The Practical Inventor's Handbook

The Practical

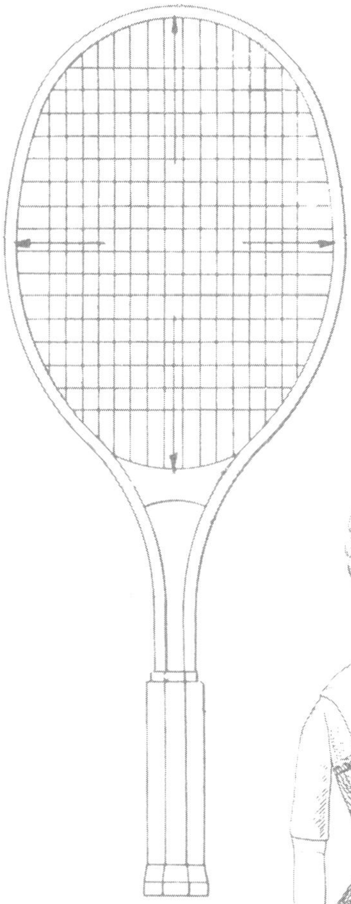

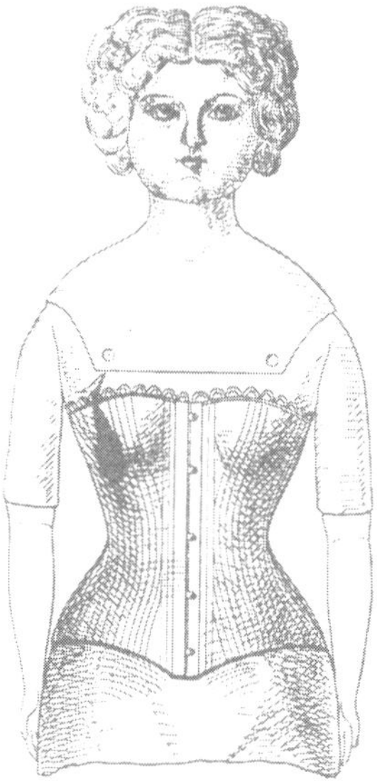

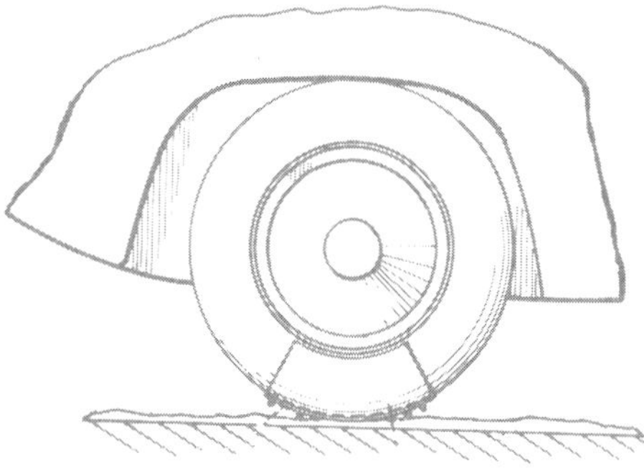

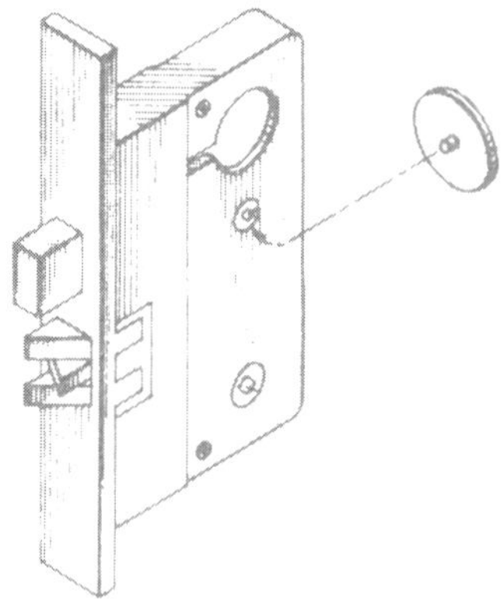

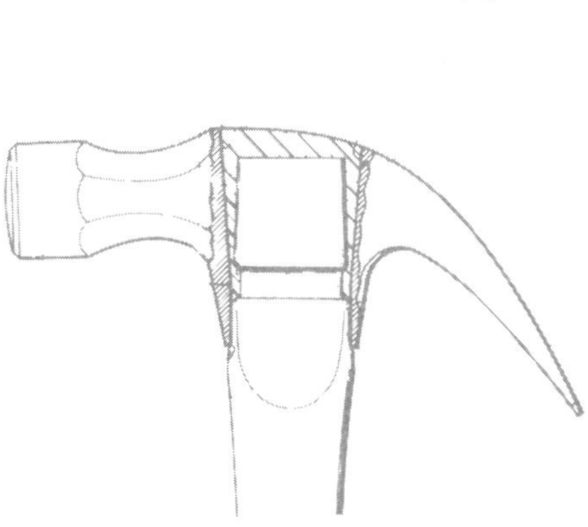

Inventor's Handbook

by Orville Greene and Frank Durr

with John Berseth

McGraw-Hill Book Company

New York St. Louis San Francisco Auckland
Bogotá Düsseldorf Johannesburg London
Madrid Mexico Montreal New Delhi
Panama São Paulo Singapore
Sydney Tokyo Toronto

Library of Congress Cataloging in Publication Data

Greene, Orville, date.
The practical inventor's handbook.

Bibliography: p.
Includes index.
1. Inventions. 2. Patents—United States.
I. Durr, Frank, joint author. II. Berseth, John,
joint author. III. Title.
T212.G59 608'.7 78-26666
ISBN 0-07-024320-4

1234567890 HDHD 7865432109

A Note on the Illustrations and Captions

Patents, being legal documents, are often long and complicated. They contain upward of several pages of text and five to ten or more explanatory diagrams. For the patents described in this book we have reproduced only the most important drawings, the one or two that best show how the invention works. Captions are edited, abridged versions of each patent's descriptive text. In some cases it seemed worthwhile to include a great deal of this original material; in other cases a few lines were sufficient to explain the drawing. In all cases the reader can obtain a copy of the full patent, complete with all illustrations and descriptive matter, for 50 cents from the U.S. Patent and Trademark Office, Washington, D.C. 20231.

*The editors for this book were Robert A. Rosenbaum, Gretlyn K. Blau,
and Celia Knight, the designer was Naomi Auerbach,
and the production supervisors were Frank P. Bellantoni
and Sally Fliess. It was set in Trump Medieval
by The Clarinda Company.
Printed and bound by Halliday Lithograph.*

Contents

Introduction

Inventions seldom bring instant success and wealth. And sometimes they can be downright hazardous to an inventor's health. In the early nineteenth century, when the Frenchman Joseph Jacquard demonstrated a model of his mechanical loom, he almost lost his life. A mob of angry workers from the big textile city of Lyon took direct action against the inventor and his machine because they were afraid of being automated out of their jobs.

More than a hundred years later, Chester Carlson, who invented the Xerox process, didn't face death, but he did experience frustration and defeat. For a long time he was unable to convince anyone that his invention was practical. One potential backer after another said No, and Carlson almost gave up in despair many times.

Yet inventors persist. They have a powerful urge to create something new, to improve something old, in some way to turn the wheels of progress forward another notch or two. To many people the art of invention is worth all the sacrifice it demands.

What makes up an inventor? In one of the few polls ever taken on this subject—Joseph Rossman questioned 710 patent holders for a book he published in 1963—a large majority (71 percent) said that perseverance was the most important character trait for a successful inventor. Optimism was also rated high on the list. Interestingly, only 21 percent thought originality to be of crucial importance. So inventors must believe in themselves and their ideas. Fortunately, most of them do, sometimes to excess.

The last third of the nineteenth century has been called the golden age of the individual inventor. There was a feeling then, especially in America, that anything could be done. The go-getter Mark Twain poked fun at in *A Connecticut Yankee in King Arthur's Court* spoke the thoughts of many of his contemporaries: "Why, I could make anything a body wanted . . . and if there wasn't any quick new-fangled way to make a thing, I could invent one." The Hank Morgans of the nation certainly weren't nobles and gentlefolk, and many of them were indeed laughable. But they and their machines came in handy when the time came to plow the fields or run a business or travel from place to place. Of course, there were plenty of more or less useless gadgets that people fooled with, too, but they were often more for fun than for work. In 1871 Mark Twain himself was granted U.S. Patent No. 121,992 for a waistband.

The fabled American ingenuity has continued into the twentieth century, but many scholars and social commentators have pointed out that the heyday of the individual inventor is over. For while the numbers of patent applications and patents granted have increased, the figures have not always kept pace with the growth of the population. In the ten-year period 1871–1880, for instance, in round figures there were 205,000 patent applications and 125,000 patents issued, mostly involving individuals. In a nationwide population that averaged around 45 million, this was one application for every 220 people, one patent for every 360 people. Closer to our own day, in 1961–1970 there were 900,000 applications and 600,000 patents, of which 160,000 patents were granted to individuals; the rest were assigned to corporations and other institutions. Taking an average population for the period of 190 million, that's one application for every 211 people, representing almost no change in 100 years. But there was only one patent granted to an individual for every 1,200 people—a 70 percent drop. Thus, things have changed, though not so much as one might think.

Corporate, university, and government research facilities are necessary today in many fields because of the large amounts of money required to make or test certain inventions. And there are many creative people, especially in electronics, chemistry, and similar fields, who find it congenial to work in a well-equipped lab, stimulated by their colleagues and bankrolled by their employers. But that life is not for everybody. According to Dr. Vannevar Bush, a brilliant engineer, administrator, and inventor, "There is the independent inventor, whose day is not past by any means, and who has a much wider scope of ideas and who often does produce out of thin air a striking new device or combination which is useful and which might be lost were it not for his keenness."

Those are the people for whom this book is written. These individual inventors face certain problems, and have certain opportunities, that need to be pointed out and explained. For instance, some areas of invention are for all practical purposes closed to most individuals. Yet there are countless fields where the horizons are as wide as they've ever been—or wider. Just as the nineteenth-century tinkers made many advances in farm tools, industrial machinery, and railroad equipment, so twentieth century inventors are developing new products and processes in sporting goods and toys, plastics and packaging, data processing and pollution control. The Energy Crisis has brought about several new trends. Many inventors have turned from the development of energy-utilization machines—labor-saving devices, they used to be called—to the invention of energy-conserving devices.

Thus there are certain lines of endeavor that offer greater chances of reward to the individual inventor. And within those broad lines there are paths that are less crooked, that have fewer and smaller obstructions to progress. In this book we hope to direct creative individuals to the areas where their efforts stand a good chance of being rewarded. The U.S. Patent Office has divided the patents on inventions into more than 400 classes and about 90,000 subclassifications. No one has the time or the mental capacity to study all or even most of these. And neophytes can get lost even in their own areas of interest. Yet it is these vast patent files that hold the building blocks to successful invention.

Technical progress is not often made in great leaps. It is more of a step-by-step process, with one innovator building on—but not copying or stealing—the work of another. And generally the ideas for something long precede its actualization. The prodigious imagination of Leonardo da Vinci spawned hundreds of great ideas in the fifteenth century. His notebooks, which were discovered about 300 years after his death, contain drawings of a flying machine, a power loom, and other marvels that have since been invented. Leonardo couldn't build them simply because neither he nor anyone else of his era had the necessary technical knowledge.

In later centuries inventors have come upon the same ideas at the same time more often than the mere chance of coincidence would allow. On the morning of February 14, 1876, Alexander Graham Bell filed his first application for a patent on the telephone. The very same afternoon, Elisha Gray filed a caveat, a kind of preapplication used at the time, giving notice that he planned to patent a device for the transmission of sounds, especially the human voice. Both men had been working on this invention for some time, and both had the benefit of their predecessors' and contemporaries' work. In subsequent years there were no fewer than 600 suits filed by Gray and others against Bell and his telephone patents, but none was really successful. And so our phone network is called the Bell System, not the Gray System. But Gray did not fade away. He went on to patent several other successful inventions and to found the company that eventually became Western Electric, which is now the manufacturing subsidiary of the American Telephone and Telegraph Company.

The Patent Office, now in Crystal City, Virginia, across the Potomac River from Washington, has more than 4 million patents on file, available for inspection either by number or according to classification. Most of them are outdated and of only historical interest. Others are of the greatest importance to contemporary workers. Every day the Public Search Room is filled with hundreds of people, delving into drawers and checking shelves, trying to find the answer to some problem or the key to a new invention. Has it been done before? What are other people in the field doing? In another part of the same office complex, government patent examiners check each application to make sure that each invention disclosed is new, useful, and practical, a genuine advance over the prior patents in the field, and nonobvious. These examiners, who have their own files of patents and of other relevant information, are assigned patent applications related to their training; thus they must be familiar with substantially all the important patents in their areas of specialization. They can often foresee future developments before anyone else, but since they are not permitted to patent anything, they cannot take advantage of this "fringe benefit" of their jobs.

This book is based on the premise that a good many alert, creative minds will be stimulated in their own inventiveness by seeing and studying patents in classifications that seem particularly suitable for investigation by the individual. There are no inventions in the chemical, electronic, or complex mechanical fields in these pages. We have chosen to describe patents that are generally in the more straightforward branches of mechanics. Included are indications of the fields where new products and processes often find a ready welcome and a wide potential market.

Almost 300 patents are discussed in later chapters, most of them granted within the last 20 years. The most pertinent drawing or drawings from each have been selected for reproduction. The captions are edited versions of the "brief descriptions" published in the patent. Very few descriptions have been reprinted here in their entirety, though they are long enough to give the reader a pretty good idea of what the invention is all about. Patents, after all, should be the best source of information about the invention to which they relate.

The first two chapters of this book explain in brief but comprehensive terms various ways to protect your ideas from infringement and of how the U.S. patent system functions, particularly with reference to the individual inventor. There are then eleven chapters, each devoted to a particular group of items. As the reader browses through these pages to examine the results of other inventors' dreams and frustrations, perhaps some of the best counsel to keep in mind comes from a recent court decision: "Invention is not always the offspring of genius; more frequently, it is the product of plain hard work; not infrequently, it arises from accident or carelessness; occasionally it is a happy thought of an ordinary mind; and there have been instances where it is the result of sheer stupidity."

Hardworking or careless, brilliant or stupid, lucky or plodding, inventors have been and are the hope of the future. Without them, we'd be lost.

CHAPTER ONE

Protecting Your Creation

In the mid-nineteenth century, a young midwestern politican did what most inventors do all the time—he got an idea. It came to him while he was traveling on a heavily loaded riverboat that ran aground on some shoals. As he watched the captain and crew struggle for hours, trying to get their craft afloat, he sketched a plan for a device that could lessen the danger of such hazards. The idea was to attach buoyant air chambers made of waterproof cloth to the sides of the boat. These "balloons" could be inflated whenever, in the words of the boatmen, the bottom of the river got too close to the top. Back home in Illinois, the politician fashioned a primitive model, drew up specifications, checked his calculations and drawings, and submitted them and his application to the Patent Office. In 1849, Abraham Lincoln was awarded U.S. Patent No. 6,469.

This was the future President's only venture into the world of patents; unfortunately, his invention never gained commercial success. Like most of his countrymen, Lincoln was a strong advocate of invention and of patent protection. His words are engraved over the entrance to the old Patent Office in Washington: "The patent system added the fuel of interest to the fire of genius."

At one time or another almost everyone, like Lincoln, is struck by a flash of inspiration that has the potential for some kind of commercial use. After the idea has been worked on, tried out, and perfected, it may take shape as a practical, useful device or process. The next step is generally to share this new wonder with others while at the same time holding onto the right of ownership. Lincoln followed the custom of his time by applying for a patent. Today's creator will probably want to follow the same course. But how is it done? What are the alternatives? How much does it all cost? These and other questions are answered in our first two chapters, which are written primarily for the independent inventor—that is, the person who invents on his or her own hook, rather than on salary.

American patent law began with the Constitution. In Article I, Section 8, the Congress was given the power "to promote the progress of science and the useful arts, by securing for limited times to authors and inventors the exclusive right to their respective writings and discoveries." In 1790, Congress fulfilled its responsibilities by passing a number of laws dealing with the issuance of federal patents. Thomas Jefferson, who himself had invented a swivel chair, a moldboard for a plow, and a good many farming tools, was at that time Secretary of State. He and two other cabinet members, the Attorney General and the Secretary of War, were appointed to a board of patent examiners. They were promptly flooded with applications. Then, in 1793, a law was passed that called for the granting of patents without any system of examination. Although this system was chaotic and often unfair—many patentees were hangers-on who drew heavily upon or simply copied the patents already registered—it lasted until 1836. In that year, Congress codified a new set of patent laws and established the Patent Office with a commissioner at its head.

The system still exists today in a form not drastically different from when it was first established. Millions of applications have been processed, scientific knowledge has grown in geometric leaps, the examination procedures have grown more complex and perhaps more stringent, but the Patent Office is still managing to serve the best interests of both inventors and the general public.

One of the major changes has been in the requirement for models. Throughout the nineteenth century, patent applications had to be accompanied by a model showing what the proposed device would look like. By the 1870s, several hundred thousand of these often beautifully executed models had been accumulated, and their public exhibition had become one of the tourist attractions in Washington. Many of them were built by professional woodworking shops in the capital. The employees were skilled craftsmen who made even some ordinary ideas into extraordinary machines.

When fire crackled through the Patent Office in September 1877, almost 100,000 of these models were damaged or destroyed. Shortly thereafter a reappraisal was made of the need for these models and of the Government's capacity to store them. From 1880 on, then, models have been required only for very few patent applications. Today the drawings are important, and several drafting studios in the D.C. area and elsewhere specialize in doing this work for inventors.

Many of the old models are still in existence. The Smithsonian Institution on the Mall in Washington has some of the most famous on display. In 1927, the penny-wise administration of Calvin Coolidge disposed of a group of about 50,000 models for $1,550 and another lot of 125,000 for $6,540. Many of these have since been acquired by collectors.

For as long as the term has existed, a patent has involved the granting of a monopoly—"a privilege," according to former U.S. Supreme Court Justice Hugo Black, "conditioned by a public purpose." Under a U.S. patent, the federal government grants to an inventor a monopoly over the rights to the manufacture and the sale of his creation for a term of seventeen years in return for a full disclosure in the patent application of just how the thing is put together and how it works. It's important to remember that a patent is not a secret: the specifications are on file in the Patent Office, and anyone can get a copy of any of the more than 4 million patents in existence for just 50 cents ($1 for certain "jumbo" patents with hundreds of pages). After the seventeen-year term has expired, anyone can make or use the patented material without fear of being sued. A patent is a contract, a trade off, in which neither side loses but both sides stand to gain.

Applying for a patent is not necessarily the only course open to those who want to safeguard their literary, artistic, or scientific creation. Other categories of protection are trade secrets, confidential disclosure, copyright, and trademarks. And among patents, there are utility patents (the type we usually mean when we say the word *patent*), design patents, and plant patents. Since each of these categories and subcategories applies to a fairly specific type of creation, the applicant does not have that much of a choice among them. It's a good idea to be aware of the various options, however, and so we will describe each of them in this chapter.

Utility Patents

It is generally known that patents are granted only for inventions. If you create something that is functional and of a mechanical, chemical, or electrical nature—as opposed to something that is literary or artistic in nature—then the only real protection obtainable is a utility patent. There's no effective substitute for it. Most experienced inventors have at one time or another applied for and been granted an utility patent.

In the following pages we present a thumbnail review of the American patent system and of how individual inventors can best take advantage of the protection it provides. The first question must be, simply, What is an invention? As of this writing there is no single definition that satisfies everyone. The statue (35 U.S.C. 100) says "The term 'invention' means invention or discovery." Judge Giles S. Rich of the Court of Customs and Patent Appeals has pointed out that even in legal proceedings the term *invention* has been used in two different ways: as a concrete noun, to refer to whatever is created by an inventor, and as an abstract noun, to refer to a standard of novelty. For the purposes of this book it is practical to refer to an invention as any useful process or device that, when related to the prior experience of the inventor, is new and at least offers significant advantage over similar processess or devices already known.

Some patent specialists and patent examiners (the people who pass on patent applications) will be dissatisfied with this definition because, under it, many developments that are not patentable can be defined as inventions. According to the Patent Office an invention isn't an invention unless it is patentable. On the other hand we like the broad definition because it does not make sense to say "That's not an invention—it's not patentable" to someone who has independently developed something new and useful without substantial help from the available literature or from others. There are indeed some reasons why it might not be patentable: (1) the same thing was created by someone else before—even just a day before—our inventor did it; (2) the thing would have been an obvious advance to anyone who had studied various carefully selected prior patents; (3) the subject matter of the thing belongs to a category that is not patentable, such as a method of doing business.

The bad news, then, is that many inventions as defined above simply are not patentable. The Patent Office has strict standards, and almost half the applications are rejected. This rejection figure seems particularly high when you consider that most applications are filed only after inventors and/or their patent attorneys or agents have searched the prior patents in the field. In about half the cases these searches result in the inventor being sent back to the drawing board or workshop. His creation just isn't sufficiently advanced beyond the previous "state of the art" to warrant going any farther forward. Other inventions get only as far as the desk of a good attorney, whose experience in the field tells him almost immediately that the developments are almost surely not patentable. After dealing with the Patent Office many experienced inventors come to believe the true meaning of the old adage, "There's nothing new under the sun."

Let us consider the several standards that an invention must meet if it is to earn a patent. First, it must fall within the subject areas eligible for consideration. According to the statute (35 U.S.C. 101), a patent may be obtained by one who "invents or discovers any new and useful process, machine, manufacture, or composition of matter or any new and useful improvement thereof." The term *process* is further defined as including a new use of a known process, machine, manufacture, composition of matter, or material. The term *manufacture* includes articles, things. As you might expect, the great majority of patents describe improvements on existing processes, machines, manufactures, or compositions of matter.

An invention that comes within at least one of these statutory classes must then meet a second set of three requirements: (1) It must be new; (2) it must be useful; (3) it must be nonobvious. To be new, the exact thing must not have been shown in any prior patents or publications. As far as usefulness is concerned, in most cases that is obvious, although an untested process has no established utility. In fact, many applications for proposed medicines are rejected for just this reason—there must be medical proof that the substance has been successful in experimental use. Finally, to be nonobvious, an invention must incorporate a reasonable advancement over the prior art. Something that represents only superior mechanical skill or power may be useful to some, but it is not patentable. Nor are simple changes in materials, size, etc.

This is not an area of blacks and whites, patents and rejects. Even experts disagree. The crucial question is, How much of an improvement is necessary for an invention to be granted a patent? Each case is decided individually, and it's often difficult to predict the outcome. Even after a patent has been issued, a court may declare it invalid in an infringement action—a suit that an inventor brings against someone he thinks has stolen his invention.

There are two other requirements that must be met before a valid patent can be obtained, although the failure to meet these requirements may not be brought to light until after the patent has been granted. First, the invention cannot have been known to or published by anyone other than the inventor in the U.S. or in a foreign country before the date of the inventor's creation of it. That is, you can't take someone else's unpatented invention and patent it as your own. The other requirement is, simply, that the invention cannot have been patented or described in a publication anywhere in the world, or placed on sale or publicly demonstrated in the United States, more than one year before the patent application is filed. Thus, you must be careful about submitting any of your patentable creations for publication in a periodical or at a meeting far in advance of the time you're going to submit your application for a patent.

Before describing the steps recommended for obtaining a patent, other kinds of protection that are available to an innovator will be discussed.

Trade Secrets

A trade secret can be any formula, pattern, device, or compilation of information that is used in a business and that gives the owner or creator an advantage over competitors who do not know or use it. Trade secrets offer little help to individual inventors, most of whom have no access to their own manufacturing or marketing facilities. But it can be a useful approach for some businessmen. Of course, the danger that trade secrets will be discovered, either independently or through spying, is substantial. Use them at your own risk.

Most trade secrets are unpatentable, but there are exceptions. A competitor who discovers a process that is one of your trade secrets can go ahead and try to get a patent on it—and succeed, if the subject matter is in fact patentable. A trade secret can be kept pretty secure if access to it is restricted to trusted employees. So, despite its drawbacks, the trade-secret approach is used by many companies that have found it suits their own purposes. Its ramifications are too broad, and its scope of interest is too narrow, for us to go into the subject any further here.

Confidential Disclosure: Submission of Ideas

Close to trade secrets, but more relevant to the concerns of the independent inventor, is confidential disclosure—that is, selling to a company a process, device, or thing that would then become one of their own trade secrets. Payment for such information usually has to be accepted in a lump sum, though some royalty arrangements have been made. There are obvious dangers in this approach, and it should probably be adopted only by those whose invention is for one reason or another not patentable. Something that the Patent Office rejects for not being original or useful or a sufficient improvement upon prior patents may be just the thing a manufacturer or promoter is looking for. Just because a preliminary search reveals that an invention has been partially disclosed in an old patent doesn't mean that the invention is destined for failure.

There is another category of items that, even when fully developed, cannot generally be protected by patents or copyrights. This includes ways of doing business, processes that require mental calculations, methods of keeping accounts, games that can be played with pieces or equipment already on the market, methods of advertising, and so on.

Once you have decided, for one reason or another, to try to sell your potentially profitable information to a company, you must establish what a lawyer would call a confidential relationship with the prospective purchaser(s). This can generally be accomplished in a letter that tells in carefully chosen, rather vague terms just how the idea relates to their business. They must also be told not to use or disclose your ideas. Be sure to mention that you're not a philanthropist—you expect to get paid if your proposal is adopted. Two samples of such letters are in the Appendix.

There has been so much costly litigation in this area that most large corporations have various form letters that they use to answer queries from submitters of ideas. What they want is a signed agreement, including a provision that the corporation is the sole judge as to the value of any submission. These are rather stiff terms, and they often confuse and embitter the inventor, but they are common to many firms.

Most businessmen want to avoid the trouble and expense of responding to lawsuits from angry inventors or idea submitters, and so they are relatively scrupulous in their dealings. There are legitimate, unavoidable conflicts, however. Many companies have large research departments, and it is possible, even probable, that someone within an organization is working on the same general idea as the outside person. These companies must be careful not to let any "confidential relationship" they establish with a non-employee jeopardize their own staff's investigative efforts. On the other hand, the independent operators must protect their own interests. They should be wary of small companies with poor reputations. They should be careful of signing away all claims to their ideas. But they should not become paranoid. If there comes a time when you think that everyone is out to steal your ideas, if you're afraid to tell anyone of a new discovery, then you've probably crossed the border between legitimate concern and irrational fear. A typical form used by a large corporation that receives many unsolicited ideas will be found in the Appendix.

In the 1920s, there was an interesting case involving the automobile industry—a long-time favorite with many inventors. An independent designer established the requisite confidential relationship with the Stutz Company and submitted to it his plans for a radically new, low-slung motorcar. Stutz had several of its designers look over the plans, but finally decided not to use them, and so returned them to the original author. Then, several years later, the famous Stutz Bearcat was introduced. It bore a strong resemblance to the auto described in the rejected plans. The designer sued and eventually won his case. Although Stutz had probably acted in good faith and not actually pirated the designs, many of the new concepts and ideas in the submitted plans had made a strong impression on the company's chief designer. Perhaps without even realizing the source of his ideas, he had incorporated them into the Bearcat.

The moral of the story is, keep careful records of what you send to a company, when you send it, and what the company replies.

Many inventors have found that it is easier to deal with corporations after the patent application has been filed. Then both parties have a measure of protection, and there are fewer grounds for expensive court action. This subject is taken up in greater detail on pages 23–26 in the section on the marketing of patents.

There is another kind of disclosure, too, one in which you must rely on the largesse or fairness of the company for your reward—the suggestion box. Most large organizations and a good many small ones welcome ideas from their employees. In general, companies are pretty fair about sharing the wealth that results from any suggestion—if they weren't, the idea flow would tend to dry up—and some clever people are able regularly to supplement their incomes in this way. The federal government is particularly interested in encouraging its personnel to think about ways to get a job done better, or quicker, or cheaper—or maybe even to eliminate the job altogether. And it costs nothing to try. Once you have submitted an idea to an organization by which you are employed, however, any patent resulting from the invention belongs to the employer.

Copyrights and trademarks are even less related to patents than are trade secrets and confidential disclosures. However, a minimal knowledge of these two protective devices is desirable for most creative people. You never know what you might come up with next.

Copyright

Another method to protect certain creations is via copyright. As has often been said, ideas are not copyrightable, but many ideas can be developed into something that can be copyrighted. Generally speaking, copyright covers literary and artistic creations. In recent years additional areas have been included under the copyright umbrella—phonograph records, fabric designs, and even distinct shapes such as are found in jewelry designs.

The first and most important step in copyright is to make sure that a proper notice is applied, such as Copyright © 1980 by Joe Writer. The © symbol is used internationally. The copyright notice must be prominently displayed so that it is readily accessible to anyone who looks for it. In a book, the notice should appear on the title page or on the page immediately following. If the notice doesn't appear in a place "that is not concealed from view upon reasonable examination," the book is not legally copyrighted.

On a work of art such as a sculpture, painting, or piece of jewelry, the notice should appear on the viewing surface—not the bottom of a statue or the back of a painting. If a complete notice would interfere with the design, the copyright owner's full name may be replaced by initials, provided that the name appears somewhere on the article. On a fabric the name can appear at regular intervals on the selvedge, even though the selvedge is often removed when the fabric is worked up into clothing or other consumer goods.

Copyright is unique in that you can place a notice of it on something without applying to the government for copyright registration. This do-it-yourself protection is advantageous for items that will probably have a short life or problematical success. However, you cannot sue for infringement unless the copyright is registered with the Library of Congress. Many owners simply wait and see how something goes, even though the fees for copyright registration are relatively small. Recent changes in the copyright law have made this course practical. If in doubt, you may want to consult an attorney with copyright experience.

Filing for a copyright is a relatively simple matter. First, request a form from the Copyright Office, Library of Congress, Washington, D.C. 20559, stating the subject matter of the item to be copyrighted. There are four broad categories: TX(text), PA (performing arts), VA (visual arts), and SR (sound recordings). Follow all the instructions on the form they send you, and return it to Washington, along with the specified number of the works copyrighted.

The Copyright Office checks all registration applications submitted to it for accuracy and also checks the copies of the work in question for proper notice. Unlike the Patent Office, the Copyright Office does not run an exhaustive search to determine whether the editions have been plagiarized, copied, or otherwise duplicated from anything that has been registered in the past.

Compared to patents, the term of copyright has always been relatively long. The law has recently been changed from a maximum term of 56 years to one that generally lasts for the author's lifetime plus 50 years. This brings U.S. law into conformity with standard international practice. In the case of works done for hire, the copyright now lasts for 75 years from publication or 100 years from the creation date, whichever is shorter.

Often, a copyright is an interesting alternative to a design patent, which is one of the nonutility types of patents and applies to a new, original, and ornamental design for an article of manufacture.

These patents run for $3^{1}/_{2}$, 7, or 14 years, depending on the fee paid. They are granted only if the object is found to be patentable after being compared with prior patents and other publications showing similar designs unearthed by a thorough search of the Patent Office files. Many applications are rejected. A copyright, on the other hand, is almost always registered.

Even in court tests, copyrights have the advantage over design patents. To be valid, a copyrighted article must only be "original"—and the degree of originality is minimal. The design patent must undergo a rigorous test in court, not only for originality but for significant differences between it and previous items of the same kind. Any similar prior design, whether a patent or in some publication, becomes an exceptionally valid reference to invalidate a design patent. Thus, anything the Patent Examiner might have missed can still be introduced into the court proceedings. Awards for damages tend to be greater in copyright cases, too, and attorney fees are more likely to be granted to the copyright owner.

It is clear that copyright registration offers significant advantages over a design patent. The latter has one point in its favor: If a number of copies of an item have been distributed without a proper notice, a valid copyright cannot be obtained. The Patent Office gives an inventor six months after first "publication" during which an application for a design patent can still be filed. (The new copyright law does, however, excuse an inadvertent omission of the copyright notice.)

Trademarks

Like copyrights, trademarks are registered, but the similarity pretty much ends there. A trademark is defined as any "word, name, symbol or device or any combination thereof adopted by a manufacturer or merchant to identify his goods and distinguish them from those manufactured or sold by others." It is not necessary to register a trademark, but registration certainly enhances its value. No trademark can be registered until it has been used on the goods and sold in interstate commerce—that is, until it has served the purpose in the marketplace of identifying the manufacturer's or the seller's goods.

To avoid unnecessary risk and expense, a trademark should not be used until a search has been made of the files at the Patent and Trademark Office. A mark is no good if it duplicates or comes too close to a previously registered trademark used to identify the same or similar goods. Even after a search, questions may arise as to a confusingly similar trademark. In 1976, for instance, the National Broadcasting Company adopted as its symbol a stylized N, of fine modern design. It was not long before it received a complaint from the Nebraska Educational Television Network, which had been using a strikingly similar logotype for several years. The case was never seriously prosecuted, since the two networks are not really in competition with each other. The only real result was to embarrass some NBC executives: They had paid an industrial design firm several hundred thousand dollars for their N; the Nebraskans had theirs made in the art department for less than a thousand.

There are several clear advantages to federal trademark registration. First, no one can claim to have been unaware of a trademark that has been registered. Second, the infringing party can be sued in federal court, where the penalties tend to be higher than in state courts. Judges tend to look with favor on the true proprietor of a trademark because they are protective of the public's right to get what it thinks it is getting. A Frigidaire made by the Quik-Buk Mfg. Co. is as much consumer fraud as it is infringement of the General Motors trademark.

A trademark offers a third advantage if it is registered. An official document that indicates exclusive ownership and right to use a trademark becomes an important part of a business. The public may buy "Spin-Easy" nuts and bolts without ever realizing that they are made by Smith & Co. If Conglomerate Industries purchases Smith & Co., an invaluable part of Smith's assets will be the trademark registrations, which show that Smith & Co. has an established right to its trademarks.

Although trademarks are registered by the Patent and Trademark Office, there is no significant overlap between them and patents, even design patents. Some confusion has been engendered by "patent medicines," which as a general rule are not patented or patentable. When the labels of these potions say "Reg. U.S. Pat Off.," as they often do, that refers only to the trademark on the label, not to the stuff inside the bottle or box.

If you obtain a patent on an invention and thereafter participate in the sale and manufacture of it, do not overlook the possibility of obtaining a trademark registration on the item you sell. A patent lasts only seventeen years, while a trademark remains in force as long as it remains in use, as long as it is renewed every twenty years. One problem is that trademarks can be lost if they become recognized by the public as generic names for the goods rather than particular names for the goods of one manufacturer. The danger of losing a trademark in this way is greatest with a patented product that dominates the market for the period of the patent. (With this kind of success you will be able to get the best legal advice, so don't lose any sleep over it right now.) Trademarks that have been lost by becoming generic include aspirin, cellophane, and escalator.

Because trademark registrations and copyrights can be identified by similar symbols, ® for the former and © for the latter, they are sometimes confused in the minds of the public. Don't make this mistake. The © with date and name must appear on the copies of the copyrightable material *before* a copyright registration is obtained. The ® can be used next to a trademark only after a T.M. registration has been obtained. Also, a "trademark use" necessary for registration is valid if the trademark appears on the package that contains the goods. A copyright notice on a package does not protect the goods inside, however; it merely gives notice that the rights to the package and/or label are being reserved. An item that is to hold a valid copyright must have the © notice on the item itself.

We have discussed the various methods of protection open to those who have created something that they would like sell or otherwise make available to a limited or a general public. Most inventors will, after careful consideration, choose the path that, if all goes well, leads to a patent. The next chapter will be devoted to a discussion of how to obtain a patent, and what to do (and what not to do) after you've got one.

CHAPTER TWO

Getting a Patent

Every week, the Patent and Trademark Office publishes its *Official Gazette,* which contains brief descriptions of the patents and trademarks issued that week. Anyone who is interested in patents should take a look at this book, which is available in many libraries. The most striking thing about it is the number and variety of patents that are issued. The average issue of the *Official Gazette* contains more than a thousand patents, everything from a new kind of thumbtack to a complicated system for turning solar energy into electricity.

The standard picture of the American inventor is of a lonely figure, spending long hours in his workshop or a company cubicle, working in isolation on some complicated problem. A study of the *Official Gazette* and other Patent Office publications reveals that there are an awful lot of (some cynics would say a lot of awful) inventors in the U.S.A., and many of them work together and are co-patentees. While there are well over a million inventors around the country, working on various projects, there's always room for one more. Don't let the sheer numbers of the system discourage you from joining it.

Let's say you have an idea you believe may lead to an invention that may be patentable. You got the idea fairly recently, and since then you have been working on it, refining it to the point where you have something that's pretty good. What's the first step you should take toward fame and fortune? Experts would agree that you should consider applying for a patent.

Actually, preparations for a patent application should begin almost immediately after an idea occurs to you. Our first advice is to keep a notebook on all your ideas and their development—not a looseleaf book, where it's easy to delete, add, or change the position of pages, but an old-fashioned, permanent-bound notebook.

Some inventors become rather contemptuous of notebooks after they have several patents to their credit. "I never have to refer to it again," they say, "so why bother?" It is true that an inventor may have a dozen or more patents and never have had to rely on a notebook to substantiate a date on any of them. In fact, however, about 2 percent of the patent applications filed end in an interference proceeding, where the Patent Office must decide which of two parties who filed substantially similar claims at about the same time should have precedence.

There is no doubt that one of the most valuable pieces of evidence to bolster your case in interference—or in any litigation in which your invention might become involved—is your record book. It is most useful in helping you prepare for the questions that it will be necessary to answer should your patent application be put into an interference where you must establish a date of invention that is earlier than the filing date of your application. In an interference you are often required to file a preliminary statement with the following information:

1. Date when the invention was conceived.
2. Date when active exercise of reasonable diligence toward reducing the invention to practice began.
3. Date of actual reduction to practice.
4. Date of first act or acts susceptible of proof other than acts of the character specified in the first three points above. These acts, if proved, would establish the conception of the invention, and would include such tangible actions as the making of a nonoperating model.

An important fact often omitted from a notebook record of a development is the degree of success (or failure) of the experiment. Don't omit this kind of information.

Many creative people make the mistake of believing that a sealed, registered letter to themselves describing their invention or new development will provide them with all the protection they need. In fact, this kind of evidence just does not stand up either in Patent Office proceedings or in court. The Board of Interference Examiners, U.S. Patent and Trademark Office, has said:

> The sealed condition of an envelope held in the personal custody of an inventor for a number of years where manipulation at leisure is possible is flimsy circumstantial evidence at best compared to the corroboration of a reliable neutral custodian (Pezzullo v. Litzenburg and White, 810 OG 896,904).

It is evident that your brother-in-law in Chicago or your friend down the street would not be neutral custodians, either.

To replace the popular practice of attempting to establish a date of invention by registered letter, the Patent Office adopted a Disclosure Document Program in 1969. It is described fully in a booklet published by the Patent Office. Under this program you may submit a disclosure directly to the Patent Office, by enclosing a

$10 fee, a self-addressed stamped envelope, and a letter, in duplicate, to the Commissioner requesting that your Disclosure be preserved for a period of two years. The Patent Office will then give the Disclosure a number, put it on the letters, and return one of them to you. Make sure you refer to this number in any subsequent patent application that has to do with the material in the Disclosure. (See Appendix for sample.)

If you have any question about this or any other Patent Office procedure, write to

Office of the Commissioner of
Patents and Trademarks
Washington, D.C. 20231

As you can imagine, that office gets tons of mail and thus must reply to most queries with a form letter or a marked copy of one of its booklets. It does not give advice, especially in regard to: patentability of your invention; whether or not something has been previously patented; any other area where the staff would get bogged down just keeping up with all the questions.

If your invention goes forward, if you are pretty sure that you have something that has a useful and practical application, then it's time to institute a search of the Patent Office files of prior patents. There are more than 4 million patents now, filed and cross-referenced among more than 500 main classes and 90,000 subclassifications. True, the subject matter of your invention limits the extent of the search, but you have to know where to look. The *Manual of Classification of Patents*, which includes an index of subject matter, can be of help here. This is a big book that is available in many libraries around the country.

The object of the search is to find out just how new and original your invention is and to determine whether you are justified in spending the money required to pursue the patent application further. How can you judge this, and do you have the time and the experience to make the search? We have come to the point where a crucial decision must be made—whether to seek the help of a patent attorney or agent, or to do it yourself.

The two authors of this book are patent attorneys. Based on our experience, we do not recommend that you go it alone. Taking an invention through all the stages that lead to a patent is a time-consuming, complicated business. It is true that some amateurs have done it and are doing it successfully—there are several books in the Bibliography that describe how it's done. There is no doubt in our minds, however, that in most cases the professional can do the job faster and better and cheaper (depending on what your time is worth per hour) than you can.

Most authorities in the field agree that it is difficult for independent inventors to get their own patents. Stacy V. Jones, who has written a regular column on patents for *The New York Times* for more than twenty years, has this advice: "Patent officials are patient with the outsider and will help him when they can, but the procedure is very technical and it is best to engage an attorney or agent." The Patent Office has its own words on this subject, stated succinctly in the helpful pamphlet *Patents & Inventions: An In-*

formation Guide for Inventors: "An inventor who prepares his own application is therefore almost certain to endanger his chances of obtaining a good patent, unless he has a great deal of experience in these matters."

If you decide that it would be best to have someone represent you, obtain a copy of the book *Directory of Registered Patent Attorneys and Agents Arranged by States and Counties.* This should be available from your local public library or bar association; the U.S. Government Printing Office also sells copies for approximately $5. Everyone whose name is on the lists in this book has the legal, technical, and scientific qualifications necessary to practice before the Patent Office. Those readers who prefer not to deal with lawyers will be pleased to learn that this listing includes the names of many who are patent agents—that is, nonlawyers who are qualified to arrange for a search, prepare the application, and follow through on the prosecution of a patent in any correspondence or meetings with Patent Office examiners. Agents are not, however, qualified to conduct any litigation in the courts or to perform any other duties that your local authorities consider to be the practice of law.

The Patent Office exercises some control over all the lawyers and agents who are registered with it as accredited practitioners. There are many regulations that registered practitioners must follow, so the client is afforded a good measure of protection. If you want to complain about any registered practitioner, write a letter to the Patent Office describing your grievance carefully, completely, and calmly. The Patent Office has the power to disbar or suspend anyone from practice before it whom its investigation determines to be violating any rules of that practice.

One thing the Patent Office does not regulate is the schedule of fees—that's between you and your practitioner. Be sure to ask for an estimate of costs before authorizing anyone to take action on your behalf. As you have probably gathered, there are three stages on the way to a patent—search, application, and prosecution—where the attorney or agent can be of service to you and will bill you for such service. You should be able to get an estimate of about how much each of these separate operations will cost. Naturally, there are going to be some variations depending on how much time must be spent at each stage. Nevertheless, it's important to set up some guidelines so that no one will get any unpleasant surprises.

Fees change, sometimes all too rapidly, but at the present time, if you shop around, you can probably get a good search on a simple invention for about $100. If the search is favorable, the attorney's charges for preparing an application on a simple device could be as much as $600. In addition to these costs, you can expect to pay about $90 per sheet for drawings (some simple inventions need only one sheet). There will be other costs six months or a year later, including the probability of time charges for work on amendments.

The Patent Office fees are substantial in themselves, though for the service they provide, the cost is quite reasonable. (Patent Office fees cover only about 40 percent of the expenses involved in maintaining the patent system.) At the time this is written there is a minimum filing fee for a patent application of $65. Once a patent application is allowed, it will not be formally granted until a final fee is paid, to cover printing costs. Right now the minimum cost here is $116.

Until recently, the Patent Office has not required any kind of preliminary search before application. Now the rules specify that the attorney, agent, or inventor must file a "Patentability Brief" within two months after filing a patent application. Such a brief cannot ordinarily be prepared without knowing the results of a search. There are other good reasons for a search, too. It would be a shame to go to the time and expense of filing a patent application if a search later revealed that the invention was too close to a prior patent to be worth a new patent. (As was mentioned in the last chapter, sometimes an experienced attorney or agent can tell an inventor this right off the bat, based on experience.)

Some inventors like to conduct their own preliminary searches. They often get good ideas, too, by examining the work of other workers in their field. The Search Room at the Patent Office in Crystal City, Virginia, just across the river from Washington and on the new Metro rail system, is without doubt the best place to conduct a search. The patents are arranged according to classification in readily available files, and it is also possible to get a suggestion of the classes you should search from a friendly examiner. Other libraries maintain up-to-date patent listings, although the patents are in numerical order rather than being arranged in their proper classifications. Some of these libraries have listings of the patents by numbers in the various clasifications, but this is an awkward way to conduct a search. (See the Appendix for names and addresses of these libraries.)

Even at this early stage, the help of a professional can be valuable. If you do not want to get involved with an attorney or agent at this time, you can go directly to one of the search services listed in the Washington, D.C., phone book yellow pages. Often these people have sufficient experience to tell you rather quickly what the chances are of your invention's being patentable. They may give a flat No, too. This is disappointing, of course, but it's a good deal less expensive to be told at this point rather than later in the game.

In this book we want to deal as often as possible with happy endings, so let's assume that a preliminary search does not turn up any invention that is too close to your own creation. The next step is for you or your attorney or agent to prepare the patent application.

In the next few pages, we review briefly the outline and construction of a patent specification, not with the object of making you an expert in the field, but so that you can understand what is going on at each step in the process. If you understand what your attorney is doing, you will be able to cooperate with him in turning out a superior job. Beyond this, we want to introduce you to the method you can use to review patent specifications that have been turned up in a search, for example.

Many technicians have said that patent specifications are not only full of repetitions, but are often difficult to understand. While this is true of some patents, which go into endless and unnecessary detail, most professionally prepared patent specifications represent the clearest and best descriptions available of the scientific processes and products they represent.

The attorney or agent who draws up the specifications must thoroughly understand the invention and its place in the overall hierarchy of patents already granted in its subject area. The inventor must carefully review every word and drawing to make sure that the explanation of the work is understandable and complete. The patent examiner, who is an expert in the subject area, will not approve the form of the specifications unless they are up to the high standards of clarity and accuracy set by the Patent Office. Patents are in fact valid fields for study to learn how certain complex mechanisms or arrangements work.

Following is the basic patent law that governs the preparation of a patent application:

35 U.S.C. 112. Specifications. The specification shall contain a written description of the invention, and of the manner and process of making and using it, in such full, clear, concise, and exact terms as to enable any person skilled in the art to which it pertains, or with which it is most nearly connected, to make and use the same, and shall set forth the best mode contemplated by the inventor of carrying out his invention.

The specification shall conclude with one or more claims particularly pointing out and distinctly claiming the subject matter which the applicant regards as his invention. A claim may be written in independent or dependent form, and if in dependent form, it shall be construed to include all the limitations of the claim incorporated by reference into the dependent claim.

An element in a claim for a combination may be expressed as a means or step for performing a specified function without the recital of structure, material, or acts in support thereof, and such claim shall be construed to cover the corresponding structure, material, or acts described in the specification and equivalents thereof.

Although it's true that these three paragraphs contain some legal gobbledygook that tends to spread a thin glaze over the eyes of most laymen, they also contain the essence of what is required in a patent application. Read them again, slowly, and you'll get the idea.

Your practitioner will generally prepare one or more drafts of the proposed patent application for your examination, criticism, and suggestions before the final copy of the application is typed. Only inventors are qualified to be applicants for a patent. If anyone else took on this role, he would perjure himself in signing the application. Any later interference or court proceedings would probably turn up this fraud, and so any patent granted would be invalidated.

Following is the suggested arrangement of the application (based on 35 U.S.C. 112), as published in the Patent Office Rules of Practice:

(a) Title of the invention
(b) Abstract of the disclosure
(c) Cross-references to related applications, if any
(d) 1. Field of the invention
    2. Description of the prior art

(e) Summary of the invention
(f) Brief description of the drawings
(g) Description of preferred embodiments
(h) Claims
(i) Signature

Some inventors are reluctant to disclose all the "secrets" of their invention to the agent or attorney. In correspondence related to another chapter in this book, this problem was emphasized by William B. Edmondson, publisher of the *Solar Energy Digest* and an inventor: "One of the main troubles with inventors is that they have an unreasoning fear their invention will be stolen and as a consequence are afraid to let anyone know about it, whereas the more people who do know about it (up to a point) the better chance the inventor has in fighting interference claims and establishing the priority of his idea." Beyond this, if anything essential to the invention is omitted from the application as filed, it cannot be added later and obviously cannot be incorporated into any of the claims.

The risks you run in making full disclosure to anyone connected with the Patent Office are minimal. Those who are registered with or employed by the Office are not likely to jeopardize their livelihoods or their reputations by stealing the secrets of any applicants. The Patent Office is quite scrupulous in defending the rights of applicants (and of examiners and practitioners, too, many of whom have been the victims of false allegations).

Your attorney or agent will also ask you to sign a power of attorney to him to act as your agent in any dealings with the Patent Office on this particular application. This document then becomes a part of the file in Washington, and thereafter the Patent Office will conduct correspondence and other business directly with your practitioner. If you file an application that does not include a power of attorney, all correspondence will be directed to you. In no case will the Patent Office correspond with two parties on the same application.

For the purposes of this discussion the patent specification can be divided into three major parts: (1) an introductory part, including items (a) through (e) on page 18; (2) the body of the application, which includes items (f) and (g), above; and (3) the claims (h).

The first of these three parts is important, but it can be skipped by those reading the patent to find out how the invention works. The second part is most important for this purpose. (In a chemical or process patent that does not have any drawings, this part begins with the Examples.) In the patents that make up the bulk of the examples that are illustrated in the later chapters of this book, substantially all the explanations are extracted from the second part.

The third part, the Claims section, is the most important part of the patent as a legal document, since it is here that the boundaries are set for whatever is to be the protected subject matter of the patent. Infringements are decided by carefully comparing the competitor's process or device with the descriptions in each of the one or more claims of the patent. Thus, it's not the invention itself that's so important to the court, but the description of the invention as set forth in the claims of the patent.

Writing the claims is difficult and tricky. It is here that the services of a patent attorney or agent are perhaps most valuable. If the claims are too broad—if, say, Edwin Land had tried to claim as his own every possible variation on the instant-print camera—the patent examiner would probably have rejected them. If the claims are too narrow—if all of Land's claims had recited all the unique features of the Polaroid camera and process—someone might have been able to copy all but one of the claimed features of the invention without infringing the patent.

Fortunately, it is permissible to draft several claims of varying scope. A skillful patent practitioner can generally get a good deal more out of an invention than can the novice through this process.

Both the Patent Office and the courts have strictly followed patent law in requiring that the application particularly point out and distinctly claim the subject matter that the applicant regards as his invention. The claims to a machine or apparatus must set forth the essential parts of such devices and indicate how these parts are combined. A claim should give a word picture of the invention, what it is and how it works. It will be disallowed if it merely describes what the invention does. There is a kind of an exception here, as set forth in the third paragraph of 35 U.S.C. 112, which is reprinted on page 18. What this means in plain English is that not every element of a combination invention has to be specifically described when the invention relates to the way in which the elements of the combination are put together. In a composition of matter, of course, all the essential ingredients must be set forth. Some of the shortest patent claims are those for new chemical compounds. An entire claim consists of the chemical name of the compound.

It is not unusual for an invention to involve a new process as well as a new machine, a new article, and/or a new composition of matter. In such cases claims in the initial patent application usually include all forms of the invention. This shows that the inventor is reserving his rights in all phases of the invention and leaves it up to the patent examiner to determine whether or not the application should be broken down into several different applications, one for each significantly different and distinct kind of invention making up the overall invention.

Once a patent application is filed in the Patent Office, it is given a serial number and a filing date, and then is assigned to an examiner who is familiar with the particular area of subject matter to which it most closely relates. The examiners take up the applications assigned to them in chronological order. They read the specifications, make sure they are understandable, and then make a search that is similar to the preliminary search, except that it is more exhaustive. The examiners have available to them foreign patents and digests and have probably made a number of searches for similar things in the recent past. If the examiner has no questions about the adequacy of your disclosures, he compares the claims with the references he has found and determines whether the claims in their broadest interpretation define a structure that can readily be reconstructed from the references or certain combinations of references. The examiner then writes an "office action" for mailing to the attorney or inventor. At this point examiners may note any deficiencies in the disclosure, reject one or more claims as being shown in or obvious over the references, and/or allow one or more of the claims. A time period, usually three months, is set within which the attorney or inventor must respond to the office action. Defects that require correction in the description (part 2, above) are usually "objected to." If the claims are too broad or if they define a structure too close to one of the references, they are rejected. Most of these rejections can be placed within two categories—the invention isn't new or original, in which case the examiner may use such words as *anticipation, fully met,* or *lack of novelty;* or the invention is not sufficiently different from inventions already patented, in which case you will see terms like *unpatentable over* and *obvious over,* followed by the designations of the patents involved.

This is the time to reconsider your application and to have a look at the patents mentioned by the examiner as being closely related to your invention. After due study of the material and consultation you or the practitioner can amend the introduction or body of the application in an attempt to overcome any deficiencies noted by the examiner. This must be done without adding anything new, however, which can be a tricky matter. Your practitioner may also amend rejected claims and add any already-disclosed additional structure necessary to differentiate your claims from those of the cited patents. Beyond this, any of the examiner's arguments can be challenged, or any of your claims that are not defensible can be withdrawn. The practitioner's charges for amending an application are usually made on a time basis, and so you should not take up any of that time unnecessarily.

Whatever is added to the claims must be based only on what was disclosed in the original application. Nothing really new can be added, especially to the descriptive sections, to get around any of the examiner's objections. All you can do is modify or amplify the claims in accordance with what has already been described. An objection can be very serious if the examiner has found deficiencies in the description (the second part of the specification). Any amendment that tends to change the scope of, or add to, the original description is called new matter and cannot be entered. Thus you understand the suggestion above that you must be sure to make a complete disclosure to your attorney or practitioner in the first place.

After a second or third consideration of a patent application, the examiner usually makes the final decision—to reject all claims, to allow all claims, or to allow some and reject others. If either you or your agent strongly feels that this final decision is wrong, you can take your case to the Patent Office's Board of Appeals. And if you fail there, you can go on to either the Court of Appeals of the District of Columbia or the Court of Customs and Patent Appeals. At this appeal stage—as at the previous stages in the Patent Office—the applicant is given the benefit of any doubt about patentability. Thus, if you can establish that there is some doubt about whether or not your invention as described in the claims is obvious to anyone thoroughly familiar with and skilled in the state of the art, you'll get the patent—all other things being equal.

The courts to which an appeal may be taken from a rejection in the Patent Office thus tend to smile on the inventor who is applying for the patent. It's a different story in the U.S. District courts, where patent-infringement cases are filed. There, the judges tend to frown suspiciously on the plaintiff in such cases. Patents are, as we have noted, a monopoly, and the U.S. legal system generally regards monopolies as abhorrent. No matter that a patent runs for only 17 years, that even during that period the public generally benefits by having access to something that did not exist before—it's still a monopoly.

Only about six of every thousand patents end up in the courts because of infringement suits, but almost half the suits are decided in favor of the defendants, the alleged infringers. Beyond the monopoly issue, there are several reasons why the inventor often finds a time in court rather unpleasant. First, the defendant has access to all of the issues raised by the patent examiner during the prosecution of the patent application. These include insufficiency of disclosure, excessive breadth of claims, indefiniteness of claims, nonobviousness of claims, and so on. In addition, an extensive search conducted by the defendant may turn up references not considered by the examiner, and thus raise new issues of obviousness and the like. The defendant may also charge that the patentee is not the true inventor, that there was fraud in the Patent Office, that there was misuse of the patent, and so on. At some points the inventor may wonder who is suing whom.

Perhaps the most difficult point to resolve in infringement cases is the old bugaboo, obviousness. A patent should not be invalidated for obviousness if it was not obvious at the time the invention was made. The problem is that most infringement suits don't reach the trial stage until long after the patent in question was issued. During the interim, it is likely that a good many related products or processes have been developed and marketed. The general state of knowledge may have, probably has, moved along to the point where the original invention is, at the time of the trial, as obvious as sliced bread. It's always easy to see something after it's been pointed out. The court thus faces a difficult task in attempting to reconstruct the state of the art at the time of the original patent. To do this for their own side, both plaintiff and defendant often rely on the testimony of experts. These people are almost always available for either side of a patent dispute, are often indispensable to a case, and are always expensive.

No inventor should be deterred from seeking a patent by the prospect of possible eventual involvement in court. First, it just isn't likely to happen. Second, there is always an area for compromise and settlement in a patent-infringement suit. Third, there have been many well-known cases where the courts have recognized the valid contributions that certain patents have made.

What should deter independent inventors, or at least make them pause to study all angles before investing money in obtaining a patent, is the state of the industry or branch of business to which their inventions relate. Who is going to manufacture your New Wonder? Who is going to sell it? Who is going to provide the capital necessary to launch any new product? Many patent holders would shout their agreement with the following plaint from an inventor in Colorado: "My experience has been interesting but frustrating, mostly because I feel that I have something here that really fills a need. But I am stymied because it seems to take a lot of money to get one's idea before the public."

An independent inventor who owns a patent is a little like someone who owns a racehorse. You can spend a considerable amount of money on your horse, or invention, and it may not ever get into a race, or market. And if it does get a chance to make a run for the money, it may never win, place, or even show. So what do you do if you think you've got a winner? Emphasize your strengths, minimize your weaknesses, and ride like hell for the finish line!

As in riding, there are certain things that can give you a leg up on the competition. A partner might help. Best of all would be an entrepreneur, a go-getter type of businessman. Like an inventor, an entrepreneur is interested in innovation, in doing something new and making a profit from it. Unlike most inventors, he has a good knowledge of business practice, has an aptitude for salesmanship and leadership, and knows where to look for capital. It goes without saying that before hooking up with a partner you should participate in the careful drawing up of an agreement that will bring benefit to both parties. Look out for your own interests, but don't try to sting the other person, lest you get stung yourself.

Let us continue by supposing that, like most inventors, you'll be doing most of the go-getting yourself. If you already have a patent, you can write freely to corporations that you think might be interested in buying outright or in licensing your invention (see the Appendix for some sample letters). In fact, it is a good idea for you to do this as soon as you have filed your application for patent. In either circumstance, the companies can deal with you freely, because your specifications are in the Patent Office, a matter of record, and no one who copies your invention can beat you to the office.

It is not unusual for inventors to license a manufacturer to make and sell their invention while a patent application is pending. One often sees the words PAT. PENDING on a new product. Inventors should not hesitate about taking and keeping royalties on an invention that has not yet been patented and that may in fact never be patented. As in the case of a trade secret or a confidential disclosure, the inventor has given the manufacturer something new, a head start on the competition. And this is true even if the invention is found, for some reason, to be unpatentable.

A good many private organizations deal in the promotion and sale of patents and inventions. Ads for some of these groups appear in magazines and newspapers and come through the mail to people whose names are on the lists of likely prospects for such services. These agencies have been the subject of some controversy, and inventors should investigate carefully before signing any contracts or agreements.

One of the potential difficulties is the matter of "front money," or payments in advance. If you bring your invention to a company that promises to take care of the whole thing, including search, market survey, patent application, appeals, and so on, it may offer a package price, with a substantial down payment required. The total cost may be quite reasonable, but at this point, you should recall the information given earlier in this chapter, that legitimate practitioners can be paid separately for each stage of the patent process. Thus, if the preliminary search indicates that you should not go any farther with the invention because of doubts about its patentability, then you pay only for the search. And so on for the various stages.

Inventors would probably be the first to admit that they are not ordinary, run-of-the-mill people. They are creative individuals, and as such they may go through periods of—well, let's call it eccentricity. Some inventors have been known to develop a strong attachment for something they have conceived. After all, it is their baby, and like many new mothers, they spend their days thinking about it and their nights dreaming about it. They look around the stores and scan the media for evidence of potential competition, and when they find none, they think they have something unique—and saleable.

Now, suppose such an individual gets together with a patent-promotion organization. The inventor answers one of the organization's ads, and it indicates interest in the invention. It offers to take care of the patent prosecution for a certain fee and further asks that a share in the patent be assigned to it as a part of that fee. "They like my invention so much that they want a piece of it!" the inventor thinks. "It's sure to make a million." Sad to say, however, only a very small part of the patent-promotion group's profits come from royalties. The mainstay of the business is the fees it receives from the inventors. Be advised, therefore, that if you exhibit certain of the personality traits outlined in these paragraphs, you should be wary of those who fuel your fantasies.

The last word on the subject belongs to the Colorado inventor we quoted a few pages back: "My one bit of advice might be for all new inventors to be especially aware of the so-called invention marketing firms and check them out very thoroughly, because the field is full of rip-off artists." Find out about them from your patent practitioner, the Better Business Bureau in their locality, or someone who has used their services. The Patent Office itself does not license anyone in the promotion field, and therefore has no control over their activities.

It has been suggested by responsible observers—including an Assistant Attorney General of New York—that the time has come to regulate idea promoters. This would at least serve to reassure the public as to the worth of the many legitimate people in the field. In general, at the moment, if someone talks in terms of getting rich from an invention—while at the same time being somewhat vague about specific opportunities *and* problems—let the inventor beware.

There are many governmental services available to inventors and other small businessmen. The Patent Office itself publishes in the *Official Gazette* a listing of inventions for sale or license. The fee for each listing is now $3.00. This publication attracts a wide, interested readership and has helped many inventors (and investors) to find a profit in patents. The federal government has an Office of Innovation and Invention, which is part of the National Bureau of Standards (Washington, D.C. 20234). You can send it a complete description of any invention you have patented that you think someone somewhere in the vast federal establishment might be able to use in his work. Or you can send the material directly to the department or agency that you believe would be interested.

The Department of Commerce has many district offices around the country that can give you advice that offers an unusual combination—it's free *and* valuable. The Small Business Administration is another source of help. Most states and many cities and counties want to promote their own economic growth, and so they have set up agencies to help businesses get established—especially new businesses. And don't forget your local chamber of commerce and the banks. All of these sources plus your own personal contacts may be all that is needed to get you started in your own business. At least one of them should be able to put you in touch with some individuals and companies who would be interested in exploring the possibility of helping to exploit the potential of your invention.

Assuming that you are not in a position to manufacture and market an invention yourself, there are some points you should know about cashing in on your creation. First, a patent can be sold outright or it can be licensed for a period of time. Patent attorneys are generally not qualified to locate a buyer or licensee for an invention; on the other hand, they are well equipped to draw up a licensing agreement. Before you enter any negotiations for any kind of sale or license, find out what your general expectations should be. You don't want to ask too much or too little.

License agreements that incorporate a royalty arrangement are preferable to those that involve an outright sale—even when the price offered seems rather high. A license can be exclusive or nonexclusive; in the latter case you can license the invention to any number of others. If you grant an exclusive license, you are entitled to (1) a down payment to cover at least part of your development costs (in some agreements this may be considered an advance against future royalties); (2) a guaranteed minimum payment of royalties after the first year or so, when the licensee would be likely to have substantial startup costs. Royalties vary, depending on the selling price of the product and how big the market is. The average is 5 percent of the manufacturer's selling price. On a small mass-market item, the royalty could go as low as ½ percent. Agreements are common in which the royalty percentage decreases after a certain number of sales per year.

Those inventors who persevere and finally get themselves in the money should know that the tax people treat patents pretty much like any other personal property. When one is sold, even one day after issuance, it's a capital gain, taxed at that lower rate. Any royalties you receive are subject to a deduction for depreciation figured on the entire 17-year life of the patent or, if you have something that's not going to last that long, over a shorter period of time.

People have been inventing things since the beginning of time. According to Abe Lincoln, man's first discovery was that he was naked, and his first invention was the fig leaf. Of course, that wasn't patentable, being simply a product of nature put to a particular and rather obvious use, but it served its purpose. Today there are probably as many inventors, per capita, as there were in Lincoln's time, though many of them now wear white coats and work in corporate research centers. Inventors are extraordinarily creative people, able to send a rocket into the stratosphere, like Robert Goddard, or to put a few new wrinkles into a piece of metal and create a safety pin, like Walter Hunt.

Creative people can't work in a vacuum. Ideas are built one upon the other. It is this principle that motivates the next eleven chapters of this book. Here are explored some productive areas of mechanical invention for the independent inventor. Examples are given of some ingenious patents in each area—and some interesting failures, too. In the fruits of past explorations will be found the seeds of future discoveries. We hope that those who study these patents will find some ideas growing as naturally as wildflowers after a spring rain. Of course, most of them will fade almost as quickly as they came. But others will take root, be nourished, and blossom. They are what make all the work and frustration—Thomas Edison's 99 percent perspiration and 1 percent inspiration—worthwhile.

In the following chapters we make various recommendations for possible inventions or improvements on inventions. By doing this we are not saying to readers that all they have to do is follow our suggestions and they'll have a patentable, marketable invention. A patent examiner might find our suggestions "obvious over" previous patents and therefore not patentable; another patent examiner might disagree, especially if he or she were confronted with a skillfully drawn list of claims.

Nor do we think manufacturers and marketing experts would jump at the chance to invest money in making and selling the inventions discussed in this book. Our purpose has been to take a look at a group of representative inventions that relate to what the home inventor can accomplish. And using these inventions as a base, we go on to discuss the many possibilities they suggest, to open doors in the inventor's mind, to encourage inventors to generate their own ideas and to follow through as far as they can.

CHAPTER THREE

The Kitchen

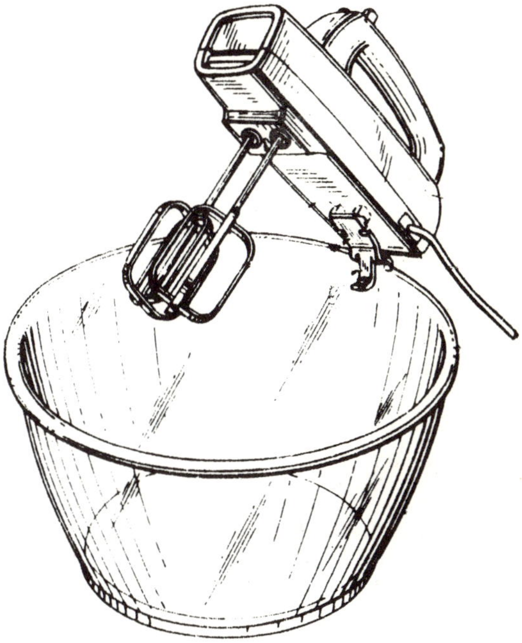

Take away all the patented inventions from the standard American kitchen and you'd be left with not much more than a stack of wood and some unformed metal. The modern kitchen is filled with all kinds of appliances, gadgets, and other innovations of every vintage and level of complexity. And there's no end in sight of the kitchen-related inventions that continue to come on the market. With good reason. The demand for new products, useful and not so useful, grows with every passing year.

The key words here form part of the contemporary American creed: We believe in things that are compact, good-looking, labor-saving, easy and convenient to use, and—well, gadgety. This seems to hold true whether the product is designed for the mass market or for the more specialized groups of consumers. The difference is that if something is destined for the mass market, it must be designed so that it can be manufactured inexpensively, by machine. It must also have a wide appeal, in many cases with a wary eye on the lowest common denominator. But there are still plenty of people who want products with individuality and "class." It is perhaps this latter group that offers the best opportunity to the independent inventor without a lot of capital to invest.

Since almost the beginning of World War II, kitchen planning has been dominated by one central fact—Mom and Sis and Aunt Jenny don't want to spend all their time chopping and baking and whipping up those old-fashioned big meals anymore. They're off at their jobs or classes or other activities outside the house. The homefires don't burn full time these days—a microwave oven does the job in a tenth of the time. The consumer wants speed, efficiency, and a minimum of scutwork.

Even in the days when many Americans had domestic help, or lots of time and the inclination to spend long hours in food preparation, there was still a desire among many for efficiency and good organization in the kitchen. Michael Shanley of Texas thought he had found it with his invention patented in 1895 (Fig. 3.1), but his hopes were illfounded. The cabinet arrangement here looks workable, but that drawer in the middle labeled Miscellaneous—it's not nearly big enough and is probably too miscellaneous.

The layout shown in Fig. 3.2 is a modern solution to the same kind of problem, on a much bigger scale and with a marketing twist. General Electric wanted to make it easy and sensible for builders to use GE products in their kitchens, and so they had Russel Sasnett of Kentucky develop the illustrated modular scheme whereby cabinets, refrigerator, dishwasher, sink, stove, and lights all fit neatly into an L-shaped area. This is a neat, workmanlike solution to an increasingly common problem of how to fit things

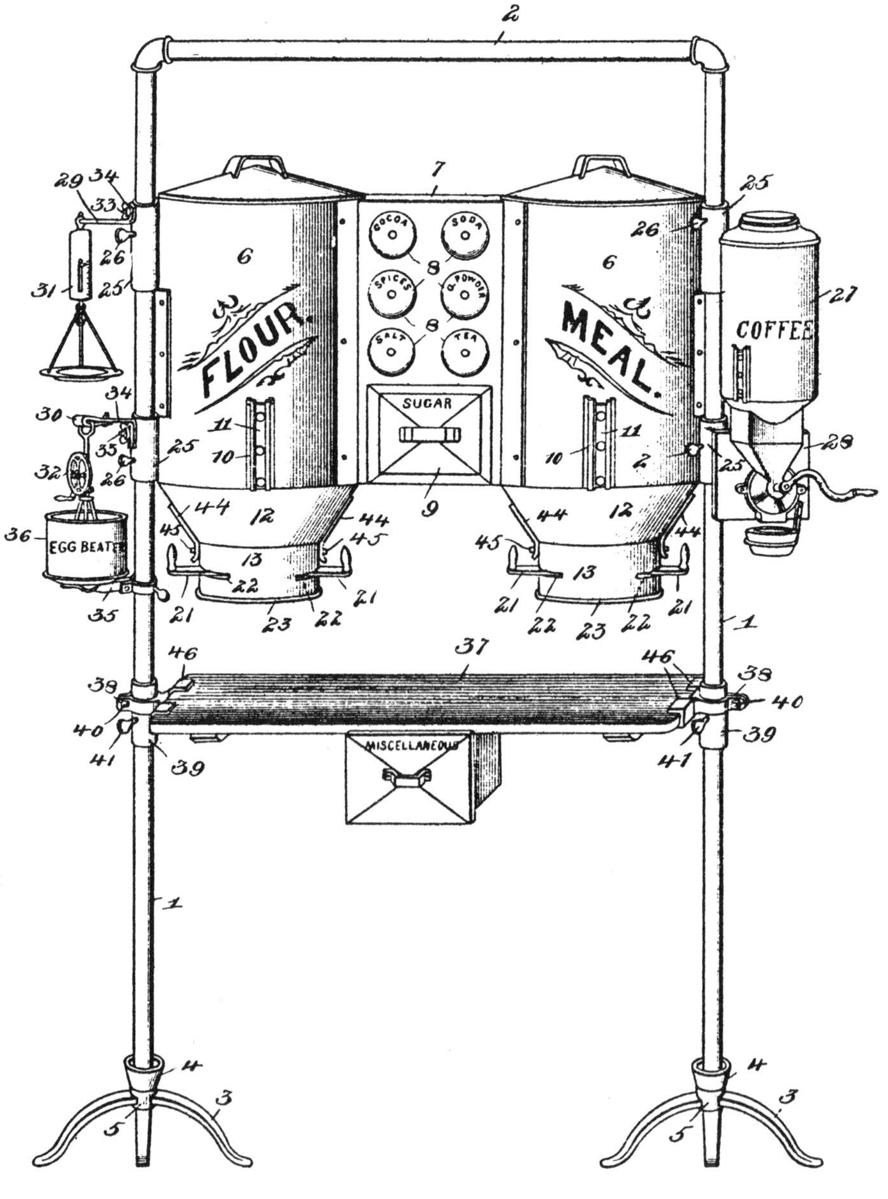

figure 3.1 A kitchen cabinet from 1895, Patent No. 541,216. Large bins hold flour and meal (i.e., cereal), medium-sized bins hold coffee, sugar, and "miscellaneous," while small drawers hold tea, spices, etc. Everything is arranged for efficiency: a coffee mill (28) at the outlet of the coffee bin, sifting arrangement (21–22) and a cap (23) at the bottom of the flour and meal bins, an eggbeater (32) inside a mixing bowl, and so on. The worktable (37) is supported between vertical standards (1).

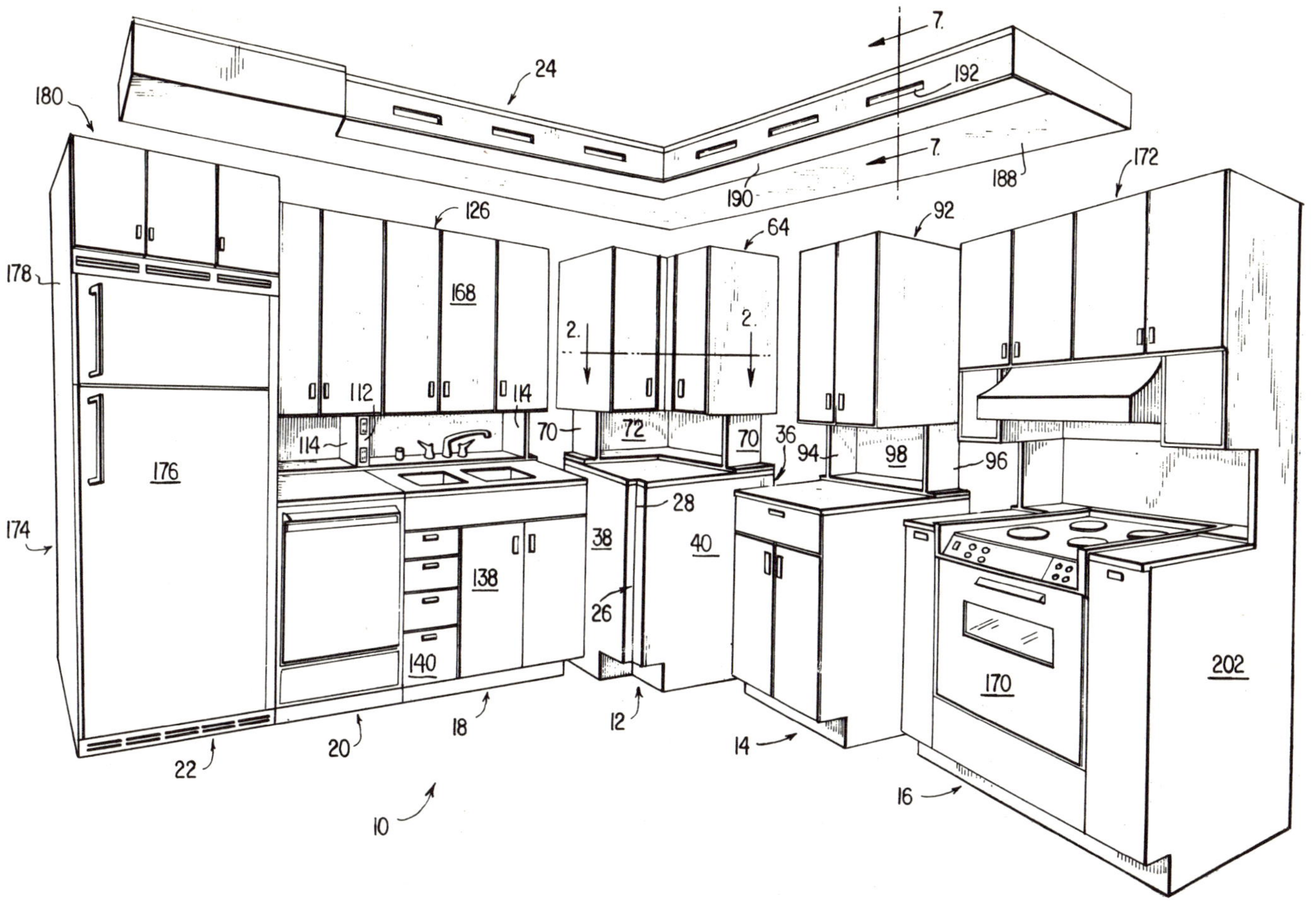

figure 3.2 Front view of a partially assembled modular kitchen arrangement, Patent No. 3,926,486 (1975). It includes the following base pieces that can be joined together to make one smooth front: a corner piece (12), storage cabinet (14), stove (16), sink (18), dishwasher (20), and refrigerator (22). These bases and their associated cabinets are designed to be equal in height to the floor-to-ceiling distance in the average new house. Any additional space is taken up by the adjustable lighting assembly (24), which goes between the top cabinets and the ceiling. The corner arrangement is constructed to allow it to be connected with the pieces next to it. A corner recess (26) makes for a smooth junction of adjacent pieces 14 and 18. In prototypes of this arrangement, corner modules have been made both two and three feet wide, which gives some space flexibility. Each unit is recessed along its back wall to allow for the passage of plumbing and electrical wiring. Outlets for counter use are mounted on an escutcheon (112) between columns (114).

into a limited area. A lot of thought went into this plan to get the most out of a small space without cramming. Kitchens and other rooms don't work out very well in everyday use if the various components don't fit together. And this is not a problem just for designers. Inventors must give careful thought to how well their products will go with other, related things.

In a very small space, a "revolving kitchen package" such as that developed by Peter Helgeson of Pennsylvania might be the answer. He assembled a combination of various appliances and work areas into a single unit that is movable so as to present "up front" the things that are needed at a particular time. In this case the idea was to have an indoor/outdoor kitchen that could be used either from a patio or from inside the house. It's a kind of variation on the old lazy susan—and thus another example of an old idea brought up to date to suit modern needs and technology. (See Fig. 3.3.)

Now let's take a look at individual appliances. There have been and will continue to be many improvements made in sinks, faucets, sprays, and the like. To avoid repetition, the discussion of plumbing has been placed in Chapter 5, on bathrooms. (But note in Fig. 3.2 how the sink was designed so that the cabinets fit snugly above it.)

An increasingly important adjunct to the kitchen sink is the dishwasher. This used to be a luxury item that often didn't work very well; now it's reasonably priced, effective, and almost a necessity. We've come a long way from the 1900 machine illustrated in

figure 3.3 Revolving kitchen package, Patent No. 3,933,400 (1976). It is made up of a grouping of selected appliances and work areas arranged side-by-side in a somewhat circular fashion around a central vertical axis, including a first large diameter grouping (18) at the top, a second large diameter grouping (20) at the bottom, and an interconnecting pedestal portion (22) between.

The diameters of the groupings 18 and 20 are substantially equal and are adapted to fit into the space between wall sections (12) and (14) as shown. Short extensions (24) and (26) of these wall sections are shaped to occupy space between the top and bottom wall sections (18) and (20) on either side of the pedestal (22) and may be glass.

The assembled package is adapted to rotate about its central axis within the opening between wall sections. The axis of rotation lies substantially inward of the wall (10) within the interior portion of the house.

Bottom grouping (20) has a preparation area (30) including sink (32), disposer, and storage cabinet, and above it in top grouping (16) is a storage cabinet (34). An adjacent bottom area is a "planning" area including dishwasher (36) having above it a pull-down refrigerator or freezer (37) and a calculator (38). The next area is a heating area including, in the bottom grouping, a surface cooking unit (40) and oven, and above, in the upper grouping (16), a pull-down upper oven (42). The fourth area, in lower grouping (20), includes another preparation area with a refrigerator, canister dispenser, etc. (44), and above it a pull-down freezer (46). Shelves (48) are conveniently located between adjacent areas at corners of the package.

To revolve, the package switches (not shown) are located in each region to run a motor and gear train mounted on the floor of the package and capable of rotating a gear (54).

The plumbing is provided through a novel coupling, which includes a movable upper element, a central drain pipe section, and a disklike device bolted to the floor for rotating movement with the package.

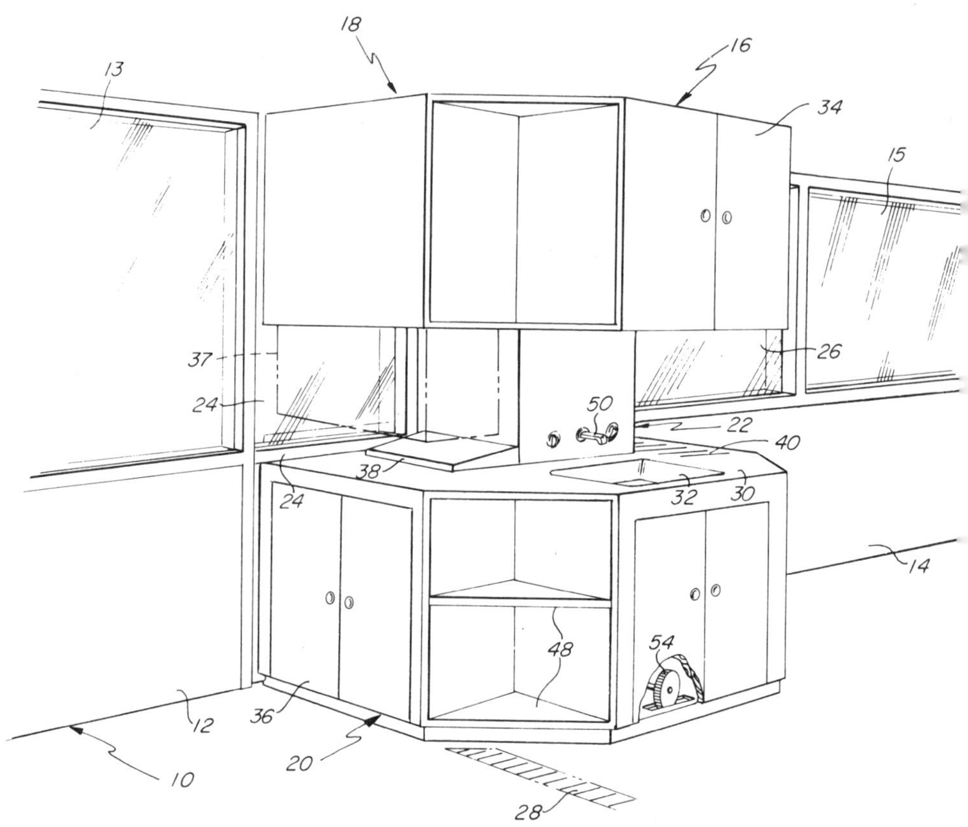

---

figure 3.4 An early dishwashing machine with an enlarged side elevation view of the dispersing wheel [Patent No. 643, 888 (1900)].

The dish-holding tank (*A*), preferably square in form, is supported on a framework (*B*), from which it can be lifted, if desired. The outlet for water (*a*) leads into the bottom of a chamber (*b*), in which reciprocates a plunger (b'). An outlet (b^2) leads upward from the bottom of the chamber (*b*) and has fitted to it a removable pipe that conducts the water back to the dishtank. The removable pipe may consist of a gooseneck (*c*), which terminates above the tank and leads through the cover (a^2). With the gooseneck-tube (*c*), there is employed a dispersing or scattering wheel (*d*), which is of peculiar construction and which is made fast to the under side of the cover (a^2), just below the opening through the cover into which the gooseneck tube c leads.

The scattering wheel is held in a hanger (*d*) and is journaled on an axle (d'). There are several, preferably four, blades to the wheel, and each blade has given to it a shape. Each blade is soldered or otherwise made secure to a tubular wheel (*e*) that constitutes the hub of the wheel. On this wheel, the main part of the blade (*g*) stands vertically or nearly vertically, and near the bottom of the blade the foot (*f*) beads abruptly to the left and extends for nearly a quarter of a circle. The surface of the foot *f* is at an angle to vertical and horizontal planes, but approaches nearly to a horizontal plane. The extreme edge extends beyond and lies partly under a spur that forms a part of the next preceding blade. The foot *f* does not touch the spurs, but between the two there is an opening for water to escape, and after passing off from the foot *f* the water will strike the oblique portion of the preçeding blade. The outer part of the vertical portion (*g*) of the blade is bent backward, so that the extreme outer edge of the blade angles backward from a plane that extends radially out from the hub. The portion of the blade which is thus bent backward is wider at the top than at the bottom, making the radial boundary line of the top of the main part, *g*, of the blade shorter than the radial distance at the bottom of the blade. This form of wheel, when acted upon by the water driven through the pipe (*c*), rotates with considerable speed and throws the water in every direction, not only sidewise, but directly down beneath it, and scatters it evenly throughout the entire tank below the cover.

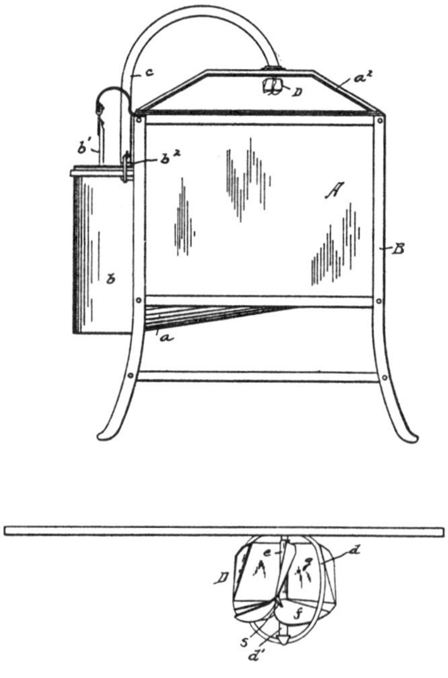

Fig. 3.4. Two water-dispersing wheels were employed then, in an effort to solve the persistent problem of how to get water to the surface of every dish.

A modern convertible dishwasher was recently patented by Richard Bergeson of Iowa, who assigned his interest to the Maytag Company. The idea here is to have a compact, centrally located sump and pump combination attached to the washing chamber (Fig. 3.5). The position of the motor is uniquely adjustable, so that it can be placed forward when the washer is built-in and to the rear when the washer is used as a portable, free-standing appliance, which would tip forward if its weight were overbalanced.

Certainly the dishwasher has become so complicated that it would be a rare and courageous inventor who would go out to the workshop and develop a totally new, improved machine. Maytag and other companies manufacture fine washers, but improvements—basic or peripheral—are not only possible, they're inevitable. So there are opportunities for the inventor who will study this patent's specifications and who will observe the operation of this or any other modern dishwashing machine.

What can be done? To take one instance, there are many people who don't like to wash dishes by hand but who have no real need for one of these large, family-size machines. For them a countertop appliance would be just fine. Fig. 3.6 illustrates one such machine, which was patented in 1969. Despite this and other efforts, no countertop model has scored significant success with the public. Yet the need is there. A small machine might be developed using the same principles and techniques as in the standard-size washers, but to be patentable, it would have to be a new departure.

The most significant change in eating habits in this century was brought about by the introduction of the home refrigerator. Where formerly most foods were eaten only in season, now supermarket cooling and freezing units offer pretty much anything consumers might want—and can afford—all the year round. Even those of us who are not long-time stockholders of General Foods can be thankful to Clarence Birdseye. His method of freezing packaged, perishable food between two metal plates was an important breakthrough. He wasn't an engineer, but a chemist, and his invention was not so much a machine as it was a technique.

Since refrigerators and freezers have been on the market for such a long time, it's questionable whether or not they offer an opportunity for innovation. Yet, the more one looks, the more chances one can see. There are always possibilities for accessories such as the one shown in Fig. 3.7. Legitimate doubts could be raised about how worthwhile these kinds of attachments are. And if they strike the public as frills, the market is gone. On the other hand, refinements of the basic appliance, such as those shown in Figs. 3.8*a* and 3.8*b*, are certainly worth consideration. The shelf and support arrangement designed by Carl Peterson of Ohio does not represent a startling departure from the norm; yet, looked at closely, it is clearly better than many designs of the past. Fig. 3.8*b* shows a refrigerator door liner with a removable shelf and stirrup. This unit was invented by a Frenchman, Marcel Lemoine, who assigned his patent to General Motors Corporation for its Frigidaire division.

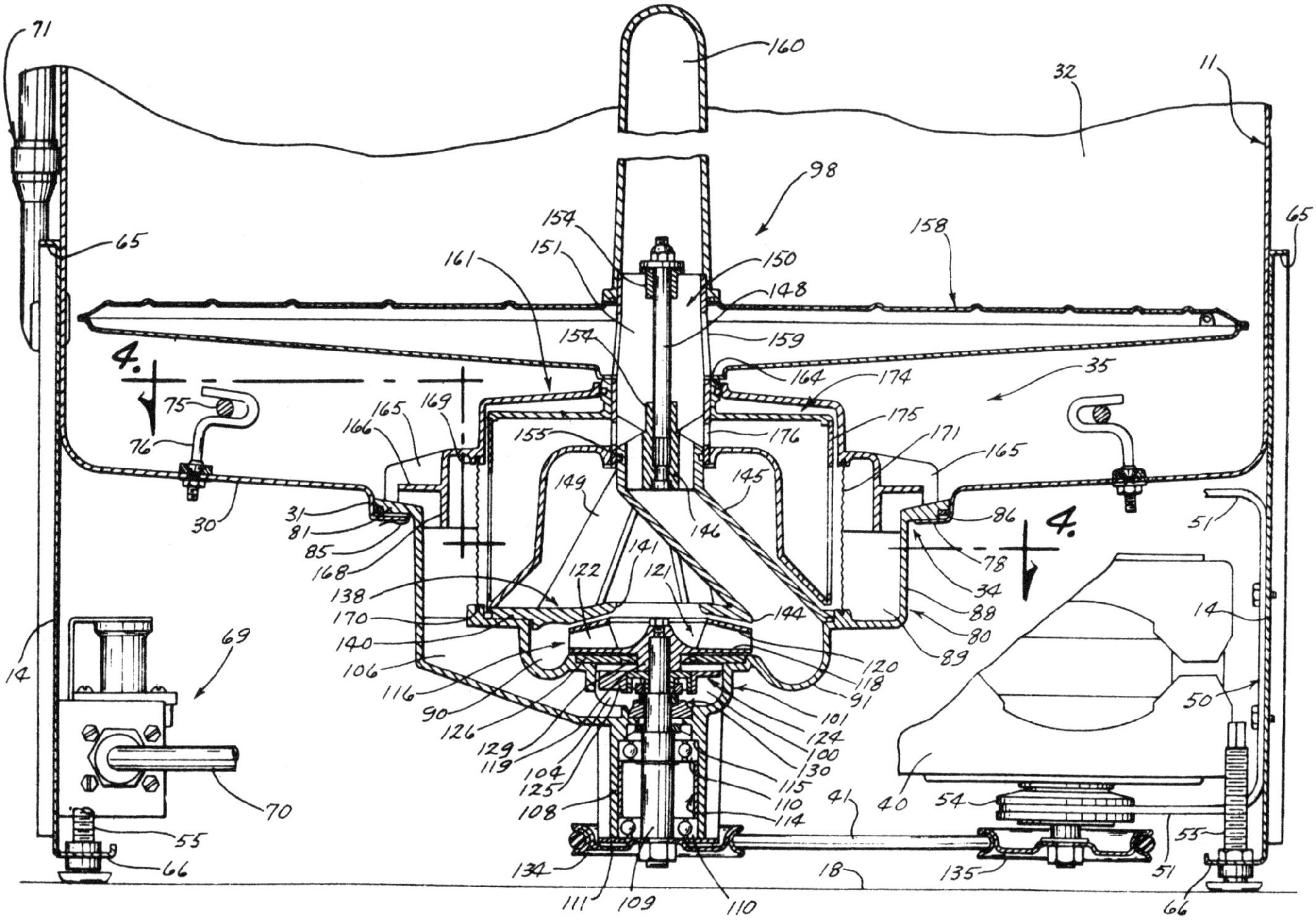

figure 3.5 Cross section of the combination pump and sump construction of a modern convertible dishwasher, Patent No. 3,905,663 (1975). It is shown in the permanently installed position.

The tub (11) is supported by a pair of side legs or supports (14) having reinforcing flanges (65) and (66) extending along the slanting top and along the bottom, respectively, and a flange (68) along the front. The bottom flanges (66) provide means for receiving the floor-engaging members, such as the two screw-in legs (55) positioned at the annular of the machine. These legs (55) may be adjusted to vary the height of the dishwasher.

The water inlet system includes a valve (69) mounted on the side support (14). A conduit (70) from the valve extends toward the rear of the machine and to the outside of the tub and connects with the inlet system (71) for directing liquid into the washing chamber (32). The inlet system along the side of the tub includes the necessary air gap. A float device may be provided for control of the water inlet system.

A heater is provided within the washing chamber (32) for raising the temperature therein and includes an elongated arcuate element (75) supported by a plurality of supports (76) attached to the fluid container bottom wall (30).

The bottom wall includes a generally centrally located recess (31) having a flange (78) defining an opening (34) for receiving the combination sump and pump assembly (35). The main sump pump housing (80) is generally annular and is assembled into the opening (34) in the bottom wall (30) from inside the washing chamber (32), so that the upper flange (81) of the housing (80) engages the recessed flange (78) of the bottom wall (30). The flange (81) is connected to the bottom wall recessed flange (78). An annular seal (86) is disposed at the joint to prevent liquid leakage from the washing chamber.

A first portion (88) of the housing (80) depending from the bottom wall defines a generally annular sump (89) communicating and facing generally upwardly toward the washing chamber. Disposed generally below the sump (89) is a first pump cavity (90) defined by a second housing portion (91) and having a volute form for receiving fluid from the sump and through which the fluid is pumped to the fluid distribution system for effecting washing of articles in the washing chamber.

Disposed below the recirculation pump cavity (90) is a radially smaller and generally annular drain pump cavity (100) defined by a third housing portion (101) for accommodating flow of fluids from the dishwasher sump (89) to an external drain. The inlet (104) communicates with the sump (89) through a conduit (106) defined by the housing (80). Also formed as a part of the main housing (80) is a depending sleeve (108) for supporting a pump shaft (109) through a pair of bearings (110), which are retained within the sleevelike housing (108) by end plate (111), spacer sleeve (114), and the shoulder (115).

A closed vane recirculation impeller (116) is mounted on the shaft (109) for operation within the recirculation pump cavity (90). The recirculation impeller (116) includes a disklike lower base (118) connected to the hub (119) and also includes an annular cover (120) defining a central opening (121) into a plurality of arcuate vanes (122). A drain pump impeller (124) is disposed within the drain pump cavity (100) and includes a plurality of downwardly extending open vanes (125).

A divider (126) is disposed below the recirculation pump impeller (116) to effectively divide the recirculation and drain pump cavities (90) and (100). The divider functions as a proximity seal and flow regulator.

Below the drain pump impeller is a

liquid seal arrangement (130) including a liquid slinger to prevent water from entering the bearings (110). A pulley (134) is secured to the lower end of the pump shaft (109) to effect rotation of the pump shaft and the pair of impellers.

The pump pulley (134) is connected to the motor pulley (135) by a round polyurethane drive belt (41). The motor (40) is operable in first and second directions for driving the pump shaft to effect, in one direction of rotation, a recirculation of fluid to the washing chamber (32), while maintaining a relatively small positive pressure head at the outlet (105) of the drain pump cavity (100) and to effect, in the opposite direction, a substantial increase in output pressure of the drain pump for draining fluid from the washing chamber and sump while maintaining recirculation of fluid by the recirculation pump, but at a significantly lower rate, as long as sufficient fluid remains in the sump.

An intermediate or internal housing (138) is disposed within the sump. It comprises a cover (140) for the recirculation pump and defines an axial inlet (141) into the recirculation pump cavity (90). The intermediate housing (138), also defines an outlet (144) from the recirculation pump at the end of the volute cavity (90) and forms an upwardly extending conduit or tube (145) from the outlet (144) for connection with the recirculation means (98) disposed in the lower portion of washing chamber. The upper end of the manifold or tube (145) includes an internal hub (146) for fixedly supporting a shaft (148) on which a spray device is mounted, as will be described. The intermediate housing (138) further includes a plurality of generally upwardly extending ribs (149) for supporting the upper end of the tube and the shaft fixed therein.

The fluid distribution system (98) includes an upstanding fluid conduit or fluid distribution hub (150) having a plurality of internal ribs (151) supporting a pair of bearings (154) for rotational support of the hub (150) on the shaft (148). The lower end of the conduit, or hub (150) engages a seal (155) disposed in a groove of the upper portion of the upwardly extending tube (145) and interlocked to this tube to prevent rotation of the seal. The main spray arm (158) is mounted on the upstanding fluid conduit, or hub for rotation therewith. Openings (159) in opposite sides of the fluid conduit, or hub (150) at the spray arm (158) accommodate the fluid flow from the recirculation pump cavity to the outwardly extending spray arm. A nozzle (160) extends upwardly from the end of the fluid conduit for distributing washing fluid up into the intermediate regions of the dishwashing chamber. The nozzle (160) is fixed to the end of the fluid conduit for rotation therewith and retaining the spray arm on the conduit.

An upper housing portion, or cover (161) is attached to the main housing (80) along the upper annular flange (81) of the main housing (80). The cover (161) is substantially imperforate and includes a generally central opening (164) for accommodating the upwardly extending fluid distribution conduit (150). A seal is disposed at the opening (164). Along the periphery of the cover are a plurality of generally upstanding radiating ribs (165) that extend radially beyond the outer horizontal flange (166) of the cover. The vanes, or ribs (165) are radially spaced around the periphery of the cover (161) to define a plurality of tunnels having, because of downwardly extending vertical flange (168), a generally 90-degree turn between the washing chamber and the sump.

The outer periphery of the cover thus defines a strainer between the washing chamber and the sump to prevent flow of large particles from the washing chamber into the sump. The spacing of the vanes (165), the radial extension of the vanes relative to the horizontal flange (166), and the 90-degree turn defined by the horizontal flange and the downwardly extending vertical flange (168) prevent flow from the washing chamber of particles or articles exceeding predetermined dimensional characteristics. This flow-control device, or strainer, thus prevents articles from getting into the drain pump cavity (100) and blocking or damaging the impeller (124) thereof.

A lower surface of the cover (161) defines a downwardly facing annular groove (169) generally aligned with a similar upwardly facing annular groove (170) in the lower portion of the sump. These grooves receive the end rings of a generally annular and cylindrical fine-mesh filter screen (171) for support within the sump. The filter screen thus effectively divides the sump into first and second portions with the inlet of the conduit (106) to the drain pump cavity (100) being outside or upstream from the filter (171) while the inlet (141) to the recirculation pump cavity is radially within or downstream relative to the filter.

The disposition of the filter within the sump between the inlets of the drain and recirculation pumps effectively establishes two fluid paths from the washing chamber. A first fluid path extends from the washing chamber through the strainer, the outer sump portion, and the filter screen (171) into the inner sump portion for conduction to the inlet (141) of the recirculation pump. The fluid is then pumped to the fluid distribution means. A second fluid path extends from the washing chamber through the strainer to the outer sump portion and from the outer sump portion through the housing-defined fluid duct (106) to the inlet (104) of the drain pump (100). From the drain pump cavity the fluid is pumped to an external drain through the drain pump outlet (105) when the pump is driven in the predetermined "drain" direction.

Within the annular filter screen is an auxiliary spray device for effecting a cleaning or backflushing of the filter screen. The device includes an arm (174) comprising two depending spray tubes (175) having an elongated slit along the surface adjacent to the screen (171) for spraying liquid onto the downstream side of the screen. The arm is fixed to and rotatable with the upstanding fluid conduit or hub (150). Fluid flows from the upstanding fluid conduit or hub into the spray arm through the openings (176) aligned with the spray arm.

Thus, with the motor (40) operating in a first direction to drive the pump shaft (109) and recirculation impeller (116) in a clockwise direction, the first fluid flow path is established and a high pressure recirculation of fluid to the washing chamber is effected. In this first direction of rotation, the drain pump is operable for maintaining a generally static fluid pressure in the drain line to prevent flow of fluid from the washing chamber while preventing backflow of fluid from the drain line into the sump. In the opposite direction of rotation, the recirculation pump is still operable for effecting fluid flow into the fluid distribution conduit, at a reduced rate, but the drain pump becomes operable for effecting a substantial increase in pumping pressure at the drain pump outlet (105) to establish the second fluid flow path and effect a relatively rapid draining of fluid from the washing chamber.

The output of the recirculation pump is effectively divided between the upper spray arm (not shown), the spray nozzle (160), the main wash arm (158), and the auxiliary filter cleaning spray arm (174). With the recirculation pump operating in the clockwise direction to produce an output of approximately 35 gallons per minute, the relative distribution of fluid from the outlet is as follows: top spray arm 10.1 gpm; spray nozzle 9.6 gpm; lower spray arm 9.6 gpm; and backwash auxiliary spray arm 5.7 gpm.

figure 3.6 Cross section of a countertop dishwasher as shown in Patent No. 3,460,878 (1969). The machine should rest on a countertop (10) backed up by a splash guard (12) between it and the wall (14). The washer is made up of a housing (16), a machine compartment, and a base (20).

The water supply and drain connections comprise a pair of hoses (22) and (24), which connect with the sink. Substantially the entire front face of the wash housing (16) is open to provide an opening bounded along its sides by the opposite, inwardly directed flanges (26) and (28). These side flanges extend upwardly and then along a gentle curve rearward, and terminate at the forward edge (30) of the top wall (32). Thus the opening includes not only substantially the entire front face of the wash housing, but also the forward portion of the top face of the housing. The tambour door (34) slides up and open, in the fashion of the roll top of a rolltop desk. In its open position, it lies below the top wall and in front of the rear wall (38).

Each of the door segments, or ribs (40), is generally channel shaped for most of its length (to conserve material and reduce weight), and is joined to an adjacent rib by the web thickness hinge line (42).

The inner face of the end of each rib (40) carries a pair of spaced tabs (44) which cooperate with the wash housing track means to form a labyrinth seal. The inner tabs (44) have angular ends so that the ribs may pivot toward each other in the more sharply curved portions of the track without the tabs interfering. The outer tabs are of rectangular shape but are obliquely disposed on the ribs so that the adjacent ends of the tabs can lap each other to form an imbricated seal.

Additional track and seal means are provided along the sides of the access opening by an additional inwardly directed flange spaced apart from the outer flange (26) a distance slightly greater than the thickness of the door ribs. Small bumps on both faces of the rib ends provide contact surfaces between the door ribs and track. The narrower track, and tighter seal, are provided only along the sides of the access opening, rather than along the entire track length, to insure that the door when closed is well sealed and free from rattles.

The base structure is essentially a shell to accommodate conduits connecting the wash housing and machine compartment components. The center part of the rear wall of the base is open to permit the hoses (22) and (24) to pass out from under the central part of the base and extend to either side through the recess (60) formed between the rear wall of the base and the back splash (12). The recess will accommodate the back splash so that the dishwasher may be pushed against the kitchen wall (14) to save space.

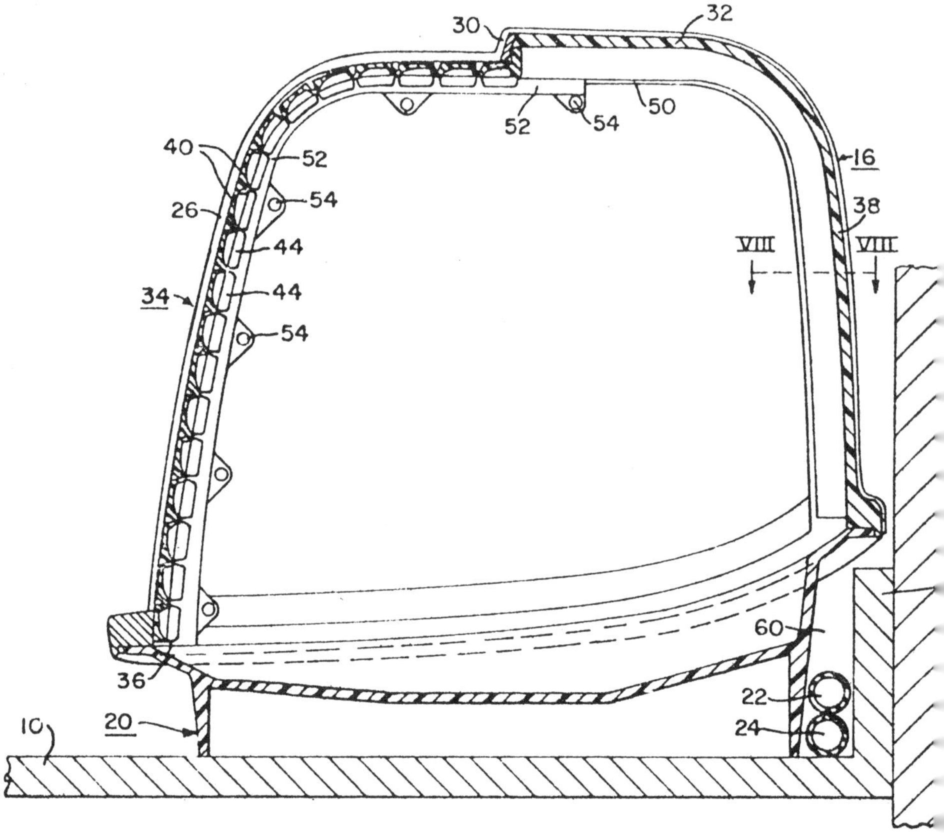

---

figure 3.7 A refrigerator door with a removable radio, as illustrated in Patent No. 3,836, 221 (1974). The radio is retained in an inwardly canted over-center position by virtue of the housing base being sloped outward to a horizontal camming rib.

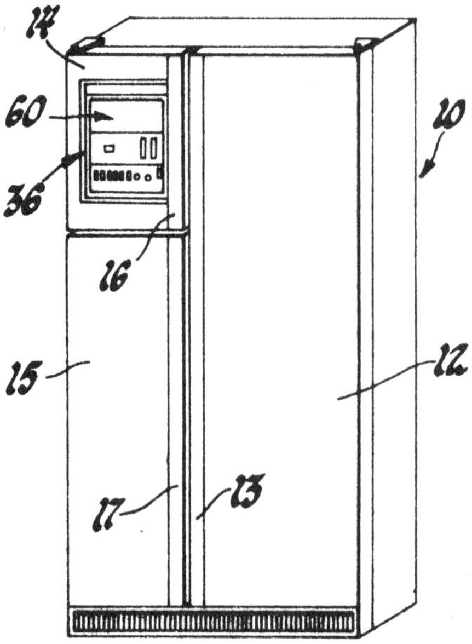

figure 3.8a Diagram of a shelf and support arrangement, generally for a refrigerator, as illustrated in Patent No. 3,857,624 (1974).

A shows the shelf (12) fully retracted into the food compartment (8). The shelf is of the wire rack type with a continuous metal rod forming one contoured side (18), an opposite straight side (19), and the rear (20). The back end (21) of each side is bent up out of the plane of the shelf, raising the rear correspondingly. The front edge (22) of the shelf is a rolled steel trim welded to both ends of the rod. The contoured side (18) includes an inwardly offset length (24) which is of slightly greater length than the length of upper ledge (16) and of sufficient depth to clear the ledge when the opposite straight side (19) is moved into position between the ledges at the opposite wall of the liner. The contoured side is shaped in this manner so that the shelf can be inserted into position in the refrigerator by placing the opposite straight side in the groove formed between the opposite upper and lower ledge while the contoured side is elevated, and then lowering the contoured side at the position where the offset and upper ledge are in registry. In the rearward position of the shelf, upward bumping of the shelf from below does not ordinarily result in dislodging the shelf because the straight side is received in and held by the groove between the upper and the lower ledge. When the shelf (12) is moved forward, both sides 18 and 19 are restrained from upward movement by the upper ledges (16).

B shows the shelf (12) in its most forward position. The shelf is held in this position by the combined effect of the support of the lower ledge (14), the restraining of upward movement by the upper ledge (16), and the back ends (21) of the sides abutting the back end (28) of the upper ledge. In this position the shelf will not tip forward even when items are placed at the very front of the shelf.

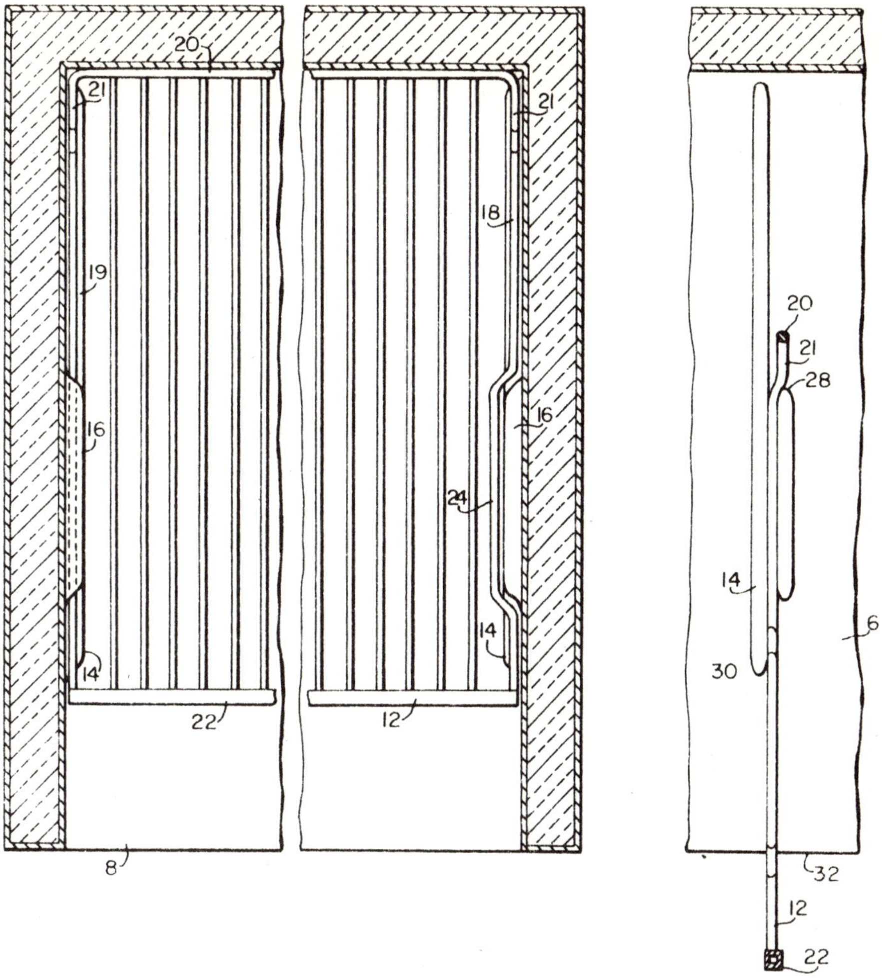

figure 3.8b A refrigerator door liner of molded plastic [Patent No. 3,709,576 (1973)]. There is a center panel (3) formed with a pair of vertical opposite side flanges (4), which reinforce the door liner (2) and support (as conventional), and the hermetic sealing material of the refrigerator cabinet (not shown). A horizontal flange (5) is integrally formed on the liner between the side flanges and is arranged to support a removable wire grid (7) or other shelf, which, when the refrigerator door is in a closed position, projects into the refrigerator cabinet. A lower support flange (5) is shown and may be formed on the panel of the liner, so that two removable shelves may be fitted. Obviously, the liner could be fitted for more than two shelves. The lower shelf is shown in position; the upper shelf is not shown.

A pair of ledges (8) is formed adjacent to each end of the support shelves (5), (6), integrally with the respective vertical flanges (4); each pair of ledges forms a horizontal groove (9), which is open at the end intended to face the interior of the refrigerator cabinet. A suitable formation such as a retainer socket (10), which may be a simple aperture formed in the back of the vertical flange (4) within the groove, is adapted to secure a stirrup (11) formed of metal or plastics material, which retains the shelf (7), as described below. In the drawing, the parts (8), (9), and (10) are visible only on the left-hand flange (4). It is to be understood that similar parts are also provided on the right-hand flange 4.

The stirrup is substantially U-shaped in plan and has side arms (12) to embrace the side walls of the shelf (7) to retain it securely on the horizontal support flange (5) or (6). The side arms of the stirrup are resilient, and each has an outer projecting peg (13), the pegs being inserted in the apertures (10) forming the retainer socket. Each side arm of the stirrup is of such a width as to fit into the groove (9) between the sides of the ledges (8), (9), (10) so as to effect a firm support at each side of the stirrup.

Thus, by inserting the side arms of the stirrup into the grooves and pushing them along the groove, the pegs (13) are urged into the complementary apertures of the sockets; the stirrup, together with the grid which it retains, is then firmly held in position on the door liner.

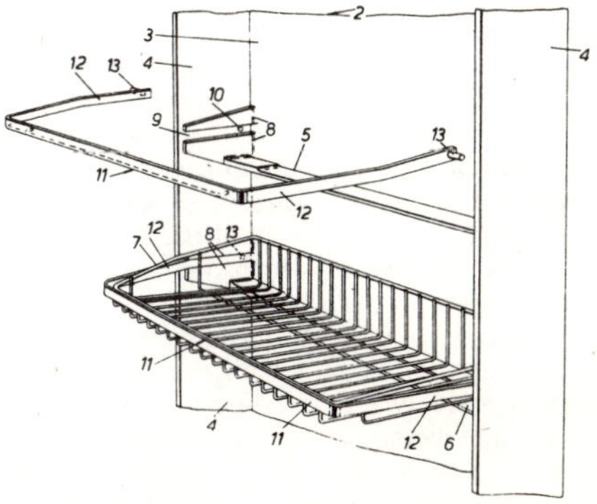

Most manufacturers are intensely competitive, always looking for an edge, a way to make their product better or their costs lower. Figs. 3.9*a* and 3.9*b* illustrate two freezer ice trays, one from 1932 and not sold anywhere now except in flea markets, the other patented in 1968 and still available in many stores. These were patented by, respectively, Ralph Chilton and Robert Strahan, both of Ohio. They are only two of the dozens of ice trays that have been invented over the years, with surely more to come in the future. Why so many? Why aren't people perfectly well satisfied with Chilton's 1932 tray? What do Americans want from an ice tray, anyway? First, they don't want watery shavings, or cubes that have to be hammered out of the tray, or ice that needs to be picked down to size. They do want an inexpensive device that gives them a large supply of ice at the touch of a button—and that seldom, if ever, breaks down. And they want it spillproof.

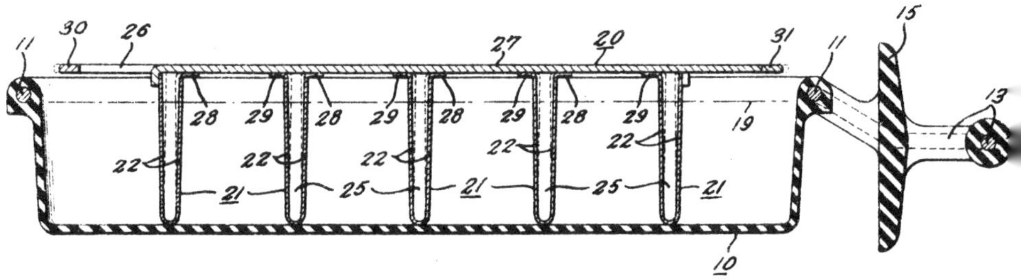

figure 3.9a A side view of a molded flexible rubber freezer tray of 1932, Patent No. 1,882,775.

A metal partition grid (20) is adapted to be inserted loosely within the flexible rubber container (10). This grid comprises a series of hollow metal partitions (21), which are preferably made from thin sheet copper plated with tin, and the double walls (22) thereof are welded or soldered together at their ends so that they will form cold air pockets extending down into the water and contacting with the bottom wall of container to effect rapid freezing of the water. The hollow partitions are held together to form a unitary removable grid by the transversely extending metal bars (26) and (27). In the form illustrated, there are two bars (26), each of which is fixed to small ears (28) on the right side wall of each of the double-wall partitions, while the central bar (27) is similarly fixed to small ears (29) on the left side wall (22) of each hollow partition (21). By this means, when handles (30) and (31), on bars (26) and (27) respectively, are pulled in opposite directions, the flexible metal walls of the hollow partitions will be slightly flexed inwardly as shown by the dotted lines, thus loosening the frozen ice bars from the metal partitions and permitting the easy removal of said ice bars without holding the grid under a water faucet, as must be done with metal grids.

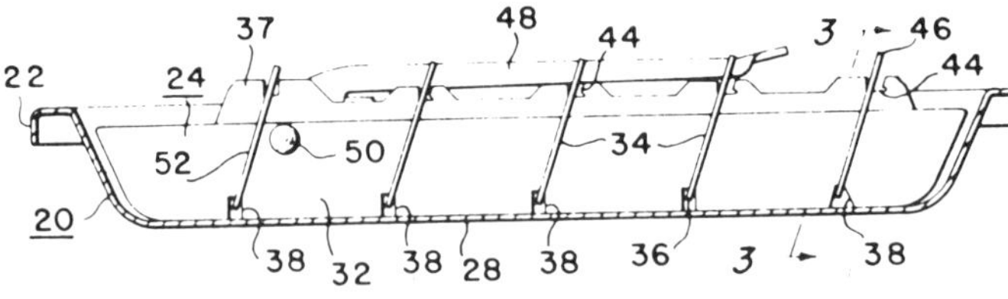

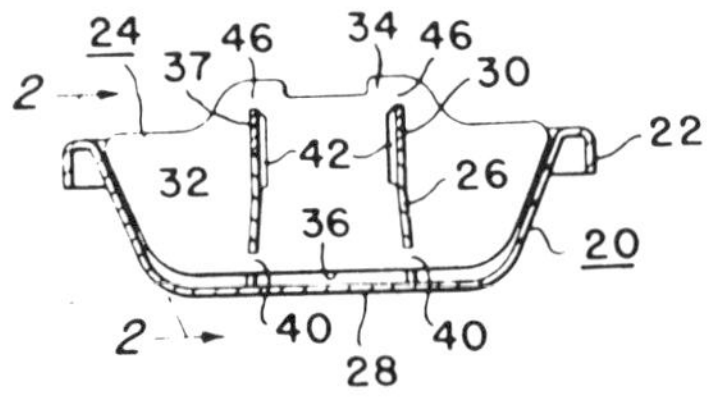

figure 3.9b Side and end views of a freezer tray, Patent No. 3,380,702 (1968). In the pan is a removable grid (24) with a longitudinal wall having a lower portion (26) resting upon the bottom (28) of the pan (20). The longitudinal wall also includes an upper portion (30) having its lower edge sliding upon the upper edge of the lower portion (26). Preferably, there is also a second longitudinal wall (32). The lower portions of these longitudinal walls (26) and (32) are inclined inwardly, so as to facilitate the release of the frozen cubes from the grid.

A lever (48) is pivoted by a rivet (50) to the lower portion of the longitudinal walls. The front portion of the lever, when lifted, pivots on the rivet and engages the upper portion of the first transverse wall (52) and through the interfitting of the first transverse wall (52) with the first notch in the upper portions (30) and (37) of the longitudinal walls to move forward the notches (44) in the upper portions to progressively move the upper portions of the transverse movable partition walls (34) and (52) to loosen the cubes from the tray preparatory to dumping or otherwise removing the frozen cubes. The bottoms of the notches (44) are sufficiently high to engage the webs (46) to hold the bottoms (36) of transverse walls (34) and (52) away from the bottom (28) of the pan throughout the movement of these transverse walls. Through this arrangement, sufficient clearance is maintained between the bottoms (36) of the transverse walls and the bottom of the pan (22) to prevent the bottoms of the transverse walls from scraping the bottom of the pan (20) during the loosening of the frozen liquid within the tray.

The grid is arranged to be freely removable except when liquid is frozen therein. This makes it convenient to wash the tray with cold water. When the transverse walls are moved forward they break the cubes of frozen liquid away from the bottom (28) of the tray and also in a shearing action break themselves loose from the ice. In this action, the bottom edges of the transverse walls are always maintained above the bottom of the tray (20). The ice beneath the bottom edge of the transverse walls also tends to prevent the scraping of the bottom of the tray. The notches (44) in the upper portions (30) and (37) are made of progressively increasing width from the front portion adjacent the rivet (50) to the rear so that the mechanical load involved in progressively breaking loose the ice cubes from the tray and grid is distributed throughout the greater portion of the movement of the handle (48) for better efficiency.

Putting aside gimmicks and relatively minor improvements, what about the basic idea of the refrigerator/freezer? Do inventors have anything to contribute here? They probably do, but there are problems, as there were with dishwashers. For one thing, manufacturers have hundreds of millions of dollars invested in machinery that can produce refrigerators as they now stand. It would take a brave corporate manager indeed to take a chance on a new type of refrigerator. What if it doesn't catch the public's fancy? What if for some unknown reason it proves to be less reliable than the compressor type of refrigerator most of us now use? Perhaps this is where the so-called little guys come in. They're the only ones with little or nothing to lose and everything to gain. The only catch here, as elsewhere, is getting together the capital and the know-how for a major manufacturing and marketing effort. That's something one person cannot do singlehandedly. So what can you do? The answer for many inventors and manufacturers has been to start small and to capitalize on the novelty of their product. Americans are still willing to pay a premium for something that's new and different and better.

figure 3.10a A design Patent for an oven, No. 39,235, issued in 1908.

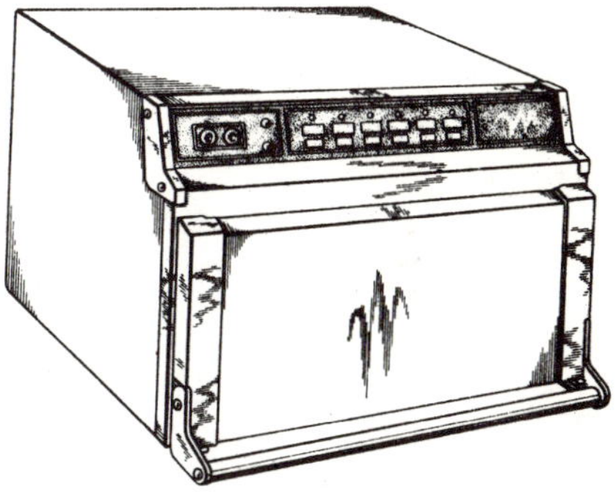

figure 3.10b A design patent from 1976, No. 239,075, for a microwave oven. This is the first of six figures in the patent, which was assigned to the Hobart Corporation.

Another major kitchen appliance is the standard stove-oven combination. Lots of improvements have been introduced over the years, from automatic timers to self-cleaning surfaces. Stoves have tended to get bigger and, in most cases, better until in recent years a new cooking device came along—the microwave oven. At first, people showed comparatively little interest in the microwave, perhaps because of price, because of fears about radiation, or a combination of these. And in fact there are still some who do not believe that the radiation question has been fully resolved, presenting to inventors a problem and an opportunity. The microwave oven has going for it the fact that it fits right in with the modern desire for speed and compact size—it looks like what it is. Two design patents are of interest here. Figure 3.10*a* is a 1908 countertop oven; round and comfortable, it fit easily into the big, warm kitchens of its day. Figure 3.10*b* shows a modern oven—sleek, pushbutton efficient, clean, and square. The microwave oven could have been fitted into a cabinet built in the style of the 1908 design, but the vast majority of buyers would have shunned it as an incongruous combination. It is important for an invention to look like what it is.

In common with most other major innovations, microwave ovens have led to the development of an array of accessories and complementary products. For instance, frozen foods should be put into a microwave shielding device before they are put in the microwave oven. This prevents overcooking. Recently, convenience foods especially prepared for microwave cooking have begun to appear in the markets. Costas Goltasos of Massachusetts designed a package for a complete meal—a new-style TV dinner—to be cooked by radiant energy. This package was designed to prevent the buildup of excess moisture in one or more of the food tray's compartments. This could spread throughout the package and result in some of the cooked food becoming soggy and unpalatable. To prevent this, the compartments in this tray are joined in se-

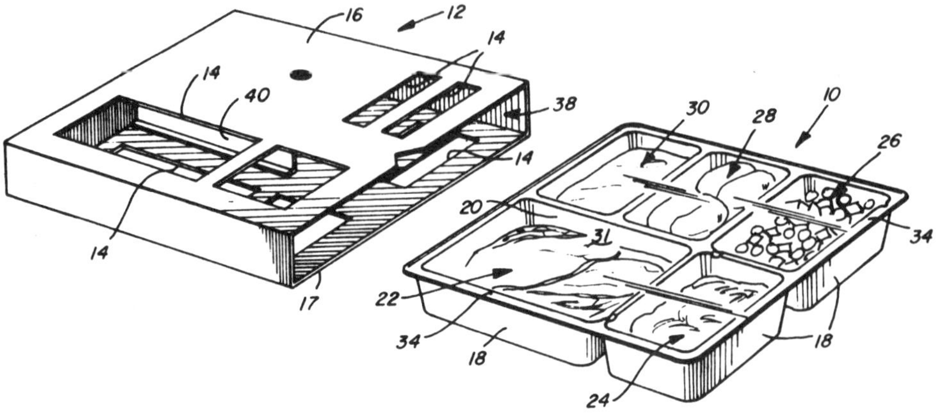

figure 3.11 A packaged meal in a tray beside its metal shielding box, as illustrated in Patent No. 4,013,798 (1977). The tray is opaque to radiant energy except for apertures (14) formed on the top and bottom walls (16), (17).

The tray is formed from a material transparent to electromagnetic radiation. It is formed from an easily moldable, inexpensive plastic and includes a bottom wall, an upstanding peripheral side wall (18) and a number of interior upstanding ridges (20). The bottom wall, the side wall, and the ridges are arranged to define a plurality of compartments (22), (24), (26), (28), and (30), which are intended to hold different components of the meal.

The seal between the compartments that bear water-laden foods is weakened so that in the event of excess pressure build-up from water vapor, the seal will rupture at its weakened portion to vent those compartments to the atmosphere. The portion of the side wall defining the vegetable compartment (26) defines a depression, or notch, which effectively reduces the width of the seal between the sealed peripheral margin of the transparent film (31) and the rim (34) of the tray at the upper edge of the side wall (18).

It is also desirable to avoid condensation within the interior, relatively inaccessible portions of the shielding box. The vent notch should be located on the tray so that it is exposed outwardly through that portion of the box in which an opening (38) is formed to receive the tray. In order to insure that the tray (10) is inserted into the shielding box (12) in the proper orientation, so that the vent notch will be located at the box opening (38), the interior of the box preferably includes a guide member (40) which is disposed within the box.

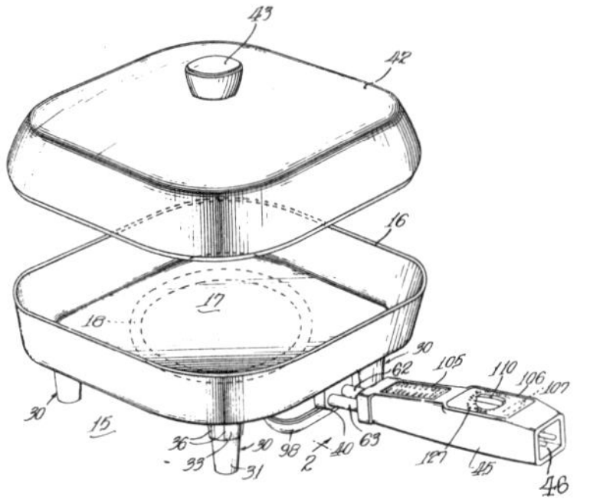

figure 3.12a Perspective view of an electric frying pan, Patent No. 2,744,995 (1956).

The vessel (16) is substantially square except that the four corners are smoothly rounded to make cleaning simple. Preferably, the vessel is cast from aluminum, a very good heat conductor. An electric heating element (18) is preferably cast within a C-shaped rib (19) on the underside of the bottom (17).

The heating element is comprised of an outer metal sheath and a length of resistance wire. It is bent into the form of a substantially one-turn ring, sometimes referred to as a C-shaped heating element having a diameter somewhat smaller than the exterior dimensions of the vessel. Having the heating element cast within the raised rib depending downwardly from the bottom does not interfere with the flat cooking surface on the upper side of the bottom.

quence and the seal between the last compartment and the side wall is weakened so that it will burst if there is too much water vapor (Fig. 3.11). In another version of this package designed for use with a standard oven, the seals for each individual compartment are set to burst under pressure, venting the vapor into the air.

Countertop appliances of all types have grown in popularity in recent decades. Small broilers, toaster ovens, and electric fry pans have all met with great success. One of the early fry pans was patented in 1956. It embodied many improvements, most importantly in being fully immersible in water for cleaning. The heat was controlled automatically, according to a setting of the thermostat in the handle. Note in Fig. 3.12*a* the square shape, which makes for efficient use of the cooking surface while avoiding the square corners that would be difficult to clean. Ivar Jepson of Illinois, the inventor, who was an employee of the Sunbeam Corporation, devoted a great deal of time to perfecting each part and detail of the pan. This didn't stop other inventors from coming along with many worthwhile improvements and variations. Indeed, Jepson and some collaborators took out new patents in 1965 and in 1968 for changes of their own (Figs. 3.12*b* and 3.12 *c*). The major addition in 1965 was a one-turn-loop sheathed heating element, which helped to make for uniform temperature on every part of the cooking surface; in 1968 the heating element was lengthened and its shape changed. Also, improved temperature control was achieved, so that foods would not have to be watched all the time during frying.

The desire for off-the-stove cookware seems almost without bounds. Crockpots are just one of several items that have been very successful in recent years. Here again, the appliance is simple to handle and to clean, and food preparation is made easier. A significant improvement in this type of cookware is shown in Fig. 3.13. The patent description sets forth some of the problems that the inventor, Robert J. Scott of Missouri, set out to solve:

It is well known to wrap an electric heater around the outside of a vessel for providing heat to partially liquid food. In normal construction, the earthenware of a ceramic vessel is glazed on the inside of the vessel . . . thus preventing the flow of electricity from the electric heater, along the electrically conductive foodstuff into the vessel. . . . This construction has a very undesirable characteristic, namely that of the

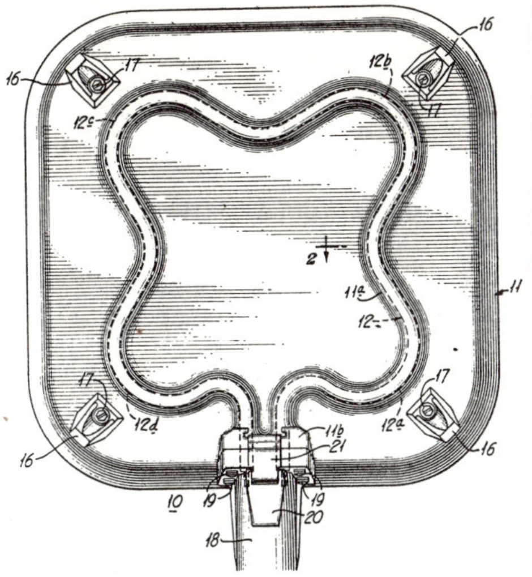

figure 3.12b Bottom view of a later electric frying pan, Patent No. 3,196,253 (1965).

There is an electric heating element in the form of a single turn loop, generally designated as (12), which is preferably of the well-known sheathed type. It has a particular lobed configuration forming an important part of the present invention, and is cast into a raised, integrally formed rib (11*a*).

The temperature control means is built into the vessel or handle, similar to the arrangement in Fig. 12*a*. The rib (11*a*) merges with a housing portion (11*b*), to which is secured a hollow handle (18), preferably formed of the same material as the legs (16). The handle is secured to the vessel (11), and specifically to the portion (11*b*) thereof, by suitable fastening means (19). Preferably, suitable sealing means are interposed between adjacent portions of the handle and portion (11*b*) to insure a completely sealed construction for immersion in liquid for cleaning purposes. Contained within the sealed chamber defined by the integral portion (11*b*) and the hollow handle, are temperature responsive control means and switch contacts for selectively controlling the energy supplied to the sheathed heating element (12) and, consequently, for controlling the temperature of the cooking surface.

There is a fifth leg (20) secured by a resilient spring element (21) to the portion (11*b*) of the housing. This arrangement permits the level of one side of the vessel to be raised to cause any liquid contained therein to move to the other side.

The single turn loop defined by the sheathed heating element is provided with a particular configuration to give substantially uniform temperature at any point over the cooking surface. This is accomplished by providing a configuration having outwardly directed lobes. By virtue of the lobe construction, a greater length of heating element is available in a single-turn loop, which permits a lower heat density per unit length, which insures less likelihood of hot spots.

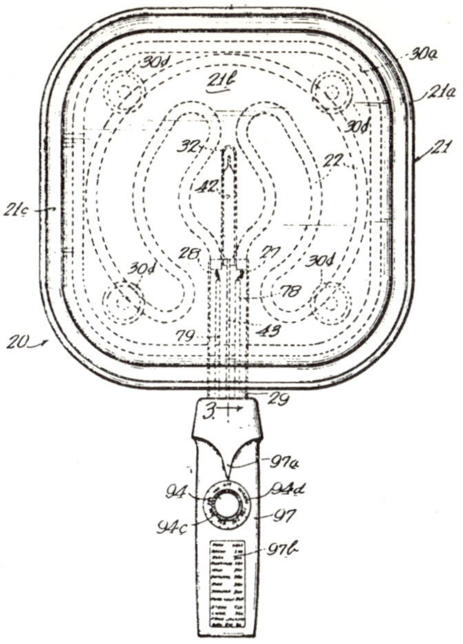

figure 3.12c Top plan view of a still later electric fry pan, Patent No. 3,384,195 (1968).

The heating element (22) is formed in the shape of a sort of double U-shaped element, as indicated in dotted lines, so as to supply heat uniformly over the entire area of the bottom surface of the vessel (21). The vessel preferably comprises, in addition to a rim (21*a*), a relatively flat bottom (21*b*) and a shallow peripheral flange (21*c*) integrally formed with the bottom (21*b*) and the rolled rim (21*a*). The peripheral upstanding flange (21*c*) is preferably flared slightly outward, and also the bottom (21*b*) and the side walls (21*c*) are preferably connected by a smooth curve.

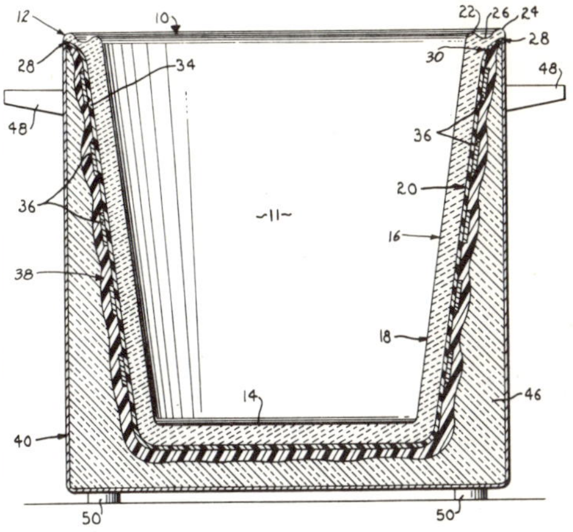

figure 3.13 Cross section of the "crock pot" illustrated in Patent No. 3,806,701 (1974).

The cooking utensil (10) has a cup-shaped vessel portion (11) and a lip portion (12). The vessel is constructed of a lightweight ceramic material that is thermal conductive, such as earthenware or stoneware. It is formed by a bottom (14) and an upwardly diverging wall (16). Bottom and wall have an inner side (18)—usually glazed—facing into the vessel and exterior second side (20). The thickness of the bottom and side wall—distance between inner side and exterior side—is usually maintained constant within the vessel. The exterior side will be unglazed and, as will be seen, accommodates and supports the electrical mating elements and the resin varnish.

The lip of the vessel has an inside ridge (22) and a higher outside concentric ridge (24) with a valley (26) located therebetween. This ridged lip construction facilitates the placing of a cover over the food being cooked or warmed within the vessel (11).

The outer and under portion of lip has a smooth arcuate contour as indicated by the numeral (30). For ease of construction, the vessel will be formed so that the ceramic material is provided with a smooth and continuous surface except for a junction edge (28), to substantially eliminate possible thermal stresses within the ceramic material.

A silicone varnish is applied to the entire exterior from the junction edge downward, including the bottom of the vessel, and will be absorbed by the ceramic refractory material. The varnish is of the type that is thermally conductive to allow heat passage when heat is applied to the vessel by the electric heater elements (36). Additionally, the silicone varnish will waterproof the exterior surface.

coat being constructed from the refractory material from which the ceramic vessel is constructed.

This refractory material is brittle, and any kind of blow might crack or chip it and thus damage the electrical connections. So the crockpot was made safer and simpler to use. Inventor Scott assigned his patent to the Rival Manufacturing Company, which gave the product wide distribution.

Almost anyone can use the crockpot described above, and this is an important point. Many inventors have fallen to the temptation of making something so complicated that only a person with technical training could use it easily. No matter how wonderful a new product is, if it's difficult to use, it's not going to find a mass market. Many of the new products of recent years have gone to the other extreme, being so obvious as not to be eligible for a patent. Many of the quickie burger and hotdog cookers that appeared in huge stacks in appliance stores during the 1970s were an example of this.

Ethnic foods are "in" these days, even to the point of having special appliances to prepare them for the table. Maria Orts of Texas has patented a tortilla warmer and hydrater (Fig. 3.14). This device makes it possible for a cook to prepare a large batch of tortillas at one time and then to serve them, hot and moist, at a number of subsequent meals. Is this appliance too specialized for most kitchens? Of course it is, but for those households where tortillas are a staple, this warmer would be used a great deal more than, say, a waffle iron. So it has its uses. Think about the different kinds of food served in your home or at the neighbors'. Is there a better way to prepare them? Does your family or someone you know have a special way of cooking a dish that might be translated into a product for a wider public? It's worth following up even specialized techniques. The United States is a huge country with such a diversity of people that there is a market of some size for virtually everything of some worth.

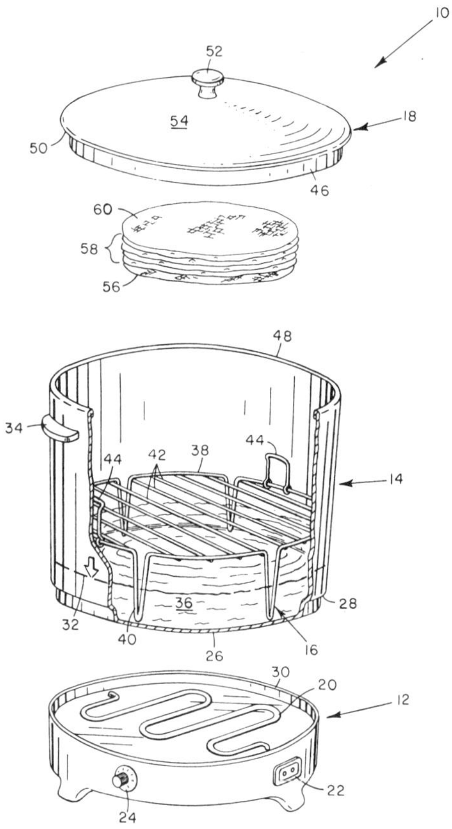

figure 3.14 Partial cross section of a tortilla warmer and hydrator, as illustrated in Patent No. 4,013,869 (1977).

It consists of a base (12), a cylindrical pot (14), a rack (16), and a top (18). The base is an electrical type having a resistive heating element (20), which receives power through an electrical connection (22). The amount of current is controlled by a thermostat (24), which in turn controls the heat given off by the resistive heating element (20).

The cylindrical pot is particularly designed to fit the base. Near the bottom (26) of the pot is a shoulder (28) that will rest against the rim (30) of the base. Upon placing the pot into position, the shoulder (28) resting on the rim (30) will hold the bottom (26) just above the resistive heating element (20). This will allow maximum heat transfer to the pot.

The pot also has a water-level indicator (32) which may be formed by indentations in its side. Handles (34) are attached to the sides.

The pot is filled with water to the water-level indicator. Above this is the rack (16) that has a circular support frame (38) with extending projections (40) that form legs. The legs (40) extend into the water (36) and rest against the bottom (26) of the cylindrical pot (14). Horizontal support bars (42) are located in a plane substantially parallel to the bottom (26) and above the water level indicator. The horizontal support bars (40) connect to the sides of the circular support frame (38) by any convenient means, such as welding.

The top (18), which is circular in design, has a downwardly extending flange (46) that fits just inside the upper rim (48) of the pot. A shoulder (50) rests against the upper rim to form a relatively good seal. A knob (52) is located in the top center of the cover portion (54) for ease of removal of the top.

In use, the pot is filled with water (36) to the water level indicator (32). The rack (16) is then placed inside the pot, which is placed on the base (12). The water is brought to a simmer. A cloth pad (56) is placed inside the cylindrical pot (14) on the rack (16). Thereafter, tortillas (58) are placed on the cloth pad. Another cloth pad (60) is placed above the tortillas. Next, the top (18) is placed securely into position and the thermostat (24) is adjusted to bring the water to a boil. The steam created by the boiling water will fill the space inside the cylindrical pot and will be absorbed by the tortillas. The cloth pads (56) and (60) ensure that the tortillas do not stick to the rack (16), and aid in the holding of moisture inside the tortillas. As many tortillas as can be conveniently held in the pot may be warmed and hydrated at the same time.

Going from the specialized to the general, one of the most popular small appliances throughout the country is the automatic coffeemaker. Over the years there have been developed almost as many ways of brewing coffee as there are varieties of coffee beans. Some methods have been more successful than others, but they have all been greeted with interest on the part of consumers. Some of these coffeemakers make more sense than others. The musical pot shown in Fig. 3.15 did not ring many bells in the marketplace. Although it's handy to have a device to tell you when something is done—the whistling teakettle was certainly a practical item—it may not be worth what you'd have to pay for the music box and accessories.

One of the great success stories of recent years is represented by the "pour-in, instant brewing electric coffeemaker," one of the best examples of which is shown in Fig. 3.16. In all of these coffee machines, cold water passes through an electric heater that brings it to the proper temperature for making coffee. Before 1972, it had been the standard practice to have a preheated metal block of rather large size as the primary heater. This was costly and cumbersome, but it was the only way that could be worked out to quickly handle water that was poured in by the user without much regard for achieving an even, controlled flow. The illustrated invention introduced several improvements:

First the gravity feed of water from the reservoir to the heating device is controlled . . . to utilize the resistance heating ability throughout the cycle; and the reservoir is superimposed upon and as a cover for a heating block having a top water flow cavity having certain baffling therein. . . . Along with effective heat transfer, relative simplicity and small size in the heating element casting is attained, decreasing the component and fabrication costs as well as enabling use of a smaller, simpler, lower-cost housing structure.

This invention, by Edmund Abel, Jr., of Ohio, was the basis for one of the fastest selling new appliances of the 1970s, the Mister Coffee machine.

Another method of brewing coffee is described in a patent granted to Alexander Daswick of California. It can be compared to the coffee bag illustrated by Fig. 8.5 in the chapter on packaging. The present invention, shown in Fig. 3.17, is a disposable filter, open at the top and closed at the bottom, held in a supporting cone. It is designed to make one cup of coffee—hot water usually would be poured through the coffee grounds in the filter directly into a cup or mug—and thus is quite economical for the solitary drinker.

Appliance accessories are one of the most important areas of investigation for the independent inventor. Almost everything has something about it that could be improved, and very often problems or opportunities for change are not apparent to the manufacturer. For instance, the portable electric mixer is an excellent convenience item that quickly found a place in many kitchens. It's a lot easier to use than the larger mixer-with-bowl appliances that were the standard for many years. And it can be used in pots on the stove or anywhere it can be plugged into an electric outlet. Carl Meyerhoefer of New York and Richard Naples of Connecticut teamed up to develop an accessory that adds to the usefulness of the portable mixer at a low cost (Fig. 3.18). Its advantages are clear,

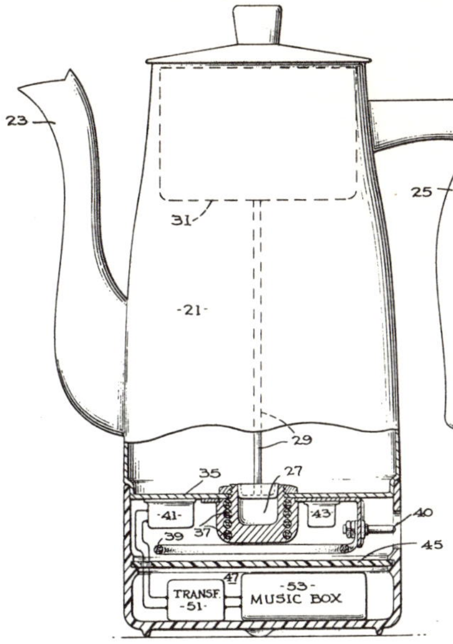

figure 3.15 Coffee percolator with timer and alarm, as shown in Patent No. 3,226, 528 (1965). The water well (27), a fountain tube (29) and a ground-coffee-holding basket (31) may be of any suitable designs. The heating and control portion of the coffee maker is separated from the fluid container (21) by a bottom wall (35). Included herein is an electric heating element (37), a current-reducing resistance element (39), power terminals (40), and first and second thermal responsive switches (41) and (43), both in a heat-transfer relationship with the content of the container (21). An insulating member (45) separates the heating and control portion from an electrified music-producing means (53), thereby protecting the latter from heat radiated by the heating element. A transformer (51) is also located in the music means section for providing the proper voltage to drive the music means (53).

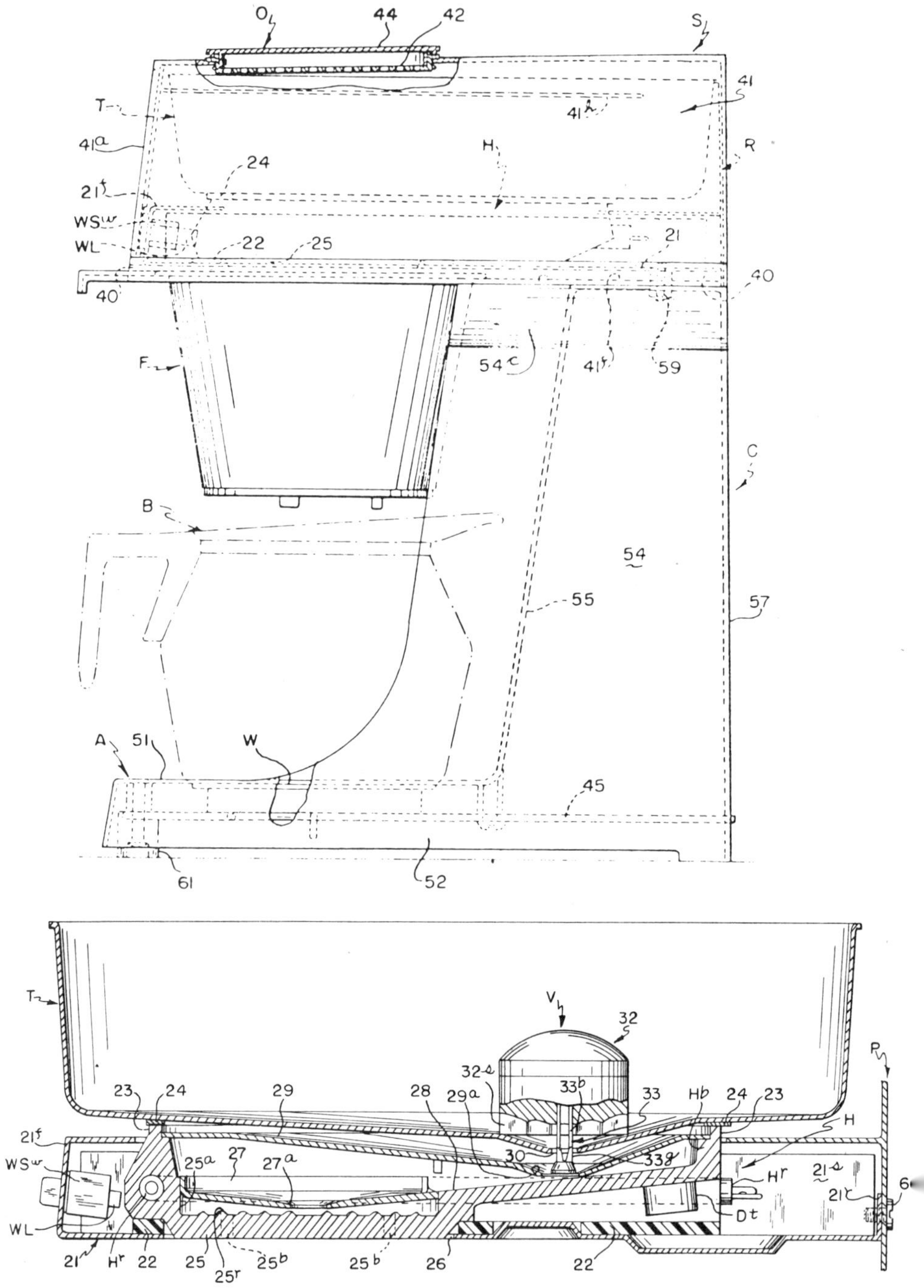

figure 3.16 Side view of a coffee brewing machine, Patent No. 3,693,535 (1973), and an enlarged side view of its water heater unit.

The principal components are a housing structure, including a pedestal element providing integrally an extended horizontal base portion (A) with a vertically extended back column portion (C), and supported on the pedestal; a top shell or hood element (S); in the base, a warmer plate (W) for a coffee-receiving beaker (B), above which a brewing funnel (F) is slideably, insertably, and removably supported in the bottom of an overhanging front part of the hood (S); and within the hood, a generally rectangular pour-in reservoir tray (T), the major part of appropriate electrical wiring for control elements, and an electrical resistance-type water heater unit (H), to discharge hot water downwardly into a funnel (F). The tray (T) has a capacity to receive a desired or rated maximum of water poured in through the hood top opening structure (O); and its bottom has an elongated central depression symmetrically sloped from the sides to a discharge bottom orifice (30) rearwardly offset on the longitudinal center line and provided with a variable orifice or valving structure (V) controlling water-gravity feed to the heater unit as hereinafter described.

The unit (H) comprises a contoured metal body or block (H*b*), symmetrical about a longitudinal center-line, and is supported on a formed sheet metal support or hood bottom plate (21), with an interposed sealing and insulating flat silicone composition bottom gasket (22).

In the top of block H*b*, a broad flow cavity, including a front circular well (25*a*) and a sloping channel 28, is circumscribed by a rim bead (23) upon which, through an interposed flat oval silicone top gasket (24), the reservoir tray, T, is supported. At the left, under well 25a, a disklike, beveled downward offset or flat boss (25), projecting into a corresponding large bottom plate opening (26), is penetrated by four equi-spaced block cavity discharge openings (25*b*). A circumferentially upwardly rim-flanged circular well baffle plate (27), dished to an edge-beveled central opening (27*a*), is press-fitted to a close spaced relation with an array of interrupted spaced concentric ridges (25*r*) integrally cast in the bottom of the well, for increased heat transfer area and flow baffling.

A further drawn sheet metal elongated oval baffle plate (29), inserted cover-like in the broad top region of the casting cavity within the rim (23), is inwardly depressed from its flat rim margins symmetrically toward the longitudinal center line and further along that line toward a circular opening (29*a*) centrally disposed beneath the tray outlet orifice (30) to direct water to the upper end of the sloping channel (28) leading downwardly to the well edge to discharge through a flange interruption to the top of the baffle (27) and into the well.

The similarly convergent bottom of the tray (T) serves not only to direct contained water toward its discharge orifice (30) centered above the opening (29*a*), but also on its sloping undersurface it guides steam condensate to the region of the orifice, to feed water to the sloping channel (28). Thus, water runs with continual opportunity for heat transfer beneath 29, along channel 28 onto the baffle disk (27), through central opening (27*a*) to flow from the center across the corrugated or ribbed bottom well surface to apertures (25*b*) at the outer well margin to discharge downward into the brew funnel (F).

The cast-in heating resistor rod (Hr) in one end of the block is formed into a nearly circular shape to surround the major circumference of the central well region from which its ends are carried in the casting out to the right parallel and adjacent the flow channel (28).

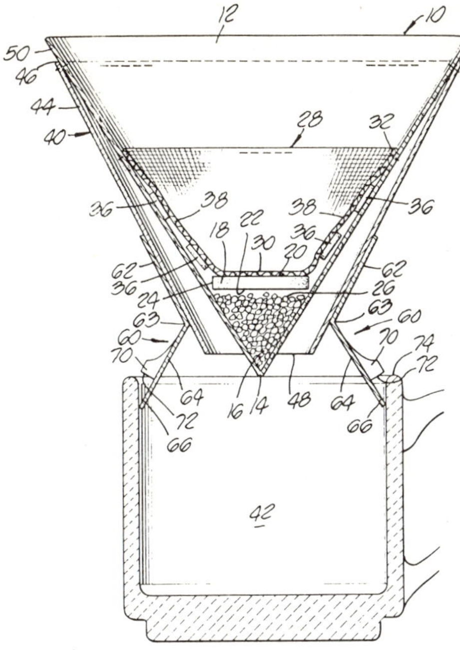

figure 3.17 Cross section of a disposable coffee brewer on top of a mug, as illustrated in Patent No. 3,971,305 (1976).

It comprises a disposable conical filter (10) having a comparatively large open upper surface surrounded by an upper edge (12), and tapering to a bottom point (14). The filter (10) is constructed of filter paper which has substantial wet strength and which is comparatively inexpensive. Within the filter (10), above the coffee grounds (16), is a primary water diverting element (18) which comprises a thin disc having an upper surface (20), a lower surface (22) and an edge (24), and having a diameter approximately equal to the cross-sectional diameter of the filter at an upper surface (26) of the coffee grounds. The element is constructed of a nonporous, buoyant material, such as Styrofoam, which will float upon water.

A retaining mesh or web (28), constructed of an inexpensive, loosely woven, flexible material such as cheesecloth, is attached, at a central region (30) thereof, to the upper element surface (20). An outer edge portion (32) of the web is attached, as by sewn stitches, to the inside surface of the filter (10), at or above the mid height thereof. The web is constructed to have a conical shape with substantially the same angle of taper as that of the filter, so that, upon assembly and with the filter vertical, the element (18) hangs initially at or near the upper surface (26) of the coffee grounds (16). A dual function is performed by the web: It retains the coffee grounds within the filter—the mesh size being sufficiently small for this purpose—and it retains the element within the filter, allowing it to float at whatever level of water is contained in the filter.

Maintained thus in a floating position by the web, the element performs the important function of diverting hot water poured into the filter and prevents it from falling directly upon the coffee grounds. Such direct pouring tends to extract the bitter acids from the grounds and also may tend to pack the coffee grounds into the bottom of the filter in a manner to cause inefficient brewing, or plugging, of the filter. For such water-diverting purposes, a number of small, thin, buoyant secondary elements (36), also made of a material such as Styrofoam, may be attached to the under side of the web (28), in spaced relationship, in an intermediate region (38) between the element (18) and the edge portion (32). As water is poured into the filter, such elements (38) float between the edge (24) of the element (18) and the adjacent sides of the filter (10) and provide additional covering for the coffee grounds.

A support (40) is provided to position the filter and its contents above a container. The support has a frusto-conical member (44) with an upper edge (46) around an open upper end and a lower edge (48) around an open lower end. The filter (10) fits downwardly into the member (44), which has a conical angle less than that of the filter, such that an upper portion (50) of the filter (10), above the maximum water level, rests against the edge (46), and is circumferentially attached thereto, as by sewn stitches.

Because of the lesser conical angle of the member (44), lower portions of the member and the filter are spaced apart so that the brewed coffee may flow from the filter freely down the outsides thereof to the point (14) which may project below the member lower edge (48).

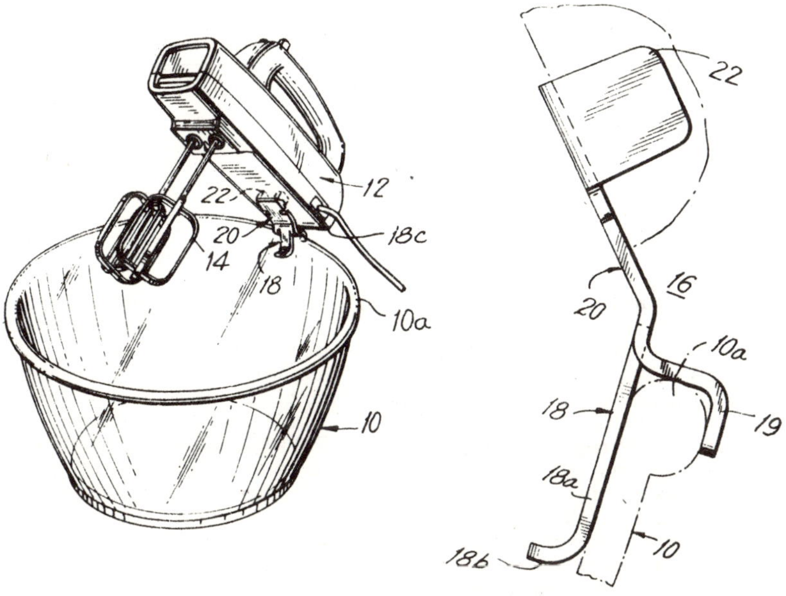

figure 3.18 Perspective view of a portable mixer held in position over a mixing bowl by means of a hanger support [Patent No. 3,827,677 (1974)], and a side view of the support.

This device (16) comprises a lower bowl-engaging portion or member (18) and an upper member (20). The lower portion is made up of a central section or tongue (18*a*), which is on the longitudinal axis of the device and which is adapted to engage the inner surface of the mixing bowl and to extend downward to a significant distance inside the bowl. The end of the tongue (18*a*) is conveniently formed at right angles to provide a smooth pad (18*b*). The pad (18*b*) serves as the third point of the tripod-like support. Corners (18*c*) of the mixer housing provide the other two points of the tripod support. A pair of ears (19) constitute the outer section of portion (18), being adapted to engage the rim (10*a*) and the outer surface of the bowl (10).

The upper portion (20) of the support device is at an angle to the longitudinal axis of the lower portion (18), being inclined radially inwardly when in use with respect to the center of the bowl (10). The upper end of the portion (20) is conveniently provided with a pair of spaced prongs (22), which are designed to be received within the apertures of the mixer when the mixer is brought into engagement with the support device. As a result of this engagement, the mixer (12) is permitted to extend over the bowl, the beaters (14) are inclined downward, and any residue of batter drips off.

yet it wasn't developed until years after the mixer itself had come onto the market. The inventors' hanger enables the user to rest the mixer on the side of the mixing pot or bowl, instead of putting it down on its side to drip onto the stove or countertop.

One way to improve two items is to combine them for easier use. An example of this is an invention by Samuel McNair of Kansas. Most kitchens are equipped with some kind of a knife sharpener, yet it seems that most people's knives are still pretty dull to the touch. McNair's invention makes it easier to keep blades sharp, being a combination storage holder and sharpener for knives. This is an interesting patent to study (Fig. 3.19). Many questions can be asked: Does it serve the need better than the

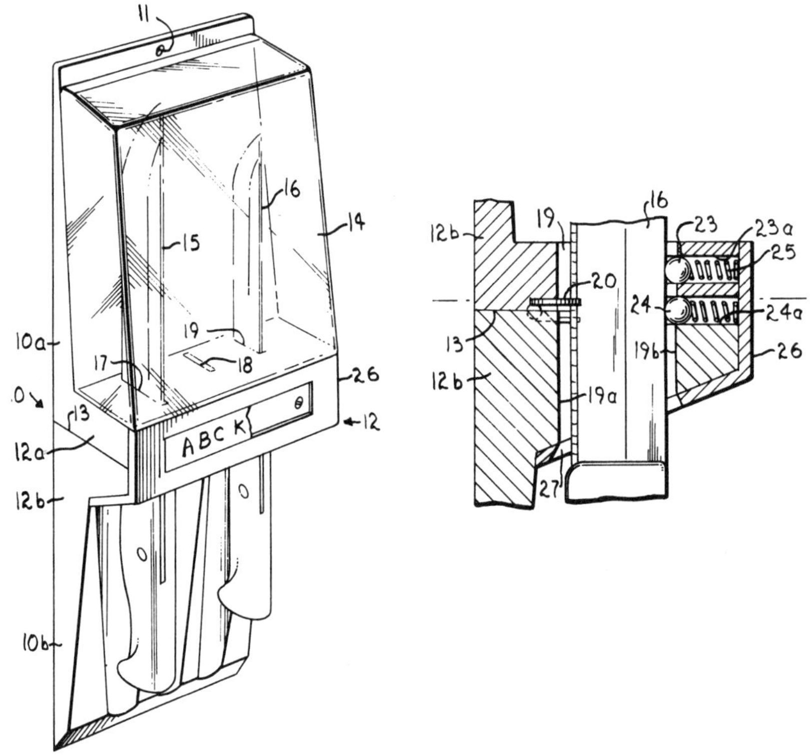

figure 3.19 Two views of a knife holder/sharpener, Patent No. 3,889,809 (1975).

A base panel (10) is adapted to be hung upon a nail or a peg in a vertical wall by means of a hole (11). The panel includes a transverse housing (12), which may be formed integral with the base panel. Panel and housing are made up of two solid plastic parts divided at line 13 so that part 12a of the housing is formed by a projection on the upper portion (10*a*) of the base panel, and the other part (12*b*) by a projection on the lower part (10*b*) of the base panel.

A transparent casing (14) is connected with the upper portion of the base panel and provides a storage compartment for the blades (15), (16) of the knives shown stored in the unit.

Housing (12) is provided with three parallel blade slots (17), (18), and (19), respectively. Each of the slots extends from top to bottom—i.e., in a direction parallel with the long axis of the panel, (10)—through the housing (12). Each slot is designed to receive a knife.

The sharpening and holding means for sharpening the knives and retaining them in the unit during times of storage are located within the housing. Each slot and its associated sharpening and holding means is in principal identical with the others.

Located adjacent to what will be referred to as the base or bottom edge (19*a*) of the slot (19) is a pair of partially overlapping discs (20). These are relatively thin circular discs of tungsten carbide. Each disc is located within a routed-out passage formed in the face of the respective housing halves (12*a*) with the discs in parallel and slightly spaced planes. The discs are loosely contained within the passages so that they are movable therein. A steel ball, of the bearing type, is located between the disc 19, and the edges of the discs bear against the ball and are maintained slightly separated by same. Each disc is generally confined in a parallel plane with respect to the other by means of the face of the opposed half of the housing (12) which forms with each passage a confined track for the disc located therein.

Extending into the slot from the edge of the slot opposite from the discs (20) are two bearing balls (23) and (24). Each of these balls is loosely confined within a cylindrical bore (23*a*), (24*a*) terminating in matching semi-cylindrical recesses formed in the opposite side walls of the slot. The bores (23*a*), (24*a*) extend inwardly past the outermost edge (19) of the slot so that with the knife removed the balls (23), (24) will be displaced inwardly of the slot edge. The balls are engaged respectively by compression springs (25). The ends of the spring opposite from the ball are retained by a cap piece (26), which is secured to the housing halves and extends along the incline entrance edge of the housing. The cap piece is slotted to coincide with the knife slots in the housing; a beveled edge on opposite sides and at the ends thereof provides a funnel-like entrance guide for the knife point as it is inserted in the housing.

As the tip of a knife blade is inserted through the entrance (27) to the slot, it eventually passes between the discs and the balls (23), (24), and upon engagement by the latter, the balls tend to force the cutting edge of the knife into alignment with the V-shaped notch formed by the discs with the sides of the knife edge engaging and supported by the discs. As the knife is pushed on into the slot, the balls will be displaced outwardly against the resistance of the springs (23*a*), (24*a*). The springs and balls thus serve to maintain constant pressure against the blade, which in turn keeps the cutting edge in contact with the discs. When the knife is fully inserted, the blade will have been engaged and honed by the discs throughout its full length. Moreover, the continuing pressure of springs will serve to hold the knife firmly within the housing until it is withdrawn again.

Obviously, upon withdrawal, the sharpening process will be effected in reverse, with sharpening continuing until the blade is completely withdrawn.

other devices that are available? Is it simple to operate and to maintain? Will it do the job well? Can it be manufactured at a cost that will make it possible to sell in the stores at a reasonable price? How can it be improved? Is there a better way? Of course, these questions cannot be answered for sure without examining a working model of the device. But this kind of examination is typical of the kind of thing inventors should do when they consider their own creations and the competition.

Gadgets hold a fascination for most Americans, and kitchen gadgets are especially popular. Most housewares counters are stocked with such items as cherry pitters, never-fail corkscrews, fish knives, vegetable choppers, salad-green driers, and so on. The public's desire for these kinds of things has not yet been satisfied, and probably never will be. Even during the Depression, pitchmen on seaside boardwalks and at county fairs were able to earn their keep selling the latest combination tomato-slicer-potato-dicer-make-everything-nicer machines. People handed over their dollar bills even though they knew deep down that the machine wouldn't work nearly as well in the home as it did in the flashing hands of the salesman.

There have been many patents granted in this area. The vegetable shredder developed by Karl Zysset of Switzerland is an interesting little machine (Fig. 3.20). In 1957, he obtained a U.S. patent on a shredder that utilizes "a single sinuous steel strip passing through a continuous slot of a wiper cup which is rotatably disposed in the housing." So good clean cuts were possible without the food sticking to the cutter. In 1963, Zysset patented the improved version illustrated in Fig. 3.21. A major change involved

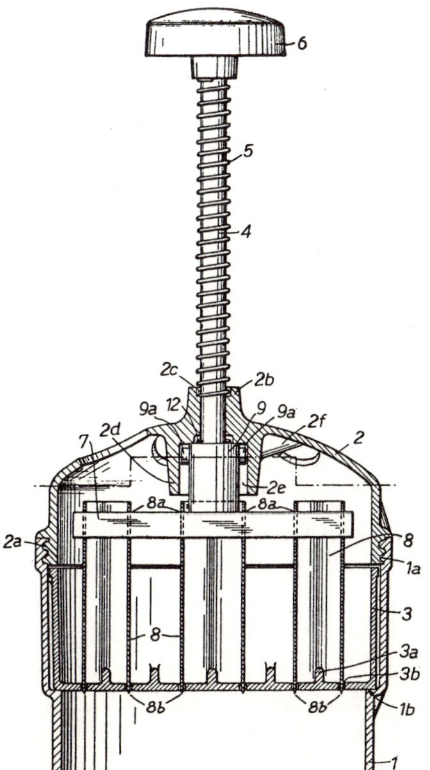

figure 3.20 A vertical cross section of a vegetable shredder, Patent No. 2,782,826 (1957).

A bell-like housing is made up of a lower portion (1) and an upper portion (2) of transparent plastic such as polystyrene. These two portions are interconnectable by means of threads 1a and 2a so as to be readily detachable from each other for cleaning purposes. Housing portion (2), on top, is formed with an integral depending boss or socket (*2d*), on the inside of which are provided six helical grooves (*2e*). A plurality of ribs (*2f*) reinforce the transition part between housing portion (2) and socket (*2d*). To the lower end of a stem (4), which is provided with a handle (6), is secured a prismatic rod (7) standing at right angles thereto. The rod (7) traverses without play a plurality of openings of a steel strip (8). The latter is formed with a knife edge (*8b*) on the bottom and, when operating stem 6 is depressed, acts as a knife for shredding the goods put in the lower open space of the appliance. A compression spring (5) is disposed between a recess (*2c*) in a shoulder (*2b*) of housing portion 2 and handle (6); after actuation of the latter, it returns the blade (8) to its upper initial position.

A wiper cup (3) serves to clean the multiple-bent knife blade during the upward stroke thereof. In the operating position of the shredder, the wiper cup bears on a shoulder (*1b*) of the cylindrical housing portion (1) and is freely rotatable when the two housing portions (1) and (2) are screwed together, but the wiper has practically no play in the axial direction. When the two housing portions are unscrewed, the wiper cup is readily removable. From the apex of a line formed by a slot (*3b*) for knife blade (8) extend a plurality of stiffening ribs (*3a*) from two sides, which brace the wiper cup.

This vegetable shredder is provided with an automatic indexing device which acts to angularly move the cutting means. This device initiates rotary movement only during the last stage of the upward stroke. The device in the present case comprises on one hand the socket 2d with helical guide grooves 2e and, on the other hand, a checking member disposed on the lower end portion of stem 4. This checking member comprises a metallic ring (10), which is freely rotatable on stem 4, a sleeve (9), which is nonrotatable with respect to the latter, and has two diametrically opposite cams (*9a*), and a torsion spring (11) coiled about stem 4 and of which the lower end (*11a*) is anchored in sleeve 9. The latter is cast or pressed as one piece of splinterproof plastic such as nylon. The checking member is prevented from axial movement on one hand through a stud (12) fixed to stem 4 and, on the other hand, through rod 7, against which abuts a metallic washer (*9b*) of the checking member.

The shredder shown operates as follows. When the knife blade (8) is depressed by means of the handle (6), the checking member is rotated by the cams (*9a*) being engaged in two guide grooves (*2e*) in the sense of separating the spring (11) from the stem (4). The friction between the operator's hand and the handle (6) is considerably greater than that between the spring (11) and the stem (4), so that the knife blade (8) cannot rotate. After the handle is released, the knife first moves straight upward under the action of spring 5. In the last stage of such upward movement, the cams (*9a*) of the checking member are engaged in the guide grooves (*2e*). Owing to the rotary movement of sleeve (9) enforced thereby, spring (11) is put tightly about stem (4), whereby the latter is coercively rotated during the remaining stage of the upward stroke. After each operation of depressing knife (8) the latter is moved through a certain angle shortly before reaching its position of rest.

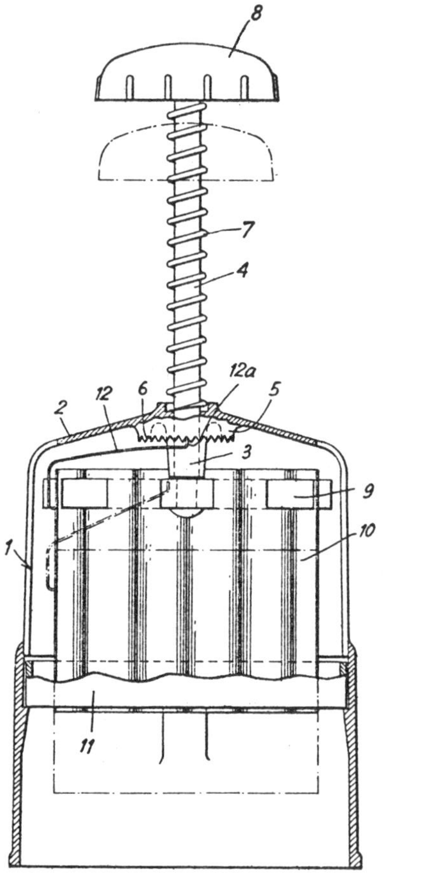

figure 3.21 A vertical cross section of an improved vegetable shredder, Patent No. 3,083,747 (1963).

The housing (1) is a press-casting of plastic; the ceiling on its underside has a boss (3) for guiding a bar (4) and a further boss (5), which is concentric with boss 3 and has radial serrations (6). Between the ceiling (2) and an actuating knob (8) screwed to bar 4 is a prestressed helical spring (7). To the lower end of bar 4 is riveted a knife-bearing traverse (9) to which is secured a zigzag knife (10) made from a rectangular piece of spring sheet steel.

The knife in use moves in a corresponding zigzag slot of a stripper (11), which is held axially immovable but rotatable in the housing (1).

When the parts and members (4), (8), (9), and (10) under the action of spring (7) move upward from their lower terminal position shown by dash-and-dot lines, and when they have moved through half their stroke, the upwardly bent free end (12*a*) of spring tongue (12) enters into engagement with the serrations (6). When said parts and members are moving further upward, the spring tongue bearing on the serrations (6) is flexed and said parts and members are indexed, i.e. angularly moved through a fraction of a revolution. In the next following downstroke, the ball of the operator's thumb rests on knob 8, and the friction between ball and knob prevents all the parts and members from being turned back; the spring-tongue lug 12a then simply jumps over some of the serrations.

simplification of the means of automatic indexing—that is, of rotating the cutting edges uniformly so as to chop each piece of vegetable thoroughly. Not only was this invention fashioned with great attention to detail, but the patent for it was drawn carefully to differentiate Zysset's machines from the many other food choppers that have been made. And they keep coming all the time. Fig. 3.22 shows a food chopper patented in 1976 by Samuel J. Popeil, who has several other patents in the field and whose company is one of the major manufacturers of gadgets. His invention includes a cutting surface and convoluted blade in the lower portion and a cushioned handle, a shaft, and a shaft guide in the upper part. It is worth study as an example of an invention by someone who has had a great deal of experience and success with the consumer.

Good design has become more important in recent years. The cheese grater illustrated in Fig. 3.23 is a model of elegant simplicity. The inventor, another Swiss, gave as his objective the creation of a grater that was "reliable and fast, clean and hygienic, easy to fill and clean, while being of pleasing and even decorative appearance." There are only a few moving parts in Jacques Buvelot's very model of a modern kitchen implement, and it can be used both in food preparation and at the table.

Sometimes a design patent will offer as good protection as a regular patent. Figure 3.24 shows such a patent for a grater. While this product offered nothing really new in the way of mechanical operation, it was certainly different enough in looks from other graters to be worth patenting for design.

Another common kitchen item, the hand can opener, has been produced in all kinds of shapes and sizes, many of which have been protected by design patents. Myron J. Zimmer is one inventor who worked on this gadget for many years. Three of his design patents, dating from 1938, 1940, and 1944, are shown in Fig. 3.25. The first two are quite similar, and had a very successful sales history. In 1956, another classic can opener design was patented (Fig. 3.26). Curiously, this design gave a somewhat bulkier appearance than earlier models, though it was certainly streamlined. It gives the impression of strength and durability. More recently, the trend has been toward electric can openers, many of them cumbersome in appearance and operation. Is there a better way? No doubt there is, though any improvements may involve the design of the cans themselves—the package more than any new opening devices. The "zip strip" on many frozen fruit juice containers is just one example of the self-openers that have been developed (see also Chapter 8, Packaging).

Sometimes a device lends itself to electrification, and sometimes it doesn't. Electric carving knives, such as the one illustrated in Fig. 3.27, work well enough and have been manufactured in fairly large quantities. However, they have not taken over from conventional blades in anything like the numbers that had been predicted. And their sales may taper off even further. Why? Probably because most people are still old-fashioned about certain things. Carving knives have a long tradition behind them; they do their job very well; many people enjoy handling them, even mak-

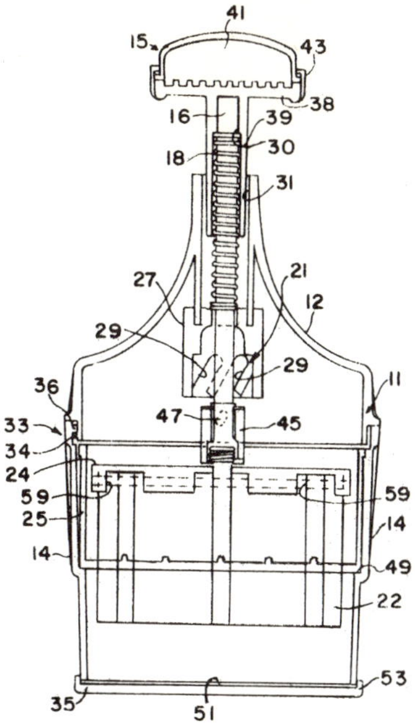

figure 3.22 Hand-operated food chopper, as illustrated in Patent No. 3,933,315 (1976). The upper housing (11) and (12) carries a handle (15) that moves the blade assembly (20). Connected to the upper housing by means of a bayonet-type locking arrangement (33) is the lower housing (14), which is closed off by a cutting-board bottom (35). After the food has been chopped, the upper housing can be disengaged by turning both housings in opposite directions. The blade assembly and stripper can also be removed so that the lower housing can serve as a container for the food.

The plunger guide (31) permits reciprocating movement of the plunger (30) for the chopping. The lower end of the plunger guide is provided with a clutch assembly (21), including a clutch housing (27) and guide slots (29).

Plunger is provided with an internal shoulder (39) such that a coil spring (18) is between the shoulder and the upper end of the clutch housing. A shaft (16) extends down through the plunger and the clutch housing (27) to be secured to the cross bar (24). A cushion (41) forms the top surface of the handle (15) to ease the impact of the operator's palm against the chopper.

At the lower end of the shaft (16), a bushing (45) is provided having a pair of clutch arms (47) extending laterally outwardly. With each down stroke of the shaft the clutch arms (47) slide into corresponding guide slots (29), and due to the slanted posture of the slots, the shaft rotates the crossbar and thc turning blades.

At the approximate mid-position of the lower housing, an internal shoulder (49) provides a convenient nesting position for the stripper (25). The internal diameter across the peripheral skirt (53) is just slightly larger than the external diameter of the lower end of the lower housing, in order to permit frictional engagement of the cutting board over the lower end of the lower housing.

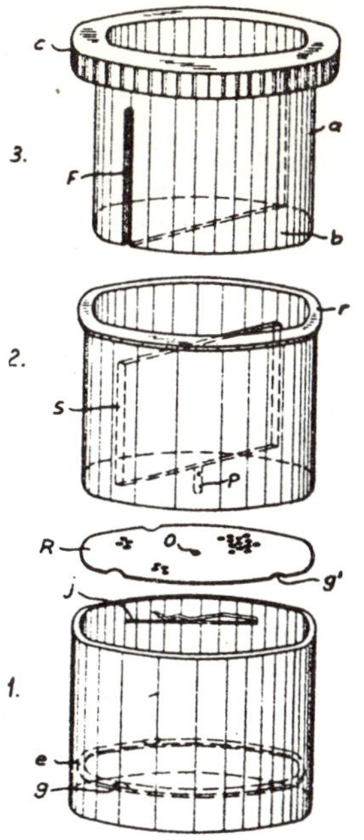

figure 3.23 A modern cheese grater, disassembled, as shown in Patent No. 3,642,045 (1972).

The housing (1) is a cylindrical body having on its inner surface a circular shoulder (e) adapted to receive a grating plate (R). The shoulder (e) is provided with inwardly projecting stops (g) which cooperate with recesses (g') on the plate (R) to prevent it from rotating.

The driver element (2) consists of a cylindrical envelope having at its upper end an outwardly projecting circular edge (r) adapted to bear upon the upper edge of case (1), and a diametrically disposed vertical partition (S). The partition is provided with a pivot (p) adapted to cooperate with an aperture (o) located in the center of plate (R).

The pivot has a lateral hole through which extends a pin (j), the function of which is to support the plate R and to scan its lower surface, when the driver (2) is rotated.

The pusher element (3) has the form of a cylinder having its bottom closed by a wall (d) and terminating at the top by a knurled crown; or knob (c).

A vertical slot (F) provided in the sidewall of the pusher element rotationally engages the partition (S).

When assembled for use, the driver element (2) and the grating plate (R) retained by the pin (j) are disposed within the housing (1). The grating plate is allowed to rest on the shoulder (e) and is prevented from rotating by stops (g) cooperating with recesses (g').

The substance to be grated is introduced into the driver element (2), and pressure is applied to the pusher element (3) to bring said substance into contact with the grating plate (R).

The pusher element is then rotated by turning the knob (c) whereby the substance to be grated is caused to rotate, while being maintained in contact with the grating plate.

During this rotational movement the pin (j) scrapes the grated substance off the lower side of the plate (R).

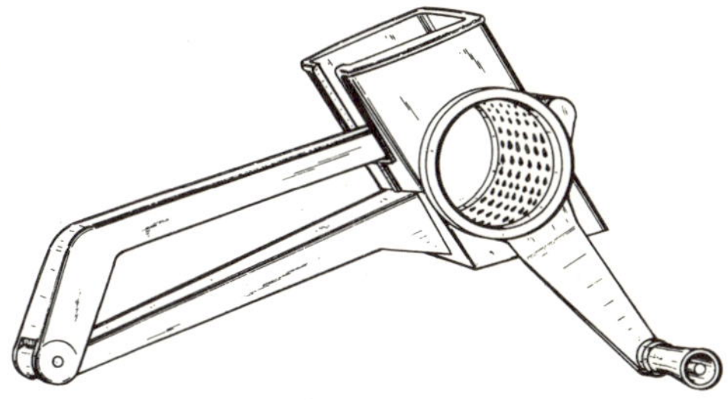

figure 3.24 The complete design Patent 235,501 (1975) for a food grater.

figure 3.25 Three design patents for a can opener filed in 1938, 1940, and 1944 by one inventor, M. J. Zimmer, Left(3.25*a*) are Figs. 1 and 2 of design Patent No. 112,628; center (Fig. 3.25*b*)), Figs. 1, 2, and 3 of No. 118,848; right (Fig. 3.25*a*), Figs. 1, 2, and 3 of No. 137.532.

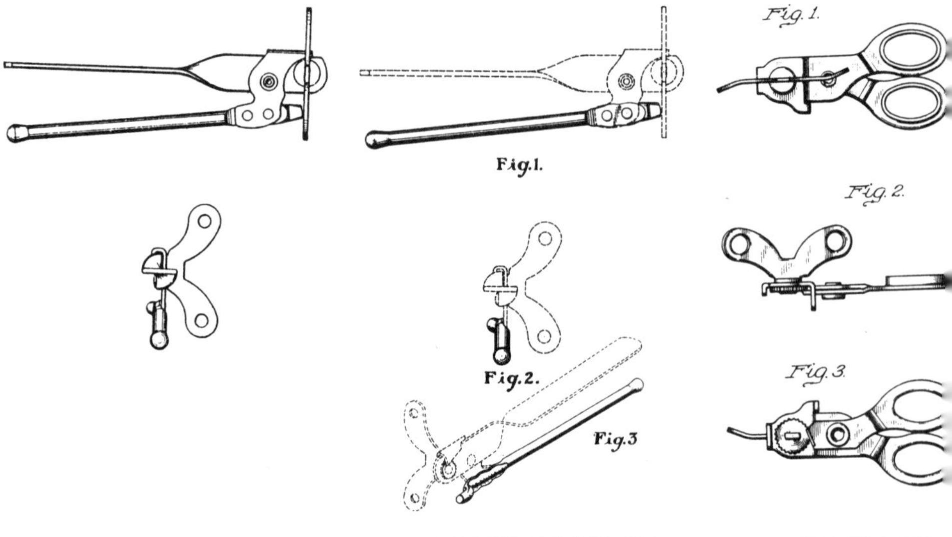

figure 3.26 Two of the drawings from a 1956 design patent for a can opener, No. 178,367.

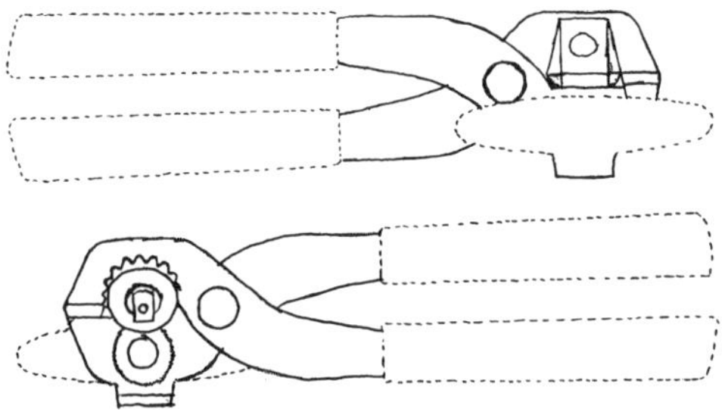

figure 3.27 Partially cutaway view of a power-operated knife, as illustrated in Patent No. 3,152,398 (1964). There is a handle or housing (10) and a pair of knife blades (11) and (12). The blades are held in close contact along their inner surfaces by a keyhole slot and rivet arrangement (13). Within the housing is a drive mechanism, including an electrical motor (14), a crank assembly, or transmission (15), a pair of connecting rods (16) and (17), and a pair of blade channels (18) and (19). Connecting rods (16) and (17) are secured to the channels by fasteners, such as (19*a*), and extend to the forward end. A bearing support (20) is secured to the lower portion of the housing, having a cover (21) secured to it by fasteners (22).

Bearing seat (20) and retainer (21) have been partially cut away to show one of a pair of hardened wear strips, such as (23), which are clamped down against the lower internal surface of housing (20) by the forward end of the bearing seat. Wear strips (23) overlie integral horizontal flanges (not shown) formed on the lower portion of the bearing seat. The lower surfaces, such as 17a of the rearward or yoke portions of the connecting rods, are slidingly supported on strips (23).

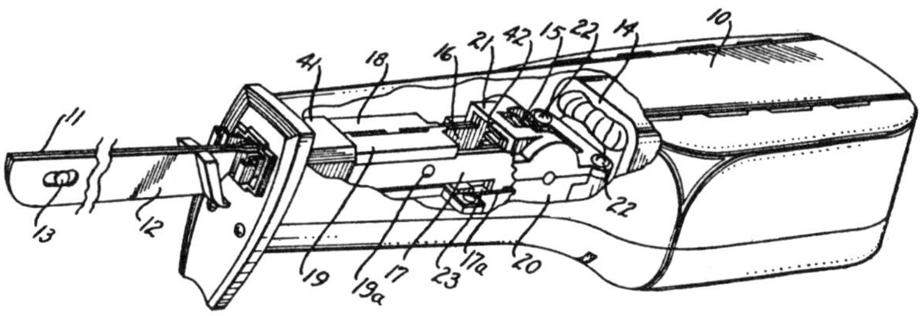

ing a kind of ceremony out of their use. The moral here is that while the American public generally welcomes and accepts new products, there are times when the old ways are difficult to displace.

We do not commonly think of table settings and napkins as an apt area for inventions, but they are—at least people keep trying, with varying success. Edmond Van Isegham of Missouri designed a prefabricated table setting that is especially useful for picnics and other large, informal gatherings (Fig. 3.28). It comes complete with plates, cups, and hand utensils folded into a tablecloth. Edith Honig of Florida invented something that she hoped would be used by airlines, railroads, and food-service companies. Her napkin is especially adapted to remain on the lap of the user and has a pocket for silverware (Fig. 3.29).

At the same time as the kitchen produces food for the table, it also generates waste—more of it every year, say the statisticians, as packaged products continue to gain in popularity. No one has

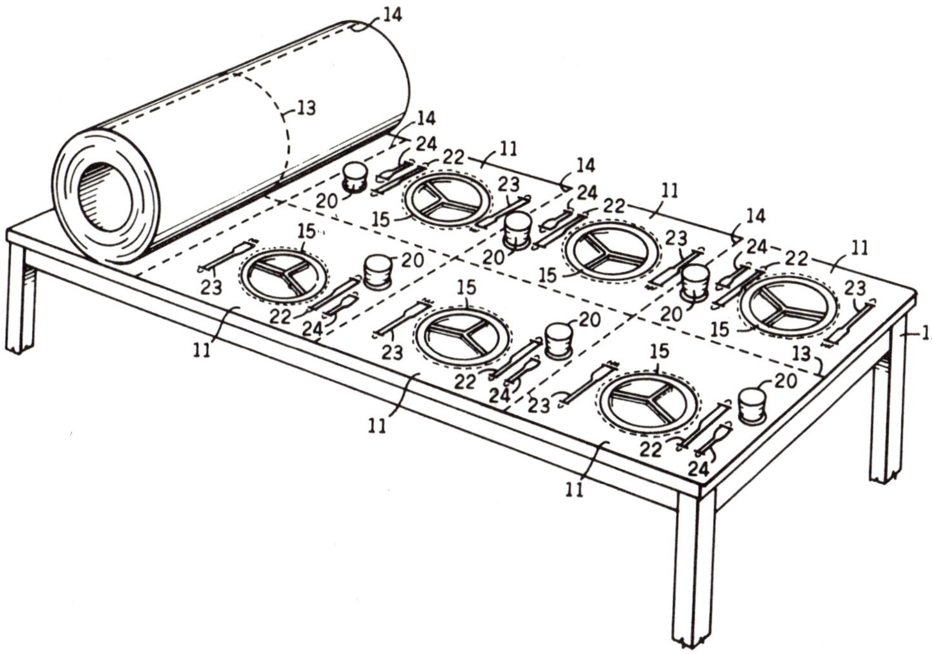

figure 3.28 Table arranged with prefabricated setting, as shown in Patent No. 3,939,976 (1976). A roll of flexible but sturdy paper or plastic material is subdivided into a number of place setting sections (11). The sections are arranged in pairs having a combined width equal to that of a conventional picnic table (12). The longitudinal and transverse divisions (13) and (14) defining each section can be perforated so that the sections can be separated from the rolled web. Each section provides a relatively recessed plate portion (15) including an inclined sidewall (16) defined by a raised rim (17). The plate is formed from the web material and is essentially self-supporting. The plate portion can be partitioned into subdivisions, and is perforated so that it can be removed from the web if desired. Each section also includes a portion that is suitable for seating a collapsible cup (20). Contact adhesive holds the cup in place on the section (11) while the web is being unrolled. Each section is also provided with eating utensils, such as a knife, fork and spoon (22), (23), and (24), which are held in place by slits.

come up with a good answer to this problem, though certainly under-the-sink disposal units and refuse compactors have made things easier for householders. There are both problems and advantages with both machines. Many people complain that garbage disposals get jammed too often, when little Freddie throws his fork down the hole and Flossie pushes in too much waste and not enough water. Then there's the noise, which in some machines is rather strident, and the apprehension that somebody just might get a finger caught down there. People's fears are more important, sometimes, than what actually takes place.

The household compactor is a comparatively recent development and does serve the purpose of getting garbage into manageable shape for removal at greater intervals than would be possible if it simply piled up in trash cans. Figure 3.30 illustrates how two inventors worked out a number of improvements for the compactor—all with the idea of making the machine easier to use and to manufacture. A ram that can produce a maximum of about 3,000 pounds of pressure compacts the refuse into a container that is sturdy, easily loaded, and readily emptied. The more trash there is in the container, the more pressure is put into the compaction. Another answer to this problem may be the large installations designed to convert trash into usable energy or to recycle it in some other way. Pollution control of this type is a growing field that is bound to see some exciting developments.

As stated at the beginning of this chapter, just about every piece of metal and wood in the kitchen is part of an article that has been at one time or another given the protection of a patent. But what about all the unique foods and recipes that have been developed? Are they patentable? What would happen if Aunt Mathilda took her secret recipe for gingersnaps to a patent attorney? Could she get it patented? Quite possibly, she could, but then the recipe would be on file in Washington, and it wouldn't be a secret anymore. If some cookie combine got wind of these gingersnaps and

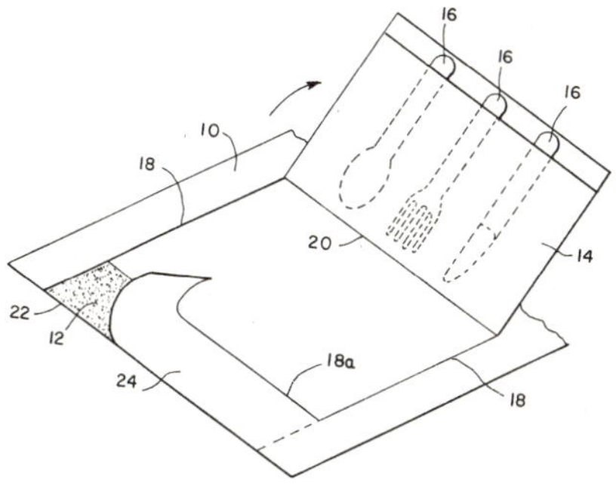

figure 3.29 Traveler's napkin with utensils, Patent No. 3,978,553 (1976).

The sheet (10) may be made from paper, cloth, plastic, or whatever material is preferred. There is at least a two-ply layer that includes the sheet itself and the layer (14) that forms the utensils' flap. The flap is formed by providing for a perforated line (18) along at least three sides. The fourth side merely defines a fold line (20), such that the flap (14) is adapted to lift upward away from the sheet (10) at that line (20). The flap is preferably formed of two opposing sheets, joined at their common base at the fold line so that there is formed a pocket to retain the flatware (16).

Along one end (22) of the sheet (10), preferably at the end opposite the fold line, there will be a layer of pressure-sensitive adhesive (12), adapted to contact the lap of the wearer. The distal ends of the border for the flap (14) will generally define the length of the adhesive side (22). The peel-off cover portion (24) may be treated with a self-release coating or be constructed of a self-release material.

If desired, the adhesive-containing end (22) may be folded backward along a fold line (18*a*) that is defined by the end from where the flap (14) was pulled upward away from the sheet (10).

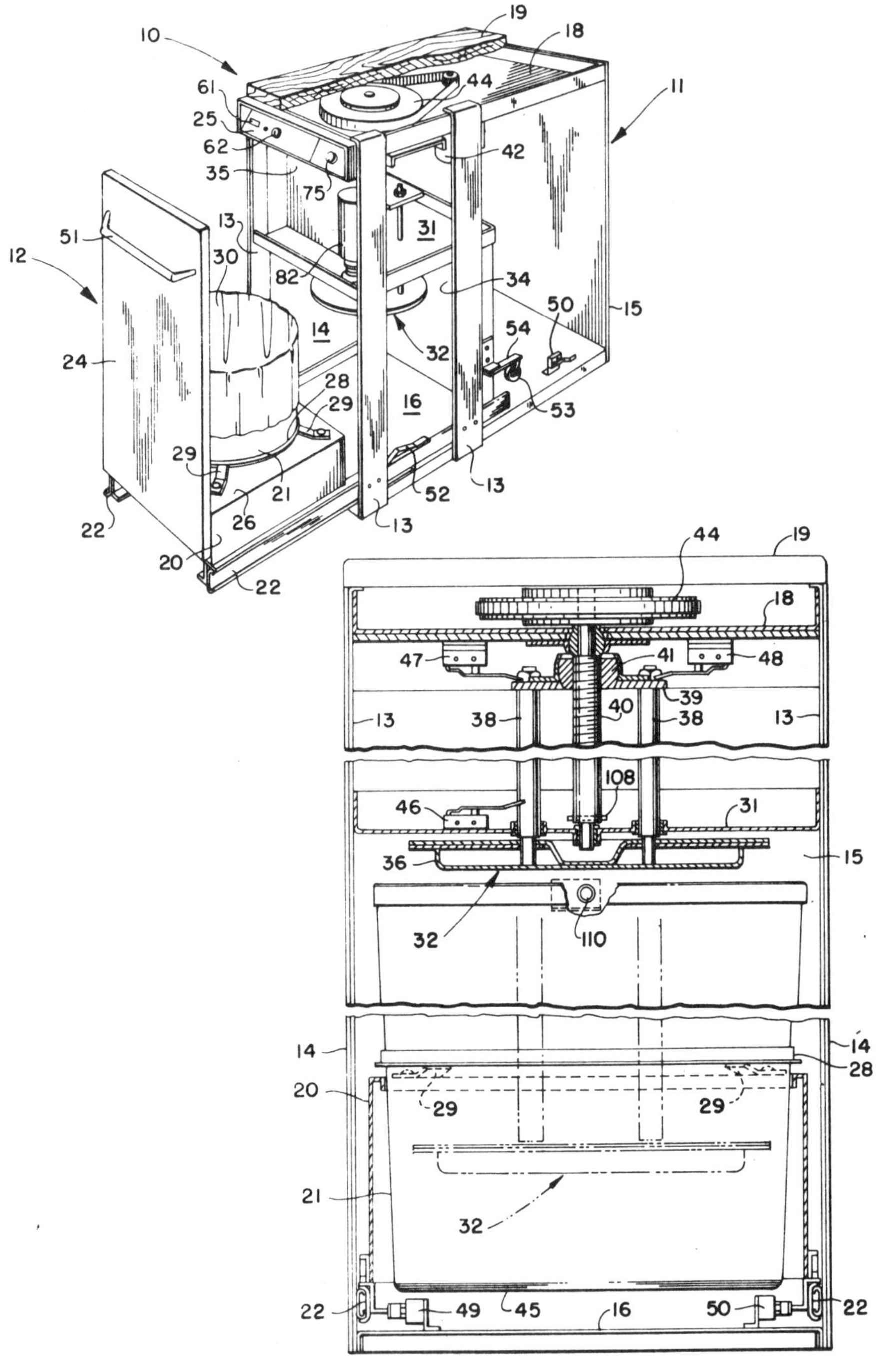

figure 3.30a and b A perspective view (3.30*a*) with parts removed of a refuse compactor, with the receptacle in position ready to receive materials. At right (3.30*b*), a front view partly in cross section. These drawings are from Patent No. 3,821,927 (1974).

The compactor (10) is made of a housing (11) and relatively movable receptacle assembly (12). The housing (11) is formed of side frame members (13) covered by exterior side panels (14), rear panel (15), bottom plate (16), and top plate (18) supporting a wood cutting board (19). The receptacle assembly (12) includes a carriage (20) and refuse container (21), the former supported on slides (22) for movement into and out of the housing (11) between a refuse-receiving position and a refuse-compacting position when fully within the housing. A front panel (24) is mounted on the carriage (20) for movement therewith and serves to close the housing (11) when the carriage is in the trash-compacting position, being then in substantial alignment with top front control panel (25).

The carriage (20) consists of an inverted box frame having an aperture in its top wall (26), adapted for receipt of the refuse container (21), the latter being of generally cylindrical configuration and supported by means of a peripheral ring (28) thereon, on spring supports (29) secured to the carriage (20). A plastic bag (30) is indicated as covering the interior of the container (21) and may be secured by means of a suitable elastic band and the like, not shown, about the cuff of the bag.

Further included within the housing (11) and supported by the side members (13) is a shelf (31) for supporting and guiding the ram assembly indicated generally at 32, an interior rear panel (34), and an interior front panel (35) thereabove a short distance inside of the front of the housing. The ram assembly (32) includes the ram pressure plate (36) supported on rods (38) passing through the shelf (31) and secured to a ram drive plate (39) which is adapted to move vertically between the shelf and top panel (18) of the housing. A drive screw (40) journaled in the shelf and the top panel cooperates with a nut (41) retained in the drive plate (39) for moving the latter, and thus the pressure plate (36), upward and downward. The drive force is supplied by an electric motor (42) via a belt and pulley (44), the latter secured to the drive screw (40) for rotation thereof.

Thus, as indicated in full line in Fig. B, with the refuse container (21) in the refuse-compacting position, the ram drive plate (39) is located just beneath the upper panel (18) of the housing (11), and the pressure plate (36) is located just beneath the shelf (31), allowing sufficient clearance for the refuse container (21) and carriage (20) to be moved with respect to the housing. In the dashed-line position, the ram pressure plate is at the lowermost position with the ram drive plate slightly above the shelf (31), any refuse in the container thereby being compacted to this position. It is significant that the lowest position of the pressure plate (36) is substantially above the bottom (45) of the refuse container (21), this distance being between one-fourth and one-half the over-all depth of the container (21). In a container on the order of two feet in depth, such dimension would be approximately six inches above the bottom of the container.

Such dimension is predetermined by the location of a limit switch (46) suitably mounted on a shelf (31) and having a leaf in the path of travel of the ram drive plate (39) and being adjustable in elevation by means of any suitable support. A further pair of limit switches (47), (48) cooperate with the drive plate for control over the electric motor (42), these including a directional switch (47) and a top limit switch (48) having leafs in contact with the drive plate (39) when the latter is at the top limit of travel.

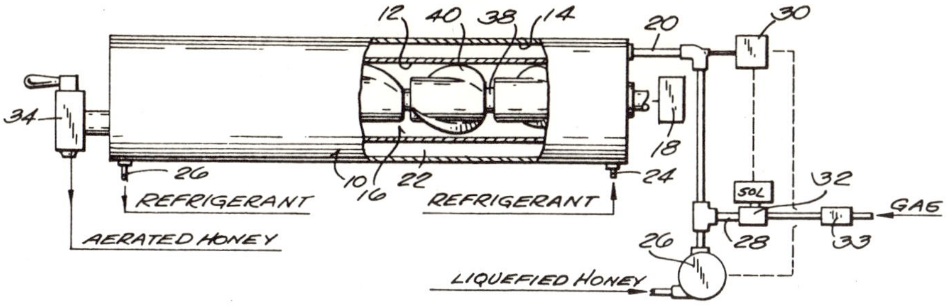

figure 3.31 Apparatus for producing the whipped honey spread described in Patent No. 4,004,040 (1977). The housing contains a closed inner cylindrical mixing chamber (12), an outer cylindrical chamber (14), and a screw-type auger (16) mounted inside the inner chamber (12) and driven by a conventional motor drive unit (18). This machine mixes gas and honey introduced through a supply line (20) into the forward end of the mixing chamber and propels the contents from the forward end toward the aft end of the mixing chamber (assisted by the pressure feed of the honey and gas). In order to maintain the mixing chamber temperature, a controlled flow of refrigerant is introduced into space (22) between the inner and outer chambers (12) and (14). The refrigerant flows through an inlet (24), flows over the outer wall of the mixing chamber, and exits through an outlet (26). To insure efficient heat transfer between the refrigerant and the mixing chamber, baffles can be provided to define a spiral flow pattern for the refrigerant.

Honey and a pressurized gas are automatically supplied to the mixing chamber by a pump (26), such as a conventional two-lobe impeller type pump, a gas inlet (28) connected to the supply line (20) downstream of the pump (26) and a pressure switch (30). The gas supply line (20) is controlled by a solenoid-operated valve (32) located in the air inlet line (28). The flow rate of the gas is regulated by a regulator (33) located in the air-inlet line (28).

A whipped or aerated honey product is withdrawn from the mixing chamber by opening a dispensing spigot valve (34). The pressure switch (30) is connected to the motor drive of the pump (26) and to the solenoid-operated valve (32). It operates through an electrical control system (not shown) in a conventional manner to cycle the pump (26) and the valve (32) on or off to maintain a predetermined pressure in the mixing chamber. If desired, the gas inlet (28) can be connected directly to the inlet end of the mixing chamber instead of being connected to the supply line (20) as illustrated.

decided to mass produce their own version, it wouldn't be difficult to change some proportions or alter an ingredient so that the taste would be the same but the recipe different. An example of a recipe that was judged to be worth patenting is the instant stuffing mix developed by researchers for General Foods Corporation. This is a stuffing you don't have to stuff, prepared from dried, yeast-leavened cornbread crumbs combined with a "hot aqueous mixture of vegetables and/or fruit and spices." It can be taken out of the box and prepared on top of the stove in about ten minutes. It carries Patent No. 3,870,803.

Another example of a recipe patent is whipped honey spread, which was created by Ruben Puta of Wisconsin. But this patent is enhanced by an apparatus designed expressly for producing the spread. Natural honey tends to crystallize and thus lose most of its appeal as a spread when it is exposed to sunlight or low temperatures for some time. Whipped honey spread, on the other hand, has a butterlike consistency and can be stored for a year or more without any changes taking place. The unique spread described in this patent is produced by putting natural, liquefied honey into a closed chamber and agitating it in the presence of a benign gas at a superatmospheric pressure for a time sufficient to uniformly distribute minute bubbles of gas thoroughout the honey. The device for mass producing this spread is illustrated in Fig. 3.31.

Inventing for the kitchen comes down finally to a question of what people need and want to make their food handling easier, faster, less expensive, or in some way better. These desires are, of course, modified by the limits of technology, the art of the possible. But there are hundreds of interesting possibilities. How about a grease-dissolving kitchen drain? A weepless onion chopper? A timing device that would help the cook get all the parts of the meal ready at about the same time? A quicker, cheaper method of doing just about anything that needs to be done in the kitchen will find a ready welcome in the vast and growing American marketplace.

CHAPTER FOUR

Around the House

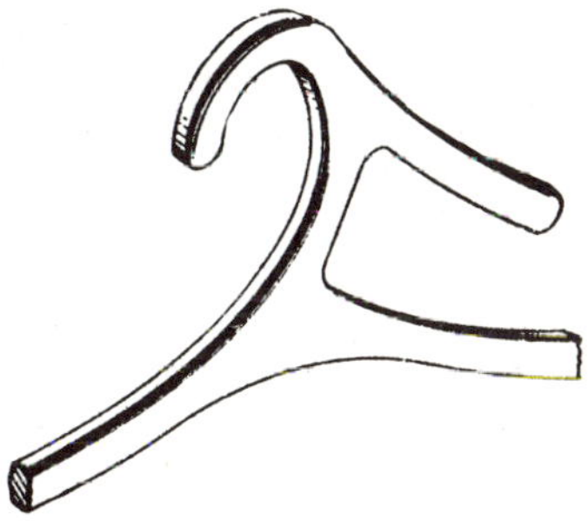

In the years since the end of World War II, the average American has gone through an almost revolutionary change in style of life. Millions of people have moved from cramped flats in the city and old farmhouses in the country to new homes on a patch of green in the suburbs. At the same time, families have become smaller and incomes have become larger—even allowing for inflation. So we've got a new, enlarged middle class, most of whom have moved into new homes in the last thirty years. They're looking for the good life, a major constituent of which seems to be the purchase of new and better things for themselves and their houses.

Many inventors and more manufacturers have done quite well for themselves by satisfying the desires of the new American buying public. In other chapters in this book it is shown how unprecedented popularity has been gained by new appliances and gadgets designed for use in the kitchen, the bathroom, the workshop. In this chapter we will consider things for the rest of the house, including the lawn and garden. It is instructive to look at some of the products that have been successful, some that have failed, and some that have never even had a chance in the marketplace. There are a great many consumer desires that have not yet been fulfilled. Even some age-old needs have not been taken care of as well as you might think.

"Build a better mousetrap, and the world will beat a path to your door," the old saying does. The hope that it's true today sustains the majority of inventors, especially the inexperienced. Those who have been around for a while and seen the weeds grow high on the

figure 4.1 Patent No. 305,102 (1884), a cat coated with luminous paint and perfumed to scare off rats and mice.

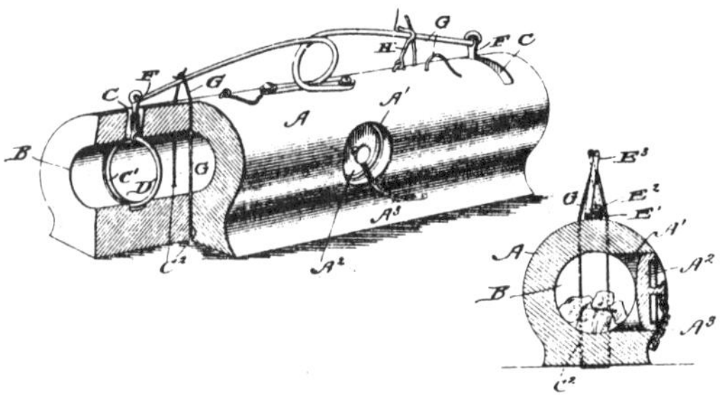

figure 4.2 Side and end views of a mousetrap granted U.S. Patent No. 690,291 in 1901. One end is broken away to show the choke ring *(D)* and spring *(E)*. A block of wood *(A)* is bored through the middle, and holes *(C)* are bored down to it, producing guideways *(C')*. Sliding in *(C')* are the choke rings, which are connected to the free ends of the spring *(E)* by means of chains *(F)*. These springs are best made from a single piece of spring wire formed into horizontal loops *(E)*, through which the screws *(E^2)* are passed to secure the spring at the top of the block, the spring being coiled or bent, as shown at E^3, in order to exert an upward tension or pull upon the free ends of the string. The bait is introduced into the bore at the center through an opening *(A')*, said opening being normally closed by means of a plug or a cap *(A^2)*, connected to the block by means of a chain *(A^3)*. A string *(G)* passes through the block and over the spring arm *(E)*, holding the spring down, so that the choke ring occupies its lowermost position, and access can be had to the bait. The string passes through the openings *(C^2)* produced in the block, the ends of the string being tied over the spring arms. While the spring is being tied around the spring arm, the arm is held in place by means of hooks pivoted to the top of the block. After the string has been tied, these hooks are disengaged from the spring arm, so that the arm is held down only by the string. Trying to gain access to the bait, the animal gnaws the string. The moment the string is severed, the spring arm draws the choke ring forcibly upward, thereby choking the animal.

path to their workshop have cause to be cynical. In fact, what *about* mousetraps? After all these years, hasn't man been able to produce the ultimate weapon for his war against rodents?

In 1884, John H. Nelson thought he had found the answer. Since mice (and some rats) are afraid of cats, he designed a kind of scaremouse, a cat statue that lit up in the dark with a glow that would terrify all but the toughest mice (Fig. 4.1). Seventeen years later, Edward Hummell of New York patented a fairly complicated trap made out of a block of wood and a length of wire. When the mouse (or rat) enters the hole of this trap to get the bait, a spring is tripped and the animal is choked to death (Fig. 4.2). A clever gadget, seemingly, but it never had any commercial success. The classic mousetrap, the one that's still sold today, is well known. Anyone who's gotten a finger caught under that vicious spring while trying to load this trap will never forget it. Another alternative of course is chemical extermination, but for many people that's too expensive and messy—some think it's dangerous, too.

So even in the well-plowed field of mousetrap development, there are opportunities for the seeds of new ideas to take hold.

In 1957, Robert Dobratz of Connecticut had a better idea, a prebaited, preset, disposable mousetrap (Fig. 4.3). His main problem was how to maintain the trap in a set position throughout the various stages from factory to mouse. This was finally accomplished by the unique configuration of latch and trip wire, as indicated. Although this seems a good solution to an old problem, it never has been widely available to the public. The time may be ripe for change.

Another persistent problem for the American householder, somewhat larger in scope than mouse-hunting, is the cleaning and simple maintenance of the house. Some people can afford to pay others to do this work for them, but even then, it's difficult to find cleaning people who will do the job the way you want it done. The majority, as always, do it themselves. Some householders like to keep things simple: They work with a broom, a mop, a carpet sweeper, and not much else. That way, there are no high electricity bills and no expensive repairs. Most Americans like a little help, though, and manufacturers are eager to do what they can.

A vacuum cleaner can make housework a lot easier. Though it's been around for a long time and enjoys wide acceptance, this kind of cleaner is still not perfect. The manufacturers in the field are quite competitive, always on the lookout for inexpensive ways to make their machines better than those of their rivals. One example of this is Sunbeam's attempt to make it easier to clean stair carpeting. This job has always been difficult because of the vacuum cleaner's natural tendency to fall down the stairs. Elmer Brunning, Donald Grover, Manfred Schild, and Robert Nickelson, all of Illinois, developed a simple, relatively inexpensive mechanism that automatically locks the wheels to prevent rotation to the rear but does allow forward movement at all times (Fig. 4.4).

Combining two or more elements with a new, patentable twist is an age-old form of invention that offers many possibilities. In 1975, Carl Parise of Nevada patented a "combination vacuum and hot water vapor cleaner that may be used as a dry vacuum cleaner *figure 4.3* A mousetrap in a package, shown unassembled, as in U.S. Patent No. 2,789,691 (1961). The mousetrap (10) has a tubular or cylindrical housing (11), which is adapted to receive a mouse at an open end (12), the other end (13) being closed by a cap or cover (14). The housing is affixed along its bottom (15) to a flat rectangular base plate (16). Mounted on the top (17) of the housing is a wire frame (18) having a pair of opposed sides (19), (20) longitudinally disposed on the housing and joined at one pair of ends (21), (22) by a transverse member (23). At the other pair of ends (24), (25) of the sides (19), (20) are coil springs (26), (27), respectively, which connect the sides to the top of the housing and bias the frame for movement in a clockwise direction.

Located within the housing is a wire loop or choker means (28), which has ends (29), (30) that extend through respective slots (31) and (32) in the housing and connect to the respective ends (21) and (22) of the frame. The frame, due to the aforementioned biasing by coil spring (26), (27) transmits to the choker means an upward bias.

A pivotally mounted latch (33) is disposable across the transverse member (23) when the latter is held adjacent the top of the housing against the spring bias, and the latch is releasably engaged by a catch on the upper end of a vertical tripping wire or releasing means to hold the transverse member in such position, i.e., a "set" condition, until the tripping wire is moved to release the catch. The tripping wire is centrally disposed within the housing and extends upward through slot (37) in the housing to permit the catch to make the above-described engagement. A transverse wire (38) is disposed through a loop in the tripping wire adjacent to the catch, and is affixed across the latter slot to the housing to provide a pivotal connection for the tripping wire.

At the lower end of the tripping wire is a hook or other bait-holding means to hold the mouse-luring bait.

The casing or container (43) is of a generally rectangular shape and of sufficient size to receive and enclose the above-described trap so that the latter will not jiggle or move around inside the casing during handling. The casing has a closed end (44) and an open end (45) with related lateral sealing flaps (46, 47) and a lid (48). Mounted within the casing is a detent means (49), which is comprised of a folded strip (50) attached at its ends (51, 52) to the inside surface (53) of the closed end (44) and thus defines an elongated protuberance extending longitudinally within the casing from the latter end. A flap (54) can be partially cut from one portion of the strip and bent so as to extend into an adjacent slit (55) in the other portion of the strip to give greater strength and rigidity to the protuberance. The attachment of the strip to the closed end of the casing can be such that the strip is removable with moderate force.

The casing (43) and the housing (11) and base plate (15) can be constructed of cardboard or other suitable material having sufficient strength yet being relatively inexpensive.

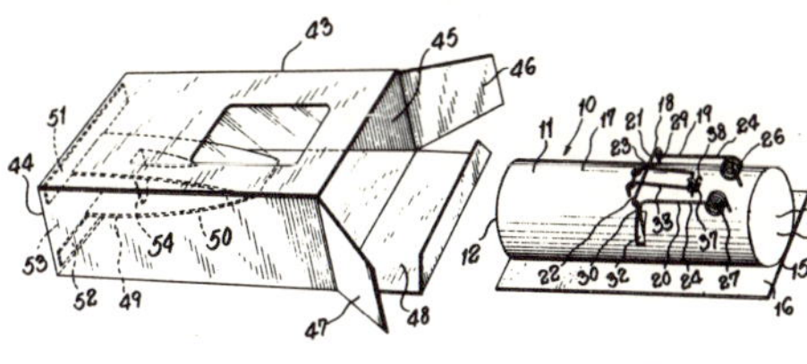

When assembled, the package is arranged with the detent means extending into the open end (12) of the tubular housing (11) and bearing against the tripping wire adjacent the bait-holding means, thereby preventing catch-actuating, i.e., trap-releasing, pivotal movement of the tripping wire in a clockwise direction.

When the container is opened and the trap is removed for use, the tripping wire is then free to move clockwise, and when a mouse enters the housing and pulls at the bait, the tripping wire is so moved, which causes the catch to release the latch, and, consequently, frame 18 is rotated clockwise by its coil springs and the choker means go up against the top of the housing, squeezing the mouse caught within its loop. The body of the mouse will lie substantially entirely within the housing. At this point, the detent means can be pulled out of the casing and discarded, the mouse-holding trap can be placed in the casing and enclosed, and the entire package disposed of as desired.

---

figure 4.4 This vacuum cleaner has a device to lock its wheels so that it will stay in place on a staircase [Patent No. 3, 820,808 (1975)]. The mechanism is employed with a cannister-type vacuum cleaner. To mount it, two holes (206) and (208) are provided in the housing (201) of the vacuum cleaner. These holes are above the wheels (112) so that the lock crank (200) may be inserted into these holes and the crank will then be held in place in the housing without the use of further mounting provisions. Holding of the vacuum cleaner on a stairway is accomplished by the lock crank, which may be formed of a single solid or hollow bar. However, for ease of assembly, it is preferred that the lock crank be formed in two sections.

The left-hand section (202) and the right-hand section (204) of the lock crank (200) are inserted through the apertures (206) and (208) in the left- and right-hand walls of the housing (201) in the vacuum cleaner, respectively. Both the left-hand section (202) and the right-hand section (204) of the lock crank are bent so that lowermost interior sections (210) and (212) of the lock crank are substantially parallel with and adjacent to the bottom (203) of the housing (201). Interior section (210) and interior section (212) are then crimped, welded, or secured by other means, to a sleeve (214), which holds them together, thereby completing the structure of the lock crank.

The locking arm sections (216) and (218) of the lock crank, which are external of the housing (201), are bent downward toward the upper surface of the polyvinyl chloride tires (114) of the wheels (112). In order to improve friction and to prevent damage to the tires (114), it is preferred that caps (220) and (222), of a resilient material such as polyvinyl chloride or other suitable material, be provided over the lowermost ends of sections 216 and 218, respectively. Although the caps (220) and (222) have generally cylindrical shapes with rounded lower ends, the particular shape of the caps 220 and 222 need not be limited to the shape shown in drawings.

The holes through which the sections are inserted, are in the housing above the front (224) of the rear wheels (112). Slightly forward of the front portion (224) of the rear wheels (112) and of the holes (206 and 208), a disengaging element or resilient nipple (228) is secured in the lower housing surface (203). The nipple (228) has a pair of enlarged flanges (230 and 232) and a reduced intermediate section (234), which allows the nipple (228) to be compressed and pressed into place in the hole (226), and to be thereby held in place in the hole (226) by the flanges (230 and 232). An upwardly projecting rigid element or stop member (236) is preferably secured to the rear of the nipple (228) in the hole (238). The stop (236) is provided to limit the amount of gripping force of the caps (220) and (222) on the tires (114) to a predetermined maximum force, in order to prevent damage to the tires (114), slippage between locking caps and tire due to over-rotation of the lock crank (200), and to provide for easier release of the locking mechanism.

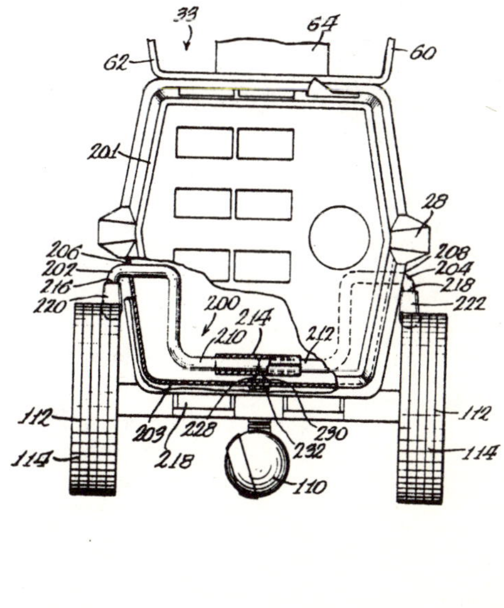

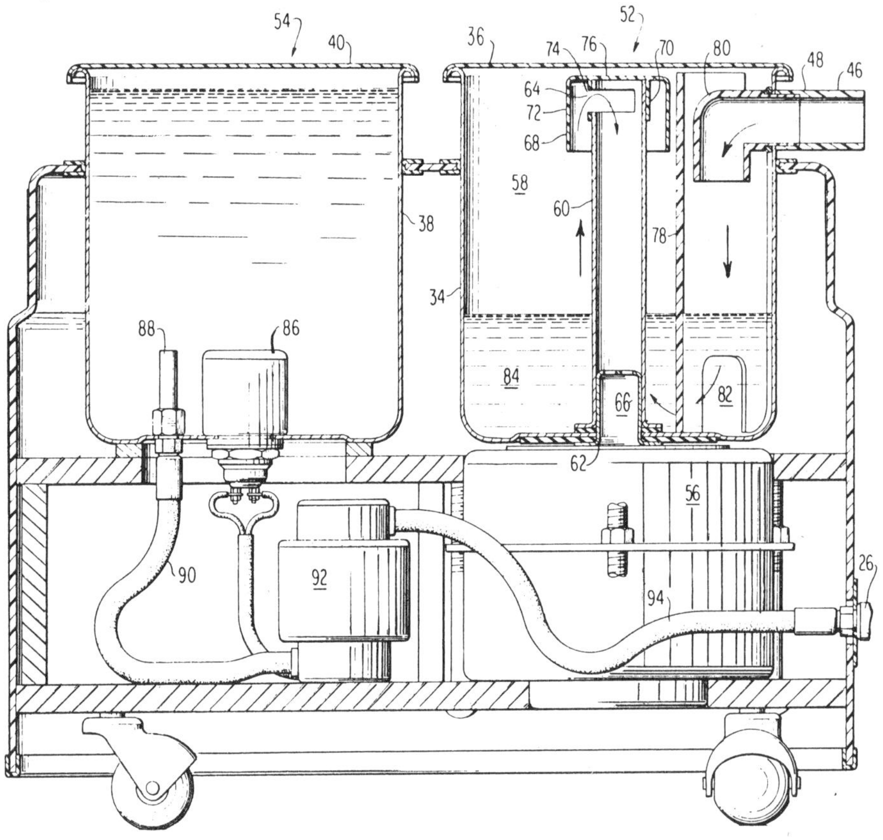

figure 4.5 Longitudinal cross section of the home cleaning system illustrated in Patent No. 3,896,521 (1975). The vacuuming apparatus (52) includes a vacuum pump (56), which must be capable of operating with very humid air, and a vacuum tank assembly (58). The latter includes the vacuum tank (34), its cover (36), and a hollow riser pipe (60) mounted on the bottom of the vacuum tank (34) over a corresponding hole (62) in the bottom of the vacuum tank (34). The riser pipe (60) has an opening (64) near its top and is open at bottom, thereby being able to snugly receive a male member (66) protruding from the vacuum pump (56). This member contains the inlet for the pump, whereby its vacuum is communicated to the interior of the vacuum tank through the riser pipe (60) and the opening (64) therein.

Removably attached to the top of the riser pipe (60) is a diffuser cap (68), shaped to reduce the amount of foreign matter sucked into the riser pipe (60). The diffuser cap (68) is composed of an inner cup-shaped member (70) and an outer cup-shaped member (72). The former is of an inner diameter sufficient to fit loosely over the top of the riser pipe and of a height sufficient to extend beneath the bottom of the opening in this pipe. It contains an opening (74) which is in size and shape approximately equal to the opening (64) in the riser pipe and which is spaced from the bottom of the inner cup-shaped member (70) by an amount sufficient to cause it to overlay the opening in the riser pipe when it is angularly positioned over it. The outer cup-shaped member (72) has a closed bottom (76), to the inner side of which the inner cup-shaped member (70) is attached, and a flat exterior surface parallel to its axis. The surface is adapted to engage a corresponding surface on a deflector divider (78), thereby positioning the opening (74) in the inner cup-shaped member over the opening in the riser pipe.

The vacuum tank assembly (58) also includes the deflector divider (78), which is positioned between the riser pipe (60) and the leg (80) of the L-shaped coupling pipe (48) extending into the interior of the vacuum tank (34). It extends substantially from the bottom to the cover (38) of the vacuum tank (34) and from one side to the other of the vacuum tank. Also, while not a part of the vacuum tank assembly (58) per se, the vacuum tank assembly is designed to be used with a layer of water (84) over the bottom of the vacuum tank (34). With a device of the overall dimensions previously stated, the water should be approximately three inches deep to start.

Thus the longest path which can be taken by foreign objects, liquids, etc., drawn into the vacuuming apparatus 52 is, as shown by the arrows, in through the L-shaped coupling pipe 48, down to the bottom of the vacuum tank (34), through openings such as (82), up along the outside of the riser pipe (60), between the inner and outer cup-shaped members (70) and (72) of the diffuser cap (68), through the opening (74) in the former and into the riser pipe (60), down the interior of the riser pipe (60), and into the vacuum pump (56). However, in practice, most foreign objects will be trapped in the water (84). The cover (38) is preferably transparent, so the user can gauge when the level of the liquid in the tank (34) has built up to a dangerous level. Thus, for the most part, only clean, moist air is drawn into the vacuum pump (56), from which it can be harmlessly expelled to the exterior of the device, preferably through a plenum chamber filled with a rock wool or other filtering material to muffle the noise of the vacuum pump (56).

The steam-cleaning apparatus (54) comprises the removable liquid tank (38), its cover (40), means (86) for heating a liquid contained in the liquid tank (38) (such as an immersion heater of conventional design), and an outlet (88), which must be located at a level above the top of means (86) if a conventional immersion heater is used in order to prevent a burnout of the heating element. The outlet is connected by a tube (90) to a liquid pump (92), and the liquid pump is connected by a tube (94) to the coupling (26) mounted on the exterior of the body (20). Means are provided for selectively activating the liquid pump (92) and the heating means (86). In the device shown, both such means are controlled by the steam switch but they may be separately controlled if it is desired to have the ability to pump cold water for cleaning certain kinds of surfaces. Also, the steam switch warning light can be designed to go on either whenever the heating means (86) is on or only when its temperature goes above a previously determined level. In the latter case, the light can be used to warn when the level of the liquid contained in the tank (38) has dropped dangerously low; in the former, observation of the user through the transparent cover (40) must be relied upon.

or a steam cleaner" (Fig. 4.5). These apparatuses may be used separately or in combination. This is a good idea, the only question being whether or not the steam cleaner will be used enough in the average house to justify its extra expense.

Built-in vacuum, or suction, cleaners have been around longer than most people think. They were placed in some houses around the turn of the century, but since then have enjoyed only modest success. It's an undeniably attractive idea, being able to remove dirt more or less at the flick of a switch, but there are some inherent problems. The size and complexity of the air ducts that have to be built into the walls constitute one major problem. In the past, these ducts have generally been used to carry dirty air to a central filter and to bring clean air into the various rooms or to one or more central units of dispersal. All this is not cheap.

In 1972, Council A. Tucker of California patented a compact, wall-recessed mountable suction cleaner consisting of a filter canister arranged in an upright position alongside a motor-fan suction unit (Fig. 4.6). The only thing left for the housecleaner to do is dispose of the dirty filters at regular intervals. This is an interesting idea, one that may well prove to be the first in a whole new line of appliances. The proposition that the way to success is by providing ways of doing things easier and faster (and maybe better) is as valid in the household field as in any other.

Many other inventions have been offered to help with the housecleaning, most of them surprisingly simple. Paul Morgan's releasable catch for a self-wringing sponge mop, patented in 1961, is still in widespread use. It made a good product even better (Fig. 4.7). The "jonny mop" itself was a great leap forward from the old-style mop and pail. It came along at a time when the use of asphalt and vinyl tile for flooring was on the increase. The new mop was ideal for cleaning these and other smooth surfaces; the products were just made for each other. Today, more and more floors in kitchens, bathrooms, and playrooms are being covered in carpet. Is there some gadget that will clean this indoor/outdoor kind of carpet as well as the jonny mop cleans tile? Not yet, perhaps, but there will be.

Inventions deal with big deals and little deals, seasonal situations and year-round problems. Take the Christmas tree, for instance. To children of all ages, it's a wonder and a delight, but to many housekeepers, it turns into a pain in the neck. Roger Price of Michigan has patented something that makes it easy to keep a tree neat and to dispose of it easily. Fig. 4.8 shows a plastic envelope with one end open and the other end closed. While the tree is on display, the envelope is spread out at the base, covering stand and trunk and catching loose needles as they fall. When it's time to get rid of the tree, the envelope can be folded out and up to enclose the whole thing in one neat, disposable package.

Getting rid of trash on a regular basis is a problem for many homeowners. The compactor and the garbage disposal are discussed in Chapter 3, The Kitchen. But even if you've got one or both of these machines, you've still got to handle some household trash in bulk. There have been many products brought out in this field, as

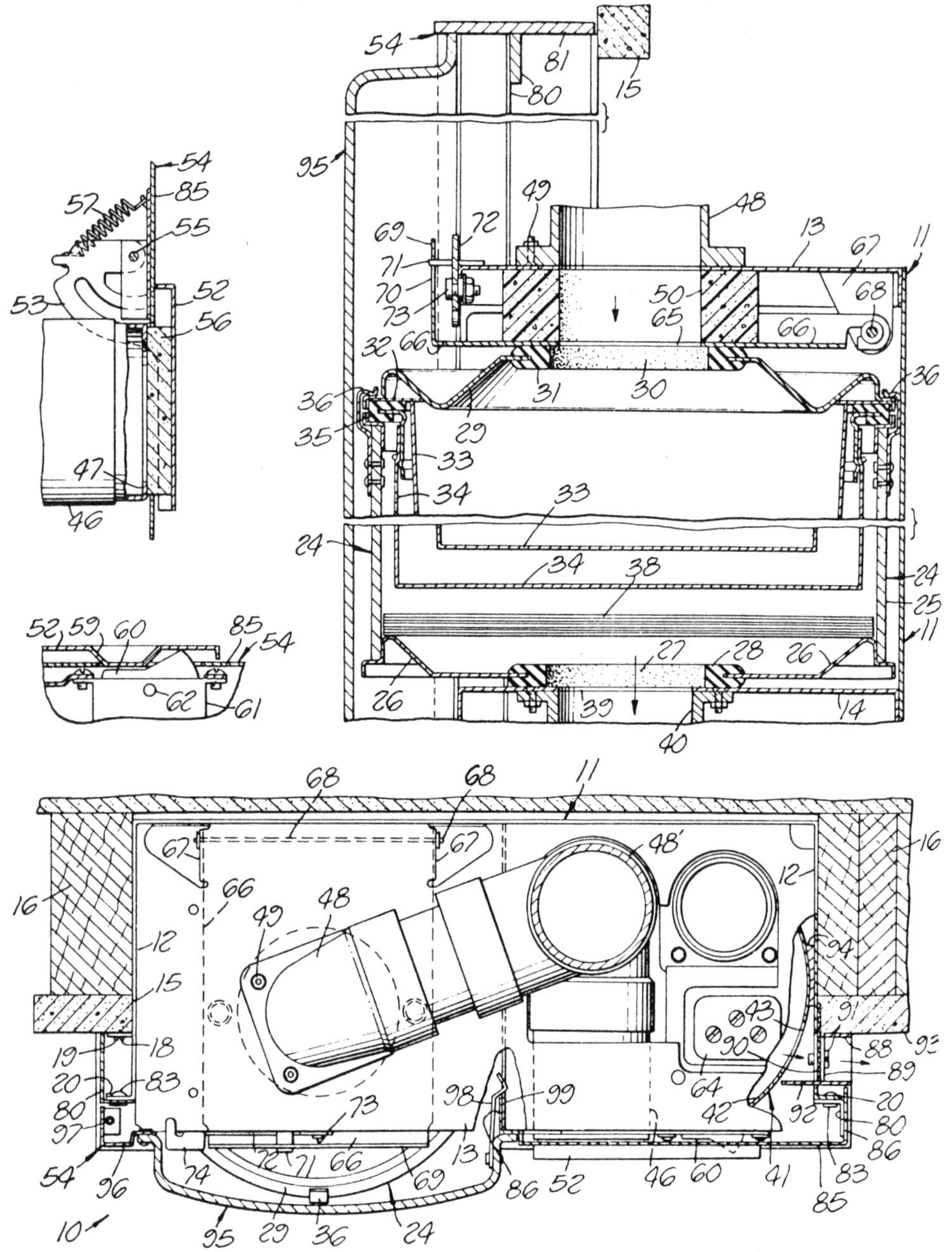

figure 4.6 Four cross-sectional views of a vacuum cleaner installed in a wall, showing portions of the cover broken away to provide a better view of the internal components [Patent No. 3,706,184 (1972)]. The cleaner (10), having a box-like sheet metal housing or main frame (11), including side walls (12, 12), an upper end wall (13), and a lower end wall (14), sized to fit in a wall opening (15), between studding (16, 16). It will be understood that wall opening (15) is sufficiently wide to have a close fit with sidewalls 12, 12 of main frame 11 but is substantially greater in height to accommodate certain horizontal air conduits extending across the cleaner outwardly of end walls 13, 14. The opposite side walls 12, 12, of the main frame include out-turned mounting tangs 18 along either side, through which fasteners (19) extend to anchor the main frame firmly to the wall. The side walls project outwardly beyond the face of the wall and include out-turned flanges (20), formed with keyhole slots to provide part of a concealed fastener to hold the appearance cover assembly in place on the main frame and about wall opening 15.

Detachably supported in an upright position along the left-hand side of main frame 11 is a unitary filter canister (24), which has its outlet gasket ring (28) in registry with a clean-air outlet (39) in the frame bottom wall (14) and with the inlet end of the clean-air conduits (40) leading into the inlet of the section fan unit (41). The latter is of any conventional construction driven by an electric motor enclosed by a cylindrical housing (42). The clean air discharging from the fan passes over the motor to cool it before discharging through an elongated, large area outlet opening (43) in the casing (42). This opening is located closely adjacent to the outer edge portion of the right-hand side wall (12) of frame (11). This outlet opening extends substantially the full height of motor casing 42 and provides for the free and unrestricted flow of clean air ready for return into the room.

The dirty-air inlet coupling and its closure are associated with the dirty-air inlet conduit 46. This conduit has an in-turned flanged inlet terminating in a vertical plane on the room side of the cleaner, the main body (48) of this conduit extending rearward, upwardly and horizontally across the exterior of top wall 14 of the main frame.

The resilient coupling between the discharge end of dirty-air conduit 48 and inlet 30 to filter canister 24, includes a resilient collar (50) secured about an opening (65) in a plate (66). The inner rear end of plate 66 is pivoted by pins (68) to brackets (67) stuck downward from top wall 13 of the main frame. The upturned forward end (69) of plate 66 is provided with an elongated opening (70) to receive a tang (71) projecting into it from the side of an operating lever (72). The latter is pivoted at 73 to a fixed support, and its left-hand end is provided with an operating thumb piece (74). It will therefore be recognized that canister 24 is released as the operator pivots lever (72) upward and clockwise about pivot (73), with tang (71) operating against the upper edge of opening (70) in plate (66), pivoting the latter upward about its pivot pins (68). When lever (72) has been rotated clockwise through (90), tang (71) will be positioned above pivot pin (73) and will be effective on the upper edge of opening (70) to lock plate (66) in open position. In this manner, the soft gasket collar (50) is compressed, releasing canister (24) so that it can be readily lifted out of its assembled position in the cleaner. The canister is readily reassembled, after which lever (72) is pivoted counterclockwise.

The unitary appearance cover assembly (54) has a rectangular, ring-like main frame (80) conforming generally to the size and shape of the side and end walls (12), (13), and (14) of the main frame. This rigid frame also includes upright angle members (80) aligned with side walls (12, 12) and projecting beyond upper and lower end walls (13, 14). Transverse plates (81) are rigidly secured crosswise of the upper and lower ends of these side-angle members (80) and serve to reinforce the frame, as well as to provide neat appearing finish strips for the cover assembly. Projecting rearward from the angle members are two pairs of shouldered rivets accurately positioned for insertion into the larger upper ends of keyhole slots in the out-turned flanges (20) of the main-frame side walls. These components function as concealed bayonet-fastener means projecting rearward from the interior side of the main frame and permitting the frame to be detachably secured to the cleaner frame by registering the heads of rivets (83) with the large end of the keyhole slots followed by pressing the rivets into the holes and then lowering cover assembly (54) until the shanks of the rivets are seated in the smaller ends of the keyhole slots.

The right-hand half of cover assembly 54 is covered by a permanently assembled decorative panel (85) extending the full height of the cover between finish strips 81, 81 and having its opposite lateral edges turned inward. The free edge of the outer flange (86) is there seen to terminate close to the out-turned flange (20) of the cleaner frame (11) and closely adjacent the edge of a channel-shaped member (88) extending the full length of the right-hand edge of cover assembly 54. The web portion of member 88 is cut away to provide a large clean-air outlet port extending throughout a major portion of the channeled member and is preferably provided with a final-filter screen (90), which also serves as a decorative covering for the air outlet. Screws (91) serve to hold air-outlet channel (88), filter screen (90), and air baffle (92) assembled to the outer edge of frame side wall (12). The L-shaped baffle (92) extends the full height of the clean-air outlet (43) from the motor fan unit and directs this clean air outward through the air-outlet port in a plane closely spaced and generally parallel to the interior surface of the room wall.

Access cover (52) for the dirty-air inlet conduit (46) overlies and conceals an opening through the panel (85) fixed to the right-hand half of cover assembly (54). Panel 85 is rigidly secured to the cover frame (80) and aids very substantially in reinforcing and strengthening this assembly.

To install cleaner the purchaser need but cut an opening in the room plaster between a pair of adjacent studs. The main frame carrying the principal cleaner components is then set into the opening with the rear wall bearing against or closely spaced from the far side of the wall. Mounting tangs (18) projecting outward from the sidewalls of the main frame will then lie flush against the face of the room wall. Screws (19) are then inserted and driven home to anchor the cleaner in place. After the electrical connections have been completed, cover assembly 54 is placed over the main frame so that rivets (83) register with the larger upper ends of keyhole slots. The cover is pressed inward, following which it is lowered slightly until rivets (83) seat in the lower end of the keyhole notches. When the cleaner is ready for operation, it merely is necessary to open the door (52) outward and to connect the usual flexible suction hose to the dirty-air inlet (46). The cleaner is placed in operation by depressing toggle button 60 counterclockwise to start the motor.

The dirty air passes along the hose into inlet (46), along conduit (48), and into the top of the filter canister. The suction produced by the motor-fan unit being conveyed into the lower end of the canister, the air passes through a series of filters and along conduit (40) into the inlet of the suction-fan unit (41), and thence over the motor within shell (42) enclosing this motor. The clean-air discharges from shell (42) through the long large area port (43) directly opposite the air-outlet port in channel member (88). This port preferably extends along a major length of channel 88, with the result that the clean air discharges in a thin, wide, sheet-like stream passing closely parallel to the adjacent face of the room wall. The deep-channel member 88 is highly effective in concealing the air return port from view, and its disposition prevents a jet of air from being discharged outward onto furnishings or persons present in the room. The entire forward-facing portion of cover assembly (54) is therefore free of air outlets interfering in any way with its decor. Accordingly, the entire front of the cleaner may be designed to harmonize with the wall covering and with the room furnishings.

After a cleaning operation has been concluded, the suction hose is detached from inlet (46) whereupon spring (57) closes door (52), sealing the inlet closed, it being noted that depression (59) in door (52) operates automatically to open the toggle switch, as described, and de-energize the motor.

---

figure 4.7 Side view of a self-wringing sponge mop, Patent No. 3,014,230 (1961) with the presser handle (19) pulled away from the mop handle socket (14). Sponge mop block (10), formed of a suitable absorbent and resilient material such as regenerated cellulose sponge, is removably attached to mop plate (11) by means of studs and wing nut. Mop handle socket (14), adapted to receive mop handle (15), is secured by welding or other suitable means to the mop plate.

Presser plate (16) is hinged to the mop plate at hinge (17). The presser plate is provided with a group of openings (18) to permit the passage of water therethrough during wringing.

Handle (19) is welded or attached by other suitable means to presser plate (16). A protuberance (25) in the form of a flange having a rectangular slot (26) extends from the rear side of mop handle socket (14). One half of the complementary U-shaped member (27) carrying protuberance (28) is shown snapped into place in slot (29) of presser plate handle 19.

Protuberance (28) has a shape complementary to that of slot (26) in flange (25). Clip catch (27) is preferably formed of a resilient material, such as spring steel, and is shaped to fit fairly snugly around flange (25) when the flange is moved into the space between the U-forming arms of the clip catch. If presser plate handle (19) is swung upward, protuberance (28) in member (27) will be pressed into slot (26) to bring flange (25) and clip catch (27) into releasable engagement.

Each wing of clip catch (27) carries at its outer end a pair of laterally extending arms (30) and (31). There are two arms (31) on each wing of member (27). Laterally extending arms (30) are received in corresponding depressions (32) on either side of slot (29) of presser plate handle (19).

Slot (26) of flange (25) is positioned so that it can be moved into releasable engagement with clip catch (27). In the engaged position, protuberances (28) on member (27) press resiliently into slot (26) of flange (25).

Each laterally extending arm (30) cooperates with the opposing laterally extending arm (or arms) (31) to snugly grasp between them the adjacent side edge of slot (29) of presser plate handle (19). When clip catch (27) is pressed into slot (29), it bends sufficiently so that laterally extending arms (31) ride over the respective side edges of the slot, and then arms snap into place when member (27) has been pressed all the way into depressions (32) in handle (19).

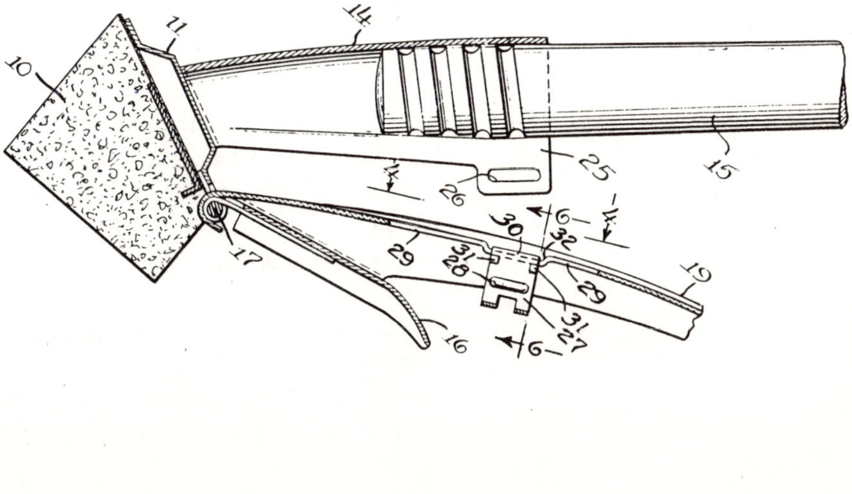

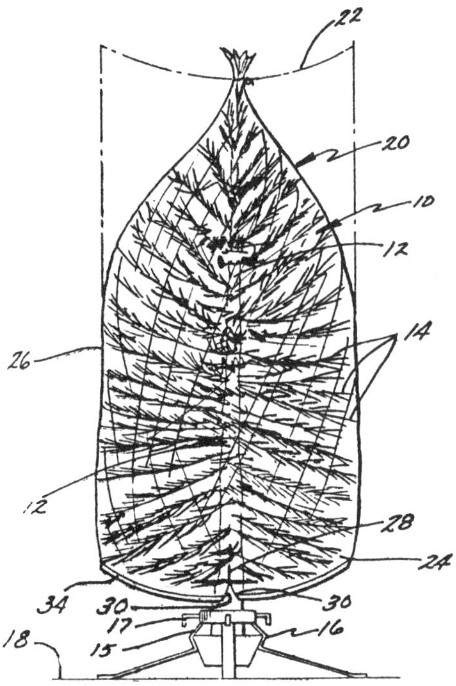

figure 4.8 Christmas tree display and disposal envelope, illustrated in Patent No. 3,924,669 (1975). The envelope (20) folded down around the butt portion (15) and base (16) of the tree (10) provides a decorative cover as well as a means of disposal.

There is a closed end (24) and sides (26) extending between the ends. The closed end includes a series of lines of perforations (28) intersecting at an apex at the midpoint or center of the closed end of the bag, providing flaps or corners (30), which are telescoped over and engage the trunk of the tree when the butt portion (15) is inserted. Alternatively, a circular aperture may be used in place of the intersecting perforated lines. Either type of aperture is formed centrally in the closed end approximately coaxially with the cylindrical axis of the bag (20), such that its size approximately corresponds to the typical butt-end portion (15) of a typical tree trunk (12).

The sealed seam (34) is an arcuate, heat-sealed seam formed within the plane of one of the diameters of the circular, cylindrical envelope (20). This seam curves outward and toward the open end of the bag between opposing sides of the cylinder from the center of the closed end. It is sealed completely from the sides (26) of the cylinder to a position adjacent the perforated lines (28) or circular aperture.

When the bag (20) is folded beneath the tree to form a decorative cover, the arcuate seam (34) supports the envelope in a dome-shaped manner, providing an attractive and decorative cover.

simple as the plastic garbage can with locking top and as complicated as the compactor. Hauling garbage out to the curb or the end of the lane for pickup is a minor chore for youngsters, a major job for some older people. Mary Larson of Florida thought of one way to make this task easier. Her "wheeled carrier for refuse containers" looks to be the result of on-the-spot research and experimentation. The carrier holds the container from underneath, and for further strength it can be attached to the side of either the metal or plastic type of can. It's lightweight, collapsible, durable, inexpensive, and simple to operate. Careful attention to detail makes this invention worth study (Fig. 4.9).

There's another waste-disposal problem that irritates some people quite a lot and others not a bit. What can be done with a cigarette or cigar ash when there's no ashtray immediately available? Figure 4.10 illustrates a solution that was offered in 1906 by Lavinia Richardson of Virginia—a lady who perhaps lost patience with those who persisted in dropping their ashes on her fine Victorian furniture. At first glance, most people might think Ms. Richardson's fish kind of silly—but is it? How many times have we seen ourselves or others flick an ash into a pants cuff or behind a sofa? Of course, it's doubtful if this invention had even a modicum of success. But many other "small" or "inconsequential" or even "silly" inventions have paid off. And right at this time a portable ashtray seems like a pretty good idea. There are some people today who dislike smoking so much that they just don't provide any kind of place to stash your ashes and butts.

Energy conservation has become as important in the home as on the road. The punishing winter of 1976–1977 brought home to many people their vulnerability to shortages of fuel. There's still a great deal of wasted heat that escapes through leaky windows, poorly insulated walls and ceilings, or up the chimney. Because of the importance of this subject, we have devoted an entire chapter to all its facets, Chapter 11, Energy. Household devices, including some that make use of solar energy, will be found in that chapter.

Electric lights generate heat, too. Is there some way to utilize it? Many people would consider this to be an unproductive area of investigation. But not Harry Noel of Indiana. He worked up a heating attachment for a bulb, illustrated in Fig. 4.11. The idea is to have two cups connected to each other. One clings to the bulb; the other can be filled with the substance you want to heat—perhaps a liquid to be vaporized or a wax depilatory to be melted. A minor issue, perhaps impractical, but it's an inexpensive way of generating a small amount of heat.

Many interesting inventions have not met with success among consumers, but they all suggest avenues of exploration that might lead to worthwhile results. "Where can we put that?" is a question often asked in American homes, especially some of the modern ones that have neither attic nor cellar. Inventors have tried to supply some of the answers here, with all kinds of racks and shelves. The products put out by the Rubbermaid Company are worth study. This company used to make mostly kitchen things, but recently they've had some success with devices that help get the most out of existing closet, door, and shelf space throughout the

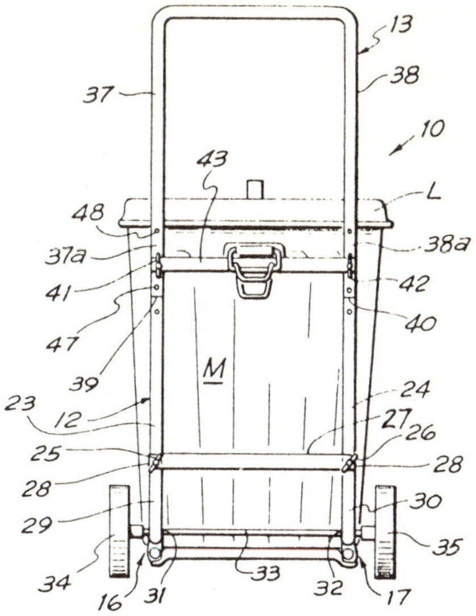

figure 4.9 Rear view of the refuse-can carrier shown in Patent No. 3,845,968 (1974). The container-base support assembly (11), preferably fabricated of bent metal tubing, comprises a U-shaped base-frame member (14) having a crossbar (15) riveted between the opposed leg portions. The outer ends of the opposed legs of the base-frame member (14) are hinged, as by hinges (16) and (17), to spaced, parallel, upstanding side members 18, 19, also of metal tubing. The hinges (18) and (19) each have side wall portions (20), (21) carrying pintles (22) journaling respective end portions of the opposed legs of the base-frame member (14) somewhat inward of the outer ends, so that they will abut against respective end portions of the upstanding side members 18 and 19 when the base-frame member is swung outward slightly more than 90 circular degrees with respect to the side members. Thus, the hinges (16) and (17) permit the base-frame member to swing pivotally.

The wheel-frame assembly (12) comprises a pair of spaced, parallel wheel-support frame members (23), (24), also fabricated of bent metal tubing, lower end portions of which are removably secured in back-to-back alignment with respective upstanding side members (18) and (19) of the container base support assembly (11), as by wing-nut bolts (25) and (26). These bolts also support an arcuate crossbar (27), also of metal tubing, outer end portions of which are arcuately flattened, as indicated at (28), for conforming fit. Lower portions of the wheel-support frame members (23), (24) are arcuately outwardly bent, as indicated at (29) and (30), respectively. Outermost end portions of the bent lower ends (29) and (30) have axles that are transversely aligned through openings (31), (32), respectively, journaling a transverse axle (33) carrying on outward-extending end portions thereof, a pair of rubber-tired wheels (34), (35) retained in place as by friction washers (36), press-fitted in place.

The handle assembly (13) is also fabricated of tubing bent into U-shape with substantially rectangular corners to provide opposing legs (37) and (38), each of which is bent at obtuse angles approximately midway along their length to define angularly-bent end portions (37*a*), (38*a*). The handle assembly legs (37) and (38) are disposed in spaced, parallel relation at a distance substantially equal to that of the distance between the wheel-support frame members (23) and (24), and the upper ends of these frame members are formed with short, reduced-diameter end portions (39) and (40) telescoped within the end openings of the handle assembly legs. A second pair of wing-nut bolts (41), (42) secure the handle assembly (13) to the respective upstanding side members (18), (19) of the base support assembly (11). These bolts also secure in place a second framework crossbar (43) of arcuate shape so as to be somewhat concave, as seen from the front of the refuse-container carrier. The second framework crossbar (45) is provided with arcuately flattened end portions (44), (45) for conforming abutment against curved outer wall portions of the handle assembly legs (37) and (38). The upstanding side members (18) and (19) of the container-base support assembly (11), and the respective opposing legs (37) and (38) of the handle assembly (13), are provided with longitudinally spaced, aligned sets of diametrically extending through openings, indicated at 47 and 48, to provide for vertically adjusted positioning of the handle assembly with respect to the base-support assembly. Centrally positioned along the second framework crossbar (43) and secured by a bolt is a container-handle attachment assembly.

figure 4.10 Two views of the pocket ashtray shown in Patent No. 820,067 (1906). At left is shown the normal position; at right the body (1) is compressed, causing the mouth (2) to open. To prevent the opening from getting too wide it should be reinforced by thickening the edges at (4).

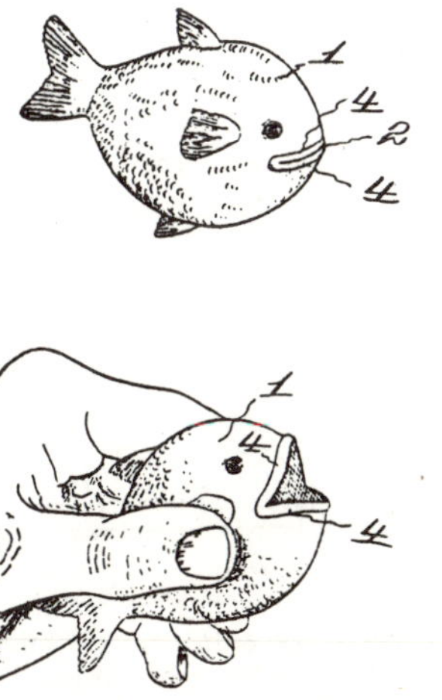

figure 4.11 Front elevational view of a heating attachment for a light bulb [Patent No. 3,151,233 (1964)]. Cups (10) and (11) are rigidly joined together at their bases and preferably formed from a heat-conductive material, such as aluminum. The cup (10) is normally disposed in an upright position and is adapted to hold the material to be heated. The cup's sidewalls abut the bulb (14) over substantially their entire extent to thus provide a substantial surface area for conducting the heat to the material carried in the cup.

The device can be mounted in a vertical or horizontal orientation. With a bulb disposed in a horizontal orientation, the edge of the recess (16) will rest upon the bulb, and cup (10) will be disposed in an upright position to retain the material (12) therein.

Circumferentially spaced openings (18) are formed in the sidewalls of the cup (11) adjacent the lower ends thereof for mounting inwardly curved legs (20) on the cup. Conveniently, each of the legs is formed as a sheet-metal stamping and comprises a pair of laterally spaced fingers (22) integrally connected at their lower ends. A tongue (24) is interposed between the fingers and is provided at its upper end with an outwardly directed offset (26), spring-stressed by the inherent resiliency of the leg into one of the openings (18); the upper ends of the fingers bearing against the outer face of the cup sidewalls on opposite sides of the opening. The legs have a radius of curvature slightly smaller than the radius of curvature of the sidewalls of the cup and, because of their inherent resiliency, will bindingly grip the bulb (14).

With the legs supporting the cups (10) and (11) in an elevated position above such a surface, there is a freedom of air movement between the cups and the supporting surface which is further enhanced by the recess 16 to thus provide an effective insulating barrier of air between the cups and supporting surface.

The lower edge of the cup (11) also lies in a single plane to support the device in an upright position on a supporting surface when the legs are removed. To remove the legs, the offsets (26) are merely sprung inwardly to clear the openings (18), and the legs are thus free to be detached from the device.

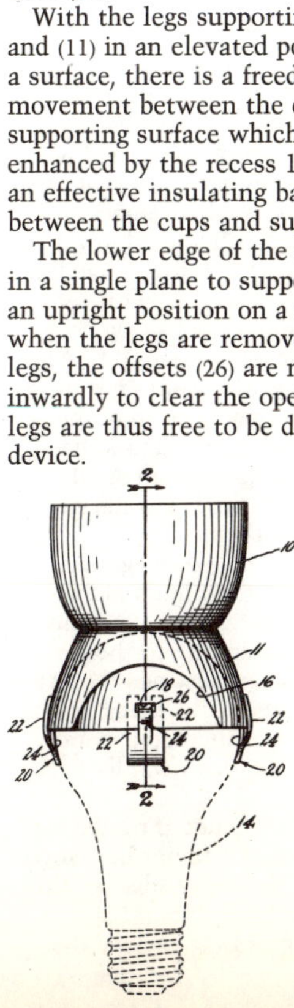

house. Most of their products carry design patents, since they have few if any moving parts. A bit more complicated is a tilt-open drawer invented by William Taylor of Ohio, illustrated in Fig. 4.12. The construction is simplified, with the drawer running on horizontal channels or tracks. The drawer has teeth that engage a slot, locking it in closed and full-open positions.

Clothes hangers have enjoyed such widespread use for such a long time that there would seem to be no room for any innovations in the field. Nevertheless, people keep trying. Design patents are also resorted to. Roger Snook of California patented the modernistic, practical hanger shown in Fig. 4.13*a*. A regular, utility patent was issued to Dayton Barnes and Robert Chiles of Tennessee for the hanger illustrated in Fig. 4.13*b*. This has a more-or-less standard design but includes several features that make it unique. (While this hanger will, of course, be used in household closets, it was designed primarily to meet the special needs of dry cleaners.) The inventors were irritated by the double strand of wire forming the neck of the hanger: "being of enlarged cross section, it often sticks in tubular conveyors and complicates ticketing." The single-strand arrangement developed by these two inventors is not only easier to work with; it's less expensive, too. Even simple things, then, lend themselves to innovation.

In general, independent inventors will have a much less difficult time if they concentrate their efforts on products that are relatively inexpensive to manufacture. It takes a lot of capital and manufacturing know-how to get something into production. As one inventor in the field has said, "It's not making the first item that's so hard, it's the first hundred or so that give you the problems." Many people have even been able to set up their own plants to manufacture, say, a new style of coat hanger. It can be done. But what about something like a sewing machine, for instance? Unless you've got backing from one of the major corporations in the field, your chances of getting anything on the market would be just about nil.

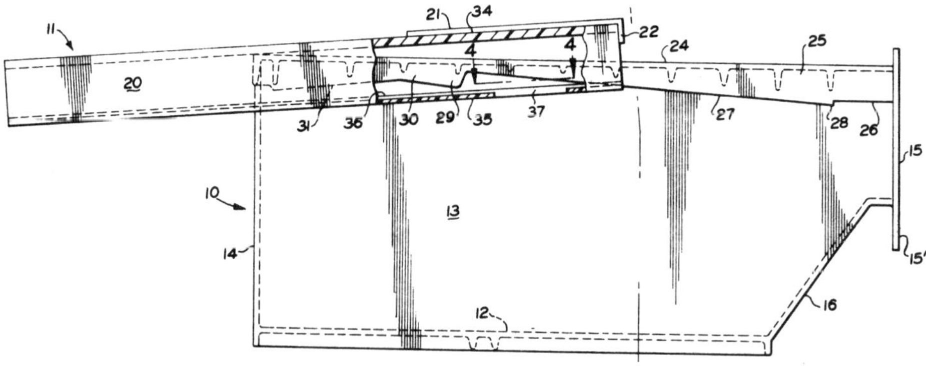

figure 4.12 Tilt-open drawer, as illustrated in Patent No. 3,722,975 (1977). The partly open drawer (10) is slidably mounted in a supporting frame, and has a bottom wall (12), side walls (13), rear wall (14), and a front wall with an upper vertical part (15) and an inclined lower part (16). The lower part (15') forms a handhold for pulling out the drawer. The carrier frame has two channel tracks (20), connected by a transverse plate (21). A flange (22) overhangs the front ends of the channels, and the upper portion of the front wall (16) of the drawer abuts the flange. Sidewalls are formed integrally with the top edges of the sidewalls and comprise horizontal flanges (24) terminating in downturned flanges (25). The bottom edges of these rails include front horizontal parts (26) and upwardly inclined parts (27) extending to the rear and having toothlike projections (28). At the rear are more "teeth," (29) and inclined edges extend back and include upwardly inclined areas (30) and downwardly inclined areas (31).

The channels (20) include inturned legs (34) and (35). Upturned retainer lip flanges (36) are formed on the inner edges of the lower legs. Longitudinal slots (37) are formed in the lower legs to receive the teeth (28) when the drawer is closed. This arrangement prevents the drawer from being pulled straight out. To pull out the drawer, the front end is first raised slightly by lifting the handhold (15') sufficiently to raise the teeth above the slots. As it is pulled out, the inclined lower edges (27) of the rails are slidably supported on the front edges of the lower legs, and the upper rear edges of the side rails engage the upper legs to limit the tilt of the drawer. When the drawer approaches the full-open position, the teeth (29) enter slots 37 and engage the stops provided by the front ends of the slots, limiting the opening, while the sliding engagement of the top rear edges of the sidewall rails with the upper legs limits the down tilt.

This does not mean that there are not some good possibilities for working with a product like a sewing machine. The year 1976 was the 125th anniversary of Isaac Singer's patent. In the intervening years, hundreds of refinements have been introduced, thousands of patents have been issued, in the field—so many that the chance for further improvements seems pretty slim. But research goes on, competition continues among the giants, and the product itself continues to change. An industrial designer now living in Florida, George Hetterich, developed a removable jacket with carrying handle for Singer. It fits neatly into the sewing machine's base (Fig. 4.14) and makes it portable.

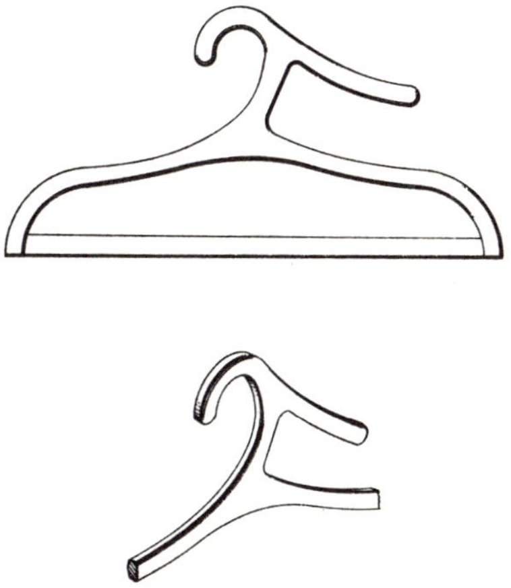

figure 4.13a Design Patent No. 179,745 (1967) for a coat hanger.

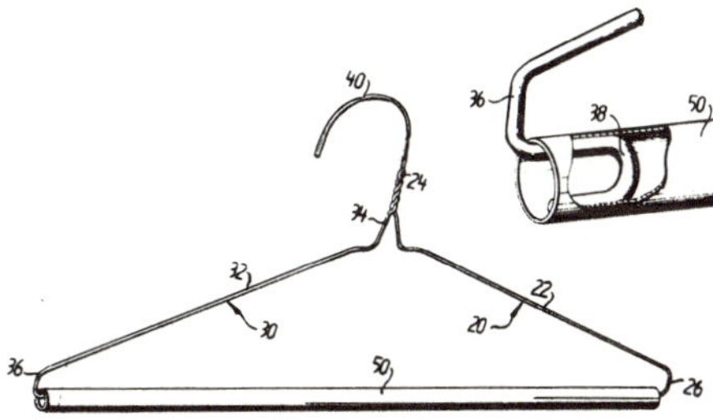

figure 4.13b A new improved, single-strand clothes hanger as illustrated in Patent No. 3,782,612 (1974). A short strand (30) is joined to a long strand (20) at the neck (24) of a hook (40). The other ends are joined below the bends 36 and 26 by a hanger tube strut (50). The bend (38) shown in the inset gives the hanger structural strength.

figure 4.14 An elevational view of a combination sewing machine and carrying case [Patent No. 3,858,954 (1975)], with the jacket (10) for the housing shown in cross section. The machine housing (10) is made up of a base or bed (14) having a standard (16) at one end.

A work-support surface (18) forms a top surface of the base (14). Extending from the standard (16) below its upper extremity is a bracket arm (20), which overlays the work-support surface and terminates in a head (22) spaced above the work support (18).

Within the head of the bracket arm, mounted for endwise reciprocation, is a needle (24) and a presser bar (26) having a presser foot (28) mounted on the free end thereof. The presser foot is adapted to cooperate in the usual manner with feet teeth (30) at the top of a feed dog (not shown), which is mounted within the base and protrudes through slots in a throat plate (32) mounted on the work support to feed the work being sewn. A thread take-up (34), for manipulating thread to the needle (24), is mounted within the head (22) of the bracket arm and is connected to a mechanism (not shown) for performing this well-known function. Any conventional sewing mechanism can be adapted to fit within the housing (10).

Extending about the periphery of the work space is a substantially continuous step recess (58). At the front and rear of the base (14), the step recess (58) is substantially equal to the thickness of the corresponding wall (50) and (52) of the jacket and defines respective front and rear slideways. Thus, the bottom edges of the walls 50 and 52 are slidably received on the recess (58) and slide into abutment with the head end facing surface 60 of the standard 16. The step recess (58) continues along the work surface and extends upward along the standard at the intersection of the post (44) with the major portion of the standard and about the surface (60), so that when the jacket is in the installed position, substantially smooth planar surfaces are formed with the exterior surface of the jacket and the adjacent exterior surfaces of the base and standard.

In order to secure the jacket to the housing, two pairs of cooperative latching mechanisms are provided. The first pair comprises a tongue (64) which extends outwardly from the open end of the jacket from a downwardly depending portion 66 of the top. The face (60) of the standard includes a slot (68) which is adapted to receive the tongue (64) when the jacket is properly positioned on the housing. The second cooperative pair of latching members comprises a block member (70) secured to the head end of the base substantially midway between the front and rear surfaces thereof. Secured to the end panel (54) of the jacket is a plastic latching member (72) having a camming surface (74) formed on the bottom thereof. The member (72) includes a first leg (76), which is secured to the end panel (54) by at least one rivet (78), and includes a downward extending leg (80) pivotably secured to the leg (76). Thus, to secure the jacket to the housing, the jacket is slidably received on the slideway (58) until the tongue (64) registers with the slot (68), and the wall (54) properly registers with the end of the base. The leg (80) is then pivotably moved outward from the leg (76), which is bent slightly until the cam surface (74) engages the bottom of the block (70). The leg (80) is then pushed back toward the wall (54) until the leg (76) is taut. The jacket is then properly installed on the housing. A carrying handle (82) may be formed on the top wall (56) of the jacket or may be secured thereto by conventional means so that the jacket and the sewing machine housing provide a carrying case.

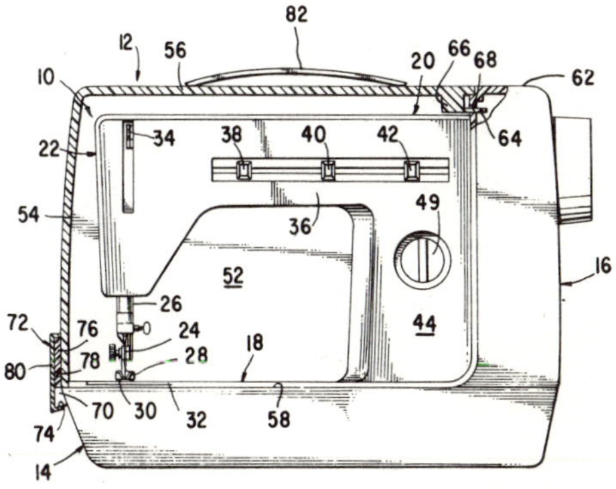

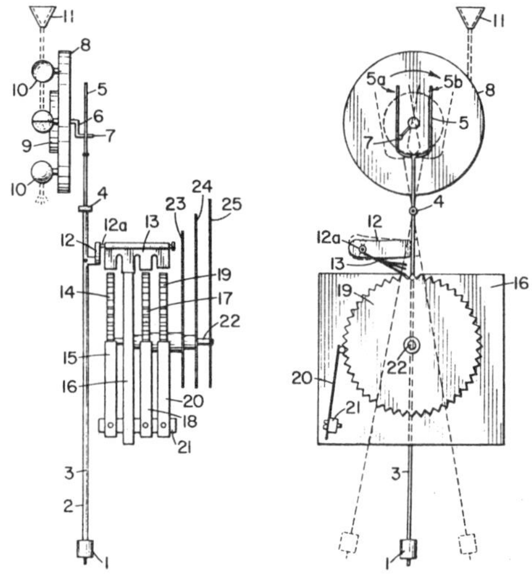

figure 4.15 A side view and a front view of the turbine clock [Patent No. 3,871,167 (1975)], showing its basic operations. The pendulum, comprised of the bob (1) and the rod (2), is mounted free to swing by means of the pivot (4). Above the pivot, the pendulum rod (2) is rigidly attached to the U-shaped fork (5) with its fingers (*5a*) and (*5b*). These fingers cooperate with the link (7), which is rigidly attached to the shaft (6). On this shaft is pivoted disc 8, which is free to rotate. This disc has several cups (10) rigidly attached to its rear surface, spaced equally around a circle whose diameter is somewhat smaller than the diameter of disc 8. The purpose of placing these cups at a smaller circle is to free the rim of the disc (8) for coupling to it a subminiature electric motor of any kind, which draws no more than 50 to 100 microamps, if and when it is so desired. Rigidly attached to the disc is the drum (9), which contains a spiral hairspring of the kind used with the balance wheel of the ordinary table clock. The inside end of this spring is rigidly attached to the shaft, and the outside end presses against the circular inside wall of the drum, and is free to slip under tension.

An outlet (11) for liquid is located directly above the center of a cup (10), and the droplets of liquid flow down into the cups. The amount of torque required to rotate the turbine wheel (8) is so small that the mere weight of a couple of drops of water in each cup will create the necessary torque to turn the wheel.

The clock is far older than the sewing machine; yet it too is being improved as time passes. The most spectacular change in recent years has of course been the LED and LCD watches, developed by the semiconductor industry at a cost of millions of dollars. Even with this kind of competition, there is still room for the independent inventor with a good idea. In 1975, Gossem Hekmat Afshar of New York was granted a patent on a pendulum clock, which is run by a turbine that can be powered by water, air, or even a tiny motor (Fig. 4.15). The invention is fairly simple in design, since its mode of operation eliminates the need for specially shaped cams, levers, and gear trains that are normally found in most clocks.

There's almost as much use for machines and gadgets outside the house as there is inside. Some of these things are for work, and some of them are for play. " 'Tis spring at last," wrote the poet, "and once more the blast of the power mower is heard o'er the land." Some of today's cutters may be pretty big and noisy, but they've got nothing on the machine patented by Thomas and William Caldwell in 1901 (Fig. 4.16). It looks capable of chewing up any lawn in the country. Yet it was one of the first inventions in an industry that has seen tremendous growth.

At this time it's questionable whether any basic or revolutionary improvements in the power lawn mower can be made by the independent inventor. It's become a big business, demanding large capital investment, and the machines themselves have been worked on and played with by scores of creative research people, looking for a better way. Yet, with all this effort, the critics still say that modern machines are too noisy and dangerous, give a poor cut to the lawn, and break down easily. If only the old reel mower, or one like it, could be powered so that it wouldn't be so difficult to push, then we'd have something, according to many home gardeners.

Donald Dean of Missouri has patented a reel-type lawn mower that makes use of opposed, overlapping spiral blades that achieve a scissorlike shearing operation during rotation. His mower blades are also self-sharpening and self-cleaning as they are worked. And the engine takes less power and runs more quietly than those now on the market (Fig. 4.17).

As far as conventional rotary mowers are concerned, there are improvements that can be made here and there. One area that has drawn the interest of inventors lately is the method of handling the grass that has been cut. Figure 4.18 illustrates two different approaches to this problem. Keith Rhodes of Nebraska sold the J. I. Case Company on his new grass-catcher assembly. Two containers with each supporting a flexible, disposable bag are the distinguishing features of this device (Fig. 4.18*a*). William H. Armstrong of Michigan invented for Montgomery Ward the grass catcher shown in Fig. 4.18*b*. It consists of a pair of frames surrounded by a flexible shroud. The frame can be attached as a cantilever to the deck of the mower; when not in use, it can be detached and folded for easy storage.

Anyone who lives surrounded by, or downwind from, a lot of trees must be prepared for a bumper crop of leaves each fall. Some people don't mind raking them by hand, but others want a better

figure 4.16 Detail view from the right side of a "Motor Lawn Mower and Sweeper" granted Patent No. 796,811 in 1905. The main frame of the machine comprises the longitudinal side bars (*A A*), connected in any desired manner and provided at the rear with a platform (*B*). The main frame is provided adjacent to its rear end with downwardly extending hangers (*C C*), in which is mounted a shaft or axle (*D*) and upon this shaft or axle are mounted the traction and lawn-rolling rollers (*E E*).

The front ends of the side bars (*A A*) of the main frame are provided with bracket castings (*a a*), connected by a tie rod (*a'*).

(*F F*) represent two caster wheels mounted in the usual yokes (*f*), the shanks or spindles of which are mounted in the brackets (*a a*), thus supporting the front end of the main frame.

(*E E*) represent the traction and lawn-rolling rollers mounted loosely on the shaft or axle (*D*) with compensating gear between them.

(*M*) represents the mower frame, comprising two side frames (*m*), suitably connected and braced and provided with the usual stationary cutter and revolving cutter or wiper, the latter being carried by the wiper shaft (*m'*), mounted in suitable bearings. The mower frame is detachably secured to the main or motor-carrying frame and is preferably carried thereby at all times.

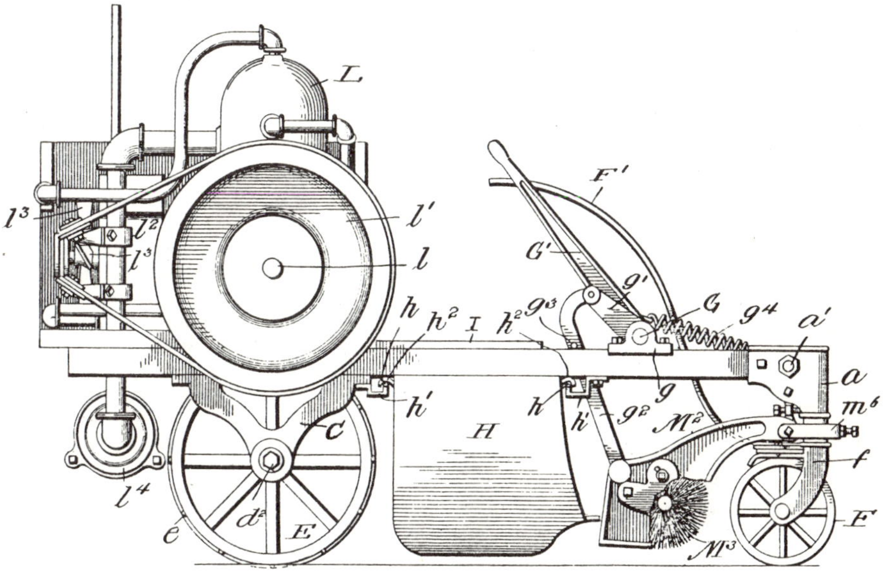

figure 4.17 Side view of a reel or spiral lawn mower, Patent No. 3,979,887 (1976). A tie piece (12) secures the side plates (10) and (11) at the front of mower, and a girt (13) secures them at rear. A horizontal plate (14) serves to support the engine (15), which is mounted thereon, and has an extension shaft (17) which is supported by a bearing (not shown) secured in the upright extension (10*a*) of side plate (10). A drive sprocket (19) is secured to shaft 17 and drives sprocket (20) by the roller chain (21). Sprocket 20 is secured to front cutter shaft (22) and has an approximate 2:1 ratio with the sprocket 19.

The front cutter shaft has bearings (24 and 25), which are fitted in flanged cups (26) and (27). These cups are pressed into the side plates (10) and (11) respectively. The front cutter has blades (23), which coact opposed spiral blades (29), which overlap the blades 23.

The rear cutter shaft rotates in bearings that have cups, which are likewise pressed into the side plates. On a girt (36) is loosely mounted a conventional handle (37), which rests on studs (38).

Slots (39) in the side plates serve for height adjustment for rear of mower. The slots (50) and (51) in front-wheel brackets serve for height adjustment of cutters. The brackets are held in place by bolts.

A deflector plate (54) with a series of baffles (55) directs the clippings to the rear of the mower and distributes them evenly, preventing slugging and windrowing of the clippings.

The resilient mounting of the cutter blades (23) and (29) for this mower with dynamic balancing (which is a well-known practice for cylindrical work), will produce a quieter mower, especially with its lower speeds and lesser power. A gear (28) is secured to the shaft (22) and meshes with the gear (35) on the torsion device.

The resilient discs (30) are bonded to the blades (29) of the rear cutter and to its shaft (31). The metal disc 32 is secured to shaft 31 and has slots (33) which receive screws (34), which are provided with conventional flat washers. The screws (34) are threaded into the gear (35), and when the proper torsional stress is set on the resilient material (30), the screws (34) are tightened to clamp the disc (32) to gear (35) to hold the initial torsion as required for the blades (23) and (29) to operate properly, regardless of varying speeds of the blades in rotation.

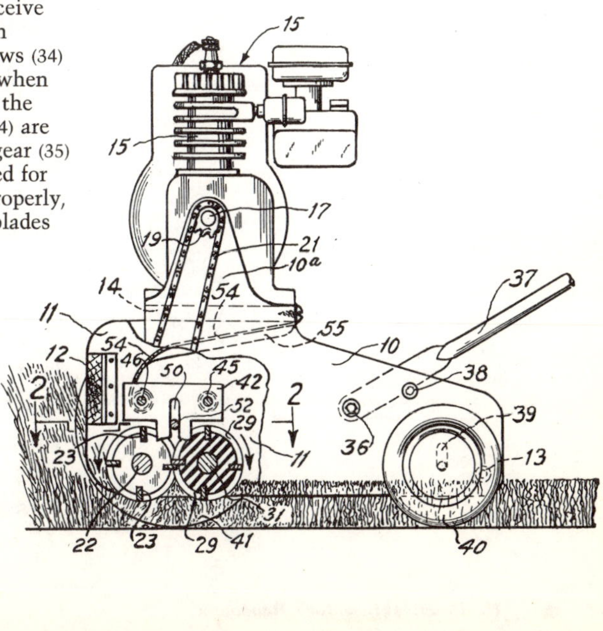

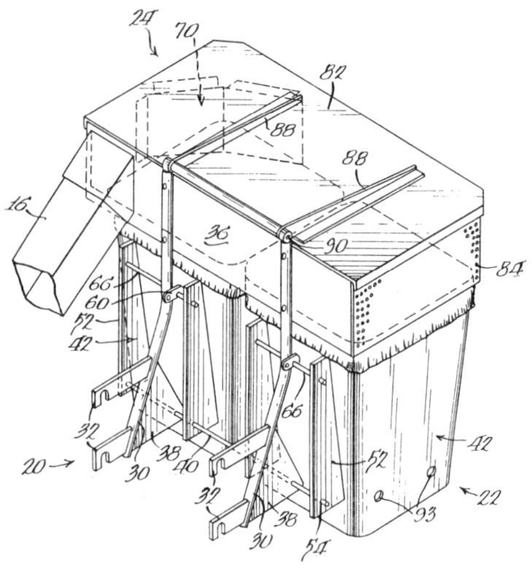

figure 4.18(a) Perspective view of a grass-catcher assembly, from Patent No. 3,974,631 (1976). The frame assembly (20) consists of a pair of transversely spaced rigid bars or angle irons (30), each having arms (32) adapted to be connected to a riding mower so that the entire grass-catcher assembly (18) is suspended on the rear end of the mower. The upper ends of the bars (30) are interconnected by a rigid plate (36), while the lower ends of the bars have gusset plates (38) secured thereto with a rod (40) extending between gusset plates (38). Rod (40) defines a generally horizontal pivot axis for container means (22).

Container means (22) consists of first and second containers (42), which are identical in construction and are both supported on rod (40) that defines a generally horizontal pivot axis. Each container has side walls (44) and a bottom wall (46), which form a container that is open at the top and is adapted to receive a disposable plastic bag (48) with the upper edge thereof draped over the upper ends of the side walls that define the open top for the container.

Each of the containers has a pair of braces (52) extending from one side wall thereof, and the braces may be secured to the side wall by welding. The lower ends of the braces have openings (54) which receive the rod so that the containers may be pivoted adjacent to the lower bottom wall about a generally horizontal pivot axis from upright to tilted.

The grass-catcher assembly also has a releasable latch between frame (20) and container (22) for normally maintaining the container in a generally upright position. A pair of latches (60) are respectively pivoted by pins on the respective bars (30). Each of the latches (60) has a recess for receiving a rod (66) extending between braces (52) on the respective containers.

Cover assembly (24) consists of a single cover for the open tops of both containers (42) and has deflecting means (70) for producing substantially equal distribution of grass clippings to the open tops of the respective containers. Deflecting means consists of a deflecting plate depending from the cover assembly (24) and located in the path of movement of grass clippings through chute 16. The deflecting plate has first and second angularly related portions, each of which has a flange extending therefrom that is secured to cover assembly (24), as by welding.

It has been found that by proper positioning of plate portions, the grass clippings and other materials will be equally distributed to the respective bags (48) located in containers (42). The two bags will be substantially simultaneously filled. Because of the suspended nature of container means (22), the vehicle moving across the terrain will produce a certain amount of shaking, which will result in compaction of the clipping and other debris.

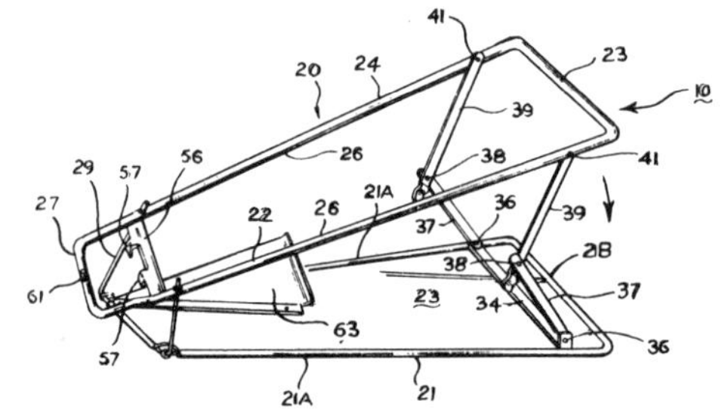

figure 4.18(b) Isometric view of a frame (20) for the rotary mower grass catcher described in Patent No. 3,732,675 (1973). The frame includes a generally U-shaped lower frame (21) having laterally spaced frame limbs (21*A*) connected by a frame element (21*B*). An upper frame (24) has a configuration generally like that of the lower. Frame 24 includes side frame elements (26) connected by an end frame element (28) and an opposed shorter frame element (27) to define with frame (20) a generally truncated pyramidal-shaped structure. Lower frame 21 has a floor plate (23), with the edges curled about the lower frame elements (21*A*).

Structure is provided for connecting the upper and lower frame elements. Floor plate (23) has a transverse extending strip (34) riveted to it, strip 34 terminating at each end in upstanding brackets (36). Lower toggle arms (37) are pivoted at the lower ends to brackets (36), with the upper end pivoted at 38 to an upper toggle arm (39), it in turn being pivotally connected at 41 to the upper frame member (26). The toggle arms are moved between collapsed and erected positions. A locking structure is provided for the toggle arms (37) and (39) consisting of a lug at the lower end of the upper toggle arm and cooperating with a slot in lower toggle arm.

The frame is adapted to be enclosed within a shroud. Structure is provided for holding the grass catcher (10) and for diverting the grass-laden current of air discharged from the mower. The cutting is diverted upward and toward the end of the shroud, being deposited upon floor plate (23).

(easier) way. There have been developed a number of attachments for power mowers to do this job, but so far none of them has worked out well. This offers a clear opportunity for inventors with ideas in this area. There are also some self-contained machines made expressly for the purpose of collecting, chopping, mulching, and compacting leaves. Such machines, of course, will be purchased only by those who've got more leaves than they know what to do with or by avid home gardeners, which means that they will never be mass produced. Jerome Lemelson of New Jersey, whose Velcro dart game is illustrated in Chapter 9, has also invented a leaf-processing machine (Fig. 4.19). It's not a simple machine, of necessity, and would not be cheap to produce. Any success would probably be achieved with a machine made as a specialty item for sale in relatively small quantities through a limited number of outlets at a fairly high price.

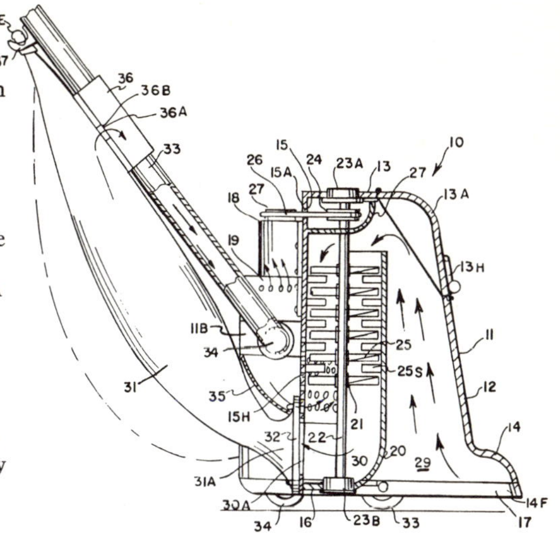

figure 4.19 Side, open view of a lawn cleaner (10), illustrated in Patent No. 3,877,207 (1975). It consists of a housing (11), supported on a frame (14F) at the lower end. An opening (17) is immediately adjacent to a scoop portion (14) of the housing and located a distance of several inches or less above the ground when front and rear wheels (33) and (34), rotationally supported by the frame, are resting on the ground. Thus the opening may be moved above dead foliage and leaves on the ground.

An elongated tubular handle (33) is pivotally supported on a rotary coupling (34) on a bracket (11*B*), which is supported by the rear wall (15) of the housing. The rotary coupling (34) extends to the inlet of a suction fan (19) of conventional design, which fan is also secured to the rear wall (15) and is rotated by an electric or a gasoline motor (18) shown secured above the fan at the upper end of the housing.

The housing (11) has a circumscribing side and front wall portion (12), a top wall (13) at the upper end thereof, having the air scoop (14) at its lower end. A cylindrical subhousing (20) is supported within housing (11) and surrounds a blade assembly (21) for mulching foliage and leaves. The housing wall (12) and subhousing (20) define a tapered passageway (29) extending from the bottom opening (14*A*) to the upper end of the housing, through which passageway leaves and light foliage are drawn and caused to flow upward into the open upper end of subhousing 20. The blade assembly (21) includes a shaft (22) extending between the bottom wall (16) of the housing (11) and the top wall (13) thereof. Ball bearings (23A and 23B) are respectively supported by the top and bottom walls (13) and (16) and the rotatable support shaft. A pulley (24) is supported near the upper end of the shaft, and a flexible belt or chain (26) extends around the pulley and through an opening (15*A*) in the rear wall, then around a pulley (27) supported at the end of the shaft of motor (18). Thus as the motor rotates, it not only drives the blades of the blower (19) but also rotates the shaft. Supported by the shaft, a plurality of spaced-apart blades (25) rotates therewith and serves to cooperate with stationary blades, disposed therebetween and secured to the subhousing (20), in chopping or mulching leaves and foliage deposited into the upper end of the subhousing. The blades may also be designed to induce an axial flow of air through the subhousing to assist the suction fan (19) in causing flow of air through passageway 29 and the subhousing as well as a container (31), which is coupled to the lower chamber (30) at the bottom of the subhousing and is adapted to receive chopped or mulched material therefrom.

The suction airflow generated by the suction fan (19) is applied to the rotary coupling (34) and the interior of the tubular handle (33). A fitting (36) is secured to handle 33 and contains a coupling (36*A*) through which air may flow from the interior of a flexible bag-shaped retainer (31) for mulched material. The lower end (31*A*) of bag (31) contains a flanged fitting (32), which is secured to an opening (31*A*) in the rear wall so that air and mulched material may be drawn from the lower chamber (30) into the bag. A filter (not shown) in the form of a screen is disposed across the opening (36*B*) in fitting (36) so that only air flows therethrough into the handle to the inlet end of the suction fan (19).

Suction is also applied from the fan through a duct (35) attached to the rear wall of the housing (11) and draws air through a plurality of holes (15*H*) in the rear wall, which defines the rear wall of the subhousing (20). Thus it is seen that air is induced to flow through the inlet (14*A*) by a combination of flows thereof, one through the tubular handle (33) and the bag (31), another through a portion of the wall of the subhousing (20), and a third by the rotation of the blades (25), any one of which flows may be used per se.

A lid (13*A*) extends from sidewall (12) to the top wall (13) of the housing (11). The lid (13*A*) is hinged at the top and contains a latch and handle (13*H*) at the bottom permitting it to be opened for the purpose of inspecting the inside of the housing, clearing material jammed or clogging the housing and, if desired, inserting material such as wet leaves, twigs, or other matter which may not be easily drawn up through the passageway (29), into the upper end of the subhousing (20).

The upper end of bag (31) contains an eye fitting (31*E*), which is held on a hook-shaped fitting (37) attached to the tubular handle (33) near its upper end, permitting the bag to be removed from the vehicle after the flanged lower end thereof has been disconnected from the rear wall (15) by loosening one or more conventional fasteners securing it thereto.

As a general rule, a machine that is made for use in or around the house must do a job a lot of people do not want to do for themselves, and it must be available at a price these people are willing to pay. (Like most rules, this one doesn't always hold up. The man who buys a $500 tractor-mower for his 50-by-100 lot isn't filling a real need: He just wants to have some fun driving that big toy around his property.)

Many, many appliances and gadgets have been invented for use outside the house, especially in the garden. The greatest boon to aching backs since Sloan's Liniment has been the Rototiller. All kinds of cutting and pruning devices have been developed, too, most of them electrical and cordless. Which brings up a question: Is there room for a cordless electric lawn mower? With all the research being done into battery-powered automobiles, the answer would seem to be yes.

There's another answer for the person who's tired of mowing the lawn—artificial turf. At the moment it seems farfetched to think of a large number of rose-covered cottages (or even mass-produced condominiums) surrounded by a blanket of green fresh from the chemical factory. But stranger things have happened. All this change just brings more openings for inventors to develop new products. For instance, you may not have to mow all that fake grass, but you do have to clean it. Anthony Salerno of New Jersey developed a turf cleaner for use in stadiums and similar places where it's important to keep the surface free of glass, dirt, and all foreign matter. His invention, illustrated in Fig. 4.20, may be the first in a line of artificial-turf cleaners for large areas and small.

In warm weather, the yard and garden are the centers of recreation. A few of the games and sporting goods suitable for use on the lawn are discussed in Chapter 10. Something that fits into the present chapter is the lawn swing invented by Romeo Gaucher of Al-

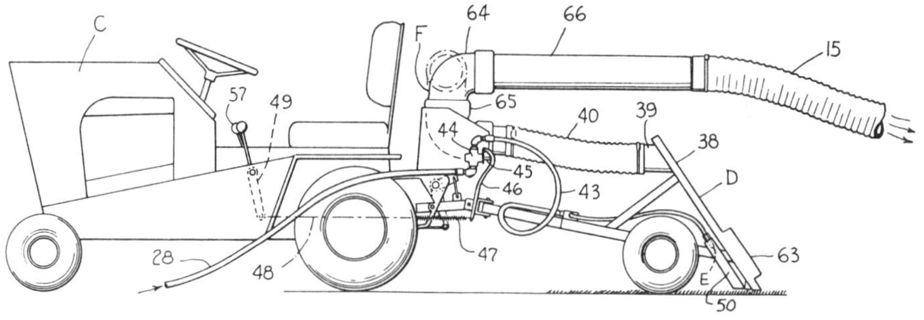

figure 4.20 Side view of a cleaning head connected to a tractor as illustrated in Patent No. 3,902,219 (1975). The device includes a tank with a vacuum hose attached. A vacuum pump communicates with the tank, causing a large volume of air to be drawn through the hose. An elongated transverse cleaning head (*D*) has a suction nozzle, which communicates with the other end of the vacuum hose, circulating air in large volume through the nozzle and the tank. A sprayer means (*E*) sprays hot cleaning medium on the turf ahead of the suction nozzle as the head is passed across the turf. A housing encompasses the sprayer and has an elongated opening closely adjacent the turf for encasing the hot cleaning medium as it is sprayed on the turf for minimizing the temperature lost during the cleaning operation. A swivel support (*F*) is carried by the motorized vehicle (*C*) for supporting the vacuum hose in order to permit the vehicle to turn under the hose as it traverses the field.

berta, Canada (Fig. 4.21). This is a clever variation on an old theme. It has hinged connections that don't need to be oiled every day in order to avoid that inevitable squeaking. In addition, these connections automatically adjust themselves for smooth operation to avoid any strain on the swing. This layout also has a table between the two seats.

On a hot day, there's nothing like a dip in some cool water to refresh the body. But not everyone has the space for, or can afford, a backyard swimming pool. Robert Gildea of Delaware has invented a water lounge complete with umbrella, refreshment table, and even a rubber duckie (Fig. 4.22). Recently, there have been several other inventions and designs that have combined several elements into one unit for convenience. It's not a bad idea.

Swimming pools have sprouted like pale-blue mushrooms all over the country. The market for accessories and improvements is a good one, too—growing all the time. Safety is the big word here, followed by cleaning. Whether a pool is in the ground or on a platform, it's got to be equipped with safety features that will prevent accidents, especially to small children. Fences, gates, and even alarm systems are possibilities. Most pool owners would give a lot to avoid the onerous chores connected with cleaning and filtering the water. One of the most successful products invented in recent years has been the Pool-Sweep, patented by Howard Arneson of California. Consisting mostly of a small motor, some plastic, and two hoses, this gadget blows all the debris that collects on a pool's surface into one spot, where it can be removed without much trouble (Fig. 4.23).

Whether at home or on the road, a picnic is a favored way to have a meal. Things can go wrong, of course. Ants, rain, and wind are among the forces of nature that picnickers must contend with. For those who like their outdoor meals not quite so rustic, Alan Hart of Colorado has developed something simple yet clever—a weighted tablecloth that resists being blown about by the wind or pushed out of alignment by a careless diner (Fig. 4.24).

Few customs are as solidly entrenched in American life as the outdoor barbecue. Each spring the stores blossom with a shiny array of new gadgets for the cookout. However, lots of people still have trouble getting the charcoal to light and the grill to work properly. There's room for an inventor to do something for his fellow man here—but the competition is formidable.

When cold weather comes, all this outdoor equipment must be moved inside the house and stored, probably in the basement or garage. Many hanging and other storage units have been developed, especially as houses have shrunk and possessions have multiplied.

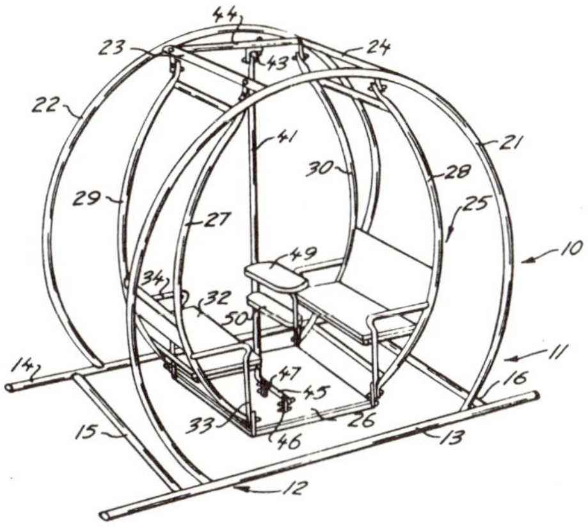

figure 4.21 Perspective view of a lawn swing, Patent No. 3,825,299 (1974). The swing (10) has a supporting frame (11), with a substantially rectangular base (12) formed of a pair of longitudinally extending members (13) and (14) and a pair of transverse members (15) and (16). The supporting frame also includes a pair of spaced-apart, hoop-like side supports (21) and (22), which extend upward from 13 and 14, spaced apart and sloping slightly inward over the base. The side supports are connected by a pair of horizontal, transversely extending braces (23) and (24), which lie in a common plane parallel to the base (12). The base members, the side supports, and the braces are tubular steel and have a welded connection so that the frame has transverse as well as longitudinal stability.

A swing proper, generally 25, is suspended from the braces (23) and (24) for swinging movement longitudinally of the supporting frame. The swing (25) has a substantially rectangular platform (26), the length of which is the same as the distance between the braces (23) and (24) and the width of which is substantially less than the width of the base (12), so as to provide clearance with the side supports (21) and (22).

The platform is suspended from the braces by means of bowed swing supports (27), (28), (29), and (30). The swing supports are of equal length, have hinged connections at lower ends with corners of the platform and at upper ends with the braces (23) and (24). A seat (32) is mounted between the supports (27) and (29) on L-shaped arm rests (33) and (34) secured to the supports (27) and (29).

A seat is mounted in a similar manner between the supports (28) and (30).

A vertical center strut (41) has a hinged connection at (43), midway between ends of a longitudinal brace (44) which extends between the transverse braces 23 and 24. The center strut has a lower leg (45) which has hinged connections at (46) and (47) with, and centrally of, the platform 26 midway. The center strut (41) has laterally extending arms on which tables (49) and (50) are rigidly mounted, upper surfaces of the tables being normal to the center strut.

figure 4.22 A backyard water lounge (10), illustrated in Patent No. 3,969,776 (1976). The bottom (14) of the body (12) is contoured to fit the human body. A drain is on the bottom, and an inlet is on the front wall. Both have adjustable valves. There is also a side table, a back rest, and an adjustable umbrella. The entire unit can be decorated, as shown, with appropriate drawings.

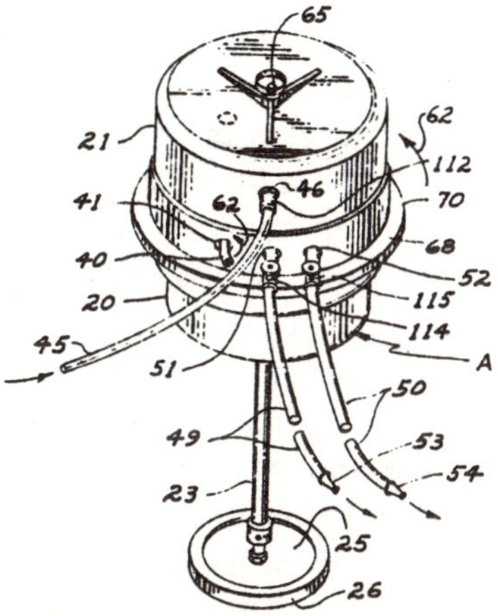

figure 4.23 Swimming pool cleaner, as shown in Patent No. 3,291,145 (1965). The main housing or float (*A*) has a bottom half (20) and a top half (21), which are rotatably mounted on each other. The top section maintains float A on the water, in which a shaft (23) is axially aligned with the two halves. The bottom of this shaft is fitted with a wheel (25) having a rubber rim (26). The power unit (*B*) is mounted on the upper half by arms (28–31). A central drive shaft (35) penetrates the top of the upper casing for free rotation. The lower half of the float is fixed to the shaft, and the top half is fixed to the power unit, so that rotation of the shaft will cause relative rotation of the lower half with respect to the upper half. The power unit has a control jet nozzle (40) that penetrates through aperture (41). A second, or reverse, jet nozzle (42) extends through aperture 43 in an opposite direction. A high-pressure inlet hose (45) passes through aperture 46 and is there connected to the power unit. Two working hoses 49 and 50 are connected to the housing and are fitted at their ends with pressure nozzles (53) and (54). The hose is connected to a fitting on the pool's side wall. A freely rotatable coupling interconnects the hose length closely adjacent to the housing. A coupling is supplied with high pressure through a pump system that draws its water from the pump's main drain and passes this water through the pump after normal filtration through the pool's filtering system. The high-pressure water flows into the power unit, causing the rotation of shaft (35), the alternative spraying of water from jet (40) or jet (42), and causes the rotation through 360 degrees of jet (40). Water is sprayed from the two working hoses (49) and (50) and from a spray nozzle (65) mounted on the top of shaft (35). These two hoses are of such length that during operation they spray generally toward the main drain. The force of the water from the nozzles causes the two hoses to move about in whiplike fashion, creating a water-jet force against the pool's bottom and side walls. The float positions the hoses at various locations so that eventually the nozzles will work against virtually every section of the pool wall. This, by hydraulic force, causes the eventual cleaning of the pool. Nozzle (65) rotates with a spray that hydraulically reacts with the side wall of the pool just about on the water line as the housing travels along this same wall.

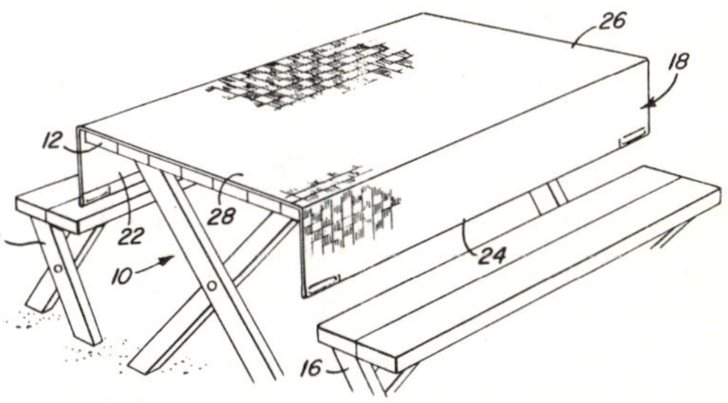

figure 4.24 Picnic table (10) spread with the table cloth claimed in Patent No. 3,960,193 (1976). There are weights at each corner of the cloth to hold it down and in proper position.

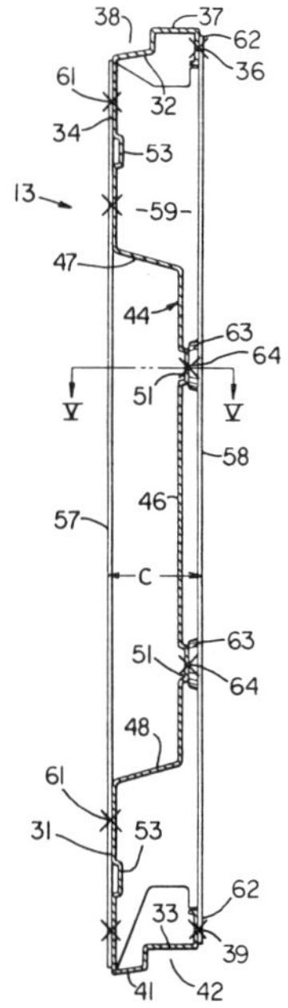

figure 4.25 Cross section of a door panel, according to Patent No. 4,008,745 (1977). Each panel (13) is of a generally channel-shaped section (31) defined by end walls (32 and 33) and interconnected by a rather wide web (34). The upper end wall (32) includes an inwardly directed flange (36) which is substantially parallel to the wall (34) and extends inwardly from the longitudinally extending edge of the panel. The flange (36) is integral with an upwardly projecting channel-like part (37), which extends partway across the thickness of the panel. The end wall (32) in addition includes a shouldered recess (38) adjacent the front side thereof.

The lower end wall (33) is of a construction somewhat similar to the upper end wall (32) and includes an inwardly projecting flange (39) disposed on the rearward side thereof and positioned substantially parallel with the front wall or web (34). The flange (39) is integral with a shouldered recess (42), which is positioned adjacent to, but rearwardly of the downwardly projecting channel-like part (41).

The section member (31) also includes a channel-shaped stiffening portion (44) formed substantially centrally thereof and extending throughout the length of the section (31). The stiffening portion (44) is defined by an offset web or wall portion (46), which is substantially parallel to but transversely and laterally spaced from the front wall (34). The wall portion (46) is connected to the upper part of the wall (34) by means of a first shoulder (47), and is connected to the lower portion of the wall (34) by means of a further shoulder (48).

The wall (46) of the stiffening portion (44) is preferably provided with one or more rib or channel-like protrusions formed therein, such as the protrusions (51) that project rearward of the section (31) and extend in substantially parallel relationship. Similar protrusions (53) can also be formed in the upper and lower front wall portions (34). The protrusions 51 and 53 improve the strength and rigidity of the section (31), particularly in its longitudinal direction, and in addition facilitate the overall assembly of the panel.

If there's to be room for the car in the garage, then space at the sides and on the walls and ceilings must be utilized efficiently. Simplicity is the key to success here, as it is elsewhere. If you've rigged up something for your own particular storage needs, think about adapting it for mass production. Many inventions have begun as conveniences clever people have built into their own homes.

The overhead garage door is quite a space saver in itself, and has been one of the more successful inventions for the home. Ever since its development, many improvements on the basic design and method of operation have been devised. Yet here, as elsewhere, people are still working on change. The original doors were made of wooden panels, which tend to be heavy and costly. Lightweight metal, fiberglass, plastic, and other materials have become commonly used for these doors. But this has caused some problems. For one thing, some of these doors have not been as strong as they needed to be. One of the latest ideas on how to change the design of the door and its panels to add strength to the entire unit is illustrated in Fig. 4.25. A channel-like center portion gives each panel the rigidity to withstand wind pressure, etc. This increased strength permits the assembled door to have end and center stiles of substantially reduced size and cross section. This invention was patented by Morris Bailey of Texas.

It must be concluded that there's no real problem connected with devising all kinds of things for use around the house. The difficulty comes in getting your invention on the market. As we discussed in the introductory chapters, there are no sure-fire answers here—in fact, anyone who offers you a plan for guaranteed success is automatically a suspect character. But there are some guidelines that generally hold true:

1. Keep your invention simple unless you think you can sell it to a large company. Large companies are the only ones willing to invest the capital to cover startup expenses on a complicated item.
2. Look at competitive items. If there are only a few, maybe the market is smaller than you think. If there are many, make sure yours is better than any of the others. If there are none, look up some of the old patents in the field and find out whether any of them reached the manufacturing stage. This previous experience may be quite revealing.
3. Make a prototype of your invention and use it around the house. Is it as close to perfect as you can get it? If not, don't try to sell it until you've made some adjustments (and tested them, too).
4. Consider whether you have the knowhow and capital necessary to manufacture and distribute the invention yourself. If you don't know the field, better not take this course.
5. Approach some companies in the field. Don't be chagrined if they don't all give you a warm welcome. Persevere.
6. Be prepared for disappointments and frustrations. Believe in yourself and in your invention.
7. Don't be afraid to ask advice of people in the field. After your invention has been patented, it's not going to be easy for them to steal it.
8. Work hard.
9. Be lucky.

CHAPTER FIVE

Bathroom Utensils and Fixtures

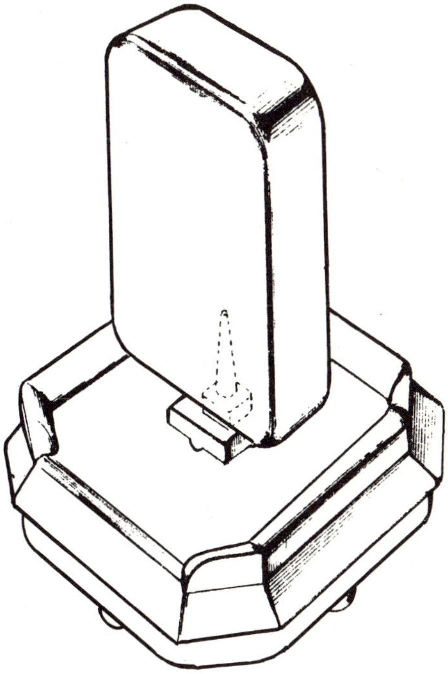

In a movie that dates from the late 1960s, *Divorce American Style,* there's a funny sequence that satirizes the nighttime-cleaning routine of a prosperous American couple. Dick Van Dyke and Debbie Reynolds race through a fantastic array of machines to make their teeth whiter, their gums firmer, their hair glossier, their skin smoother, their muscles firmer—their whole bodies cleaner, healthier, and happier. In laughing at Dick and Debby, the movie audience is also knowingly laughing at themselves and the American obsessions with clean bodies and gadgets. If the trend holds, the bathroom of the future will be a combination of a do-it-yourself massage parlor and a Turkish bath.

The country has come quite a way from the days, not so long ago in some areas, when just about all a household had were a wooden tub in the kitchen and an outhouse in the back. Of course, inventors have been working on products for the bath ever since colonial days, but never have they enjoyed as much success as they do today. Our pampered generation would have felt right at home in ancient Rome, where the bath and accouterments held an important place in the community structure. In modern America, security, comfort, and happiness have become the rights of all people—perhaps the Declaration of Independence will be modernized to guarantee the "possession" rather than the "pursuit" of happiness.

Things for the bathroom continue to pour into the marketplace at an astonishing rate. Recently, shower nozzles and heads, massage sprays, and other watery attachments have enjoyed great

figure 5.1 A cross section of a Teledyne spray nozzle, Patent No. 3,801,019 (1974). The lower housing unit is designated 20. At its upper end is an externally threaded neck (22).

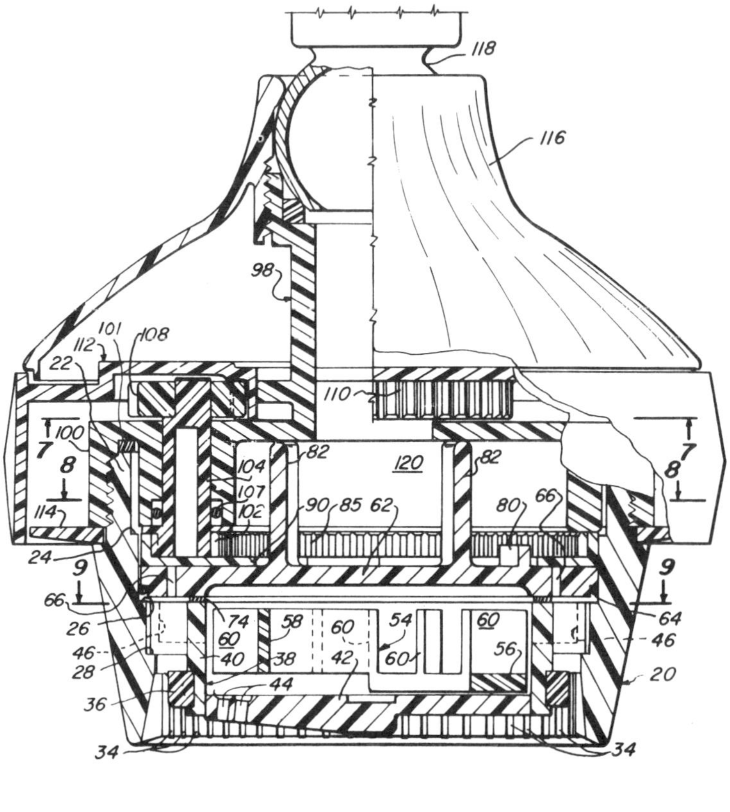

The internal central passage through lower housing (22) is formed with three radial shoulders (24), (26), and (28), which provide seating shoulders for axially locating other elements of the shower head within lower housing (20). Two sets of diametrically opposed axial grooves (30) and (32) extend downward respectively from shoulders 24 and 26 to rotatively orient other elements of the assembly. The lower end of the central passage through housing unit (20) is formed with a series of axially extending slots (34). In the assembled shower head, a washer (36) is seated to extend across the inner sides of grooves (34) to complete definition of the grooves as one group of spray-discharge orifices.

Washer (36) is seated within an annular groove formed on the exterior of a spray-cup assembly (designated, generally, 38) having a tubular main body (40) and an end wall or orifice plate (42) seated within and extending across the open lower end of the tubular body (40). Three like groups of discharge orifices (44) are bored through end wall 42 and lie in a symmetrical relationship within an annular band concentric with the central axis of the unit. At the upper end of the main body (40), a pair of like flow-carrying troughs (46) extend partially around the outer circumference of member 40 in symmetrically disposed relationship. The adjacent ends of the two troughs terminate short of each other to provide a pair of axially extending flow passages through the troughs. The exterior areas of the troughs are seated upon the lowermost shoulder (28) of lower housing (20) to axially locate the spray cup (38) within housing (20), the cup being rotatively oriented within the housing by a pair of projecting lugs received within locating slots of the housing. When the spray cup (38) is seated within the housing, washer (36) is radially compressed against grooves (34) to cause the grooves to define one group of orifices, while orifices (44) in end wall (42) define a second group of spray discharge orifices.

A rotary valve member, designated generally (54), rests upon the inner or upper side of end wall 42 and is retained by the inner wall of main body (40) for rotation about the central axis of the unit. Valve member 54 is one piece, molded preferably from glass-reinforced nylon, and includes a flat, generally C-shaped base plate portion (56), which lies in a radial general plane and extends for approximately 180 degrees about its central axis. A semicylindrical portion (58) is integrally joined to the opposite ends of portion (56) and extends angularly around the remaining 180 degrees of rotor (54). The lower edge of semicylindrical portion 58 is coplanar with the top or upper surface of flat portion (56) so that the latter has its lower surface spaced downwardly from the lower edge of semicylindrical portion (58). Radially extending blades (60) are integrally mounted upon portions 56 and 58 in symmetrical relationship to the central axis of the unit. The relative angular extent of plate (56) may vary. However, the 180-degree extent shown in the drawings represents a preferred form.

Plate (56) of valve rotor (54) rests on the inner surface of end wall (42) of the spray cup assembly and is so located as to cover, at all times and rotative positions, at least a portion of orifices (44), the annular band within which the orifices lie corresponding in general to the annular path traversed by end plate (56) upon rotation of valve rotor (54). Radial blades (60) are so located as to be struck by water discharged through tangential passages, and the valve rotor is thus driven in rotation at a rate which varies with the rate of flow of water through tangential passages of the spray cup assembly.

A flow-directing plate, designated generally (62), overlies the upper end of spray cup 38 and is employed to direct and control the flow of water to the various discharge orifices. Flow-directing plate (62) is seated upon the second or middle radial shoulder of lower housing unit (20), and an O-ring (64) seated between the periphery of plate (62) and housing (20) provides a seal against the flow of water around the outer periphery. Plate (62) is formed with a first pair of segmental openings (66), which are located in diametrically opposed relationship to each other, a second pair of diametrically opposed segmently shaped openings, and a third pair of segmental openings. Radially projecting tabs on plate (62) are engaged in axial slots in housing 20 to rotatively orient plate (62) relative to spray cup 38 so that openings (66) are vertically aligned with and communicate directly with flow passages (48) in spray cup (38). The openings in plate (62) are aligned and communicate with troughs (46) of spray cup (38), while other openings are located radially inwardly of the inner wall of main body (40) of the spray cup. A gasket (74) is seated between the lower side of flow-directing plate (62) and the upper end of the spray cup.

When connecting tube (98) is threaded onto lower housing unit (20), compression tabs (82) of flow-directing plate (62) are engaged by the underside of connecting tube (98) so that, as tube (98) and lower housing unit are threaded into each other, the plate is forced down against the top of spray cup 38 to clamp gasket (74) and also to force spray cup downward so that its troughs (46) are seated on lower shoulder (28) in the housing. It is not necessary that the plate be firmly seated axially against shoulder (26), since O-ring (64) is radially compressed to form the necessary seal around the outer periphery of plate (62). During assembly, an annular ring (114) is clamped between the lower end of connecting tube (98) and a shoulder on lower housing unit (20), ring (114) being primarily for cosmetic purposes and providing a stationary member upon which a scale for indicating the rotative position of control ring (112) relative to the housing may be located. An upper housing unit (116) (stationary supply form) is threaded upon the upper end of connecting tube (98), and when threaded onto tube 98 frictionally clamps a swivel ball fitting (118) to the unit to provide a means for mounting the assembly upon a stationary supply pipe. The skirt of upper housing unit (116) also serves to axially confine control ring (112).

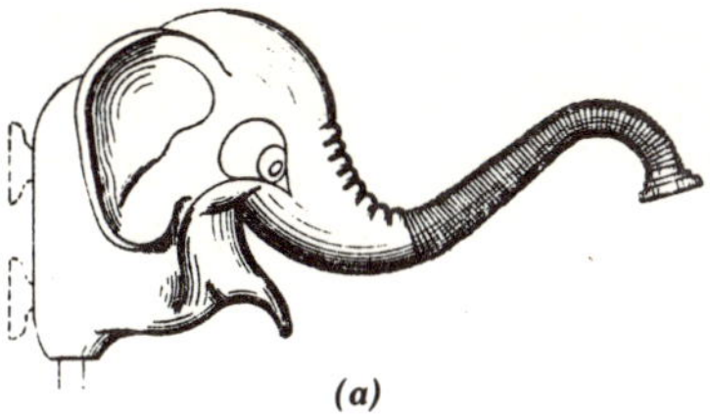

(a)

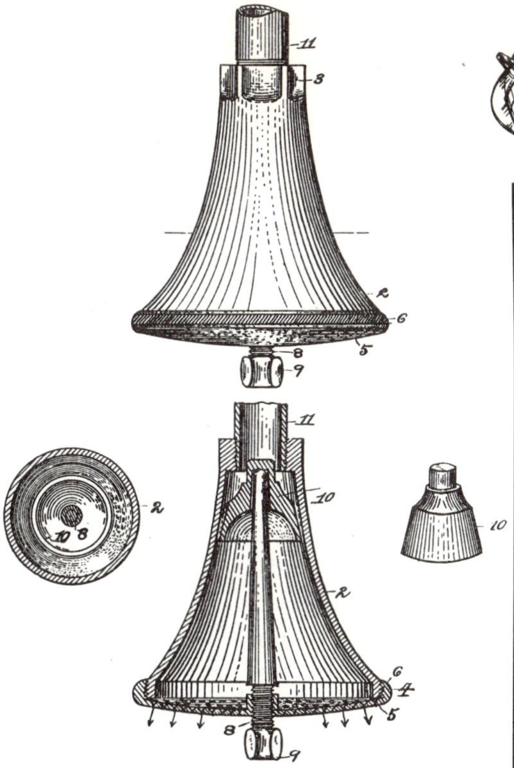

(b)

(c)

figure 5.2(a) A side elevation of a bath shower head, design Patent No. 173,979 (1955).

figure 5.2(b) Four views of a showerhead, Patent No. 1,295,250 (1919). Figure A is a side view of the showerhead secured to a supply pipe, and Fig. B is a section centrally through the head. Figure C is a transverse section of the head on line 3-3, Figure A. Figure D is a perspective view of the spinning-valve member.

The invention comprises a hollow bell-shaped head (2) having flat, wrench-engaging faces (3) at its upper and smaller end and screwthreads (4) externally at its rim or larger end. A perforated bottom member (5) is removably secured to this screw-threaded rim by a screw-threaded flange 6, and a central integral boss in this bottom member has a screw-threaded bore in which a screw stem 8 is held upright for vertical adjustment. The exposed end of the stem is provided with a corrugated finger grip (9), and the upper end of the stem supports a flaring or cone-shaped valve (10) in rotatable relation to the internal flaring sides of head 2, opposite the end of supply pipe 11. Member 10 serves to spread the water uniformly within the head and promotes the discharge of water at the circumference of perforated member 5 where ordinarily a stronger flow would occur at the center of said member. Sleeving of the spreader 10 loosely upon the stem also permits it to spin or revolve under the pressure of the water with an advantageous effect on the distribution, while the working of the spreader (10) and the amount of water discharged is regulated by screw stem 8.

figure 5.2(c) Three views, perspective, side, and front, of a showerhead, design Patent No. 235,997 (1975).

popularity. The newest nozzles produce a pulsating spray that can be fine-tuned by the user to provide a great variety of effects. One of the most active companies in this area has been Teledyne, makers of the Water-Pik. (This is of course one of the most successful bathroom machines introduced in the last twenty years. It was developed by John W. Mattingly and first patented in its present form in 1966. Since then, there have been many improvements—and some competition. See page 85 for further details.)

Another success story is represented by the patent illustrated in Fig. 5.1. It is an improvement on another patent issued a year earlier to the same inventors, John Trenary and David W. Smith of Colorado. This later invention offers a choice of either pulsating or nonpulsating spray or a timed combination of both. The earlier nozzle had met with some dissatisfaction from some people who often wanted a straight shower. In this nozzle and in its predecessor there are two internal flow passages. The water rushing through one of these passages drives a rotary valve, which cyclically blocks the discharge of water through the second passage. By varying the water pressure and the size of the opening for the second passage, a wide range of pulsating frequencies can be achieved. Like having your own masseuse (or masseur) on call. The unit can be mounted on the wall or held in the hand.

Many different kinds of shower nozzles have been patented over the years, some for their design, others for a unique arrangement of moving parts. Fig. 5.2 shows parts of three of these patents: *(a)* is a design patent issued in 1955 to James Crenshaw of Missouri; *(b)* is a regulating shower head patented in 1917 by Leon Yeiser of Ohio; *(c)* is a design by Allan B. Johnson of California that has been manufactured by Teledyne. Comparing the nozzles, it is evident that the modern one is more complicated and thus more expensive. A fairly simple product has been turned into a kind of luxury item available to, and sought by, a mass market. This is contrary to the general rule that a new version of an old item should be simpler and less expensive to make than its predecessors. But if people are willing to pay for a more complicated machine, inventors and manufacturers are ready to supply them with all they can use. This is a product area that has probably not reached its peak, and thus offers opportunities to inventors with a better idea.

The sybarite of the shower needs more than a pulsating nozzle to feel luxurious—balms and oils, soaps and powders, are part of the routine, too. Several inventors have had the idea of an improved dispensing arrangement for use in the shower. Figure 5.3*a* shows a device made by Alan Toper of California, which thus far has not had commercial success. Figure 5.3*b* is the design patent for a simple arrangement issued in 1974. It was developed by Dwight Wooters for the Rubbermaid Company, which has one of the best track records as a marketer of accessories for kitchen, bath, and household.

Other similar products have come on the market without benefit of patent protection and have done quite well. The question is, Why do some products gain fame and fortune while others remain obscure? Perhaps it has less to do with their intrinsic worth, one against the other, than with good distribution by a promotion-minded manufacturer. Not that even the most skilled marketer can sell a piece of junk; the product must have worth. But many inventors would testify that a good sales push can make all the difference. "We know we have a fantastic product," Mr. Toper wrote of his dispenser, "but we just don't know how to get it on the market."

All the soaps and shampoos used in the shower can make the floor dangerously slippery. Bathroom falls are not uncommon, especially among older people, and various devices have been developed in attempts to cope with this situation. For one thing, manufacturers of bathtubs and tiles have made their products less glossy smooth. Also on the market are various kinds of rough-surface strips that can be pasted down on shower floors and tub surfaces. These are not 100 percent effective, however, and many other solutions to this problem have been advanced. In a 1976 patent, Paul Kollsman of New York proposed a new type of contoured surface for tubs and shower floors (Fig. 5.4). He envisioned a great number of alternate raised and depressed segments bounded by sharp edges to break down any slippery liquid film.

A bath can, of course, be the simplest of pleasures. All you need is a tub, some soap, and plenty of water at the right temperature. But inventors are not generally in favor of things as they are. They're always looking for a better way. In 1891, M. H. Welles found one answer to the old problem of how you get at your back when your arms aren't long enough. The back rest he patented has a friction attachment for back rubbing in the tub (Fig. 5.5). Sounds nice, though the man in the drawing doesn't look quite as comfortable as he could be. A modern arrangement was patented by Lars

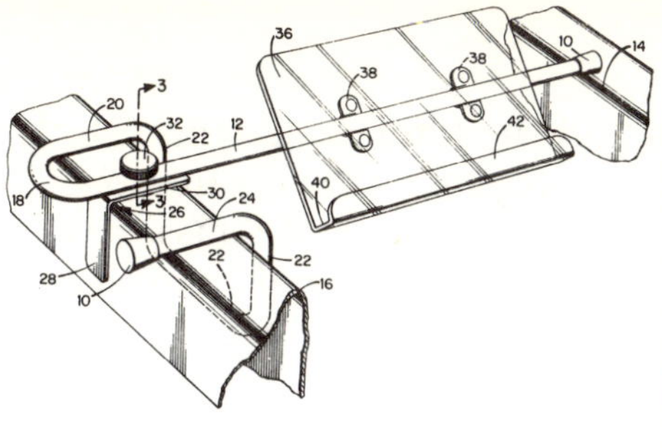

figure 5.7 A bathtub bookstand, Patent No. 3,950,793 (1976). It is preferably made from 3/4-inch-diameter hollow aluminum tubing which has been bright-dipped and clear anodized. The initially straight hollow tubing is bent into the shape shown and is capped at both ends with rubber or plastic caps (10), for purposes of safety and to prevent marring of the bathtub finish.

Horizontal beam (12) has an extension (18) which is bent so as to have a first horizontal portion (20), a vertical U-shaped portion (22), and a second horizontal portion (24). Horizontal portions (20) and (24) prevent the horizontal support beam (12) from rotating about its longitudinal axis. Vertical U-shaped portion (22) is positioned against the inner side of outer tub rim (16) to prevent the horizontal support beam (12) from moving outwardly along its longitudinal axis and to provide a long bearing or clamping surface to support horizontal beam (12).

In order to clamp the horizontal beam (12) to the outer tub rim (16), a movable clamping jaw (26) is slidably mounted on the underside of the junction between beam (12) and beam extension (18). Clamping jaw (26) is an L-shaped element with a vertical jaw (28) and a horizontal slide (30), and is preferably provided with pads on its surfaces which contact the bathtub. Clamping jaw (26) and U-shaped portion (22) clamp outer tub rim (16) between them.

The horizontal slide (30) is longitudinally slotted to accommodate the shafts of a threaded bolt (32) and an internally threaded inverted T-nut (34) into which the bolt (32) is engaged. The head of the T-nut has an upwardly extending projection which rides in the slot of the slide (30) to prevent the T-nut from rotating. When the head of bolt (32) is manually tightened, the clamping jaw (26) is immobilized. When the head of the bolt (32) is manually loosened, the clamping jaw (26) is slidable toward or away from the vertical U-shaped portion (22) which constitutes a fixed clamping jaw.

In order to securely hold a book, a magazine, cosmetics or the like, a planar book support member (36) is attached to a horizontal support beam (12). Preferably, the attachment is adjustable so that the angle of book support member 36 can be conveniently varied by the bather. Two roughened metal straps (38) encircle the support beam and are bolted to the support member. The two straps frictionally hold the support member securely in position, yet permit it to be tilted. Also, the support member can be moved laterally along the beam under manual pressure for centering, reversing, or other purposes.

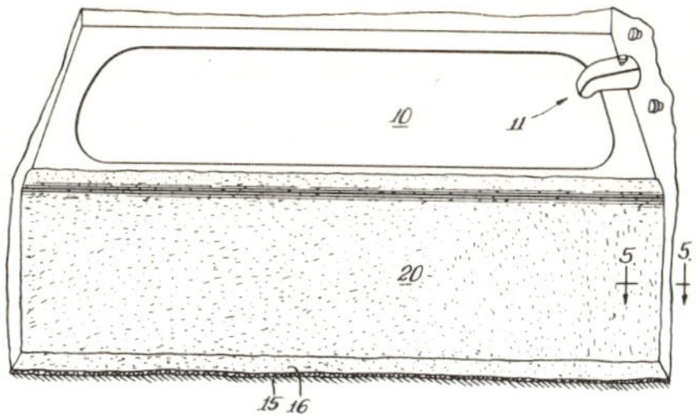

figure 5.8 A bathtub cover, Patent No. 3,950,795 (1976). A conventional bathtub (10) with faucet (11) is covered with a decorative textile (2) on its outer wall (shaded area). The back side of the fabric (20) is coated with an adhesive covered by a peel-off paper or protective film that remains in place until the cover is ready for application. The purchaser can cut the fabric to fit the tub, expose the adhesive by peeling off its protective covering, and then press the fabric down to make a firm bond.

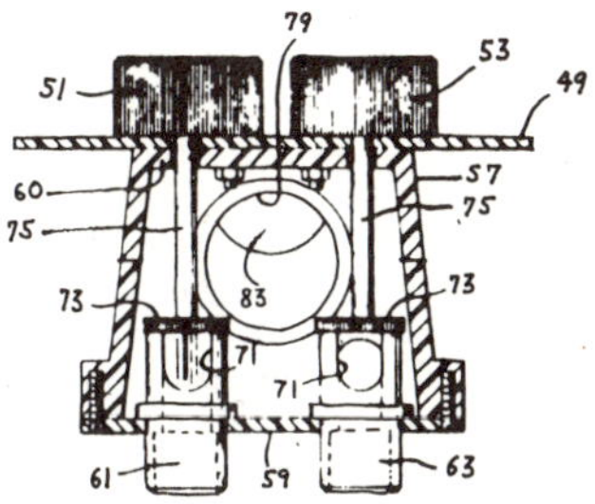

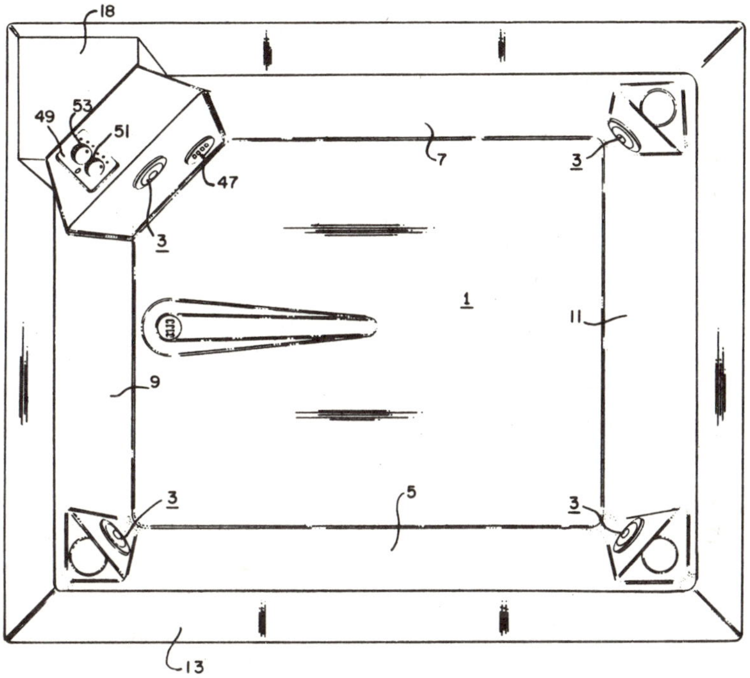

figure 5.9 Hydromassage tub with jet head assembly (right, above) as described in Patent No. 3,874,374 (1975).

A tub (1) is of substantially rectangular shape, having a jet head assembly (3) installed in each corner thereof. The tub's side walls (5, 7) are preferably of slightly greater length than the end walls (9, 11). A peripheral apron (13) surrounds the tub walls to create a peripheral chamber, which not only houses the jet head assemblies, but also a pump (17) in one corner (18) and all piping associated with the system.

Each jet head assembly may be of conventional design. All four assemblies are supplied in parallel with water from a pump, the water being returned to the pump via a return opening (47) in the corner of the tub. Air control to the assemblies is provided from a slightly sloping panel (49) installed in the corner of the tub that houses the pump, and above the proximate jet head assembly. Two control knobs (51), (53) are installed in this panel, one in control of the air to the jet head assemblies at one end of the tub, and the other in control of air to the other two jet head assemblies at the opposite end.

These two control knobs are components of an air control unit comprising a housing (57) formed of side walls closed at one end by a floor (59) and at its other end by an end wall (60) upon which the panel (49) rests and is retained by appropriate fastening means.

Installed in and extending through the floor of the air control unit are a pair of air tubes (61), (63), a portion of each extending outside the housing for coupling to an air line (67), (69) respectively, from which to supply air to two of the jet head assemblies and a portion of each extending within the housing.

The portion of each air tube within the housing has an opening (71) adjacent its edge. Telescoping this end of each air tube is a rotatable cap (73) having a matching opening, which by rotation of the cap as a valve, can be made to move into and out of registry with the associated opening in the tube, and to varying degrees. Each tube and associated cap thus constitutes a valve assembly for determining air intake.

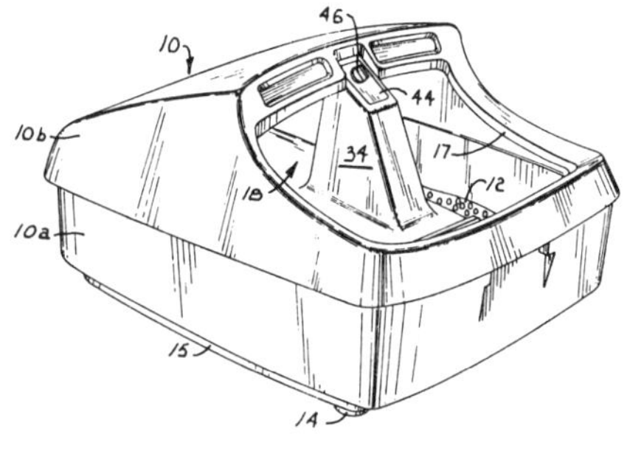

figure 5.10 Perspective view of a heating and massaging foot bath, as illustrated in Patent No. 3,965,495 (1976).

The foot tub (10) includes a lower portion (10a) and an upper portion (10b). The casing is preferably made of a suitable high-grade molding plastic. The lower casing portion (10a) has a floor (11) and upstanding side and end walls, which provide a generally rectangular tub for containing liquid. Two embossed sections (12), raised slightly above floor (11), are shaped generally to the contour of the human foot.

The casing is supported on four short legs (13) which extend below floor (11) near the four corners thereof. Each leg (13) has a pad (14) on its lower end. A continuous, peripheral, downwardly extending flange (15) projects below the floor (11) inwardly of its periphery and outwardly of legs (13).

The upper casing (10b) is secured on top of the lower casing (10a) and is constructed to overhang and cover approximately the forward half of the tub. Casing portion (10b) is open directly above the rearward portions of foot-receiving sections (12) in order to accommodate the legs of the user. An inward and downward turned flange (17) extends around the open area to help retain liquid in the tub.

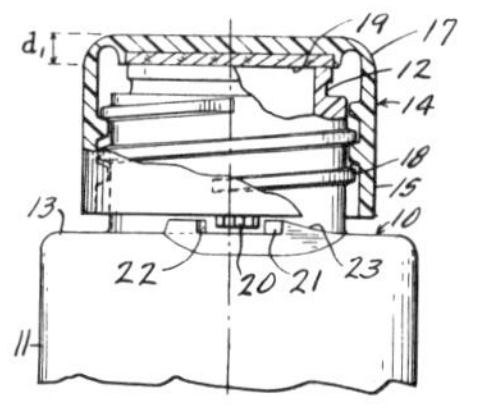

figure 5.11 A child-resistant closure, as illustrated in Patent No. 3,894,647 (1975). The cap (14) and the container (10) have a lock that consists of a radially/outwardly directed tab (20) at the lower edge of the cap skirt (15) and a pair of stops (21, 22) formed on the shoulder (13) of the container. When the cap is turned, it must be rotated a sufficient distance for the tab to ride up an incline (23) on the container's shoulder and then to snap by its own resiliency down behind the stop (21). This first stop is the closed position relative to the container neck (12). The second stop is angularly spaced beyond the first stop (21) only a distance sufficient that the tab will drop down between the stops and thus prevent the cap from being turned beyond this normal closed position.

In order to compensate for tolerance variations in the mass manufacture of this product the cap has the annular shoulder (17) so that the liner (19) is brought into a seal with the end of the threaded neck (12) whether the tolerances result in the locking tab (21) reaching the normal position (distance d_1) or a longer distance. The shoulder stretches downward to compensate for variations.

Feet are among the achingest parts of the human anatomy. Dr. Scholl and others have made fortunes because they figured out ways to help people's feet feel better. Samuel McNair of Kansas has patented a hydrotherapy foot tub with heat and massage capabilities (Fig. 5.10). There have been many other foot tubs, of course, and in his design, assigned to the Dazey Corporation, McNair attempted to improve on them by developing a machine to do all of the following: heat the water quickly and evenly; eliminate potential burn hazards; have a single easily accessible on/off switch; provide adequate ventilation for the vibrator motor to prevent overheating.

Surely none of the inventions described here, nor others that are on the market, has put a cap on new developments in this area. There's still plenty of room for improvement, for innovation. Despite all the work that's been done, the jet machines for the bath still tend to be cumbersome and expensive, or cheap and ineffective.

Another category of products made for use in the bathroom concerns the medicine cabinet and all the different things it holds. Safety is an important element here. In the past, children were able to get at medicines and, in some cases, to take toxic overdoses. Several solutions to this problem have been put forward. The medicine cabinet itself can be redesigned to provide a locking mechanism only adults can open. More important, bottle tops have been invented to be child-resistant. Gary Van Montgomery of Indiana patented one of these closures, designed to be leakproof and to be manufactured in large quantities at relatively low cost (Fig. 5.11).

As might be expected, many patents for plumbing parts and appliances have been granted over the past near-200 years. The current emphasis is on economy in several ways, as costs of materials and plumbers' time continue to rise and as water supplies dwindle in various parts of the country. Not only builders and contractors, but also average homeowners, are looking for ways to keep expenses down. There's a ready market here—even, it must be admitted, if the new is not quite as sturdy or effective as the old. This is not to say that cutting costs always means cutting corners, too.

Most modern faucets are single—that is, hot and cold water come out of the same tap at a temperature selected by the user, who moves the tap control left or right or back and forth. William Schmitt of Wisconsin has patented a valve for this kind of faucet that contains fewer parts and fewer seals between the mixing chamber and the rotatable stem (Fig. 5.12). Here, less offers more. There are some other improvements in this invention, too, but the basic idea was to make the valve less complicated and easier to maintain. As a result, it will probably last longer than the older-style valves.

Another important trend in this area is toward the availability of faucets and other appliances that are easy for the do-it-yourselfer to install. The homebuilding and home maintenance industry is changing from one dominated in many parts of the country by out-of-date building codes and regulations to one that is open to new ideas and ways of getting the job done.

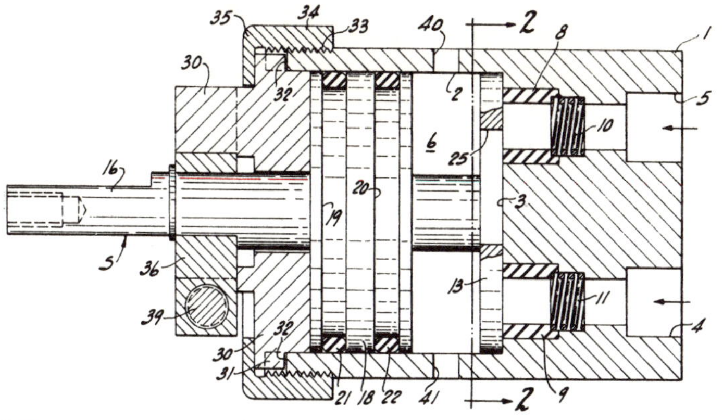

figure 5.12 Longitudinal, cross-sectional view through the mixing valve, as shown in Patent No. 3,916,951 (1975).

The cartridge assembly includes a cylindrical cartridge (1) having a large bore (2) extending from one end and terminating in an inner end wall (3). An inlet passage (4) and an inlet passage (5) for hot and cold water, respectively, extend from the mixing chamber (6) that is defined by the bore (2) and outwardly of the cartridge where they can communicate with sources of hot and cold water (not shown).

Each of the passages (4) and (5) has a cylindrical seal (8) and (9), respectively, which are urged by their springs (10) and (11), respectively, into sealing engagement with a valve disc (13) of the stem *S*. The stem includes a handle in the form of an operating portion (16) which extends outwardly from the cartridge. The stem also includes a radially enlarged cylindrical portion (18) which is also of considerable axial length and around the periphery of which are formed two sealing grooves (19) and (20). O-ring seals (21) and (22) are located in these grooves and form a rotatable seal with the bore (2). The stem also includes the valve disc (13) at its inner end, which abuts against the inner end wall (3) of the cartridge and is rotatable in respect thereto. This valve disc has a pair of openings in the forms of kidney-shaped ports (24) and (25), each of which is registerable with the inlets (4) and (5), depending on the direction and the extent of rotation of the valve disc (13).

The enlarged portion (18) of the stem is held captive in the bore (2) by the cartridge top (30), which is insertable partially into the bore. A pair of ears (31) extending from the top (30) engage in slots (32) of the cartridge, to prevent relative rotation between the top and the cartridge. The top (30) is releasably clamped in the cartridge by a removable cap nut (35) threadably engaged as at 33 with the threaded end of the cartridge body. The radially inward extending flange (34) of the cap nut (35) extends over the top (30) to hold the top captive. An inward-extending stop (36) is carried by the top (30) and has stop surfaces, which are engaged by an adjustable stop (39) secured on the stem so as to limit the rotation of the stem in either the hot or cold water directions.

Outlet ports (40) and (41) extend from the mixing chamber (6), by means of which the water is directed to the spigot or the like (not shown) in the conventional manner.

The conservation of water is becoming more important all the time. Even in regions where rainfall is generally plentiful there can be devastating droughts—as there were in California and the Pacific Northwest in 1977. From the desert country of Israel has come a quick-acting water tap designed to alleviate part of this problem. The tap is comprised of (1) a vertical ball valve where the ball is freely movable above a valve seated in a housing adapted for connection to a faucet and (2) a water spout consisting of a spherical top part fastened in the housing below the valve. The basic idea here is a familiar one—people are lazy and careless, and the easier it is for them to do something, the more readily they will do it. Sometimes it's even too much just to turn off a faucet. (If you think that's farfetched, think of the times you've seen water gushing into the sink in a public bathroom, with no one in sight.) But if a tap can be opened and closed with the flick of a wrist, most people will use only the water they need. And this is a much more convenient system than the spring taps in many public places, where in order to keep the water flowing you've got to keep one hand turning the faucet handle to keep it open. The device illustrated in Fig. 5.13 can be opened and closed by a horizontal movement of the hand, and can be attached to most existing faucets without much trouble.

Another place where there is considerable potential for water conservation is the toilet. Almost half of the water consumption of many households is flushed down the soilpipe. Most planners agree that the toilets of the future must be designed to use less water. California has already enacted a law designed to speed this development. There are several possibilities. Dr. Zeev Raz of Ben-Gurion University, Israel, has been working on a simple mechanism involving two levers—one to release a fairly high volume of water, for solid wastes, and one to release a low volume of water, for liquid wastes. Some householders in this country have taken a fairly simple action, placing a brick or two in the toilet tank to cut down the volume of water necessary to fill the tank.

That's somewhat the same principle used by Walter Klein of Virginia in his invention (Fig. 5.14), an inexpensive, flexible plastic bag with an adhesive that holds it in place during flushes. Filled with a gallon of water, the bag saves that much water on each flush without diminishing the water pressure necessary for proper operation. Klein did not think of this as an item to be sold at retail, but

figure 5.13 Two cross sections through the quick-acting water tap of Patent No. 3,888,461 (1975).

It comprises a cylindrical housing (1), a movable spout (2), a ball valve (3), and a removable lower seat (4). The top part of the housing is provided with an internal screw thread (11) for connection to a faucet or to the spout of a mixing battery. In a continuation of the threaded part, the housing is formed as a cylindrical hollow space (12) which contains a ball valve (3) and terminates at its bottom in a conical valve seat (13), which permits closing of the tap when the ball rests thereon. The housing is enlarged at its bottom to form another cylindrical hollow space (15), the lowermost portion of which is internally threaded (14), and which serves to contain the upper spherical portion (21) of the spout (2). The spherical portion is continued downwards by the actual spout portion in the shape of a tube with a bore (25), to the end of which tube an annular grip (23) is attached by screw thread or other bonding, which may also serve for connecting a hose.

A finger (24) with a rounded end projects out of the top of the sphere (21) in continuation of the bore (25). It is in contact with the ball valve (3) in vertical position of the spout (2) and disappears below the valve seat (13) when the spout is inclined sideways.

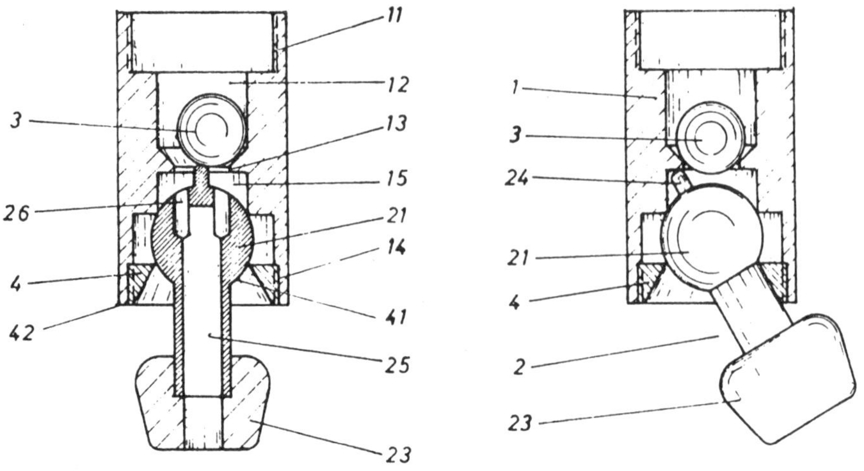

The bore (25) does not reach to the top of the sphere but ends somewhat below it. Four water passages (26) are drilled into the sphere around the finger (24) parallel to, and entering, the bore (25). These passages connect the space (15) of the housing with the outside through the bore.

The spout is held in position by a flat, centrally perforated disc (4), which is screwed into the housing from below. It possesses a stepped inner seat (41) for the spherical spout portion (21), which also serves as seal against penetration of water to the outside. To enable the sideways tilting of the spout, part of the bottom surface of the disc is cut away in the shape of an oblique recess (42), which is of a width somewhat greater than the outer diameter of the spout tube.

Water passes through the tap when the spout is in coaxial alignment with the housing and the ball valve is lifted off its seat, and then the ball valve drops onto its seat (13) when the spout is tilted, entering the recess (42), thereby closing the tap.

---

figure 5.14 Two drawings from Patent No. 3,877,081 (1975) illustrate a water-saving device and its installation in a toilet tank.

The water container preferably consists of a flat, transparent bag (10) formed from any suitable plastic material, the plastic preferably being approximately 2 to 4 mils in thickness. The bag may be manufactured by any one of several known methods, but for ease of manufacture may consist of tubular stock material, heat-sealed or otherwise welded along one end, as at 12, to form the bottom of the container.

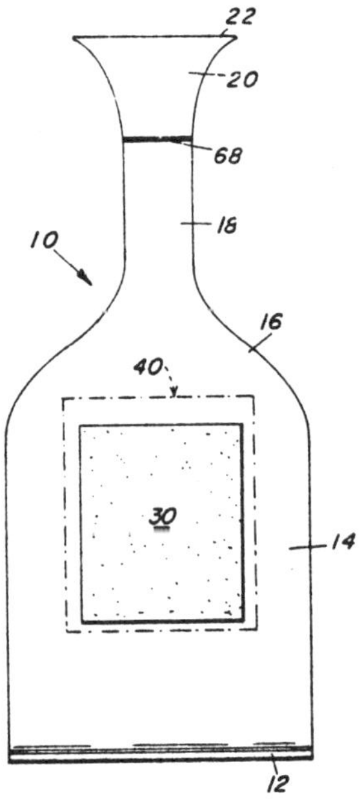

The side wall (14) tapers as at shoulder portion (16) to form an elongated neck portion (18), which, in turn, is flared at 20 to produce a funnel-shaped opening (22) by means of which the container can be filled. The neck portion (18) may, if desired, be sufficiently long to act as a hose, whereby filling of the bag from a faucet is facilitated, but in any event should be at least long enough to allow a knot to be tied in it to seal off the container after it is filled with water.

Although the container (10) can be placed in any one of several locations within a conventional flush tank, the preferred location is adjacent the left-hand end wall (50) of the flush tank (52). In this location, the bag will be supported not only by the wall of the tank, but by the water inlet pipe (54) near the left-hand wall (50). This pipe supports the filler tube (56) as well as the inlet valve mechanism (58), the ball float mechanisms (not shown), and the like. The container is located approximately midway between the front and back walls of the tank. The adhesive patch (30) is shown in position, securing the container to the end wall (50) with the container full of water and the neck portion (18) knotted as at 64, so that the container will retain the entrapped water. The adhesive patch (30) and the inlet pipe (54) cooperate to hold the container firmly in position so that it will not shift unduly as the flush tank fills or is emptied by the operation of a ball valve 66 or equivalent device.

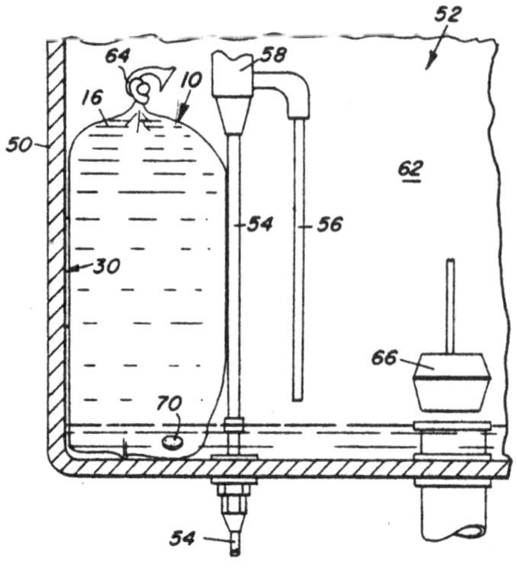

A tablet or powder (70) of a vegetable or other harmless dye may be included in the container to dissolve in the water when the bag is filled. If the bag should develop a leak, the dye within the container will be released into the flush tank to provide an indication of such leakage. In the event the bag is damaged, the adhesive patch (30) serves the additional function of retaining the plastic container in the tank so that a flushing operation will not cause it to be carried into the outlet valve where it would interfere with the operation of valve 66, the float valve, lift wires, or the like, or otherwise prevent proper operation.

as something that "can easily be folded and packaged in a business-sized or similar envelope, thus allowing it to be easily distributed by utility companies, commercial organizations, or governmental agencies at a very low cost." It's such a simple, straighforward idea that one might wonder why this device was patentable. But, like many other simple things that did not exist before some clever person thought them up, Klein's bag was indeed a patentable invention.

In the last several years, the water in many American toilet tanks has turned blue. This is not the result of pollution, of course but of householders' attempts to keep their bathrooms clean and germless with less work. It's not a new idea. In 1940, Edward Ahern of Connecticut patented a "cleansing and deodorizing device for water closets" that looks pretty good from the patent specifications but did not find favor with either manufacturers or consumers (see Fig. 5.15). Again, it looks as if timing and marketing push spelled the difference between success and failure. When the big household-products companies decide that a product's time has come, they look for or develop the best invention they can, and then promote it with vigor and skill—and money. Naturally, they don't score 100 percent, but they do pretty well (in 1967, Procter & Gamble had sales of just under $2.5 billion; in 1977 their sales came in just over $7 billion).

A more recent entry in the blue-water sweepstakes was created by Glenn N. Taylor of Illinois for the Colgate-Palmolive Company (Fig. 5.16). One of the points of interest in this patent is the unusually long list of deficiencies in prior versions of the same idea that it enumerates. Perhaps this was done for good reason. A consumer survey taken in 1975 found that although many people liked the idea of an automatic bowl washer, they did not think that the blue-water cleaners then on the market did a particularly good job. Mr. Taylor's device is somewhat unique in that it is designed to be attached outside the tank. This makes refilling and maintenance easy, but it's not the most attractive accessory we have seen. It is likely that other inventors will be heard from before long. Anyone who's got some ideas in this area should find some interested listeners.

Inventions come in many different categories—there are necessities, conveniences, novelties, luxuries, and some in-betweens, too. There is also some switching back and forth as some luxuries have become necessities and vice versa, as times and tastes change. Take toilet seats, for example. The regular hardwood or plastic models have been standard for so long that any improvement would seem to be just a novelty. But, according to Donald Gallegos of New Mexico in the descriptive matter of his 1976 patent, "The discomfort one experiences when sitting on a cold toilet seat has for many years been recognized as objectionable." He's probably right. We've been unhappy all these years without ever doing anything about it. Gallegos's solution to the problem is a detachable heating pad that can be plugged into any nearby outlet (Fig. 5.17). Washable and soft, the pad is also inexpensive.

Hot or cold, a soft seat is nicer than a hard one. David Harrison of

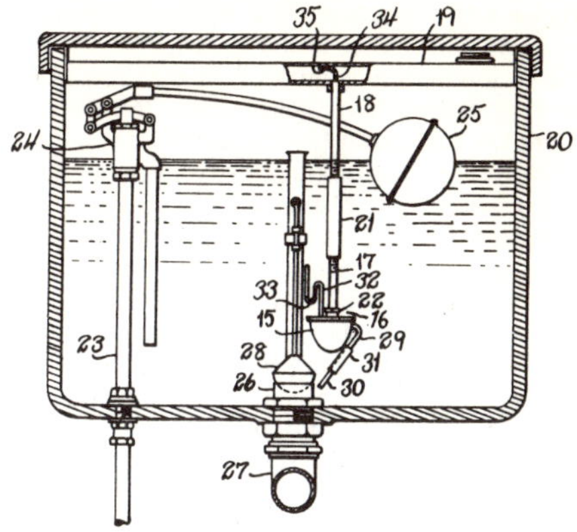

figure 5.15 Sectional view of a toilet tank illustrating the cleansing device described in Patent No. 2,199,094 (1940).

A mixing chamber (15) is adjustably connected to the lower end (16) of a solution feed pipe (17), the upper end (18) of which is connected to a container (19) of any desired form, herein shown as of box-like form, adapted to be suspended in the toilet tank (20). The two parts of the feed pipe are connected by a coupling sleeve (21) to provide for lengthening or shortening the feed pipe, and the lower end (16) may be held in position by a lock nut (22). The toilet tank is provided with the usual inlet pipe (23), a valve (24), a float (25) to control the valve, and a seat (26) at the upper end of the flush pipe (27) to receive a ball valve (28) adapted to be raised in the usual manner.

Extending into the mixing chamber (15) and downward therefrom is a siphon tube (29) having a discharge end (30) of any desired diameter directed toward the flush pipe (27) and adjustably connected to the siphon tube by a coupling sleeve (31), so as to shorten or lengthen the siphon tube (29) as required. Opening out of the top of the mixing chamber is a vent pipe (32) which may be formed with a trap (33). Extending upward through the feed pipe (17) is an air pipe (34) which extends into the solution container (19) to admit air thereto as the solution is drawn off. Preferably, the upper end (35) of the pipe (34) is flexible, so as to bend if the feed tube is shortened.

When the mixing chamber (15) is empty, the solution will flow down the feed pipe (17) until the lower end of the feed pipe is submerged and the flow stopped. Then, as water rises in the tank (20), water will pass through the siphon tube (29) and mix with the solution, the air in the chamber escaping through the vent pipe (32).

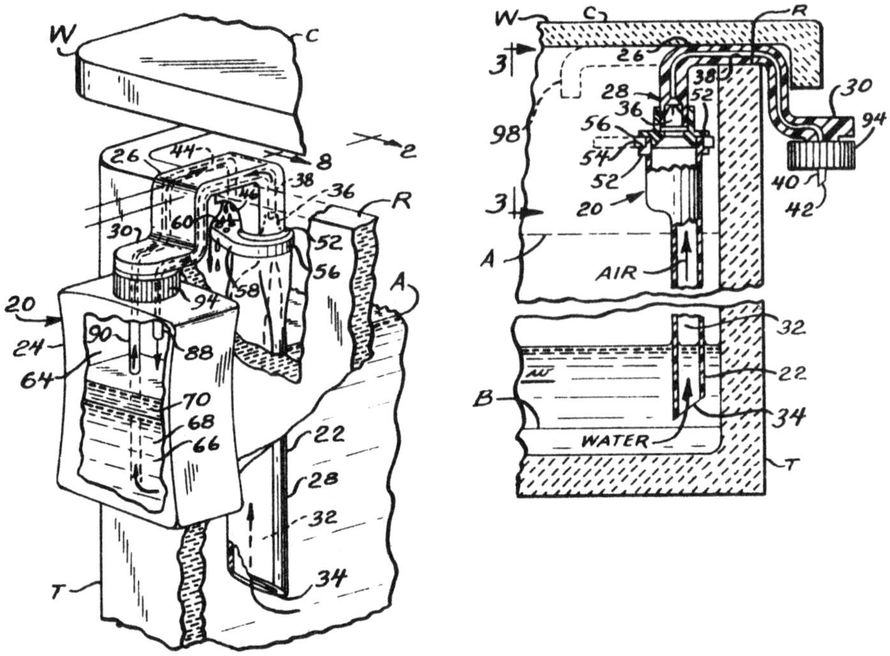

figure 5.16 A perspective view, partly broken away (left), and a fragmentary elevational view (right), of a toilet additive device, Patent No. 3,953,902 (1976). The tank (T) has a supply of water that reaches. level (A). A cleaning device, generally designated 20, is also shown having a connector (22) and a container (24).
The connector (22) has a central portion (26) adapted for placement on a rim (R) of the tank at its upper end, a first end portion (28) depending from one end of the central portion (26) for placement inside the tank, and a second end portion (30) depending from the other end of the central portion (26) for placement outside the tank (T). When the connector (22) has been initially placed in its proper position with the first end portion (28) inside the tank and the second end portion (30) outside the tank, a cover (C) for the tank (T) is placed over the tank and the central portion (26) of the connector.

The first end portion (28) has an elongated air chamber (32), which is vertically disposed in the tank (T). The first end portion (28) also has an inlet opening (34) at the lower end of the first end portion and communicating with the air chamber (32). The inlet opening or opening means (34) is located intermediate the first and second levels (A) and (B) respectively, of the flushing liquid in the tank (T), and is preferably located slightly above the lower second level (B).

The first end portion (28) has a one-way valve (36), such as a flap valve, permitting passage of air from the air chamber (32) into a first lumen (38), while preventing passage of fluid from the first lumen into the air chamber (32). The first lumen (38) extends through the connector (22) and through a first tube section (40) depending from a lower end of the second end portion (30) of the connector. The first lumen thus communicates between the valve (36) and a first aperture (42) adjacent the outer end of the first tube section (40).

The first end portion (28) has a pair of spaced flanges (52) located above the first level (A) of the flushing liquid and defining an annular groove (54). An annular portion (56) of a ledge (58) is received in the groove (54) such that the ledge (58) may be rotated about the first end portion (28) of the connector (22). Thus, the outwardly extending ledge (58) may be moved between a first position below the outlet opening (46) and a second position spaced from a location below the outlet opening (46). The ledge (58) may have a plurality of recesses (60) in its upper surface for a purpose which will be described.

The container (24) has a chamber (64) for retaining an additive liquid (66), a cleaning solution such as a liquid having a surfactant, which cleanses the water closet without foaming. The solution may also include a bleach to remove stains from the bowl of the water closet. As shown, the liquid (66) may include a first liquid (68) having a first density, and a second liquid (70), having a second, lesser density, such that the second liquid (70) floats above the first liquid (68), as shown. The second liquid (70) preferably has a color which contrasts with the color of the first.

In use of the device, the connector (22) is initially placed on the tank (T) by removing the cover (C) and positioning the central portion (26) of the connector over the rim (R) of the tank (T), as previously described, after which the cover (C) is placed over the tank and connector (22). As will be seen below, it is no longer necessary to remove the tank cover for attachment or replacement of containers. After initial placement of the connector, a container (24) containing an additive liquid (66) is secured to the second end portion (30) of the connector (22) by puncturing the sealing film of the container and by turning the threaded rim (94).

When the water closet is flushed, air is permitted to pass through the inlet opening (34) into the air chamber (32) of the first end portion (28). As the water is subsequently filled into the tank, the water level in the tank rises toward the first level (A). During this time, the water passes through the inlet opening (34) and into the air chamber (32), thus forcing the previously admitted air toward the upper end of the air chamber (32), through the one-way valve (36), the first lumen (38), the first tube section (40) and into the upper end of the chamber (64) in the container (24), as indicated by the direction of the arrows. Since the inlet opening (34) is located adjacent the lower second level (B) of the flushing liquid, a substantial volume of air is forced into the container (24) where it is compressed, while the one-way valve (36) prevents passage of the compressed air from the container (24) to the air chamber (32), particularly when the water level lowers in the tank.

The compressed air in the chamber (64) forces a portion of the additive liquid (66) through the second tube (90), the second tube section (48), the second lumen (44), and the outlet opening (46), where it falls onto the ledge 58. A portion of the additive liquid which passes from the outlet opening (46) is retained in the ledge recesses (60), while the remainder of the liquid flows over the sides of the ledge (58) and into the flushing liquid below. Thus, a portion of the additive liquid (66) is introduced into the water supply in the tank (T) responsive to flushing of the water closet, where it serves to treat the water retained in the tank and the water closet bowl (not shown) during subsequent flushing of the water closet. As shown in the drawings, the outlet opening (46) is preferably located at a level above the liquid in the container (24) to prevent possible siphoning of the additive liquid from the container into the tank.

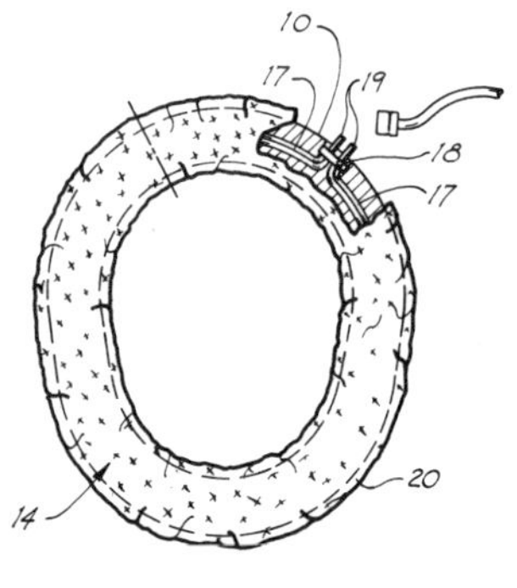

figure 5.17 A toilet seat warmer is described in Patent No. 3,968,344 (1976). The seat is of standard design and construction. The cover is generally indicated at 14, and conforms in general shape to the seat. The pad is relatively thick, made of some resilient material such as polyurethane foam. It is provided with electrical heating elements in the form of insulated resistance wires (17). In the present embodiment of the invention, three resistance wires are shown, but it will be understood that the number may be varied. The ends of the resistance wires are soldered to a terminal block (18) provided with terminal posts (19) for plugging the heating wires into any convenient wall outlet or the like.

Mississippi came up with an improved model of a soft seat (Fig. 5.18). His patent, like many of those filed by employees of manufacturing companies, deals extensively not only with the invention itself but with the most economical and efficient means of producing it. As has been noted elsewhere, an invention isn't much good if it can't be manufactured economically. Any independent inventor who can think of a better way to produce either his own invention or something that is already on the market should be able to get a fair hearing for his plans.

A few paragraphs back, it was noted that one manufacturer had some success with a clever idea for soap—molding it in the shape of a hand microphone. Over the years many other variations on the ordinary bar of soap have been patented by inventors and designers. It's an excellent illustration of how many different ways there are of looking at one minor area of interest. In Fig. 5.19*a*, the bar of soap is joined to a suction cup that can be temporarily attached to almost any surface. This solves the problem of where to put the soap when you want to keep it handy. In 5.19*b*, the soap is embedded in a sponge that, with its handle, also serves as a soap dish. A new kind of soap holder is shown in 5.19*c*, and an easily mounted soap dish is illustrated in 5.19*d*. Two variations on essentially the same theme are presented in 5.19*e* and 5.19*f*. They involve attaching a cord or rope to the soap bar to make it easier to handle, especially in the shower. As is evident, soap dishes and other soap holders are often quite wasteful of the soap. The old-fashioned ones, especially, often collect water that has nowhere to go. If a bar of soap is placed in this kind of receptacle, the soap just melts away.

All of the rope soaps illustrated can be hung out to dry when not in use, thus cutting down on waste. During the shower or bath they can be placed around the neck or other part of the body, within easy reach. There are problems, of course. The cord may get in the way of the soap bar or may be irritating to the body. Perhaps a softer cord is the answer. The "bathing appliance" in Fig. 5.19*f* offers two loops, one for each hand, so that the soap can be drawn back and forth against and across the back and other areas of the body that are not readily accessible using just one hand.

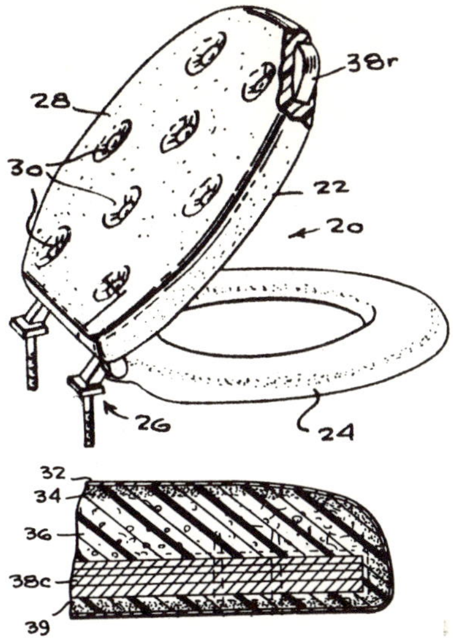

figure 5.18 Perspective view of the soft toilet seat described in Patent No. 3,863,277 (1975).

The seat (20) includes a cover (22) and a seat ring (24) hinged together by a suitable conventional hinge assembly (26). The upper surface (28) of the cover (22) may be embossed or otherwise decorated with various designs (30) or may be smooth. The cover and the seat ring are essentially identical in construction except for shape and surface ornamentation.

The outer skin (32) is composed of a film that may be one of many well-known chemical coatings. Positioned within and contiguous to the outer skin of the film is a core of polyurethane foam having a high-density, soft, tough skin (34) and a soft, flexible, foam interior (36). The outer skin of film (32) and the tough skin (34) of high-density polyurethane provides a gas- and liquid-impervious barrier surrounding the flexible foam (36). For strength, and in order to permit the attachment of the hinge (26), a rigid insert (38) is provided, which is surrounded by and contiguous to the flexible foam (36).

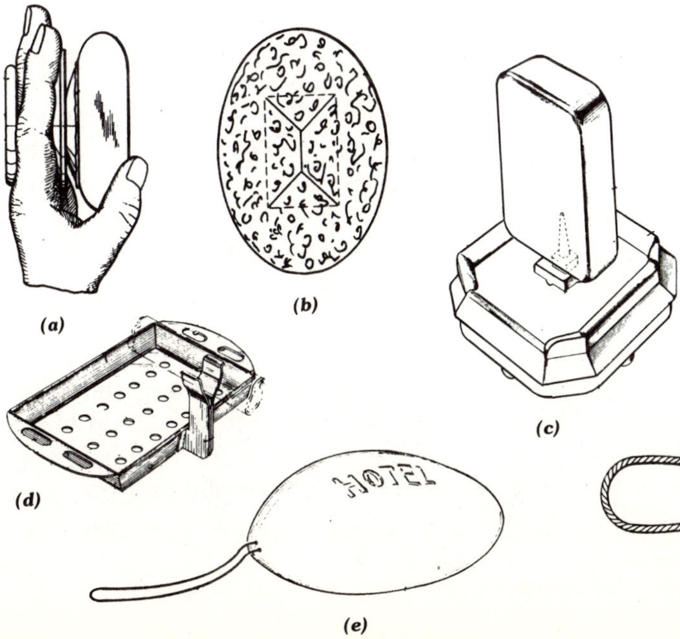

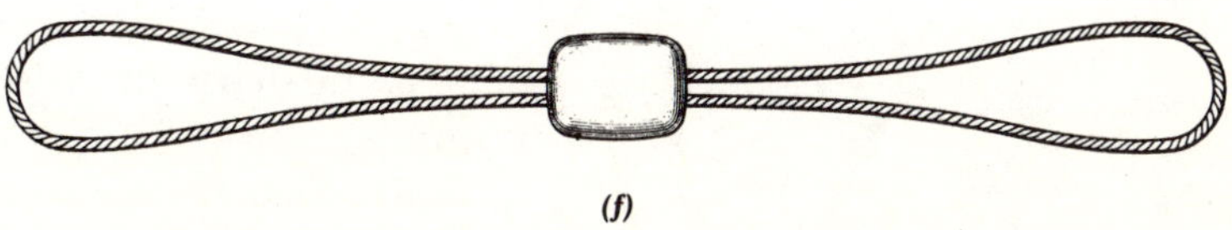

figure 5.19 Six of the hundreds of ways of handling soap that have been patented over the years. (*a*) Patent No. 3,679,111 (1972) has the soap attached to a suction cup. (*b*) Patent No. 3,601,493 (1971) puts the soap in the middle of a sponge. (*c*) Patent No. 3,164,356 (1965) is a soap dish that holds the bar upright, impaled on a stud. (*d*) Patent No. 3,194,526 (1965) is an easily mountable soap dish with drain openings. (*e*) Patent No. 3,796,665 (1974) is a bar of soap attached to a rope, for convenience in the shower. (*f*) Patent No. 3,160,523 (1965) is an earlier version of the rope soap, this one with double loops, to make back rubbing easy.

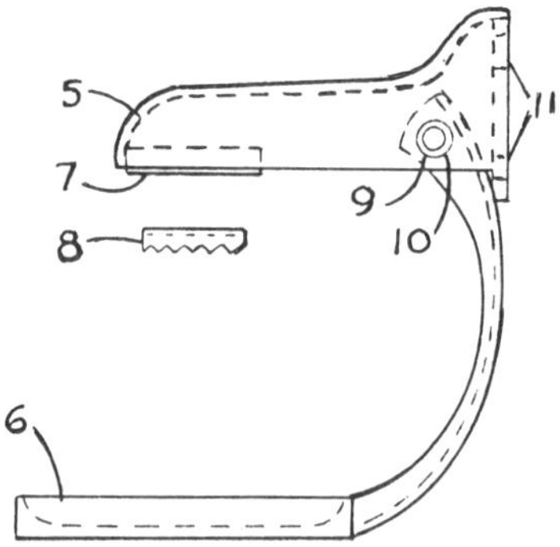

figure 5.20 Side view of the magnetic soap holder/toothbrush holder/soap-drop catcher combination described in Patent No. 3,472,391 (1969).

There is a wall bracket (5) with a generally elongated convex shape, the front portion of which is rounded off. The back portion of this wall bracket curves out to a semicircular shape with a flat back, which is used to support the unit to the wall. The back of the wall bracket is provided with openings (11) in the event it is desired that the wall bracket be secured to the wall with screws. The wall bracket can also be secured to the wall with cement glue. In the front portion of the wall bracket there is secured on the underside a circular-shaped iron magnet (7). A cap-shaped magnetic metallic disc (8) with jagged edges is pressed into a bar of soap. The bar of soap is thereby attracted to and held by magnetic force. The wall bracket also contains an opening on either side of the back of its elongated portion through thich a pin (9) with slightly rounded ends is inserted.

The pin contains two spaced exterior grooves to help anchor it to the openings on the sides of the wall bracket. The pin serves a dual purpose: It serves as a means for connecting the soap-drop catcher (6) to the wall bracket (5), and acts as a bar upon which to hang toothbrushes by means of the holes in the toothbrushes' handles. The pin is firmly anchored into position with the placement around the pin of a C-ring (10) on either side of the wall bracket. The C-rings serve as a means of keeping the entire assembly in place.

The base of the soap drop catcher 6 is rectangular in shape in its outer periphery. The inside of the base is slightly curved or dish-shaped to facilitate its cleaning. The soap drop catcher flanges upward from one side of the base in a generally arced position. The flange is adapted and curved to enter the wall bracket and contains two openings which correspond to and complement those in the wall bracket. The pin is placed first through the opening in the wall bracket, then through the openings in the flange of the soap-drop catcher and then through the other opening in the wall bracket.

When the entire assembly is in place and secured to the wall, the soap drop catcher can be tipped upward if desired for cleaning, but the wall bracket will remain stationary.

Have soapholders and similar devices been invented to death? Judging from the material on file in the Patent Office Search Room, they have been. On the other hand, most people are using soap pretty much as they have been since the nineteenth century and before. None of the inventions have really caught on. Whether this means that there is no need for something new here or if there is simply nothing proposed so far that has taken the public's fancy is a question that can be answered by individual readers for themselves.

Bathroom shelves are generally messy little areas, with open toothpaste tubes and combs and bottles and packets of who-knows-what strewn about. Inventors are often trying to tidy things up—except in their own workshops, of course—with some new-fangled holder or container. A holder for both toothbrushes and soap seems a good idea to reduce clutter, with the soap held upright by a magnet (Fig. 5.20). The soap dries neatly, and drippings are caught by the tray below. Gino Bolognesi of California invented this holder in 1969. Since then, many others have had ideas along the same lines. A Texan, Lorraine McLeod, developed a modern toothbrush holder for which she was granted a design patent (Fig. 5.21).

One problem that remains is the handling of toothpaste and other tubes and their tops. Charles Miles of Indiana invented a gadget on which tubes could be hung in such a way that their contents could be dispensed "without the necessity of removing the tube from the storage position." It's nothing fancy, but it looks practical and easy to use. There seems to be a need for some kind of device like this. Why, then, isn't it widely available? Perhaps it needs further improvement (see Fig. 5.22), or perhaps it needs the right kind of push.

Toothpaste tubes are obviously just one of the elements in contemporary tooth care. Many inventions have been patented in this field. Carroll Conklin of Ohio came up with an interesting idea for a disposable tooth cleaner. A number of thick plastic bristles on this small device can, if properly applied, give the teeth a vigorous and thorough cleaning. Adjustable to fit any size finger, this disposable item is considerably freer of germs than a brush or other utensil that is used every day (Fig. 5.23).

An improvement on the good old toothbrush was patented in 1977 by Frederic Huish of California. The sulcus toothbrush has round-tipped, soft bristles that are particularly adapted to remove bits of food that lodge in the area between teeth and gums. The innovative design illustrated in Fig. 5.23*b* offers two double rows of relatively long sulcus bristles that diverge from each other at an angle of about 30 degrees. This arrangement makes it possible to clean upper and lower gum crevices at the same time.

As noted earlier in this chapter, one of the most successful Twentieth-century inventions for the care of teeth has been the Water-Pik. Figure 5.24 shows some of the key details of the 1966 patent on this device. The Water-Pik was developed as an efficient way to clean the areas around teeth and gums and thus to keep both healthy. In recent years, dentists have been paying more and more attention to the gums. A mechanical gum massager was

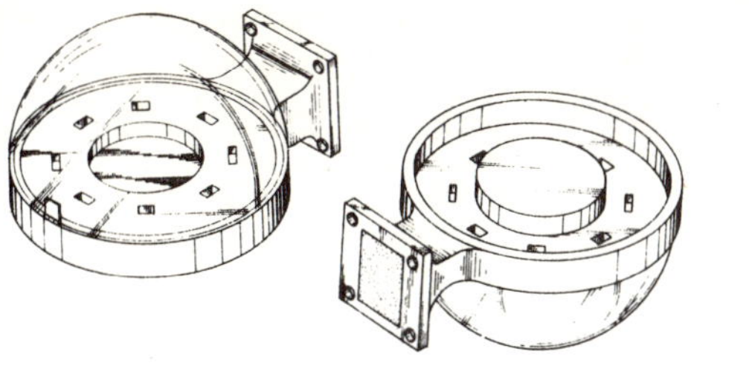

figure 5.21 Part of Design Patent No. 234,481 (1975) for a toothbrush holder.

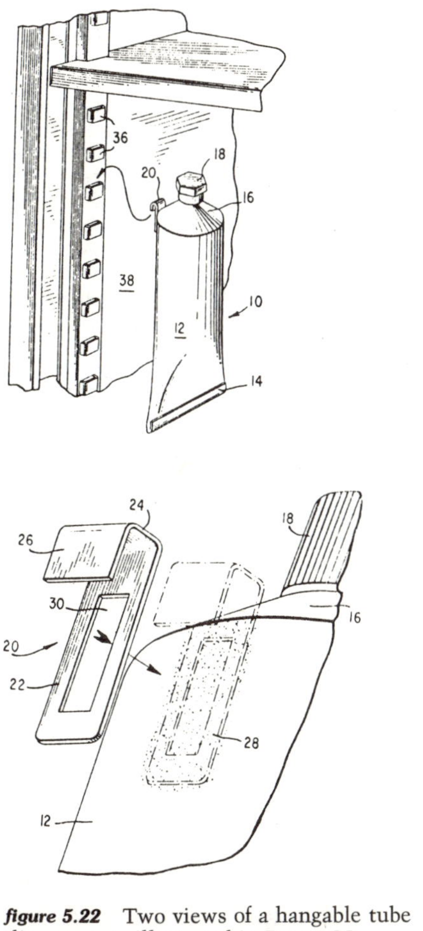

figure 5.22 Two views of a hangable tube dispenser, as illustrated in Patent No. 3,190,500 (1965). There is a base (22), a bight (24), and a lip (26).

The base may be flat or slightly curved to compatibly embrace the side wall of the tube whereon a quantity of suitable adhesive (28) may be disposed adjacent the shoulder (16) to receive and then to secure the hanging means (20) to the body portion of the tube (12). The bight is spaced longitudinally beyond the juncture of the conical end closure (16) with the body portion (12). The element (20) does not extend as far as the plane normal to the axis of the nozzle (18) and of the body portion (24) and intersecting the mouth of said nozzle. However, in certain applications extension beyond said plane may be desirable. An opening (30) may be provided in the base member (22) of the hook (20) to facilitate the application of the adhesive and to secure an improved bond between the hook and the tube wall.

figure 5.23(a) Disposable tooth cleaner, Patent No. 3,798,698 (1974). The body is formed of some material such as ethylvinylacetate and includes a central platform with an upper surface (16), evenly spaced nubs (14), and two strips (18) and (20). The cleaner can be fitted snugly around the finger of the user by adjusting the protrusion (42) and its enlarged head (44) in one of the holes (40). Thus the strips (18) and (20) are locked in place.

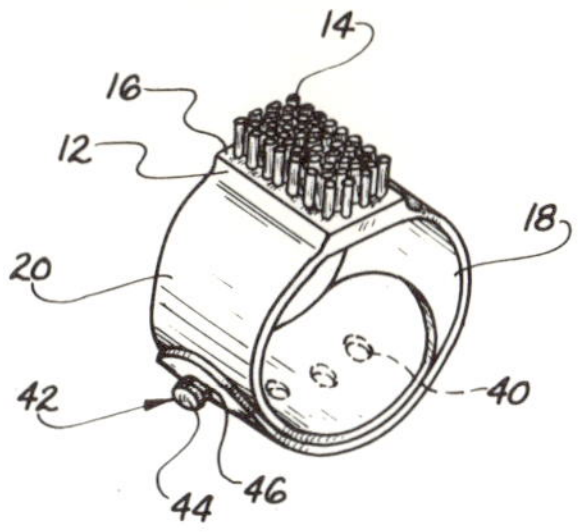

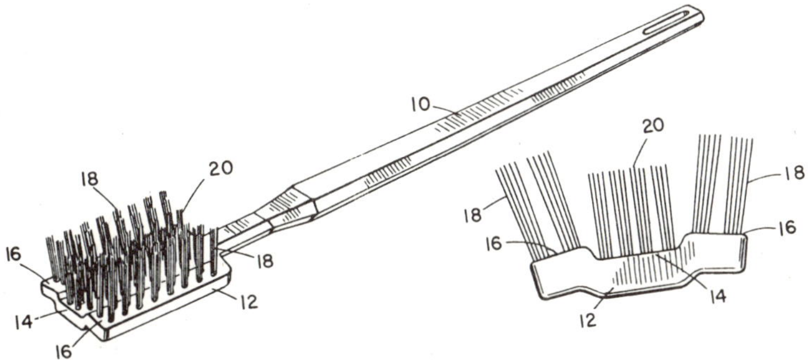

figure 5.23(b) The double sulcus toothbrush described in Patent No. 4,010,509 (1977). It has the usual handle (10) and head (12), which supports the bristles. The inset shows how the sulcus bristles (18) are enbedded in each shoulder at an angle.

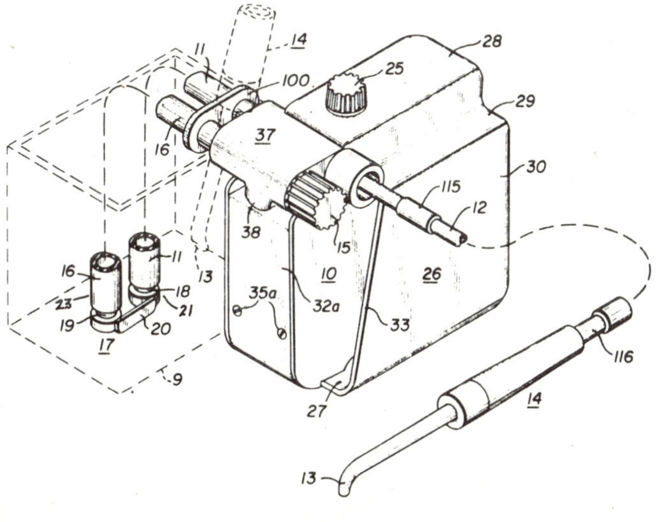

figure 5.24 The famous Water Pik, Patent No. 3,227,158 (1966). A motor-pump assembly (10) is provided with a flexible intake conduit (11) and a flexible discharge conduit (12). A jet nozzle (13) is secured to the conduit (12) by a coupling (14) and may easily be manipulated to direct a high-velocity jet produced at the nozzle (13) against the teeth and the gum tissue. The pump unit (10) is provided with an adjusting knob (15) which controls a bypass from the discharge side of the pump to a flexible conduit (16). The conduit (16) is made of a resilient flexible material which expands upon application of internal fluid pressure and provides a damping action. The conduit (12) has a relatively inexpansible flexible material to minimize damping action in the discharge line. The ends of the conduits (11) and (16) are attached to a fitting (17), which may be placed in a cup or other liquid container as indicated at 9. The fitting (17) includes two coupling sleeves (13) and (19) of molded plastic material, over which the ends of the conduits (11) and (16) respectively, are secured. The sleeves are made integrally with a yoke (20), which extends below the ends of the sleeves and spaces the ends from the surface on which the fitting rests. The sleeve (18) has its lower end open, as indicated at 21, and when the fitting rests within a liquid container, the yoke (20) holds the open end spaced from the bottom of the container and allows liquid to enter the opening freely. The sleeve (19) is closed at its lower end and is provided with a restricted opening (23), which limits the discharge of liquid through the bypass tube (16).

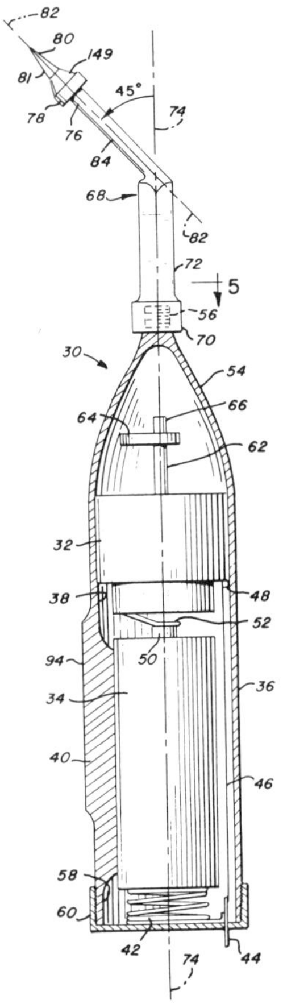

figure 5.25 Side view, partly in section, of a mechanical gum massager, Patent No. 3,967,617 (1976).

The vibratory mechanism (30) has a drive means that includes an electric motor (32), actuated by a dry cell battery (34), both of which are enclosed in an elongated tubular casing (36) suitable for holding in the user's hand. Both motor (32) and battery (34) are rigidly secured to interior wall (38) of the casing. In the case of the battery (34), three equally spaced flanges (40) receive the battery when it is slid into place within the casing.

Positioning and negative contact spring 42 holds the battery in place. A slide switch (44) is connected through lead (46) to one terminal (48) of the motor (32). Positive terminal (50) of the battery is connected with the other side of the motor (32) through a leaf spring contact (52).

Casing (36) is formed of a suitable nonconducting material such as plastic. Its upper end (54) is necked down to terminate in a threaded nub (56). Its bottom end (58) is closed by a cap (60), through which the slide switch (44) extends.

The upper end of the electric motor (32) carries a rotatable shaft (62) extending outwardly therefrom. A suitable speed of rotation for the motor (32) is in the range of about 55 to 70 c.p.s. An eccentric flywheel (64) is carried at the outer end (66) of the rotatable shaft (62). The vibrating shank (68) has a flange (70) at the end of its lower portion (72), which is threaded onto vibratory mechanism (30) through the nub (56). Other suitable means of attaching casing (36) and vibrating shank (68) may be employed, and if desired, these two members may be integrally formed, as, for example, by a molding process.

Lower shank portion (72) is positioned with its longitudinal axis generally coinciding with the longitudinal axis (74–74) of the vibratory mechanism (30). The relative length of the lower shank with respect to the overall length of the vibrating shank may be other than is shown in the drawing, so long as the shape and dimensions of the shank (68) as a whole, and the material of which that element is formed, are such that the free end of the shank traces a generally elliptical path. The free end (76) of the shank (68) carries a stimulator element (78), which has its operative portion (80, 81) facing upward and to the left in the drawing.

When the electric motor 32 is actuated by closing the switch (44), the rotation of the eccentric flywheel (64) about the shaft (62) tends to produce a substantially circular movement in the shaft that is transmitted to the motor (32). This cyclical movement is in turn transmitted through the motor (32) to the tubular casing (36), since the casing is naturally held in the user's hand with less than complete rigidity. As a consequence, the nub (56) on the casing and bottom end portion (70) of the vibrating shank (68) tend also to rotate with circular motion.

Since the vibrating shank (68) has considerably smaller transverse cross-sectional dimensions than the casing (36) and is thus not a rigid body, its outer end (76) does not necessarily trace the same circular path as is followed by its bottom end portion (70). The fact is that because the shank portion (72) has a larger transverse dimension in one direction than the other, the circular motion of its bottom end portion (70) will be damped in the direction of the larger dimension when the motion is transmitted upward through the shank portion (72). The reason for this is the greater rigidity of portion (72) in the direction of its larger transverse dimension.

As a result, the upper end of shank portion (72) is caused to trace a generally elliptical path in a plane perpendicular to the longitudinal axis (74–74) with that elliptical path having a major axis extending perpendicularly to plane (83), and thus also to the direction (82) in which operative portion (80) and (81) of stimulator element (78) faces during use of the gum massager. This generally elliptical motion is transmitted through shank portions (84) and (76) to stimulator element (78).

---

patented in 1976 by Paul Krolik of Illinois. It represented an improvement over prior devices in that it provided for an elliptical, rather than a circular or a back-and-forth movement on the gums (see Fig. 5.25). In addition, this massage is effective because it is applied continuously in one direction. Pressure can be varied smoothly from strong to weak, so that the gums are not exposed to hammer blows.

Both the Water-Pik and the gum massager illustrated are fairly complicated devices. Figure 5.26 is an illustration of a very simple and inexpensive device that is also designed to help keep teeth healthy. The enemy that is being attacked here is plaque, the accumulations of bacteria in the mouth. Research has indicated that the tongue is a source of these bacteria and, thus, of plaque. The answer provided by the device described in this patent is a tongue brush, or cleaner, to prevent the accumulation of bacteria in the crevices of the tongue. So now, in addition to brushing and "pikking" and massaging, those people who want to be smiled upon by the tooth fairy must also clean their tongues. Will this catch on? Well, perhaps not right away.

There are some inventions that are ahead of their time, that

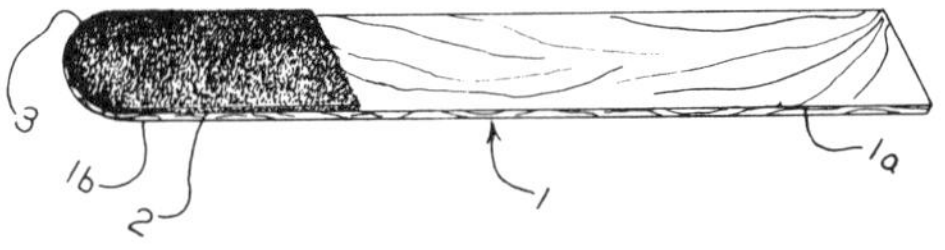

figure 5.26 Tongue cleaning stick, as described in Patent No. 3,943,592 (1976). It is a flat piece of wood or plastic (1) with one end (1*a*) used as a handle and the other end (1*b*) as a platform for a piece of Velcro tape (2).

aren't quite right, that for some reason(s) fail to catch on. In 1954, Alexander Samko of New York was granted a patent for a disposable razor. This invention, illustrated in Fig. 5.27, did not become a commercial success. Yet in the 1970s, disposable razors have taken over a good share of the market, pushed by competition among the manufacturers. Why didn't this happen in the 1950s? Probably because the existing technology was able to produce a blade that gave only one or two shaves, on the average. Today, the double-bladed new razors can be used quite a number of times, depending on a beard's toughness. The customers don't mind throwing away a razor they've used for a week or two or longer.

Figure 5.28 shows one of the patents Gillette was assigned for its Trac II razor and blades. It was developed by Roger L. Perry of Massachusetts, who has been with Gillette for more than twenty-five years and has thirty patents to his credit. About a dozen patents by Perry and others are part of the Trac II system. They relate not only to the intrinsic nature of the blade, but also to the techniques necessary to achieve its mass production.

A visit to a home-improvement shopping center or to a hardware or similar store is bound to impress anyone with the number and variety of things available for the bathroom. Is the field glutted, and are the opportunities for new products limited to the large and powerful marketing organizations? The answers would be Yes, if we could look at this field and say that everything is just about right as it is, except for a few fripperies. But if there are some things that could stand improvement, if there are still areas of need, then the bathroom is an excellent area for investigation by the independent inventor. The big companies can't do it all. Often they get bogged down by their systems and committees.

The money is there, and the market is active. Any invention that is (1) useful and (2) handled intelligently after development stands a good chance of success.

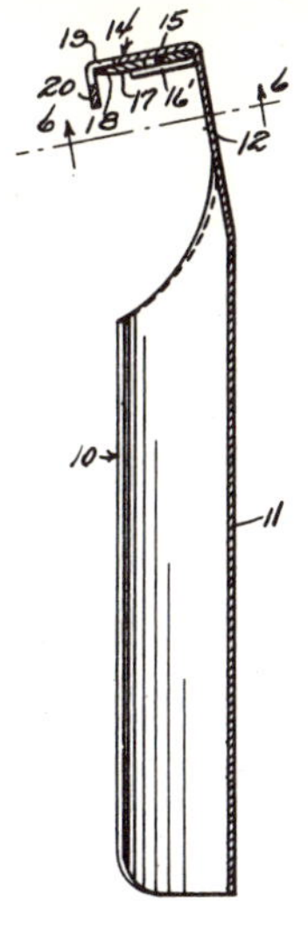

figure 5.27 Side view of a disposable razor, Patent No. 2,684,529 (1954). The razor (10) includes a handle (11) and a flat section (12), from which projects a holder (14). This latter includes a top wall (15), and arranged below the top wall (15) is a pair of spaced tongues or clips (16) or (16'). These coact with the top wall (15) to retain or hold therein a razor blade (17), with a sharp cutting edge (18).

The holder (14) is provided with a slot (19), which enables the cutting edge (18) of the razor blade to perform its normal cutting function. Thus, by gripping the handle (11) and moving the razor along the face or other part of the body being shaved, the portion of the cutting edge (18) exposed by the slot (19) will effectively cut the hairs from the face. Arranged transversely with respect to the top wall (15) and projecting therefrom is a guard bar (20), which helps to maintain the blade in its proper place in the holder (14), and also to prevent the user from being accidentally cut by the blade.

figure 5.28 Two views of a modern Gillette razor, Patent No. 3,832,774 (1974).

There is a handle (10) and a blade (12). The handle includes a grip portion (14) that is 4¼ inches long and of generally square cross-sectional configuration with a slight taper along its axial length. Grooves (16) along a portion of the length of the grip (14) facilitate its handling by the user. At the upper end of the grip is a neck (18) which extends forward and upward from the upper end of grip portion (14) at an angle of 120 degrees to the axis of the grip so that the end surface of the neck portion (18) is offset a distance (*A*) of 0.37 inch from the axis of the grip. Secured to the end surface of the neck is a transversely extending support (20) that has two outwardly directed flange or rail portions (22), the upper surfaces of which define a reference plane (24) that is perpendicular to the axis of the neck. This plane (24) is spaced a distance (*B*) of 0.55 inch from the junction of neck and grip and a distance (*C*) of 0.195 inch above the end surface of the neck. Two laterally extending slots (26) are formed in the base of support head (20), one on either side of the point of attachment to the neck. A spring plate (28) functions as a resilient latch member and cooperates with the forward rail (22) of the coupling structure (20) to secure the blade unit (12) on the handle (10).

The cooperating blade unit has a transverse length of 1.56 inch, a width (*D*) of 0.426 inch, and a depth (*E*) of 0.230 inch. It includes a base or platform (30) molded of high-impact polystyrene on which are supported two blade elements (32 and 34) and an interposed spacer member (36) so that the shaving edges of the blade elements are maintained in spaced, parallel offset relation. Cap member (38), also molded of high-impact polystyrene, has depending pins (40) which extend through apertures in the blades, spacer, and base, and are cold-headed to secure the blade elements and spacer to the base.

The base defines an upper platform surface (42) on which the leading blade member (32) rests and includes a front wall (44) and a rear wall (46). Five webs (48) extend between the front and rear walls, and the lower surfaces (50) of these webs define a plane that is disposed at an angle of 35 degrees to the platform surface (42). Apertures (52) are aligned with apertures in the blade (32) and permit shaving debris to flow through the base (30) away from the shaving zone. Extending downward, perpendicular to the surface (50), are two spaced wall extensions (54), each of which has at its lower end an inturned flange (56) that projects inwardly 0.018 inch to define a recess in the form of slot (58) that has a height of 0.026 inch. The space between flanges (56) is 0.244 inch. A notch (60) in the front flange has a width of about 0.04 inch and a depth of 0.04 inch.

Web extensions on the forward side of the wall (44) support a transversely extending guard structure (64) that has a surface spaced about 0.23 inch above surface 50 (dimension E), and when the blade unit (12) is attached to the handle unit (10), that surface is offset about 0.6 inch from the front surface of the upper end of the grip portion (14).

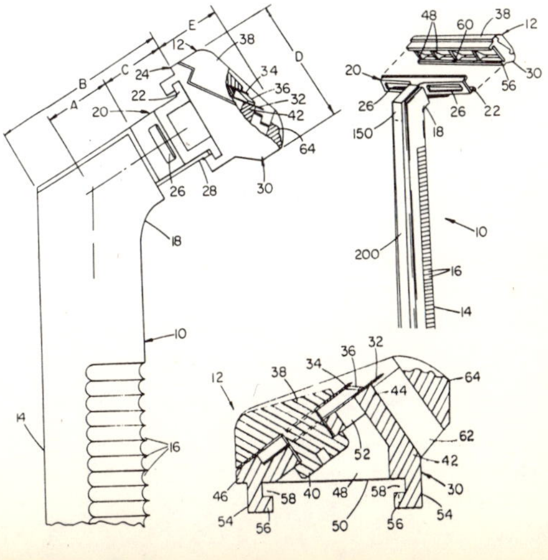

CHAPTER SIX

Security Devices

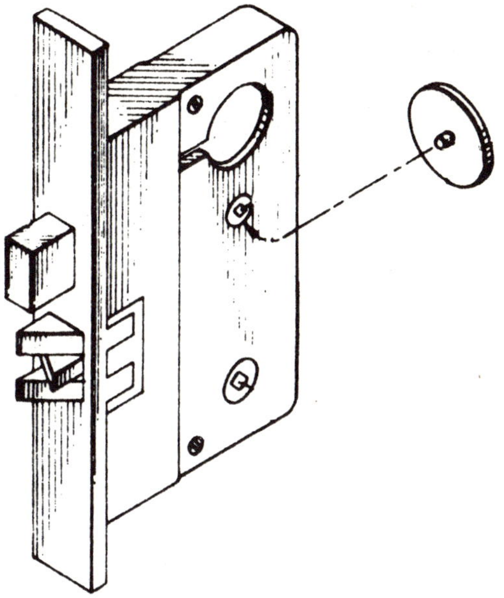

Personal safety and protection of property have been important to people for thousands of years. The problem of security was around for a long time before the modern American urban crisis made it a national obsession. Primitive man wanted to bar unwanted visitors from his cave just as much as modern householders want to keep intruders out of their houses. One early inventor discovered that you could close a cave opening if you had a stone of the right size and shape. You simply rolled the stone over a length of wood that had been carefully concealed in the ground at the entrance, just below the surface. The cave's occupants would know the location of the wood and could remove it, thus dislodging the stone—opening the door, you might say—and permitting them to enter. The door could be closed again by reversing the process.

Early people had to face many hazards, and their survival is due to the inventiveness they applied to solving their problems. The lock and key date from antiquity, having been known in Egypt sometime before 2000 B.C. In the early days, the key was generally quite large and made of wood. It also served as a handle for the door. Because of the keyhole's size and accessibility, it was important to conceal it or to protect it in some way so that it could not be picked. Some were even double-locked with another key.

In the fifteenth century, the lever tumbler lock was invented, and over the next 300 to 400 years it was improved at a slow, steady pace. Then, as the Industrial Revolution gathered force, the pace of development quickened. The most famous name in American locks, Linus Yale, took out his first patent on the pin tumbler lock in 1857. The Yale lock soon became the standard—it was secure, inexpensive, easy to install, and its keys were of reasonable size and could be duplicated from the original without too much trouble. The pin tumbler system remains the most popular today. Anywhere from three to eight pins descend and align along each shear line. Generally speaking, the more pins there are, the more difficult it is to pick the lock. The average good lock of this type has five or six pins.

The burglary business is booming these days, especially in many older American cities. Locksmiths are making a lot of money, too. As thieves get more sophisticated (or simply bolder), the defenses that are thrown up against them get stronger and more complicated. There was a time, and there are still places, where it was and is possible to use inexpensive locks. A fairly common example of this is the lockset, where the keyhole is right in the doorknob. These are simple to install, inexpensive to make, easy to find in the dark, and a cinch to get through. The current trend is away from this sort of lock and toward more secure types, especially those with some kind of dead bolt.

The underlying principles of lock engineering have not changed all that much in the last 150 years. Yale's very first patent was issued in 1844 (Fig. 6.1). He went on to patent many other locks and improvements on them, and the company that bore his name manufactured millions of locks for use all around the world. The Yale lock always had competition, but for many years it set the standard that the others tried to beat. In 1924, an important breakthrough was achieved by Samuel Segal of New York when he patented a lock with a bolt "adapted to interlock with a suitable keeper fastened to the door frame or other support. The casing shall be adjustable in any direction without disturbing or changing the position of the spindle or other means adapted to actuate the lock mechanism and operated by a key from without the door" (Fig. 6.2).

figure 6.1 Linus Yale's first patent for a door lock, No. 3,630, dated 1844.

Figure 1 is a view of the front of the lock. Figure 2 is a vertical longitudinal section. Figure 3 is a plan of the interior of the lock, the back plate being removed. Figure 4 is a vertical transverse section. Figure 5 is a plan of the pistons. Figure 6 is a plan of the key. Figure 7 is a section of same.

The improvement in this lock consists in having a cylinder or circular rim (*C*) on the inside of the front plate of the box fastened to it, forming a cylindrical socket, pierced from the outer to the inner circumference with round apertures on lines radiating from the center to the periphery. In these apertures are placed cylindrical pistons (*D*), which pass through the apertures and enter corresponding apertures in a revolving cylinder or tumbler (*E*) having a hub or arms that turns in the cylindrical socket for locking the bolt. The pistons are kept pressed inward toward the center [against other pistons (*F*)] by springs (*G*) attached to the front plate of the box.

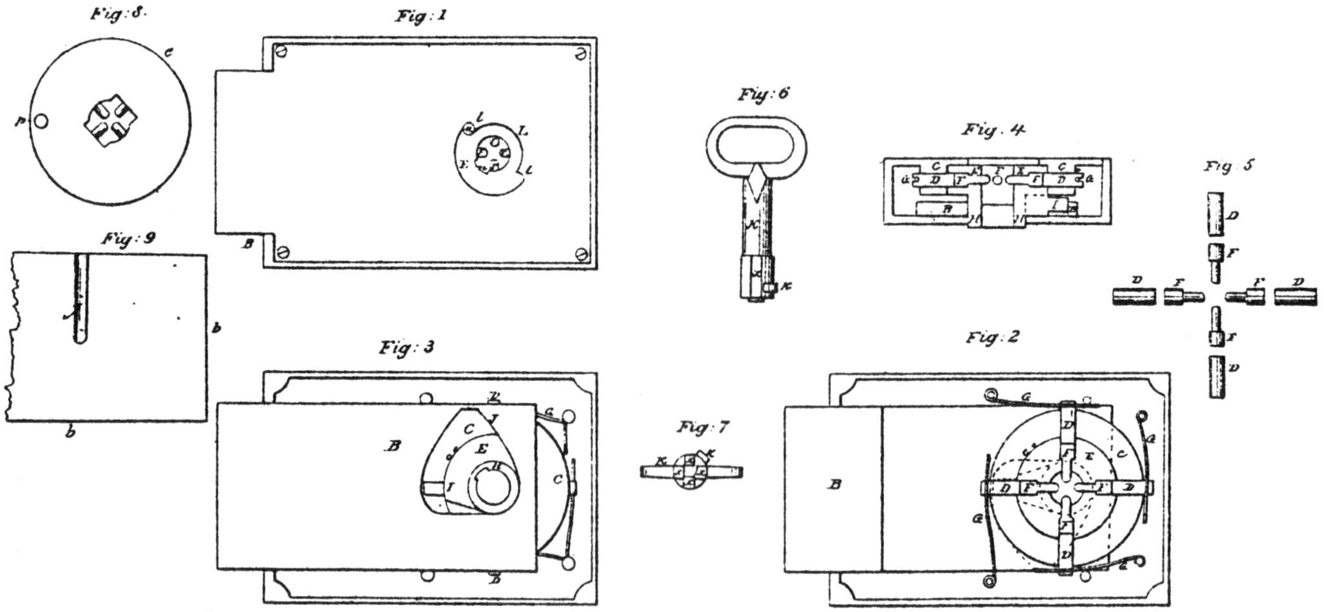

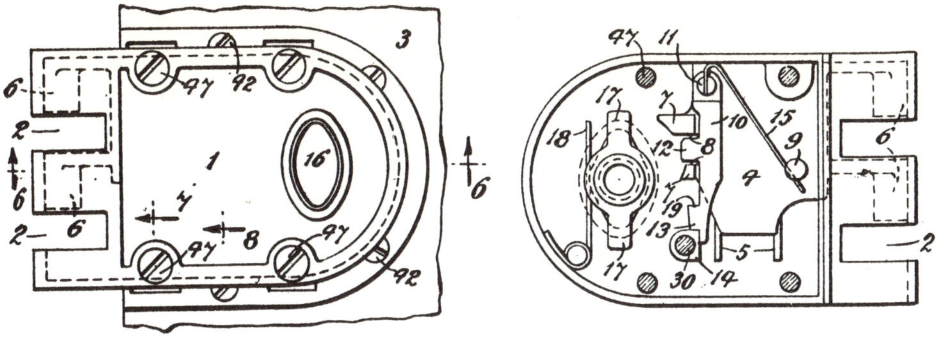

figure 6.2 Samuel Segal's 1924 lock, Patent No. 1,482,766. A is a front elevation of the lock casing applied to a door, and B is a rear elevation of a plate fastened to the door and to which the lock casing is removably secured.

The casing of a lock is indicated at 1 and is provided with horizontal slots (2) at that end thereof which extends beyond the edge of a door (3) when the lock is in mounted position on the door. A locking bolt (4) slides vertically on tracks (5) within the casing and is provided with hooked ends of fingers (6) normally concealed by the casing and adapted to span or extend across slots (2) when in locked position, and to interlock with the perforated ears of a suitable keeper (not shown) fastened to the door frame. The locking bolt is provided with lugs (7), spaces (8), and a pin (9). A tumbler pawl (10) is pivoted to locking bolt (4) as at (11). The pawl has an arm (12) normally engaging a post (14) fastened to the outer wall of the casing. A spring (15) is secured to the upper end of the tumbler pawl and engages the pin (9). The action of this spring tends to swing the pawl on its pivot (11) in a clockwise direction whereby the arm will extend between the lugs, and the shoulder (13) will engage the post (14). This holds the locking bolt in raised and unlocked position in an obvious manner.

Segal's dead bolt was improved upon over the years by the inventor and his associates, and it came to be known as the lock that was difficult, though not impossible, to pick or jimmy—if it was properly installed, of course. Many kinds of changes have been proposed for the basic dead bolt lock, but all the possibilities for improvement have not yet been exhausted. Figure. 6.3*a* illustrates a crossbolt deadlock invented by Gregory Colombo of Minnesota. This lock is designed for use on doors that are adjacent to glass panels or some other weak spot that can be broken through by a thief, who can then reach in and open the lock from the inside. Colombo's invention includes a crossbolt that can be operated only by a key from either side of the door. A key is also required for the removal of the lock from the door. Some locks of older design were easy to remove with just a screwdriver.

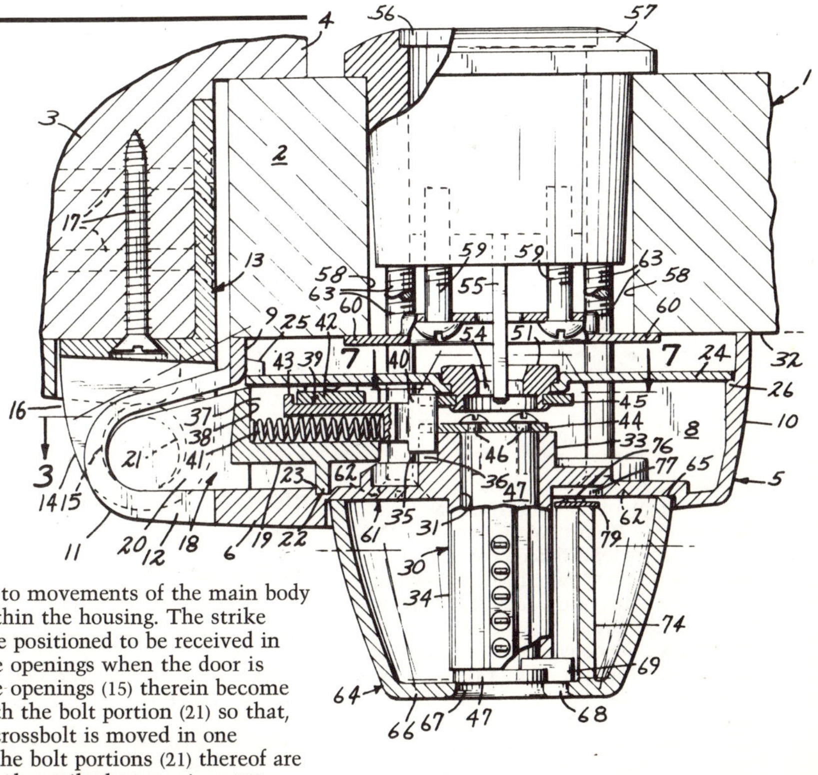

figure 6.3(a) Fragmentary section of a crossbolt deadlock attached to a door. The drawing is from Patent No. 3,760,619 (1973).

A conventional hinged door (1) is shown fragmentarily. The lock is mounted on the free edge portion (2) of the door opposite the hinged edge of the door (not shown), the door being mounted in a conventional door frame (3).

The deadlock includes a housing (5), an outer or front wall portion (6), and a peripheral flange forming top and bottom side walls (7) and (8) and opposite end walls (9) and (10). The former provides an outward-projecting nose portion (11) that is cut away to provide a pair of vertically spaced slotlike openings (12) opening to the interior of the housing. The nose projects laterally outward from the edge of the door and over part of the door frame when the door is closed. A strike (13) includes a pair of vertically spaced lugs (14), each of which is received in a different one of the openings when the door is closed. The lugs have axially aligned openings (15) and are integrally formed with a mounting flange or base portion (16) that is rigidly secured to the door frame.

A crossbolt (18) is mounted in the housing for reciprocatory movements in a generally vertical direction, and comprises a main body portion (19), a pair of vertically spaced arms (20) projecting laterally outward from the body portion (19) within the nose portion of the housing, and a pair of axially aligned bolt portions which move transversely into and out of the slotlike openings (15) that are responsive to movements of the main body portion within the housing. The strike lugs (14) are positioned to be received in the slotlike openings when the door is closed. The openings (15) therein become aligned with the bolt portion (21) so that, when the crossbolt is moved in one direction, the bolt portions (21) thereof are received in the strike-lug openings (15). The body portion is formed to provide an elongated lug or rail (22) that is longitudinally slidably received in an elongated guide channel or groove (23) in the housing wall (6), whereby the crossbolt (18) is guided for sliding movements within the housing (5) between locked and unlocked positions relative to the strike (13).

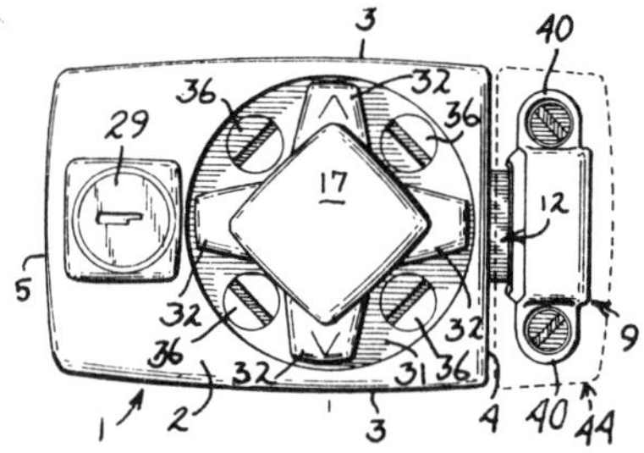

figure 6.3(b) Partial view of the lock with concealed mounting fasteners, mounted on a door and door frame, as illustrated in Patent No. 3,580,017 (1971). The dead bolt housing (1) has walls (2), (3), (4), (5). There are a number of openings in the plate through which pins extend and are riveted or peened in place. The strike (9) is normally close to the inner end wall (4). The housing is mounted on a door, and the strike is mounted on the door frame.

The outer surface of the front wall provides a circular recess (31), which receives the axially inner portion of the knob (17). As shown, the knob has four "wings" (32) that are disposed in the recess (31). Behind this recess, and within its confines, the housing is formed to provide bosses (33), having axial openings (34) that are countersunk, as indicated at 35, for reception of anchoring screws (36). The rear cover plate (6) is provided with openings (37) aligned with the openings (34) for passage of the screws. The screw heads are exposed to view for installation or removal of the deadlock only when the knob (17) is turned to a position where the bolt is between its fully extended and fully retracted positions.

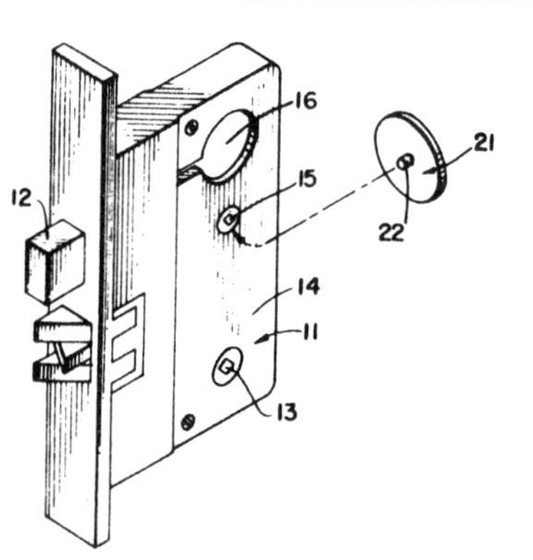

figure 6.4 A side perspective view of a protective guard for a mortise-type lock, as illustrated in Patent No. 3,741,596 (1973). There is an extended dead bolt (12). The usual doorknob spindle hole (13) is shown near the bottom of the lock (11). Side (14) is to face the exterior of the door, which is not to have the conventional thumb and finger knob, for operation of the dead bolt, so it is this side that is to receive the protective guard. The spindle hole having the hub to operate the dead bolt is depicted by the number 15. The cylinder-receiving hole (16) is located directly above the spindle hole.

The protective guard (21) has a stub that is inserted in the spindle hole. The result is to present a face where the disc covers not only the spindle hole but the area around it.

Russell Waldo of Minnesota found another way of meeting this problem, as shown in Fig. 6.3*b*. The screws that fasten the lock and corresponding strike elements to the door and adjacent frame are concealed by a knob that is mounted on the housing. The anchoring screws are exposed only when this knob is manipulated to remove the dead bolt to an intermediate position, neither fully retracted nor extended.

Mortise locks generally combine a mechanical lock with a dead bolt mechanism. And they're better looking than locks that are simply screwed onto door and frame. Usually this kind of lock has spindle holes that are readily accessible from both sides of its housing, so that it can be used on either the right or left side of the door. Only one spindle hole is generally used, leaving the other one empty, a passage to the inside. A burglar is often able to figure out the location of this hole, get to it by drilling or even with just a sharpened screwdriver, and then turn the hub and retract the dead bolt. Richard Ohno of Connecticut invented a device that can be used to block off the dead bolt spindle hole and thus prevent unwanted entrance (Fig. 6.4).

The search for a pickproof lock has been a long and frustrating one. In the nineteenth century, English manufacturers proudly offered a reward to anyone who could open their tough locks without a key. And each one of them eventually had to pay off, often to someone who then went into the lock business himself. Locksmiths state that any lock with a keyhole—in effect, an entrance into the inner machinery of the lock—can be picked. Obviously, the better the lock, the more skill and time will be required to open it. The state of the art continues to advance, and locks get better and safer all the time. In 1976, Norman Epstein of New York patented a pickproof cylinder lock based on the novel principle of utilizing the tumbler pins to bring a corresponding number of elements such as balls into a straight-line chain that transmits a force longitudinally to a pivoting part that produces the action that opens the lock (Fig. 6.5). The inventor owns his own lock company, and is manufacturing his pickproof lock.

Many people don't feel secure in their own homes until the doors have been manually bolted or chained. Perhaps this goes back to colonial days, when the latchstring was out for family and friends, but was pulled in at night or when no one was expected. Many variations on the bolt have been worked out. Two of the most recent, both the products of independent inventors, are shown in Figs. 6.6*a* and 6.6*b*. The first illustration is of a door bolt device patented by Kenneth Westcott of Illinois. It is an economical product, stamped out of metal and with no very close tolerances, and it is designed to resist tampering. The first and second plates can be easily connected to door jamb and door end, respectively, and have nonhorizontal slots that align when the door is closed. The key is used inside the door, and dovetails with these two plates to complete the locking arrangement.

Jose Diaz of New Jersey developed the sliding bolt shown in Fig. 6.6*b*. It consists of "a sliding bolt housing adapted to be remounted on a door, a key-releasable lock assembly adapted to be mounted

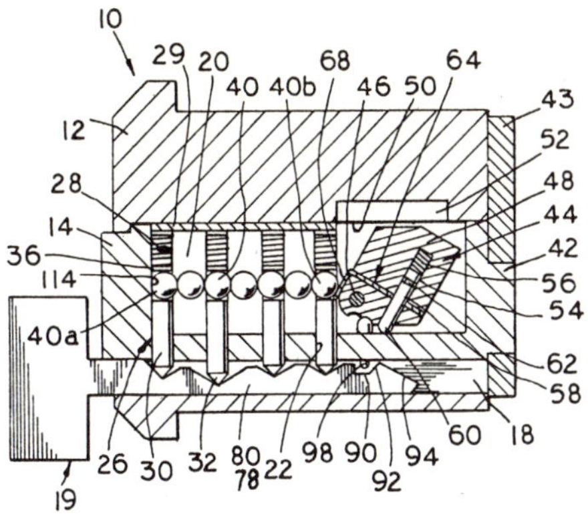

figure 6.5 Cross section of the pickproof lock described in Patent No. 3,968,668 (1976). The lock (10) comprises a hollow cylindrical casing (12) within which is mounted a lock cylinder (14). This cylinder has in its front wall a keyhole communicating with a slot (18) into which a key (19) may be inserted. In contrast to conventional tumbler locks, keyhole and key slot are not centrally located in the cylinder but are offset from center. In the locked arrangement the keyhole is adjacent to the underside of the lock cylinder (14). A longitudinal slot (20) opens through the side of the cylinder opposite the keyway (18). A row of bores (22) is formed in the cylinder along one edge of the slot, and a second row of bores is formed along the opposite edge. These bores extend diametrically through the interior of the cylinder to the keyway.

In each of the bores, a tumbler pin (26) is pushed down by coil springs (28) seated against a retainer (29). The pins vary in length. Each has a cylindrical body (30) having a tapered bottom (32). Upstanding from the body is a narrow post terminating in an integral flat disc (36). This disc and the flat upper surface of each pin body provide a space for mounting a small metal ball (40), which is the prime element in the locking and unlocking process. Each pin mounts one of the balls, and in each instance its ball is held against motion axially of the pin by its confinement in the recess in the side of the pin. Each ball is confined between the post of its associated pin and the wall of slot (20), remote from this post. The balls are thus held with their centers in a common plane parallel to the longitudinal axis of the lock cylinder.

The end plate (42) is secured to an external latch (43). When the lock cylinder is rotated, it moves the end plate and thus the latch to open the external latch mechanism. When the lock is closed, the cylinder is locked to the hollow casing (12) by a lock-operating lever (44), which is mounted by a pivot (46) within the cylinder. The lock lever has a projecting upper part (48), which in the locked position extends through a slot (50) in the cylinder and into a registering slot (52) in the casing, thus blocking rotation of the cylinder. Lever (44) has a bore (54) within which is mounted a helical spring (56) and a retaining pin (58). The spring biases the pin downward through bore 54, so that the rounded tip (60) of the pin bears against the inner surface (63) of cylinder (14) thus pushing lever (44) to pivot to the locked position. The retaining pin (58) is formed with a circumferential slot (62), which cooperates with a leaf spring (64) to lock lever (44).

When the proper key is inserted in the lock, the balls are raised by pins (26) into a horizontal chain. The end ball (40*b*) turns lever (44) counterclockwise to a lock-open position, in which the upper part (48) moves out of slot (52) and out of engagement with the casing (12), and the cylinder is released for rotation. The leaf spring (64) ensures that the cylinder will be kept locked until the balls (40) are brought into proper alignment.

When the lever (44) is in locking position and the balls (40) are nonaligned, the edge of the leaf spring arm is located in the circumferential groove 62 of retaining pin 58, thus locking the pin rigidly in extended position and in turn locking the lever. A key (19) is used to open the lock. Since the tumbler pins (26) are mounted in the two rows of bores (22) and (24), to present two spaced rows of pins, in order to actuate these two rows of pins, the key is made in double-shank form, having a pair of parallel shanks (78) and (80) connected by a crosspiece. When the key is inserted in the keyhole and slot, the inclined leading surfaces (94) of the key shanks engage the tapered tips (32) of the tumbler pins of each row, raising each pin in the keyway slot. The dwells and elevations on the key shanks are in registry with the tapered tips of the pins of each row, elevating each pin to the proper level, the balls being carried upward with each pin.

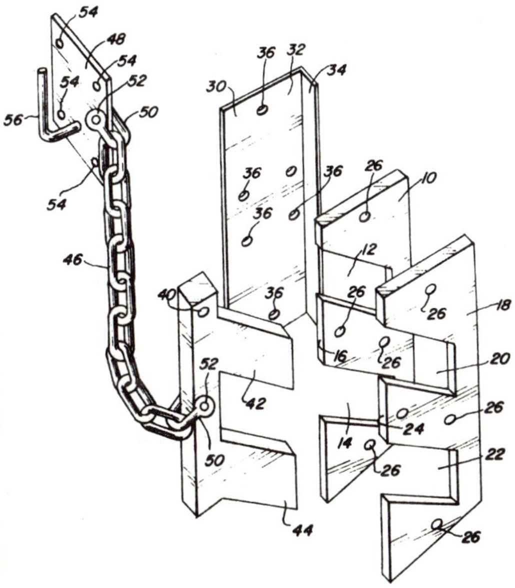

figure 6.6(a) A perspective view of the door bolt device described in Patent No. 3,761,118 (1973). The device is comprised of a first plate (10) defining a pair of nonhorizontal rectilinear slots (12) and (14), which open to the end (16) of plate (10). A second plate (18), which is identical to the first plate (10), comprises nonhorizontal rectilinear slots (20) and (22) which open to end (24) of plate (18). Preferably, the plates (10, 18) are formed of steel stampings and define apertures (26) for suitable connection to a door jamb and door, as explained below.

A bolster plate (30) is provided, having a planar coupling portion (32) and a perpendicularly extending flange (34), and defining a number of apertures (36) for permitting connection of bolster plate (30) to a door jamb. Flange (34) is greater in width than the thickness of first plate (10).

A key (40) is provided and is preferably formed of a steel stamping being greater in thickness than the thickness of either plate (10) or plate (18). This key carries a pair of nonhorizontal extending projections (42) and (44), which are complementary with and substantially conforming in shape to slots (12), (14), (20), and (22). In this manner, when plates (10) and (18) are aligned so that slot (12) is aligned with slot (20) and slot (14) is aligned with slot (22), projection (42) will enter the opening comprising slots (12) and (20), while projection (44) will enter the opening comprising slots (14) and (22).

A chain (46) couples the key to hanger (48) and is connected to them by means of U-shaped connectors (50), which pivot about pins (52). Bolster plate (30) and first plate (10) are aligned and fastened to a door jamb (60) so that the coupling portion (32) and first plate lie in a plane perpendicular to the walls to which the jamb is attached. Flange (34) extends perpendicular to the end of the first plate, toward the outside of the premises.

Second plate (18) is fastened to the free end of the door and is placed at a horizontal level so that when the door is closed, plates (10) and (18) will be aligned and have the ability to receive projections (42) and (44) of the key. Because the width of flange (34) is greater than the thickness of the first plate, a person trying to tamper with the bolt device from the outside will not be able to do so when the key is in place in the nonhorizontal slots defined by the plates. Any force on the door tending to open it will only tend to strengthen the relationship between key member (40) and plates (10) and (18). Although the projections and slots could be at various angles, an angle of 40 degrees with respect to the horizontal has been found satisfactory.

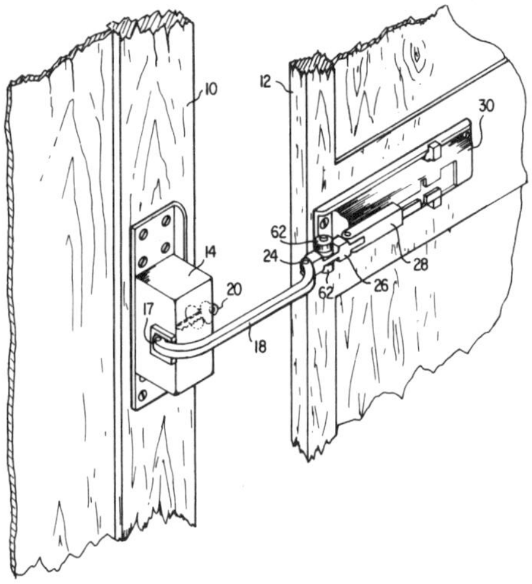

figure 6.6(b) The safetv lock described in Patent No. 3,869,886 (1975), installed on a door that is partially ajar. The tumbler-lock housing (14) engages and secures a lock plunger pivotally attached to end (17) of hardened steel door restraining arm (18). Key (20), shown partially in phantom, engages a tumbler-lock mechanism (22) and is actuable from either side of the door to disengage the lock plunger. At its opposite end (24), arm (18) is pivotally attached to roller link (26), which in turn is pivotally attached to sliding bolt (28), engaged in sliding-bolt housing (30).

Located along the vertical face of the housing (14), most remote from the door opening, is an opening for receiving a lock-plunger guide, an essentially U-shaped element that extends with its open end facing the interior of the building.

on the adjacent door frame, and an articulated, hardened steel arm having on one end a sliding bolt . . . and on the other end a lock plunger . . . thereby permitting the door to be opened only slightly." Thus the householder can look outside without opening the door all the way.

Window grilles are another important security device for the home where burglary is a possibility. The problem with many grilles is that they not only lock out criminals, they lock in the occupants of the house. That's fine, except in case of fire or other emergencies when the doors are not accessible. Jean Astié and her brothers, Gabriel and Georges, all of Florida, have invented a device that allows for the opening of a window grille from the inside when necessary. The brothers Astié learned ornamental ironworking from their father in France and have built up a good business since their arrival in the United States. They developed the lock in response to requests from their customers. It has been approved by local fire and building departments and is produced in conformance with the South Florida building codes. Distribution on a wide scale has not yet come about. One problem: The lock may freeze during cold northern winters. (See Fig. 6.7.) Construction is not complicated. The grille has a socket to receive a lock shaft projecting through the wall. A lock plate attached to this shaft on the inside of the wall holds the unit in place. The grille can be opened simply by rotating a handle connected to the inside end of the lock shaft.

Stephen Lajcak of Colorado has patented a sash lock for swinging or sliding windows and doors that consists of a cam rod mounted in the window opening. "The rod is inset to provide a lip extending longitudinally and spirally whereby it flatly engages a window edge when rotated by an operating lever at one end thereof to lock the window in closed position" (Fig. 6.8).

These inventions are practical and useful advances in an important area of security, but they do not represent any revolutionary breakthroughs. One area with promise of providing some important new developments is that of electronic locks. In 1974, Raymond Kambic patented an electric lock with a two-part bolt that has a bell crank connected to it pivotally. The bolt can be moved only by the crank, which in turn is activated by a solenoid (Fig. 6.9). The bolt spring protects the device against tampering from the outside.

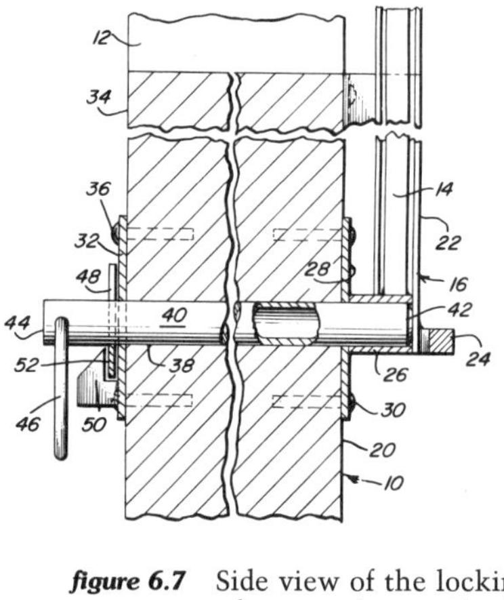

figure 6.7 Side view of the locking arangement for a window grille illustrated in Patent No. 3,913,957 (1975).

The window opening (12) is blocked on the outside by a grille (14). A bracket assembly (16) is mounted by fasteners on the exterior surface (20) of the wall structure and includes parallel spaced vertical slides or tracks (22) interconnected below the window opening by a bridging member (24). The grille is slidably received in the tracks by being inserted upwardly from below the bracket assembly.

A tubular socket (26) is secured to the window grille centrally along its bottom edge and projects axially into abutment with an exterior anchor plate (28) positioned centrally below the window opening and aligned with another anchor plate (32), secured to the interior surface. The socket is axially aligned with holes centrally formed in the anchor plates and a horizontal opening (38) in the wall structure, when the grille is in proper position as shown.

The grille is retained in position by an elongated, tubular lock shaft (40) having one axial end (42) received in the socket. The lock shaft is slidably and rotatably supported in the opening (38) and by the anchor plates (28) and (32) with its axial end (44) exposed on the inside of the wall structure. A handle (46) is in the form of a ring element and is pivotally connected to the lock shaft adjacent to the end. By use of the handle ring (46), axial and angular displacement may be imparted to the lock shaft.

A circular lock plate (48) is concentrically fixed to the lock shaft to limit its insertion into the wall structure by abutment with the anchor plate (32). A planar hook element (50) is secured to the anchor plate in radial overlying relation to the lock plate (48). The hook element thus forms a recess (52) through which the lock plate is rotatable but prevents axial displacement of the lock plate inwardly and withdrawal of the lock shaft from the socket (26).

The lock plate (48) is provided with a radial slot that is wider than the hook element (50) with which it may be aligned. Thus, the plate may be rotated by the handle ring (46) to the single release position, permitting axial withdrawal of the lock shaft. When the lock shaft is withdrawn from the socket, the grille (16) will drop down out of the bracket assembly.

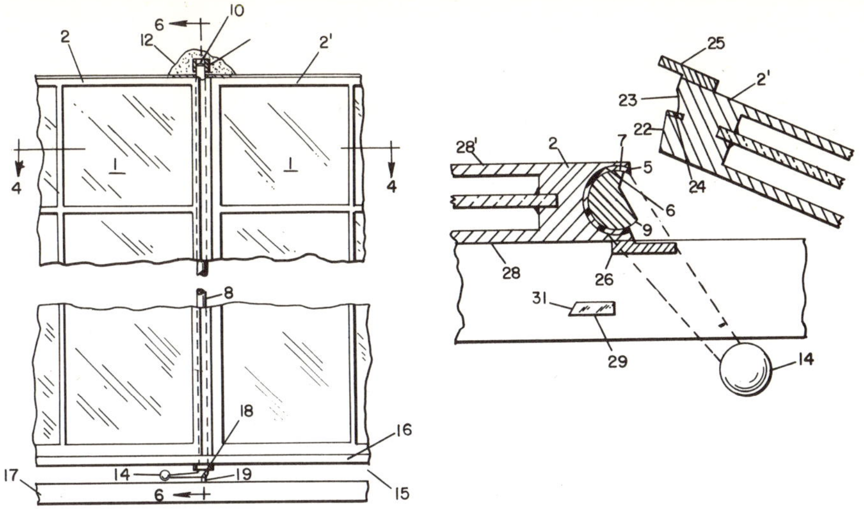

figure 6.8 Sash lock for windows illustrated in two drawings from Patent No. 3,977,711 (1976). At right is a sectional view along line 4–4 of the main drawing, with the sash lock cam rod (8) unlocked and one of the casement windows partially open.

The sash (2) has a concave channel (6) extending from top to bottom and lined with a plastic semitubular bearing (7). A cam rod (8) having a curved side wall (9) snugly interfits the ends of the bearing. The extended upper end (10) of the cam rod is received in a bearing in upper window frame (12), and the extended lower end (13) is formed with an operating lever (14) that projects laterally of the space (15) between the window stool (16) and sill (17). A recess (18) is provided in the underside of lever (14), in alignment with the axis of the cam rod, and receives an upright pin (19) suitably anchored in sill (17) and functioning as a pivot point for the lower end of rod (8). Thus, the upper and lower ends are so supported that the rod may be rotated within bearing (7) when lever (14) is manually arcuately moved.

The cam rod is inset longitudinally to provide a spirally extending flat face terminating in a flat flange. The inner side edge of window sash (2′) has an arcuate-shaped channel (23), which terminates in a flat face (24) so that cam rod interfits channel, and the flange thereon flatly engages face (24) when the cam rod is rotated and twisted clockwise to the closed position. To ensure proper alignment of the sashes, flat stop plates (25, 26) are attached to the inner face of sash (2′) and the outer face of sash (2). A beveled stop (29) is attached to the top face of window sill (17) so that when lever (14) extends to the right and the cam rod (8) is in the untwisted and unlocked position, sash 2′ may be swung open. When sash (2′) is moved to closed position and the cam rod is rotated and twisted, the lever (14) is moved clockwise and rides up the beveled top face of stop (29) and snaps downwardly when clearing the end (31) so as to be retained in that position by the end. During this movement the cam rod (8) will be rotated and twisted with its curve wall (9) rotating relative to bearing (7) and the spirally extending flat face progressively upwardly engaging and pressing against the flat flange (24) in channel (23) of sash (2′) and successively and continuously forcing this flange and its sash closed.

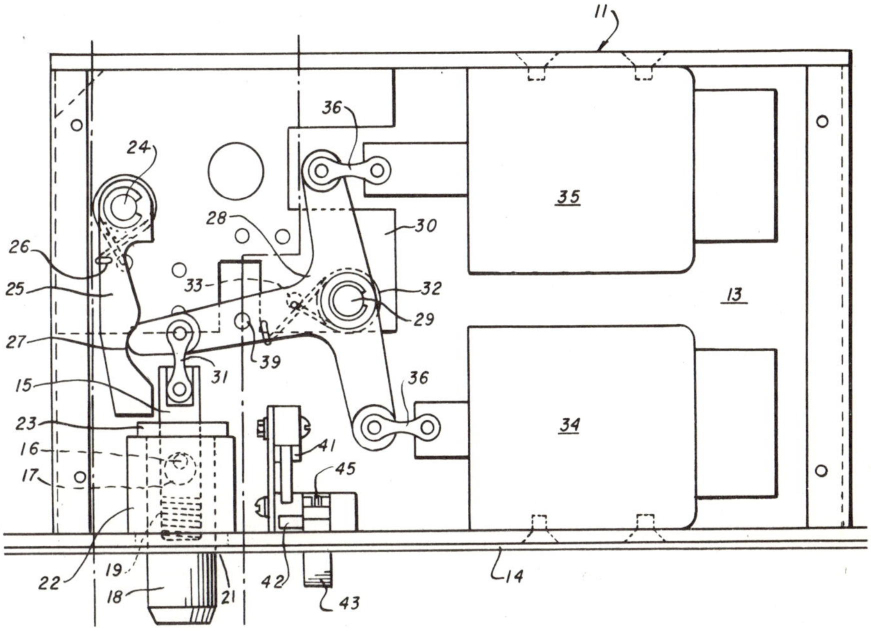

figure 6.9 Front view of an electric lock, as described in Patent No. 3,792,888 (1974). It is a two-part bolt made up of a bolt pin (15) having a cross pin (16) extending into apertures (17) in the walls of an inner bored portion of a dead bolt or outer part (18). Such pin and slot connections (16) and (17) permit but limit longitudinal movements of the inner and outer parts (15) and (18), and a compression spring (19) is disposed in the inner bored portion of the dead bolt between the latter and the lower end of the bolt pin disposed therein. The face plate (14) is provided with an aperture (21) accommodating the dead bolt, and a bolt guide or collar (22) is secured in any suitable manner to, and extends upwardly from, the casing means in vertical alignment with the aperture 21. The upper end of that bolt guide cooperates with a flange (23) on the upper or inner end of the dead bolt to limit downward or extended movement of the bolt, wherein the bolt is illustrated in its extended position.

Pivotally mounted at its upper end on a cross stud (24) is a dead-lock lever (25), the lower end of which is urged in a counterclockwise direction to active position into the path of retracting movement of the dead bolt (18) and (23) by a spring (26). The inner vertical edge of this dead-lock lever is shaped to provide a cam surface (27) with which the outer end of the central arm of a bell crank (28) cooperates.

This bell crank is pivotally mounted on a cross shaft (29) supported by a vertical bracket (20), and its central arm is pivotally connected near its outer end by means of a link (31) to the bolt pin or inner part (15) of the two-part bolt. A bell-crank spring (32) has one end engaging in an aperture (33) in bracket (30) and its other end terminating in a bent portion adapted to engage either the lower or upper edge of the central arm of the bell crank. This bent end of the spring (32) engages the lower edge of the central arm to urge the bell crank in a clockwise direction. if it is desired to move this in a counterclockwise direction, it is necessary only to remove the ends of the spring from their engagement with the bell crank and the bracket and reverse the spring so that the longer bent end engages the opposite edge of the central portion while the other end extends into the aperture (33) from the opposite side.

The most successful lock-without-a-key has of course been the combination lock. It was first introduced in the West during the nineteenth century; the Chinese had used it before then in slightly different form. Many inventors had a hand in the improvements that were made to the combination lock in the early days—most notably James Sargent and Linus Yale, Jr. Perhaps the most significant advance was the time lock, which has proven effective for use in bank vaults and other sophisticated installations where its high cost is far outweighed by the value of what it is protecting. Right now, it is expensive and a bit complicated for use in the home. Perhaps an inventor will someday find a way of getting around these problems.

One of the best patent drawings of an old-style "permutation padlock" is shown in Fig. 6.10. It was invented by Edgar W. Wilson of Colorado around 1885. His object was "to simplify and cheapen and make more positive and efficient in operation this type of lock." A modern combination padlock is shown in Fig. 6.11. It was designed for the Presto Lock Company by Irving Feinberg and Henry Heine, both of New Jersey, and gave the lock owner the option of setting and changing the combination. Other locks of this type had been on the market, but the inventors found fault with all of them. So they eliminated the obtrusive means of changing the combination—to keep it simple, to make it easy to miniaturize, and to use the shackle for setting the combination, but in a way that would eliminate the possibility of making inadvertent changes.

There are all kinds of applications for combination locks—padlocks, locks for safes, and in circumstances where it's better to have a changeable combination than a loseable key. Most combination locks aren't cheap; the moving parts make for a comparatively costly device, just as they make for something that is not

figure 6.10 Side elevation view of a permutation padlock, Patent No. 327,046 (1885).

(*A*) is the shell or casing within which the tumblers are concealed. The inner wall of this shell is recessed vertically upon opposite sides, as shown at (*a*), for the purpose hereinafter explained. At its open end this shell is internally threaded, as seen at (*b*), while its opposite end is provided with two holes coincident with the recesses (*a*).

(*B*) is the staple, having a long leg (*c*) and a short leg (*d*). Each of these legs is notched upon its inner face, as shown at (*e*), there being four notches in each leg, and the notches of one leg are on the same horizontal plane as the notches in the other legs. Each leg is provided with a shoulder (*f*) to limit its movement within the shell, and the long leg is provided near its end with a stop (*g*) to prevent its being entirely withdrawn from the shell. This top is preferably formed by a wire coiled around the leg of the staple and having its ends passed through an opening therein. The opening in the end of the shell is countersunk inside, as shown at (*h*), to receive this stop when the staple is withdrawn to its fullest extent.

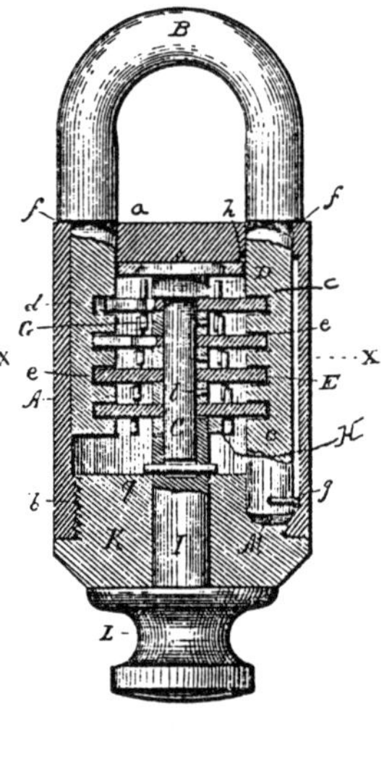

(*C*) is the tumbler-spindle, provided with a spline (*i*).

(*D*) is a tumbler rigidly secured to this spindle, and preferably resting against the inner face of the end of the shell, although the shaft may extend through the same and be stepped in the end, as shown in dotted lines.

(*E*) are locking tumblers, each notched on its diametric opposite edges, as shown at (*F*), and having a central opening to permit its sleeving on the spindle. These cutaway portions or notches correspond in shape to the shape of the staple legs. Pins (*k*) project upon either side of the tumbler, which latter may be of any desired number and sleeved upon the spindle (*C*). These tumblers are spaced by the collars (*G*), each of which is provided with a pin or feather (*l*), which engages the spline (*i*) in the tumbler spindle (*C*), and prevents the collars from turning on the spindle.

(*H*) is a locking plate, having an opening (*m*), to receive the tumbler spindle, which opening is extended to form a rectangular slot (*n*), the object of which will soon be apparent. A slot (*o*) is also formed in one end of this plate. When the tumblers are all sleeved on the spindle, as shown, this locking plate is sleeved on top of them, when the slot engages one of the pins on the top tumbler.

The operating spindle (*I*) has formed on its end a lug (*J*), projecting from one side of the opening (*p*), which opening is adapted to receive the end of the tumbler spindle and form a guide bearing therefor, while the lug (*J*) is adapted to pass within the slot (*n*) of the plate (*H*), thus locking the parts together. The operating spindle passes loosely through the cap (*K*), and is prevented from accidental withdrawal therefrom by a pin (*q*), passed through it.

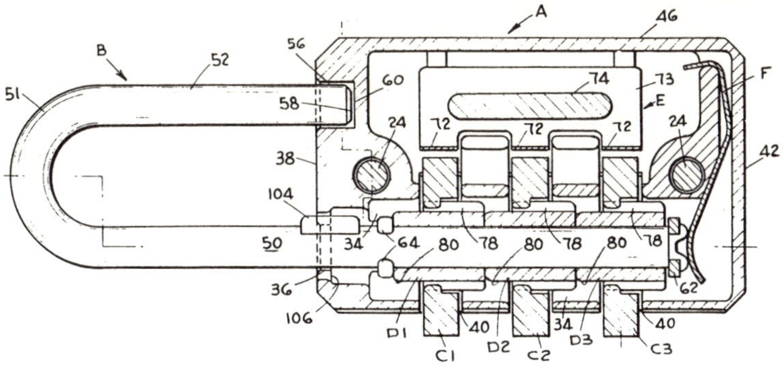

figure 6.11 Cross section of a combination padlock with the shackle closed, as illustrated in Patent No. 3,720,082 (1973). It is composed of a body (*A*), a shackle (*B*), three dials (*C1*), (*C2*), (*C3*), and sleeves (*D1*), (*D2*), (*D3*). Each dial has ten numbers, so there are 1,000 different combinations available. For more combinations the number of dials can be increased.

The shackle has the usual J-shape, and comprises a long portion (50) and a short portion (52). The legs are connected by a curved portion (51). The long portion (50) extends through the opening (36) with a small amount of clearance and into the cavity (34). The body A is provided with a recess (56) to receive the end (58) of the short portion (52) when the shackle is in closed position. The base (60) of the recess serves as a stop to limit the extent that the shackle may be moved inwardly to the closed position shown.

The sleeves (D) are mounted on the long portion (50) of the shackle in abutting, end-to-end relation. Each of the sleeves is provided with a central bore through which the long portion of the shackle extends with a slight amount of clearance. The sleeves are secured to the long portion of the shackle. The shackle may be rotated with respect to the sleeves and the body. The sleeves are rotatable on the long portion of the shackle, and the sleeves move linearly with the shackle as a unit.

easy to "pick." For this reason, combination locks have not been used very much in certain applications. Wallace Atkinson of Virginia saw that it was not economical to use combination locks for luggage unless the construction could be simplified and thus made less expensive. He developed a lock that has a number of adjustable dials (Fig. 6.12) "rotatably supported in side-by-side relation on a housing enclosing a fence plate movable back and forth in one direction radially of the dial members and coactive with a bolt plate reciprocatively movable between latching and unlatching position along an axis perpendicular to the axis of movement of the fence mechanism and having a coupling formation on one end."

A padlock generally offers good enough protection as long as the two things it is joining are themselves secure. Consider, for instance, the conventional hasp and large, U-shaped staple. A sustained attack with a sledgehammer or other such tool would cer-

figure 6.12 Perspective view of a combination lock for use on luggage, as illustrated in Patent No. 3,973,419 (1976).

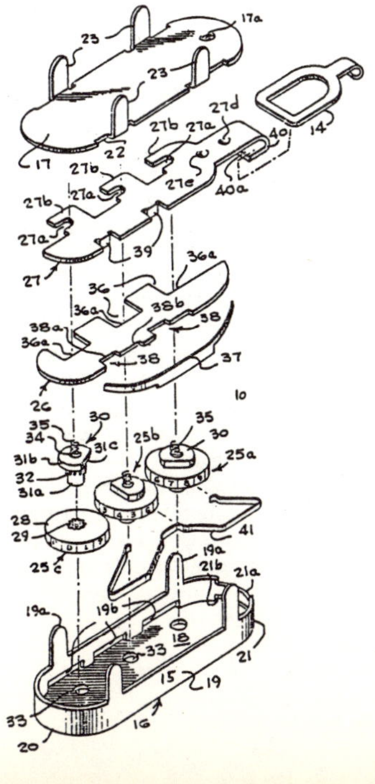

The lock (10) comprises a lock casing (15) formed from a stamped cover section (16) and a mounting plate section (17). The cover section has walls (18), (19), (20), (21), and a bolt opening (21*a*). The mounting plate (17) may have fastening tabs (23) that extend through the wall of the luggage and are bent inward for retaining the lock housing (15) on the luggage. There are three side-by-side dial wheel assemblies (25*a*), (25*b*), (25*c*), each of similar construction and with an outer dial wheel (28). Each wheel has an axial bore (29) and a hub (30). A collar (31*b*) butts against the confronting face of the adjacent dial wheel (28). A number of eccentrically located, axially elongated, interlocking formations, shown as ten centrifugally spaced, axially elongated pins of lugs (32), project axially from the collar (31*b*) to be received in the notches of the bore (29) against relative angular movement at any of ten selected positions corresponding to the ten dial numbers on the periphery of the dial wheel. The relatively smaller diameter body (31*a*) of hub (30) is of sufficient axial extent to project entirely through the axial bore and into one of the three journal sockets (33) in the front wall (18) of the cover section (16). The three journal sockets are positioned to define the axes of rotation of the three dial assemblies. The rearmost ends of the hub are provided with rear opening central sockets (34) in which are partially nested coil springs (35) that extend through recesses (27*a*) in the lug formations (27*b*), projecting laterally from the main body portion of the bolt (27), and bear against the rear cover plate (17). The coil springs (35) resiliently bias each hub into fully coupled relation in the bores of their associated dial wheels.

The fence plate (26) has recesses (36) along one edge for accommodating collars (31*b*) on the three hub members, and including a straight hub contact surface (36*a*) along a side of each of the recesses (36), paralleling the longitudinal axis of the bolt (27) and bearing against the periphery of collar (31*b*), causing the fence plate to move into locking position when the contact surfaces bear against the cylindrical portions of the collar peripheries and permitting the fence plate to be moved by biasing spring (37) laterally, relative to the longitudinal axis of bolt, and toward the hub centers when the flats (31c) on the hub collars are positioned exactly parallel to the contact surfaces. The opposite lateral edge portions of the fence plate have stepped recesses (38), each with a deep recess (31*a*) and a shallow recess (38*b*) to coact with forwardly turned locking lugs (39) on the corresponding edge portion of the bolt.

When the fence plate is in the locking position, the deeper recess portions register with and receive the locking lugs of the bolt (27) and restrain the latter in the retracted (locked) position. The bolt is loosely restrained in this position by registry of the detent nose formation (27*d*) in the detent socket formation (17*a*) of the mounting plate (17). The bolt is curved to define a hooked formation (40) on one end, projecting through the opening (21*a*) in the housing end wall (21). When the dial wheels and hubs are adjusted to the particular combination of the lock, the flats (31*c*) on the hub collars (31*b*) all face the contact surfaces (36*a*) of the fence plate paralleling those surfaces. This permits spring-forced movement on the fence plate toward the hub axes to the release position, aligning the shallow recess portions (38*b*) of the stepped recesses with the locking lugs (39) along an axis paralleling the longitudinal axis of the bolt, to accommodate sliding movement of the bolt along that longitudinal axis to the release position. The detent nose (27*e*) registers with the detent socket (17*a*), and the hook formation (40) is at a position out from the end wall (21) of the lock casing, to permit the zipper-pull loop (14) to be withdrawn from the hook formation (40) for manual operation of the zipper. A specially shaped leaf spring (41) in the interior chamber has end portions and a medial portion bearing against the peripheries of the dial wheels (28) to restrain them resiliently in the positions to which they are manually set.

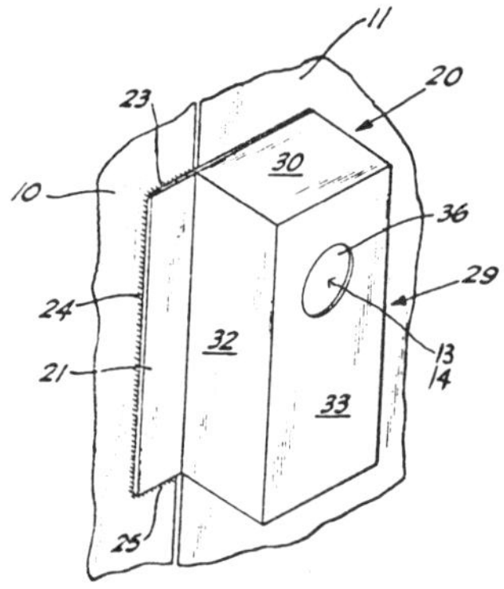

figure 6.13 Burglar-proof lock, Patent No. 3,718,014 (1973). It is designed to shield staple (13) and padlock (14) from a sledgehammer or other violent attack.

Backplate (21) is adapted to completely replace a hasp on door (10). This could be achieved by welding the plate to the door as at (23), (24), and (25). The other side of backplate (21) has a slot (22) having the same general dimensions as slot (15) in hasp (12). A staple (13) extends through the slot in the backplate.

Housing (29) entirely encloses slot (22) and staple (13) to shield them. Typically, each of plates (21) and (30-33) is made of hardened steel, and they are connected together to form a structure that is strong enough to defy sustained attack. The bottom of housing (29) is open to permit access to the staple. Finally, front plate (33) has a key opening (36) aligned with and facing the slot and staple.

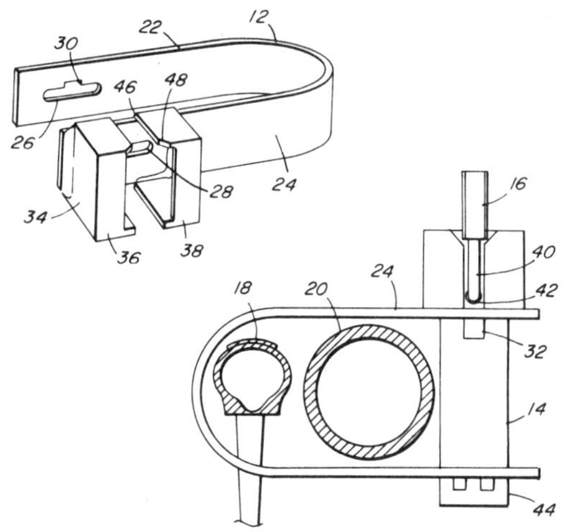

figure 6.14 Two views of the bicycle lock described in Patent No. 3,800,570 (1974). It is comprised of a U-shaped shackle (12) and a bolt (14) lockable across the open end of the shackle by means of a conventional padlock (16). Used as a bicycle lock, the shackle (12) typically has a length of eight or nine inches, the legs being spaced apart approximately 3½ inches—sufficient to accommodate a bicycle wheel and a post.

The shackle is formed with two parallel legs (22) and (24), with a slot opening (26) and (28), respectively, to receive the bolt. Each has a width slightly greater than the bolt, and a slot (26) is also formed with a cutaway portion (30) to accommodate a boss (32) on the bolt.

Box (34) is formed by short sections of channel pieces (36) and (38) welded to the outer face of the end of the leg at each end of the slot (28) to form a relatively deep socket substantially enclosing the padlock when its hasp (40) engages the end of the bolt through an opening (42). The bolt is also formed with a pair of shoulder stops (44), which cooperate with the leg (22) to prevent the bolt from passing entirely through slot (26). Alternatively, the end of the bolt may be bent at a right angle to form a stop. The length of the bolt is such that the end with the opening (42) extends a short distance through slot (28) so that the opening just clears the other side of leg (24) when the bolt is in place. The hasp of the padlock passes through the opening to lock the bolt in position against retraction.

Boss (32) prevents the bolt from being pushed through slot (28) by forcing together the legs of the shackle in an attempt to move the padlock clear of enclosure (34). Thus, even if sufficient force could be brought to bear to close the shackle legs, the boss would prevent displacement of the bolt. The cutaway portion (30) of slot (26) allows the boss to pass through when the bolt is being withdrawn from the shackle after the padlock is unlocked and removed.

tainly break off one of these components, making the lock useless. Fred Delgadillo of California recognized this problem and designed a boxlike structure to fit over the locking apparatus. The back wall is designed to attach to or engage the staple (in the above situation), and the front wall has a hole big enough for the key to pass through and open the lock. So the box cannot be moved unless the padlock is disengaged, and the box is tough enough to resist most attacks that are within the capability of the ordinary thief. (See Fig. 6.13.)

Bicycle locks offer another example of a protective device that is not always thugproof. A bolt cutter or hacksaw can often make short work of many types of bike locks. And since bikes are high on the list of stealable items for small-time criminals, there is a ready market for improved locking devices. One of the best on the market carries a stiff price tag—more than $25. Stanley Kaplan of Massachusetts has developed a U-shaped shackle and movable bolt adapted to extend between the legs of the shackle, to which it can be attached by a padlock. The arrangement is such (see Fig. 6.14) that the padlock is not accessible to a cutting tool. The shackle and bolt are strong enough to withstand most attempts at theft.

Figure 6.15 illustrates a much more complicated invention, by Anthony Aiello of California. It is a permanently anchored vertical post to which a bike can be locked and with a locker for safekeeping of packages and other personal property that the cyclist might be carrying. This invention operates more or less like the public lockers in bus stations. After you insert a coin, you can lock up and take away the key. Since this stand is made for outdoor installation, there's an alarm set to go off if anyone tampers with bike, locker, or coin box. This is a fairly complicated arrangement, and it would not be easy to install. It would be especially difficult to get the tamperproof features working properly, since they probably would be subject to vandalism. Yet the need is there. In many areas the standard "picket fence" type of bike rack is not much utilized because a bike can't be locked securely into it except crosswise, taking up virtually the entire rack. So this is an area worth thought and experimentation. The bike boom described in Chapter 10, Sporting Goods, is here, and many people would like to ride their two-wheelers to schools and stores and workplaces. But they are not going to do that if there is not a secure place to tie up.

Another item likely to attract thieves is the unattended outboard motor. For some of these engines, all it takes is some muscle to lift them up and carry them away. Several different kinds of

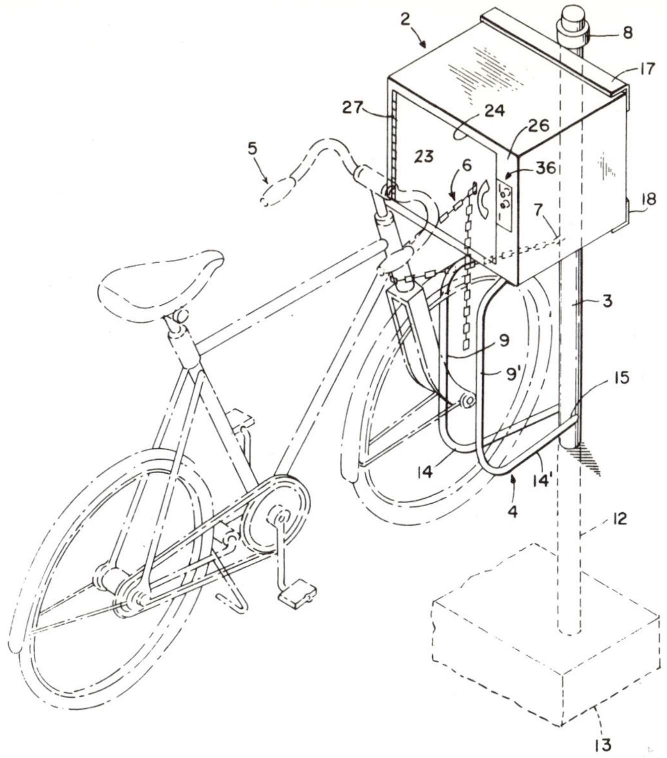

figure 6.15 A bike locker in use, as illustrated in Patent No. 3,827,773 (1974).

A locker is mounted on a pedestal and provided with a pair of wheel braces (4). One end (7) of chain (6) is securely welded either to post (3) as shown, or to receptacle (2). A cycle may be secured by inserting the front wheel of the cycle between the parallel bars (9) and (9′). The chain is then draped through the cycle and one of the links locked to the locker or post. The lower end (12) is embedded in the ground and anchored by an appropriate concrete block (13). Such construction will prevent the post and attached locker from being carried away by vandals.

The locker is provided with a door (23) to close an opening (24) in the front wall (26) of the locker. The door is appropriately hinged at (27), the construction being of the type that will prevent tampering by vandals.

The lock mechanism (36) comprises a conventional lock structure.

locking devices for outboards have been marketed—they are generally locking tubular covers for the actuators that control movement of the motor's clamps that hold it to the boat. But here again the lock is exposed and can be broken off without much trouble. Adrian Quinley of Alabama has offered a solution to this problem in an invention that rearranges the format of the tubular cover so that the lock is in a position where it cannot be tampered with easily (Fig. 6.16).

One of the biggest prizes in the war between thieves and the rest of us—and sometimes it's difficult to tell who's on which side—is the automobile. So far in the 1970s, the average annual take of car thieves has been almost $2 billion. Car makers try new locks and new ways of preventing ignition jumps and other tricks of the trade, but criminals keep on figuring out new ways to beat the system. Experts have said that if someone really wants your car, there's very little you can do to prevent the theft short of standing guard around the clock. Most of the time, of course, the bum doesn't want *your* car—he wants any car that can be stolen and sold or stripped with the least effort and the greatest profit. Many people, especially those who park in high crime areas, play the percentages—they install a protective device or two and hope that potential thieves will move on to a car that looks easier to take.

There are many locks and other security attachments for automobiles. Some of them are sold and installed by the auto manufacturers and their dealerships. Other are available through auto-accessory stores and other independent outlets. Security comes in all sizes, types, and price ranges. For one thing, there is a broad range of electronic alarms available. Depending on circumstances, these can be effective. But if you sometimes park in a relatively deserted area, and if the thief knows something about disconnecting

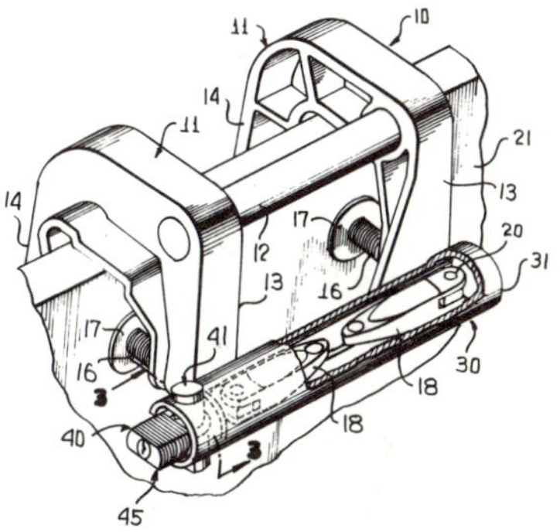

figure 6.16 Outboard motor mounting bracket clamped into position by the antitheft device described in Patent No. 3,848,441 (1974). The illustrated mounting bracket (10) has a pair of U-shaped yoke members (11) which are suitably interconnected by a bar (12). Each yoke member includes a forward leg (13) and a rear leg (14). Each forward leg has the rear end of each clamping member (16) provided with a clamping pad (17), while the forward end is provided with an actuator (18) pivotally connected to the clamp by a pivot pin (20).

The antitheft device, or latch assembly (30), includes a tube (31) which can be telescoped over the actuators (18) when they are in alignment, as shown. It is preferable that the actuators be arranged in opposed relation, as shown, so as to require a minimum length of the tube.

figure 6.17 Parking lock for transmissions, as illustrated in Patent No. 3,610,004 (1971).

As shown, the device is in the locked position with the piston pin in engagement with the lockwheel where it remains due to the force from the coil spring (19). The transmission output shaft cannot turn, and the vehicle is stationary. At this time the contacts of the switch (31) are open so there is no current flow through the coil (25), and the blocker pin biasing spring (24) is held under compression by the blocker pin that is in engagement with the piston. The engine is not running, so no pressure exists in the apply line or pressure line. The valve element (47) of the lock control value is biased by its spring so that the pressure line (49) is blocked and so that the apply line (43) and chamber (14) are opened to the sump (57) through the exhaust (67).

The plunger (73) of the solenoid valve is held down by its spring so that the orifice (71) is closed; the solenoid (81) has no signal since the ignition switch is opened and the lock switch unit (40) is closed, indicating that the system is locked.

To prepare to move the vehicle, the ignition switch is closed and the engine started; pressure from the pump (55) appears in the pressure line and current is passed through the switch (89) into the coil of the solenoid valve, since the lock switch unit is closed. This causes the armature (77) to be moved and plunger (73) compresses the spring to open the orifice (71). There can now be no appreciable pressure established in the chamber (61). The valve element (47) of the lock control valve remains in the position shown, and no fluid pressure is generated in the apply line. The piston pin thus remains in the lock position.

The parking lock switch unit (40) is then open. The signal current through the solenoid coil (81) is interrupted and the spring (75) moves the plunger into blocking engagement with the orifice (71). Pressure builds up in the chamber (61), moving the valve element to the right to compress the valve spring (69) to block exhaust (67) and permit the fluid from the source to pass from pressure line (49) into the apply line (43) and up into the pressure chamber (14) behind the piston section. Fluid pressure then builds up in this chamber and moves the piston to the left to compress the spring (19) and close the contacts of the switch (31). This action removes the end (7) of the piston from the locking wheel. The piston pin moves until it has compressed the spring and bottoms against the end of the bore on the closure member. In this piston position the blocker pin (23) is pushed into the chamber and into blocking position in front of the piston section by spring (24). The piston pin will now remain disengaged from the lock wheel until the blocker pin is removed from the chamber (14) even when there is no pressure in the chamber.

The engagement of the piston pin end with the lock wheel is accomplished by closing the locking switch (40) so that current will flow through the coils (25) and (81). The ignition switch is on and there is pressure in line (49), and, when the circuit through the solenoid coil (81) is established, the armature moves the plunger (73) to compress spring (75) and open orifice (71). The pressure in chamber (61) drops as it is opened to exhaust (72). When this occurs, the valve element (47) moves by the force of spring (69) and the chamber is exhausted through exhaust (67).

The switch (39) of the lock switch unit (40), when closed, causes a current signal to pass through the switch (31) held closed by the piston pin and on through the coil (25). The current flow through the coil causes the blocker pin (23) to move out of the chamber (14) and compress the spring. As soon as the blocker pin is removed from the chamber, the spring (19) forces the piston pin end (7) into engagement with the lock wheel to lock the transmission output shaft. This piston movement allows the contacts of the switch (31) to open.

Current flow through the coil is stopped when the contacts of switch are allowed to open. The blocker pin is held out of the chamber by the piston section (11). When the ignition switch is opened to stop the engine, pump (55) stops, and there is a decay of the pressure in the pressure line (49) so that the pressure in the chamber (61) drops. The opening of the ignition switch interrupts the signal current flow through the coil of the solenoid valve so that the spring can move the plunger to close the orifice (71). However, since the engine is not running, there is low-level pressure in this chamber and the spring will keep the valve element (47) in the position where the apply line and the chamber are opened to the exhaust.

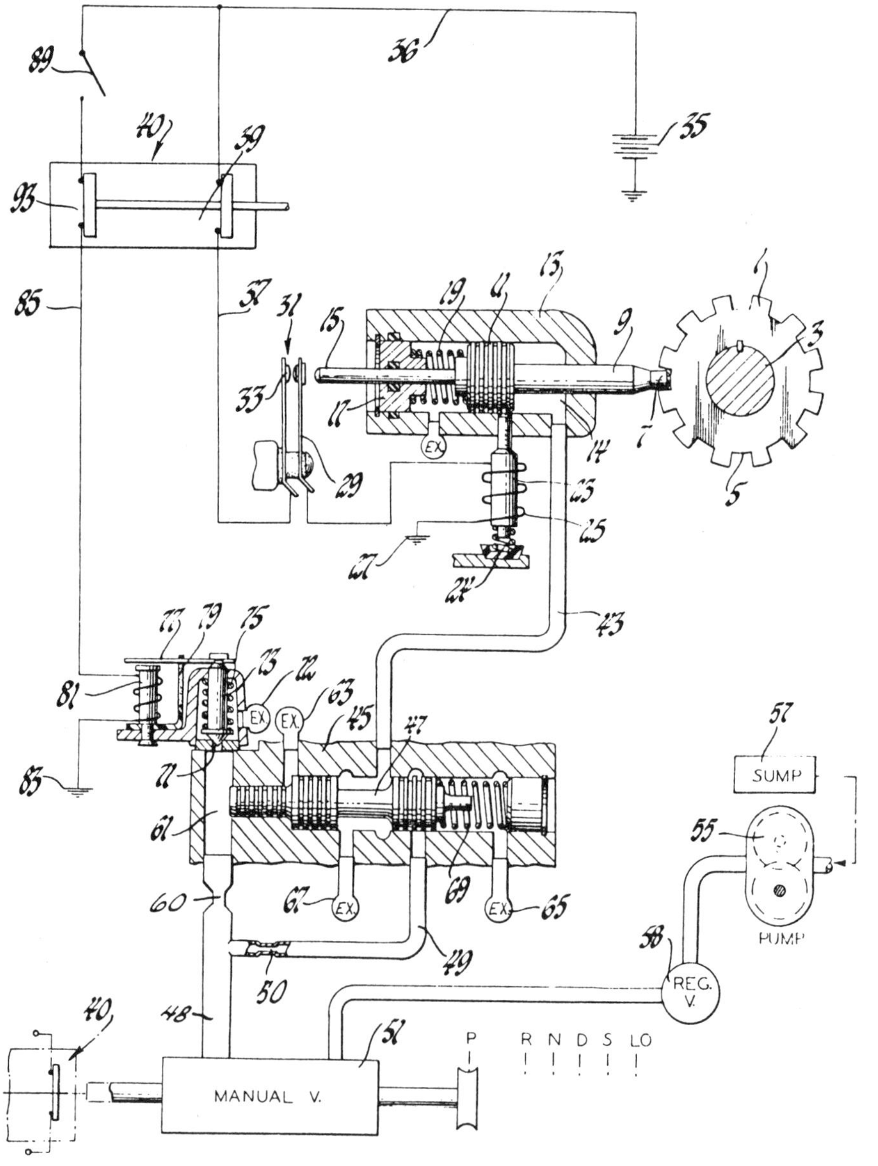

them, these alarms do not provide perfect protection. Other locks include the Krook-Lok, which locks the steering wheel to the brake. Philip Neese of Indiana has invented a transmission lock, the patent for which he has assigned to General Motors. He describes it in part as follows: "The parking lock has a piston with a rod that engages and locks a locking wheel fixed to a transmission output shaft. . . . A hydraulic control valve is operative to direct pressure fluid to the piston to an unlock position [and] a blocker pin automatically moves to keep the piston in this position unless there is a transmission lock signal." (See Fig. 6.17.)

In some situations, the greatest threat to an automobile is the bent coat hanger burglar. The lock on the typical American car door is opened either with a key or, from the inside, by lifting the rod that extends upward, ending in a knob on the windowsill. Now, even with all the windows tightly closed, a practiced and patient thief can manipulate a wire through the space between door and jamb and wiggle it around so that it lifts the knob of the lock. Bert Richman and Harvey Appell of California think they have found the way to stop this common kind of break-in. They have invented a button to be attached to car doors in place of the buttons supplied by the manufacturer. Before their buttons are installed, "an elongate body is mounted on the sill of the vehicle's door. The body has a vertically extending recess that tapers downwardly and inwardly and merges into a bore in the body. . . . The replacement handle (knob) includes an elongate portion that on the upper end develops into an enlarged head" that is mostly tucked into the recess when the door is locked. It thus becomes impossible to grip the thing with the bent wire, and the burglar moves on to some more likely vehicle. (See Fig. 6.18.)

The CB radio boom of recent years made millions of dollars for manufacturers of the equipment, for publishers of books and records about CB—and for auto burglars. Radios, tape decks, and similar equipment are vulnerable to theft even by amateurs. To combat part of this problem, Lyle Anderson of Pennsylvania developed a mounting assembly for a receiver and transmitter that was designed to deter theft. Early CB installations were easy to accomplish—a few screws and brackets put in place and the job was done. It was, of course, just as easy to remove the equipment. The invention shown in Fig. 6.19 is interesting because it acknowl-

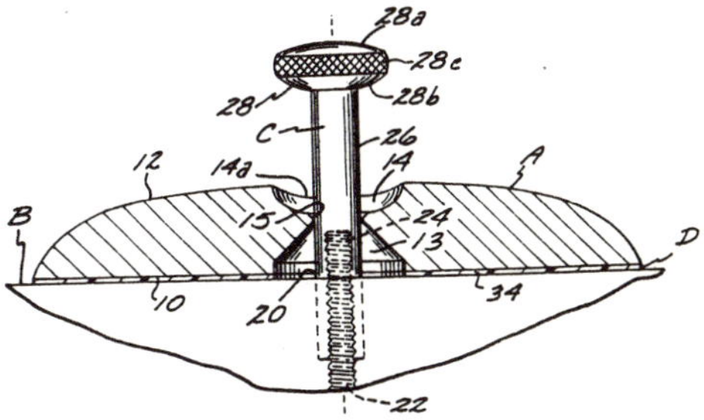

figure 6.18 Auto door lock protection accessory, as shown in Patent No. 3,915,485 (1975).

Sill (*B*) on the door of an auto has an opening (20) through which an end of a rod (22) extends that has threads (24). The rod by vertical movement is used to actuate a lock on the door.

The protective body (*A*) is used with a handle (*C*), which includes a shaft (26) and a head (28). This head is of circular transverse cross section, and is defined by an upwardly convex surface (28*a*), a downwardly convex surface (28*b*), and an intermediate cylindrical surface (28*c*) situated between surfaces (28*a*) and (28*b*). Surfaces (28*b*) and (28*c*) are of such size as to fit within the recess (14) when the handle is in the down position. The longitudinal side edges (14*a*) of the recess are concave, and permit the surface (28*c*) of the handle to be engaged between the thumb and forefinger of a user (not shown) when the handle is in a lowered locking position.

Body (*A*) has the flat surface (10) adhered to sill (*B*) by pressure-sensitive adhesive means.

Due to the upward and inward tapered configuration of cavity (13), the rod (22) may be off-centered relative to the cavity, with the cylindrical portion (26) of handle (*C*) being vertically movable relative to the body without binding as a result thereof.

figure 6.19 Combined antitheft and mounting assembly for a CB radio, as shown in Patent No. 3,994,148 (1976). The device includes a U-shaped frame (14) that includes a bottom plate (16) that supports side members (18), which project from one side of the plate (16), and a lug (20), which projects from the other side of the plate. The lug projects down at a central location flush with the front edge of the plate (16). The spacing between the side members (18) is such that the radio unit is nested in relation to the U-shaped configuration formed by the frame. Each side member has a drilled hole adapted to pass a fastening device such as a flathead machine screw into the tapped hole formed in the side of the radio.

The lug is employed as a support for a key-operated lock assembly (24). The lug has an annular bore into which the lock assembly is received and secured in place by a nut (26) at the back surface of the lug. Spaced outwardly from the nut is a radially extending latch plate (28), which is secured against a shoulder and the end of the lock assembly by a nut (30).

The housing has a depth greater than the depth of the side members so that the housing encloses, and thereby prevents access to, the fastening means (22). The housing (32) includes a top wall and is screwed into a plate (42). The bottom wall supports a lock box (44) with an open front to receive the lug when the frame is supported within the compartment. The thickness of the lug is substantially increased as compared with the thickness of the side walls of the lock box. Any forced entry into the interior of the lock box, housing, or frame will endanger the CB unit to destruction. The lock box further includes a top wall (46) that defines a slotted opening (48) that receives the lock arm (28) to form an interlocking relation between the frame (14) and the housing (32).

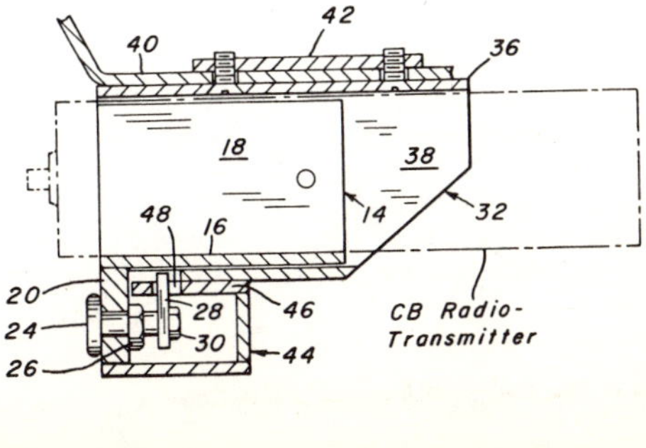

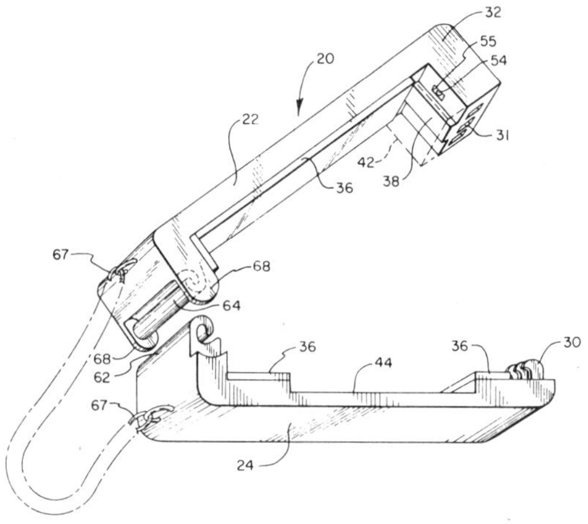

figure 6.20 Ski lock and chain attachment with hinge uncoupled, as illustrated in Patent No. 3,999,409 (1976).

The lock (20) comprises two rigid jaws (22) and (24) hinged at one end and adapted to snugly encircle a pair of skis at a point of least combined cross-sectional area. A bracket-receiving slot (44), shown here in jaw (24), adapts the ski lock to receive a fixed bracket (not shown), in addition to the skis, to thereby secure the skis to the bracket.

Felt pads (36) line a portion of all ski-contacting interior faces of jaws (22) and (24) to preclude injury to the skis. A spline (38) on an interior face of jaw (22) serves to engage a keyway of a spacer block (42). The spacer block normally is interposed in a receiving area adjacent to the spline as shown in broken lines. Jaw (22) is adapted to receive spacer blocks of varying thicknesses to enable the lock to accommodate skis of varying widths.

A reset lever (54) enables the owner of the lock to readily alter the lock combination to any setting. This lever is operably in a recess (55) in an internal face of bracket (22) to preclude tampering with the lock mechanism. Internal mechanism of lock (32) prevents alteration of the combination unless the lock is opened.

edges that a CB can be taken by a determined crook even if you've locked it in tight. There's a lock-and-key arrangement so that the equipment can be removed if the vehicle is going to be parked where there is clear danger of theft—a deserted area, crime-ridden neighborhood, etc. In normal situations, the radio can be left as installed, since the lock will provide sufficient protection.

Skis are another piece of equipment that can be easily stolen if no protective devices are used. At most ski areas there are racks where people can leave their skis while they take a break at a rest or refreshment area. Both professional and amateur thieves steal from these racks, the latter often exchanging a cheap pair of skis for a more expensive set. Sterling Bell of Texas has patented a compact, portable device that can lock a pair of skis to a rack or some other base without doing any harm to the skis (Fig. 6.20). It is fairly simple to use and to manufacture, and is easy to carry around.

There are, of course, hundreds and probably thousands of different uses for locks and other security devices. The more things that come onto the market, the more things there are to steal. And, seemingly, the more people come together in cities and in suburbs, the more they prey upon one another. One urban security need that has not been met is a riot-protection door or other device that will keep looters out of a store. After New York City's experience in the 1977 blackout, many security experts went back to their drawing boards in an effort to find something that filled the bill yet did not cost as much as the armor at Fort Knox.

Security is a growing business, with no end in sight. The thoughtful, alert inventor will see many opportunities for profit through creative change. And while the mass market for locks is difficult to exploit without a lot of capital, there are many smaller, local needs that can be handled by the independent operator.

CHAPTER SEVEN

Everything for the Office

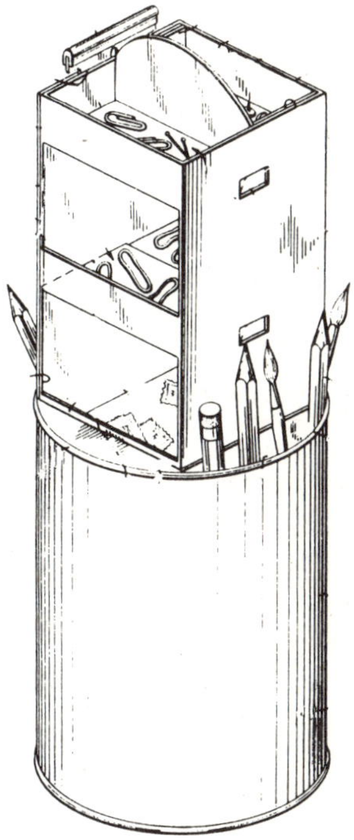

People's need to communicate—almost as basic as their need for food, clothing, and shelter—has inspired thousands of inventions, from communications satellites to paper clips. Businesses and home offices have an insatiable appetite for all kinds of equipment to help in the formation, transmission, and storage of words, both written and spoken. This process churns along on many different levels and for countless purposes. The programmer of a supercomputer can push some buttons and see data printed out at a blinding, mindless speed—a torrent of paragraphs. The young person writing a love letter may take an old-fashioned fountain pen in hand and carefully shape each word. Both are using inventions, the first a fantastic combination of complex new devices, the second a relatively simple hand instrument.

It should be no surprise that the independent inventor has only a limited chance of making any significant discoveries in the lofty world of computer science. IBM and other corporate giants invest billions of dollars in their research and development programs. The individual who doesn't have a free-flowing reservoir of capital to draw from simply can't afford to buy or lease the hardware that is essential for most experimental work. Even so, the giants can't cover all the expanding territory, and occasionally a lone inventor scores a breakthrough. With the special knowledge necessary to operate in this field, and by selecting a special area of interest, individuals do have a chance. Many ambitious people have left large companies to go on their own and work on new ideas. Some have achieved great success. Where the technology is expanding, creative individuals have an opportunity to make their mark quickly.

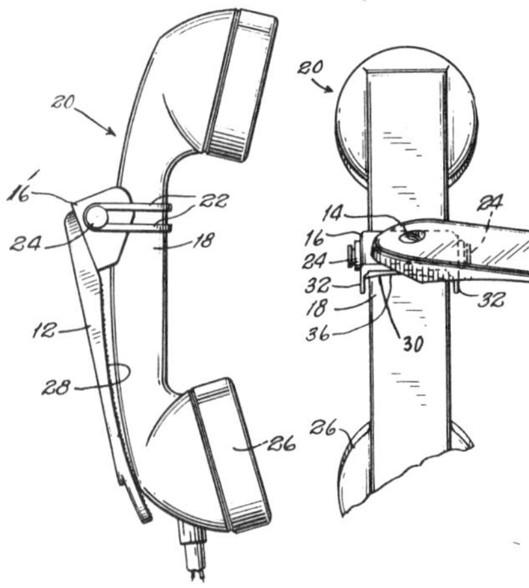

figure 7.1 A side view and a back view of a telephone with supporting device attached. The drawings are from Patent No. 3,141,072 (1964). The support comprises an arm (12), attached by pivot pin (14) to a mount (16), which is fastened to the neck (18) of a telephone instrument (20) by a band (22), of any suitable elastic material such as relatively soft rubber or such as a coil spring, extending about the inner side of the neck and looped onto integral lugs (24) formed on opposite sides of the mount.

The arm extends toward and overlies the transmitter (26) of the telephone instrument; and its underside can be given a suitable high-friction covering (28) to oppose sliding on the user's shoulder. The band may be continuous or endless, as illustrated, or it may be a noncontinuous length of elastic or spring material provided with end loops for engagement with the lugs.

Conditioning of the device for use involves two pivotal movements of the arm—first, to the left, then outward, or to the right. This first swinging action of the arm depends upon a cam coaction of the mount with the neck of the instrument. To afford such cam action, the mount is specially formed. It comprises a transverse top plate (30) having integral side wings (32), which extend downwardly close to opposite sides of the neck. At an intermediate point thereof, the element (16) is formed with a camming web between and integral with the wings, serving not only a camming purpose but also the purpose of bracing the wings against possible deflection.

The telephone business is another field that is dominated, necessarily so, by large companies. Bell Labs, the major research arm of AT&T, is one of the U.S. Patent Office's most constant correspondents. It is of course impossible to compete with them on such big-ticket items as million-dollar switching machines and complicated circuitry. But each year dozens of inventors who are not employed by any phone company are granted patents in this field. Depending on the outcome of the current dispute between Uncle Sam and Ma Bell, those two ancient adversaries, as to how wide the Telephone Company's monopoly should be, there may be even more phone-related inventions to come. In future, it is possible that Bell System transmission lines may be open to use by a greater diversity of equipment than now. Phone instruments themselves, for instance, are now being offered by some independent manufacturers in a wide range of styles and prices.

At last count there were more than 120 million phones in the United States, used daily, often endlessly, by just about everybody. Even after all these years, the system is still not perfect. Many businessmen have found to their pleasure that there is a large market for accessories of all kinds. Figure 7.1 illustrates a patent for a seemingly simple device that enables busy people to make phone calls while keeping their hands free for jotting down notes, going through papers, or whatever. Patented in 1964, it is still a good seller.

Right now, many business phones serve two purposes—they transmit the human voice and, often at night, they transmit data from one computer or computer terminal to another, beeping cheerfully at each other for hours at a time. In general, home phones now serve only one purpose—to transmit voices. Many experts believe that the time is drawing near when both home and office phones will accomplish many different jobs, from watching the house to paying bills automatically. Some of these changes will be accomplished with existing equipment; many of them will require a new invention or two. Will Bell Labs be the major factor in this field? Will independent inventors and companies be given a chance to play an important role? The answer to both questions would seem to be Yes.

One recent invention to help the telephone do double duty is illustrated in Fig. 7.2. Patented by Robert Conrad of Pennsylvania, it functions as a communications link between sick, old, and disabled people who stay at home—shut-ins, they used to be called—and a central monitoring station. The subscribers call in daily within a set time period. If they don't, the machine alerts an attendant who sets in motion a checking process to find out what is wrong. If the subscriber needs help, it can be on the scene very quickly. This automatic system enables a small staff to monitor a large number of people at the other end of the line.

Talking isn't everything; lots of people still write. New kinds of pens and pencils are almost as much a Patent Office staple as they have been in the past. Advances are still being made in ballpoints and mechanical pencils as well as in the increasingly popular felt tips. Even the fountain pen is being modified.

The ballpoint boom began slowly enough, in 1941, when the Hungarian inventor Laszlo Biro was granted his first patent (Patent No. 2,258,841). This provided him with minimal protection, however, because the pen's action was quite close to that of a standard fountain pen. At first, the Biro was marketed principally in South America. It was seen and admired there by a flamboyant American promoter, Milton Reynolds, who proceeded to steal the show in 1945 with a line of ballpoints that he advertised as being able to write at high altitudes and underwater. His publicity campaign was brilliant, and Reynolds got rich. His pens were sold at around $12.50 each at first, but after a couple of years they became lost in a flood of inexpensive ballpoints that sold for under 50 cents apiece. Biro continued to work on the pen, and his Patent No. 2,400,679 of 1946 covers the basic features of the ballpoint, though he added and refined various features with new patents in the next two years.

It would seem that after twenty-five years of mass distribution the ballpoint pen would have reached the point where almost all possible improvements have been made. Some new patents have been issued recently, however, and there will probably be more in years to come. Figure 7.3 shows part of the patent that was issued to Hartley Sears and Clarence Schrader of California. Their invention is a plastic cylinder with a tapered tip that is inserted into the end of a ballpoint cartridge to prevent leakage. A string extends in the center of this cylinder, providing tiny passageways into the cartridge as the ink is used. Any ink that flows backward will be partially blocked by the string, and its surface will come into contact with air brought in through the passages. This process will dry up the ink and stop the backflow. This device sounds complicated, but it is not. In fact, it couldn't be. Complexity means expensive to

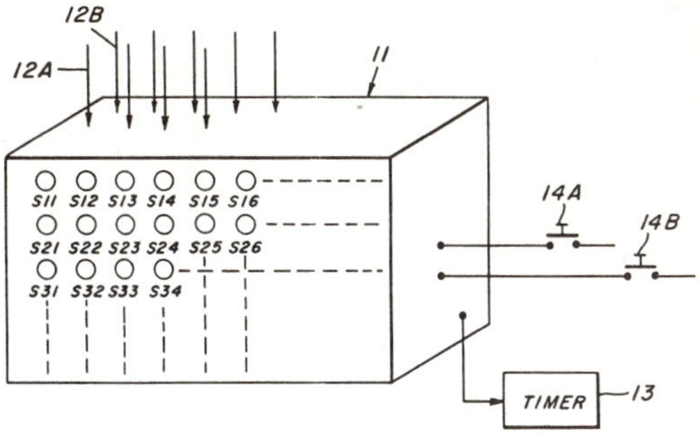

figure 7.2 Perspective view of a monitor (11) of the type suitable for communication with individual subscribers, according to Patent No. 4,011,409 (1977). It includes an array of light indicators arranged as columns and rows and identified as S11, S12, S13 . . . and S11, S21, S31. . . , respectively. Such light indicators are energized by telephone link responses from subscribers that are coupled to the monitor by lines (12*A*), (12*B*), etc. A timer (13) is coupled to the monitor to provide an attendant with a signal such as an audio or a light signal for resetting one or more of the light indicators. For this purpose, reset switches (14*A*), (14*B*) are coupled to the monitor.

figure 7.3 A fragmentary, exploded longitudinal sectional view, partially in elevation, of the end of an ink cartridge and plug element designed to prevent leaks from a ball-point pen [Patent No. 3,698,825 (1972)]. The ink plug (10) of this invention fits within what is normally the open outer end of a conventional tubular ball-point pen cartridge (11), which defines an ink reservoir that is cylindrical for most of its length. The solid body of the plug is made of plastic; there is a cylindrical part (12) that connects, through a shoulder (13), to a second cylindrical portion (14) of smaller diameter. The latter section of the plug is larger at its base than at its outer apical end. Each end has a 45-degree beveled edge, (15 and 16 respectively). The outer end wall (17) and the shoulder (13) are sloped at a 10-degree angle. These contours facilitate removal of the plug from the mold.

Extending axially through a central opening (18) in the plug is a length of nylon string (19), about 0.030 inch in diameter, composed of individual strands that are in turn made up of small filaments. The plug and string are securely attached. A short piece of the string, around one-sixteenth inch long, may extend beyond the end wall (17) of the plug. At the opposite end, the string is flush with the end wall (24).

The plug is installed in the end of the cartridge (11) simply by forcing it. The outer end of the plug portion (14) is of slightly smaller diameter than the entrance (25) to facilitate this installation; the bevel (16) also helps. However, next to shoulder (13) the plug is of larger diameter so that it is firmly gripped by the tubular cartridge body. The shoulder acts as a stop, positioning the plug axially when it is forced into the end of the cartridge.

The plug does not hermetically seal the end of the cartridge. Air still may enter through the interstices provided by the strands and filaments of the string (19). The string must have a diameter adequate to allow air to enter the cartridge rapidly enough to replace the volume of ink removed at the ball point as the cartridge is used.

In the event that the ink within the cartridge (11) flows to the end occupied by the plug (10), it will be prevented from leaking outwardly around the periphery of the plug because of the tight engagement between the outer wall of the plug and the inner surface of the cartridge. Also, the ink cannot flow between the string (19) and the wall of the opening (18) in the plug because of the complementary interengagement of the string and the plug. However, the ink will be able to commence flowing through the many small passageways within the string (19) at the axis of the plug. As the ink flows through these capillary openings, it is exposed to air and quickly dries so that it no longer is fluid. Long before reaching the outer end wall (24), the ink solidifies within the interior of the plug, forming a barrier that blocks any further outward movement of the ink. As a result, there is no leakage of ink from the cartridge.

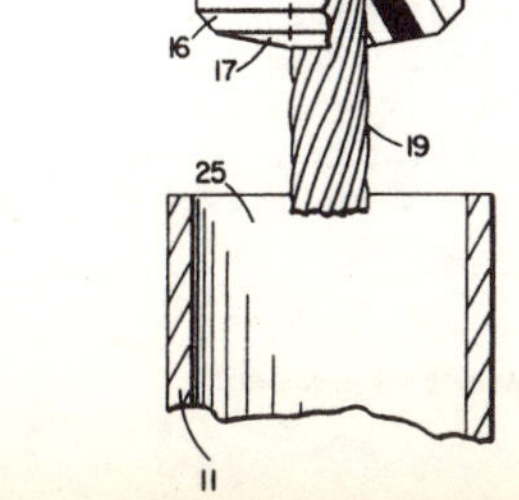

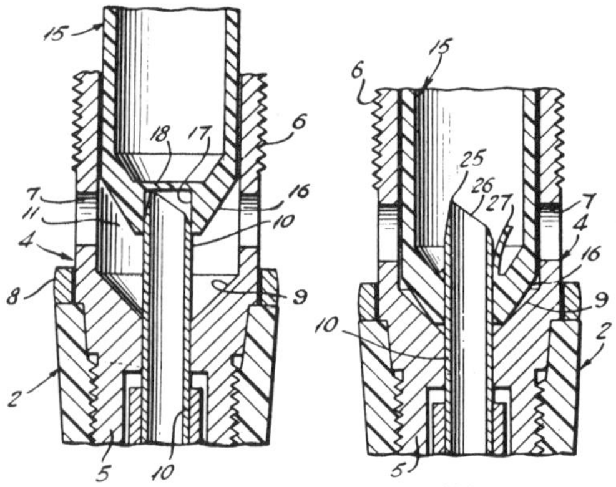

(a) (b)

figure 7.4 Two drawings of an enlarged longitudinal section of a cartridge for a fountain pen, (*a*) prior to piercing and (*b*) after piercing. As described in Patent No. 2,802,448 (1957), this type of cartridge and the pen designed to receive it are virtually leakproof. The rear or barrel end of the feed section of the pen incorporates a cartridge-receiving cavity and a cartridge seat and a rearwardly projecting piercing tube adapted to penetrate the front end of a cartridge housed in the barrel. The cartridge seat (9) is actually formed in the coupling and is preferably of conical, concave form. Piercing tube (10), protruding rearwardly from the center of the cartridge seat, extends part way back in the coupling member cavity or hollow interior (11). The position of the piercing tube relative to the coupling member is fixed.

The forward end of the cartridge is shaped to form a convex conical seat portion (16) and incorporates a bore (17) the base or inner end of which is normally closed by a restricted diaphragm area or wall (18).

The front end of the cartridge is a sliding fit in cavity (11), and the piercing tube is a snug fit in the cartridge bore (17). Accordingly, with the barrel removed, the cartridge can be inserted to its (*a*) position in which it is supported in the feed section both by the cavity wall and by the piercing tube. In other words, it cannot teeter or get out of alignment as the barrel is slid over it. Screwing the barrel home serves to telescope the cartridge bore and piercing tube, causing the latter to penetrate the diaphragm wall and place the tube in communication with the ink supply. (The details of the feed section construction and the control of the ink flow through the tube and to the nib form no part of the present invention and, hence, will not be described.)

The ink cartridge is made of polyethylene, or like material, which is readily pierced by the sharpened end of tube (10). As will be noted, the tube is shown sharpened on its inner edge (25) (Section *b*) rather than on its outer edge, and the end (26) of the tube is cut on the bias. In the result, the diaphragm wall 18 is not completely severed but is partially cut out and forced back as the tube and cartridge are telescoped, leaving a more or less hinged flap (27) which, being still attached, cannot float around in the cartridge and possibly clog the tube outlet.

As the barrel is screwed home, the cartridge seat portion (16) comes to rest against the feed section seat (9), thus effecting a second seal against the escape of ink around the piercing tube and into the barrel. A spring, with a cap abutting the rear end of the cartridge, serves to maintain the cartridge in close engagement with its seat.

As illustrated in (*b*), the angle of conicity of seat (9) is preferably greater than that of seat portion (16) of the cartridge, so that, under the influence of a spring, a squeezing action is applied against the periphery of the forward extremity of the cartridge. By this means, any tendency of the polyethylene to "flow" and thereby loosen the original snug fit of the tube in the bore is overcome, the sealing fit being maintained indefinitely.

manufacture, and in the world of mass merchandising, that just cannot be.

Ballpoints have been the popular pens, but fountain pens have managed to hold onto a share of the market, mostly in the medium- and high-price ranges. Their continued use is due in major part to the improvements that have been introduced over the years. One example is an invention by Donald H. Young of Connecticut, who assigned his patent to the Waterman Company. This patent was important enough to be contested in the courts by another of the giants in this field, the Parker Pen Company. The judge eventually decided in favor of Waterman and upheld the patent. This invention's significance was that it made ink cartridges less messy to handle and therefore more acceptable to a public interested in convenient and trouble-free operation (Fig. 7.4).

The first Eberhard Faber died in 1879, but the lead pencils with eraser tips that he introduced to the United States are still manufactured and used on a wide scale. Not until 1970, however, did a clever inventor, Charles Francis of Washington State, come up with a solution for people who erase almost as much as they write. Figure 7.5 shows a replacement eraser that is inexpensive to man-

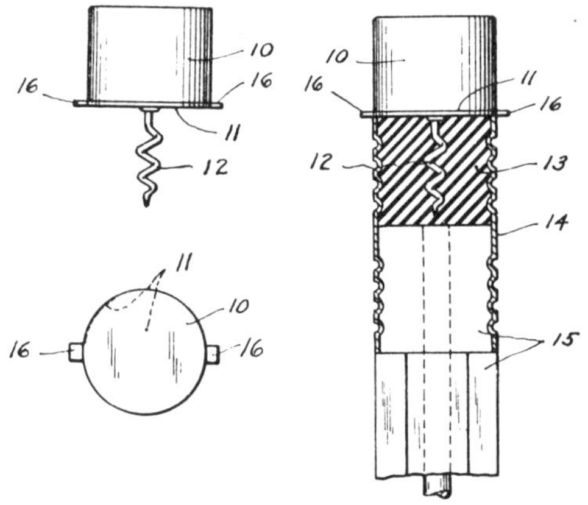

figure 7.5 Three drawings of a replacement eraser for pencils, taken from Patent No. 3,523,321 (1970). The rubber eraser (10) is preferably of about the same size and shape as the usable part of an ordinary lead pencil eraser. It has at least one flat end, which is cemented to a small disk-shaped mount (11) of relatively stiff strong material, such as metal or plastic. The mount has a rigidly connected anchor (12) protruding from the side opposite the eraser. The anchor is of small diameter and is helically shaped like a cork screw and terminates in a pointed outer end. It can be threaded into a stub or base part (13) of a worndown eraser in a ferrule (14) on an end of an ordinary lead pencil (15). The eraser must be close to and in alignment with the pencil and in substantially the position occupied originally by the worn-away part of the pencil eraser.

Preferably, one or more short tabs or lugs (16) are provided on the periphery of the mounting disc (11). These outwardly protruding lugs or tabs prevent the pencil from rolling on a flat surface and are effective even though the flat surface is slightly inclined relative to the horizontal. Also, they tend to prevent the pencil from falling or being thrown out of the pocket. Also they serve as protruding parts with which the fingers can engage when the user is screwing the anchor into an eraser stub (14).

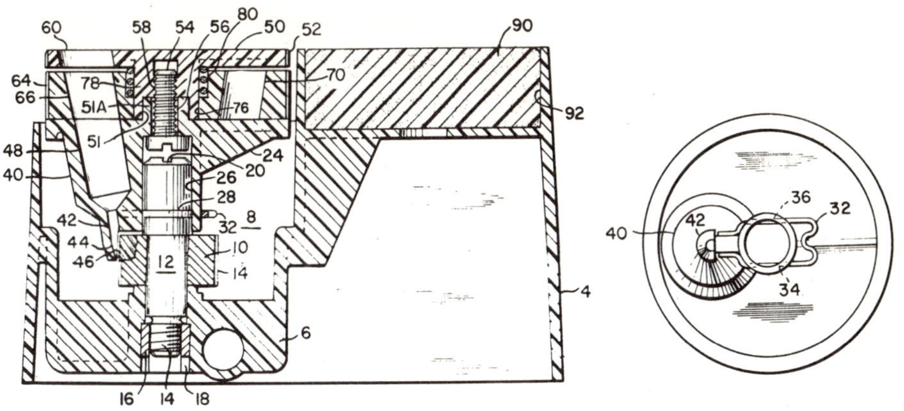

figure 7.6 A vertical section of a pencil pointer, Patent No. 3,952,784 (1976), with a plan of the pencil holder (inset). The pencil pointer has a housing (4) with a base member (6) at the bottom of a receptacle (8) within the housing. A cutter (10) is mounted by a pressed fit on a vertical rod (12), which at its lower end has threads (14) engaged by a hexagonal nut (16) mounted in a hexagonal recess (18). Rod (12) is provided with a screwdriver slot (20).

A pencil holder (24) has a bore (26) which receives the upper end of the rod for the rotatable mounting of the pencil holder. The rod has a peripheral groove (28) which is engaged by a spring clip (32), which is received in slots (34) and (36) in the carrier (24) (inset).

The holder has a guide (40), which has a reduced-diameter portion (42) for guiding a piece of lead. Reduced-diameter portion has a tapered lower end (44) with an opening (46) to expose a piece of lead to cutter (10). The guide has an enlarged diameter portion (48) for receiving the largest diameter lead holder contemplated.

A guide member (52) is fixedly secured to pencil holder (24) by a screw (54) which passes through an opening (56) in the holder and is threaded to the guide member as indicated at (58). The guide member has a hexagonal recess (51) which receives a hexagonal boss (51A) on the holder. The guide member has an opening (60) which is coaxial with guide (40).

A guide member (64) between holder (24) and guide member (52) has a plurality of openings which are adapted to register with opening (60) and guide (40). Opening (66) has a diameter equal to the diameter of enlarged portion (48) of guide (40), and other openings have successively smaller diameters. This guide member has a hub (76) which rotatably engages the exterior of boss (78) on guide member (52).

A compression coil spring (80) surrounds boss (78) and abuts against the lower face of member (52) and against hub (76) to urge guide member (64) downwardly against holder (24). The guide member has downwardly extending projections spaced 90-degrees apart and adapted to engage recesses (84) spaced 90-degrees apart on the upper face of carrier (24).

A piece of foamed resin (90), for example, foamed polyethylene or foamed polypropylene, is positioned in a cavity (92) in housing (4) to provide for the cleaning of the lead after it has been pointed.

ufacture and simple to attach. Perhaps the next step toward making lead pencils more economical would be the development of a secure holder for the rather large nubs that most of us throw away. Depending on the size of a person's fingers, up to one third of a sharpened pencil may end up in the trash can. For sharpening pencils, a new product is illustrated in Fig. 7.6.

The ideal mechanical pencil has long been sought but has never been found. In fact, mechanical pencils are probably less popular now than they were forty years ago. In 1976, two Japanese inventors, Hidehira Kageyama and Takahika Suzuki, were granted a U.S. patent for a push-button pencil that incorporated two notable improvements over pencils of the past—first, its new design made the lead less prone to split and break under the pressure of writing; second, its resilient construction made writing softer and more comfortable (Fig. 7.7).

Soft-tip (felt-tip) pens and markers are the fastest growing segment of the writing instrument business today. They used to be made primarily for artists and illustrators, but in recent years more people have become interested in adding a little flair to their handwritten notes. Generally speaking, felt tips are less messy than ballpoints, too, although there are undoubtedly some ballpoint manufacturers who would think that statement little short of treason. A good many patents have been issued in the felt-tip area, several to Japanese connected with the Pentel Company. One of the most recent, shown in Fig. 7.8, introduces an improved nib. Made from a synthetic resin, this nib is constructed with a core and a number of subcore elements in such a way that the ink can flow to its point without any dripping, splashing, or clogging.

Is there anywhere to go from here? The answer is undoubtedly

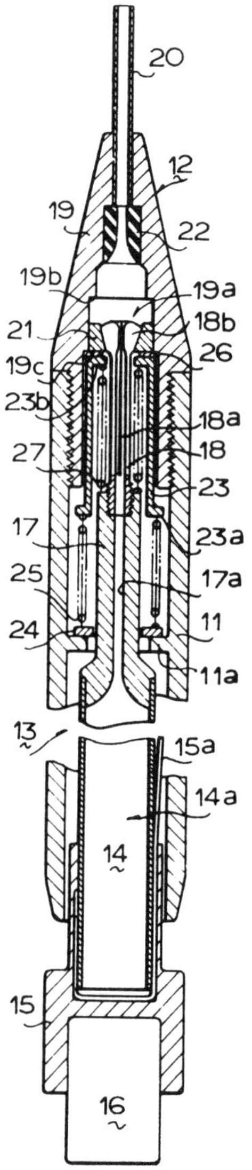

figure 7.7 A longitudinal, sectional view, partly cut away, of a propelling mechanism for an automatic pencil [Patent No. 3,947,133 (1976)]. The numeral (11) generally designates a holder or outer tubular member having an annular projection (11*a*) on the inner surface thereof. The holder is formed of a metal or plastic material, and on the front end there is detachably mounted a guide (12). In addition, the tubular member encloses a lead-holding means (13) in the interior thereof.

This holder includes a tubular housing (14) forming a chamber (14*a*) for storing one or more lead rods or pieces. A cap (15) is attached to the rear end of the housing, and has an eraser (16) extending out from the rear end. The eraser, in effect, forms a push button. Preferably, a resilient wire member or spring (15*a*) is provided, which has a pair of portions acting to resiliently engage the end of the housing, so as to prevent the cap from slipping off. One of the portions extends from the end of the cap to provide a pushing portion for the end.

An inner tubular member (18) is connected to the front end of the housing through a connecting member (17), which is coaxial with the outer member (11). The interior of the inner member (18) is connected to the chamber (14*a*) via a path (17*a*) formed in the connecting member (17) for receiving a single lead rod supplied from the chamber (14*a*). Furthermore, the inner member has a plurality of slits (18*a*) extending in the axial direction from the front end to the central portion thereof to form a plurality of chucks (18*b*), which are biased outwardly by the resiliency of the member to form a space through which the lead rod projects.

The guide means (12) includes a tubular holding member (20) securely fixed within the top of member (19). The holding member is arranged coaxially with the inner member (18), so that the lead rod in the inner member can project through member (20), and provide a protruding portion for writing. The tubular member has a cylindrical recess (19*a*) provided for movably receiving the chucking portions (18*b*) and a ring (21). The front end of the recess (19*a*) forms a stopper for restricting the forward movement of the ring but allows for forward movement of the chucking portions. A friction ring (22) provides the proper frictional engagement with the lead rod to control the axial movement of the lead rod.

In the cylindrical space defined by the inner tube (18), the outer tube (11), and the tubular member (19), there is enclosed a generally cylindrical or annular member (23) having a pair of flat surfaces (23*a*) and (23*b*) at both ends. Member (23) is movable relative to the outer and inner members (11) and (18), and is biased in the forward direction by a spring (25), which is placed between a washer (24) abutting the projection (11*a*) and the first flat surface (23*a*). In the ordinary state, member (23) is pressed into engagement with an annular surface (19*c*) formed at the rear end of the recess (19*a*) via a washer (26). An additional spring (27) is placed between the second surface (23*b*) and the front end of the connecting member (17), and acts to press the inner member (18) and the other members integral therewith backward against the movable member (23). The spring force from the spring (27) provides the chucking force by movement of the inner member relative to the ring (21).

The forward movement of the lead rod, registered within the chuck (18), is accomplished by pushing the button (16) forward. This causes the holding means (13) to move in the same direction with the lead being chucked and the ring (21) frictionally coupled with the chucking portions (18*b*), until the front end contacts with the first stopper (19*b*). Accordingly, the lead rod is kept in the chucked condition and moves forward by overcoming the friction between the lead rod and the friction member (22). The ring (21) stops its forward movement upon contacting the first stopper (19*b*), while the inner member continues to move forward. As the result, the chucking portions (18*b*) are released from the chucking engagement with the ring and moved outward due to their resiliency, thereby allowing continued movement of the lead rod in the axial direction.

After the push button is released from the pushing operation, the holding means moves backward under urging of the spring (27). However, the lead rod is kept in position by friction between the lead rod and the friction member. The lead rod is again chucked when the inner member returns to its original position, where the chucking portions (18*b*) are pressed inward by the ring, which is pushed forward by the washer (26). The position of each of the movable members is the same as they were initially with the exception that the lead rod has been moved forward by one step length. Accordingly, one can move the lead rod the requisite length by repeating the pushing and releasing operation. When the push button has been pushed continuously, the leads can be moved forward or backward, since they encounter only a slight friction.

When writing, a larger force is impressed on the outer member (11), in order to push it against the surface of the paper. A counter force will act to push backward the lead rod and the holding means (13). Although the holding means is held in a stable position, restricted by the washer (26) and the ring (21), it is moved backward by compressing the spring (25), which pushes the chucking ring (21) through the movable member (23) and washer (26), when the lead is subjected to a force acting backward. After the large force is removed, the holding means will return to its original position by the action of the spring.

As has been stated, in propelling the lead according to this invention, the lead rod and the holding means will move backward, when subjected to a force pushing the outer member toward the surface of a paper sheet. Accordingly, the lead rod does not receive a force greater than a predetermined value corresponding to the righting movement of the spring, thereby eliminating breaking of the lead rod. In addition, the backward movement of the lead rod, i.e., the decrease in the projecting length of the lead rod from the top of the holding member (20), provides a good writing feel.

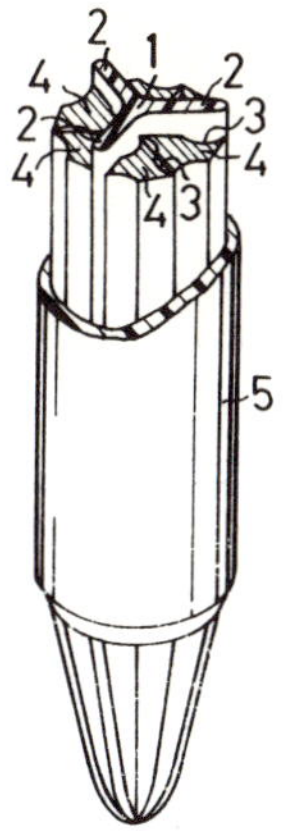

figure 7.8 A partially cut-away perspective of the pen nib described in Patent No. 3,830,576 (1974). In the drawing, a core member (1) comprises a number of legs (2) radially extending with substantially equal angles, which also extend longitudinally. Each of the legs has a substantially constant thickness from a center body portion to an end portion, and is made of synthetic resin. The legs may be three, as shown, or six. Thickness and length of the legs are about 0.05 to 0.3 mm and 0.1 to 1.0 mm, respectively.

Though the legs have smooth surfaces on their longitudinal sides, they may be provided with a number of minor grooves to facilitate ink capillary action. And though they have smooth surfaces on their longitudinal sides, they may be provided with a number of parallel minor grooves.

Each subcore element (4) made of synthetic resin has a substantially triangular prism shape, each triangular side of which is provided with groove or grooves along the longitudinal direction thereof. The subcore elements are positioned between the adjoining legs (2), and may be preferably positioned in pairs between the adjoining legs.

A tubular shroud (5) encloses the outer peripheral surfaces of the legs (2) of the core member (1) and the subcore elements to form an integrally combined body. Ink-feeding channels (3) are formed or left between the legs and the subcore elements, between the tubular shroud and the subcore elements, and also between adjacent pairs of subcore elements when that preferred structure is employed. The tubular shroud is a thin-walled tubular housing which is as thick as 0.2 to 0.5 mm and is made of synthetic resins such a polyvinyl chloride, polyethylene, or polyasetal.

Such materials enclosed by the tubular shroud as described above are cut a predetermined length to form a tapered writing tip portion terminating at a writing point.

The core member (1) has a relatively small area at the writing point compared with the known writing instrument, which has the star-shaped core member. The cross section of the writing point to make contact with writing surfaces includes parts of the subcore elements adjacent to the center body portion of the core member. Therefore, the capillary ink-feeding channels are still formed at the writing point between the center body portion of the core member and the parts of the subcore elements. Moreover, as the subcore elements (4), each provided with a plurality of grooves, are confined by the core member (1) and the tubular shroud, when a writing pressure enough to bend the flexible nib is applied to the writing point of the nib, the capillary ink-feeding channels between the relatively rigid core, the relatively flexible subcore elements, and the shroud are somewhat varied to allow the ink to move toward and away from the central body portion of the core. Thus, the ink flows transversely from the ink-feeding channels to the other channels and is homogeneously supplied to the writing point. Furthermore, as the capillary ink-feeding channels are varied by slight movement of the subcore elements due to flexion thereof, the subcore elements function for cleaning of the capillary channels by themselves so as not to blockade the latter.

Furthermore, each side of the legs (2) having a constant thickness holds firmly the subcore elements there-between. Therefore, ink never splashes even when the tapered tip portion of the subcore elements contacts paper or the like, and smooth writing is insured.

Yes, though there is less user dissatisfaction with this kind of writing instrument than with most others. There is in fact room for some kind of improvement in almost anything. Even a paper clip can be redesigned so that it can better fulfill its function. Figure 7.9 shows part of the patent for a reversible locking paper clip invented by John Langwell of New York. It is effective, easy to use, and inexpensive to manufacture. It is also difficult to describe. The entire Detailed Description from the patent is quoted in the caption as a good example of how complicated, involved, and even convoluted some of the wording in patents can be. It's almost like a foreign language. Most inventors find that they need an interpreter, their patent attorney or practitioner, to describe their invention in terms that satisfy the patent examiners and at the same time protect against infringement.

Besides paper clips, there have been many other devices made to hold papers together—file folders, three-ring binders, clipborads, and an array of stapling machines. Now, it's easy enough to slip papers into a folder or to tuck them under the metal jaw of a clipboard, but getting them into a ring binder requires an accessory, some kind of hole punch. For years, the most common device has been the spring-operated train conductor's ticket punch. Even this old workhorse wasn't perfect, however, and in 1968 a patent was granted to James Hendricks and Spencer Wernham of Illinois for an improved version (Fig. 7.10). The inventors improved the catcher for the punch by (1) making it of a transparent plastic so the user could see through to *exactly* where the hole would be punched,

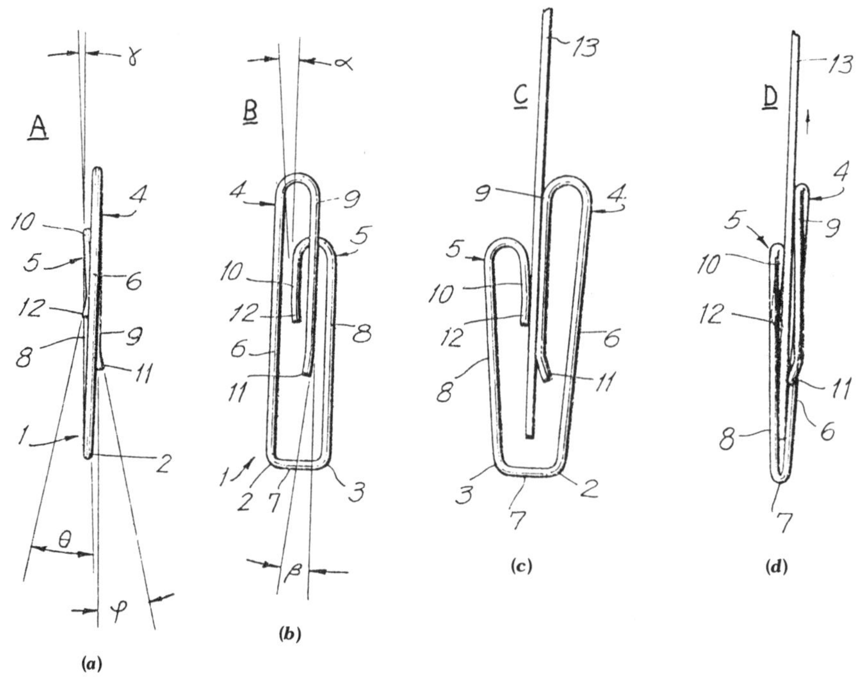

figure 7.9 Four views of a reversible locking paper clip, Patent No. 3,840,943 (1974). From left to right: (*a*) front elevation; (*b*) side elevation; (*c*) side elevation with sheets of material being held by clip; (*d*) side elevation with clip in fully reversed position.

As seen in the drawing, the clip is generally designated by numeral (1) and is constituted by a single length of wire which is formed with right angle bends (2) and (3) from which extend U-shaped sections (4) and (5).

The sections overlap one another. Section (4) includes a leg (6) which is perpendicular to base (7) and integrally formed therewith at bend (2), while section (5) includes a leg (8) perpendicular to base (7) and integrally formed therewith at bend (3). Section (4) has a free leg (9), and section (5) has a free leg (10). Section (4) is considered to define a plane of the clip as seen in (*a*).

At the free end of leg (9), there is formed a bent portion (11), and at the free end of leg (10) there is formed a bent portion (12). The bent portions extend upward at an angle to the plane of the clip as shown in (*a*) so that the bent portions are divergent. In the plane of the clip, the bent portions are convergent as shown in (*a*). Hence, each bent portion is inclined in two directions with respect to its associated leg.

Section (4) has a greater length than section (5), and accordingly leg (9) has a greater length than leg (10). Leg (9) extends approximately midway between legs (8) and (10), and leg (10) extends approximately midway between legs (6) and (9).

When the clip is in the configuration as shown in (*a*) and (*b*), it is a simple matter to insert sheets of paper (13) or the like between the sections (4) and (5) in much the same manner as a conventional paper clip, the sheets being inserted downward into the clip in (*a*) and (*b*). The divergent relation of bent portions (11) and (12) permits unobstructed entry of the paper between sections (4) and (5).

After the sheets (13) have been at least partially inserted between the sections (4) and (5) of the clip, the sheets can be locked in the clip by turning the clip through an angle of 180 degrees so that the clip is reversed. This is achieved by holding the base (7) and turning the clip in a direction to raise leg (6) above the plane of the drawing while lowering the leg (8) below the plane of the drawing. When the clip has been turned 90 degrees, it occupies the position as shown in (*c*) when viewed from the side. Herein it can be seen that, due to the angular relation of the bent portions (11) and (12), they offer no obstruction or resistance to the turning of the clip up to this point. After the clip has reached the 90-degree position, it "springs" into the fully turned position under its own resilience. After the full 180-degree turn of the clip, it occupies the position as shown in (*d*) and herein it can be seen that the bent portions (11) and (12) face into the sheets. As a result, they lock the sheets and securely prevent movement of the sheets in the direction of the arrow in (*d*). However, the bent portions offer relatively little resistance to further introduction of the sheets into the clip in a direction opposite to the arrow.

In the course of turning the clip, the sections (4) and (5) initially turn or pivot on legs (9) and (10), respectively, as fulcrums. Due to the parallel relation of the legs (6), (8), (9), and (10) and the outward direction of the bent portions (11) and (12), the force applied to the legs (9) and (10) during turning is applied substantially uniformly along the length of the legs. This is insured by the relative positions of legs (9) and (10), leg (9) being midway between legs (8) and (10) and leg (10) being midway between legs (6) and (9), and also by the location of leg (10) centrally along the length of longer leg (9). As the clip is being turned, the legs (9) and (10) apply counterbearing forces, and due to the proximity of legs (9) and (10), these forces have virtually no tendency to laterally shift the face sheets of the stack, as is experienced with conventional reversible clips where the fulcrums are individual points or where the fulcrums are laterally spaced by a substantial distance.

The spacing between the free ends of bent portions (11) and (12) as measured longitudinally along the clip is advantageous in that in the locked position of the clip, the sheets are gripped at longitudinally offset positions which laterally stabilize the clip on the sheets and resist lateral misalignment of the sheets.

The perpendicular relation between the legs (6) and (8) and base (7) not only enables the clip to clear a thicker stack of sheets, but it also provides relatively rigid corners (2) and (3), which rigidifies the clip in this region and insures turning of the clip around legs (9) and (10), especially when the sheets are made of heavy material and are stacked to a great depth.

It is an important feature of the invention that the bent portion (12) of short leg (10) has a smaller angle of divergence as compared to the angle of divergence of leg (9) so that the sharp edge of portion (12) will bite into the sheets before the edge of portion (11) will do so. The angle of divergence is so adjusted that the edge of bent portion (12) will grip the sheets approximately when the clip reaches its 90-degree position of turning.

Upon further turning of the clip, the leg (10) will serve as a primary retaining leg and leg (9) will slide with respect to the edge of bent portion (12) which now engages the sheets. The leg (9) undergoes a positive spring back to the locking or free position, depending on the direction in which the clip is turned, and assures a perfect and positive positioning and locking of the clip every time.

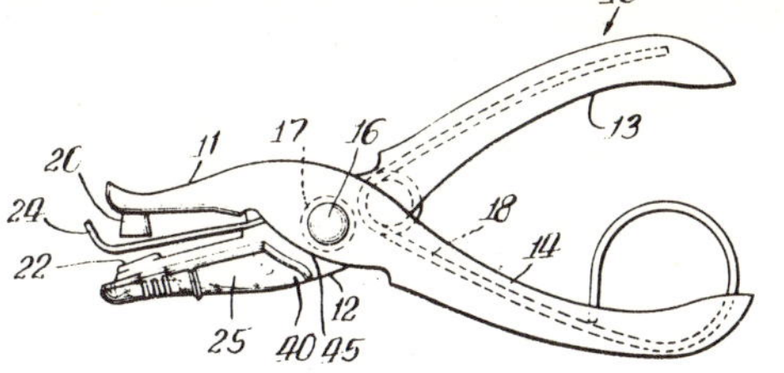

(a)

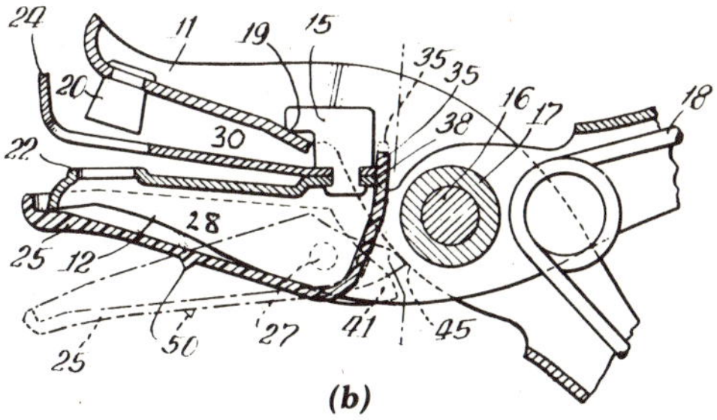

(b)

figure 7.10 Two views of an improved paper punch, Patent No. 3,392,447 (1967)—side elevation and cross section. The punch (10) includes a male jaw (11) and a female jaw (12), each of which has appropriate handle portions (14) and (13), respectively. The jaw pivots about a pintle (16) having a central sleeve (17). An appropriate spring (18) is provided between the handles (13) and (14). The male jaw has a die (20) that is staked thereto and the female jaw has a die seat (22), so that relative movement of the jaws causes the die to fit into the seat to punch sheet material placed between the jaws. A tongue (24) between the jaws acts to clear the paper from the die after the punching action and is staked to the female jaw by post (15). This post has a hooked portion (19) that forms a stop against which the spring (18) biases the male jaw.

A transparent catcher (25) is pivotally connected to the female jaw by snapping the two pimples into dimples on either side of the jaw. The walls (30) and (31) of the catcher (25) are tapered slightly inward so that they have to be forced apart to fit over the female jaw. Therefore, when the pimples (27) and (28) snap into the dimples, the walls press tightly against the walls of the jaw to hold the catcher firmly in place.

A type of leaf spring (35) is formed from an integral portion of the catcher (25). With the catcher in place on the jaw (11), the leaf spring rests against the wall (38) of the jaw and the tongue (24). The portion (35) not only acts as a leaf spring, but when the catcher is mounted to the female jaw, it functions to close off the back of the catcher to provide an enclosed compartment to receive the perforated material subsequent to the punching action.

In operation, the material to be punched is placed between the jaws. Because the catcher is transparent, the punch operator is able to view the seat (22) for the die (20) through the bottom of the catcher. This enables him to accurately judge the spot on the sheet that is to be perforated. The rib (50) and the groove portions also serve as gauges to determine the proper distance for inserting the sheet material into the jaws of the punch prior to commencing the perforated action, and may be viewed by the operator with the sheet material inserted in the punch.

The spring (35) contacting the wall (38) of the female jaw acts to bias the catcher into a closed position against the bottom of the jaw. Therefore, when the material is perforated, the waste material falls into the catcher. When it becomes necessary to empty the catcher, the operator using minimum force and effort simply pivots the catcher about the pimples (27) and (28) to an open position against the bias of the spring. The fins (40) and (41) extending from the catcher butt up against the walls (45) and (44) of the female jaw and act as stops—dotted lines in (*b*). Should the operator continue to apply pressure in an attempt to move the catcher past the limit of the open position, the force will cause the pimples (27) and (28) to jump out of the dimples in the jaw, thereby disconnecting the catcher from the jaw. This prevents the operator from inadvertently applying so much pressure to the catcher that he fractures the spring.

---

and (2) designing it so that clamping and unclamping would be easier, eliminating the problem of sticking.

Another idea for a punch is to combine it with a ring binder. This doesn't represent any startling technological breakthrough, but it is something new and original, more convenient and useful than what had existed before. This gadget has gained popularity, especially among students and others who want to put new sheets in place in the binder as soon as they receive them (Fig. 7.11).

Many other combinations of devices have been tried, some with success. Many more are still to be worked out. Generally, such mergers are most successful between things that are generally used together—like the hole punch and the binder—but there is room for new and original combinations, too. In Chapter 13, Tools, there are several examples of combinations that have worked very well.

"Let's get organized!" Most people, on hearing those words from a spouse or parent, boss or colleague, run for the nearest shelter. They know that getting organized often involves several hours, days, or even weeks of chaos. Inventors have often tried to help the hapless, sometimes with good results. Desks seem especially liable to the disease of disorganization. The nineteenth-century way of meeting this situation frequently took the form of rows of cubbyholes, like those found in rolltop desks. Ideally, cubbies offer a place for everything. Practically, they are often cumbersome and seem to attract all kinds of papers and doodads that really should have been tossed in the wastebasket.

Many desk organizers have been patented and manufactured. Figures 7.12 and 7.13 illustrate two different inventions that can

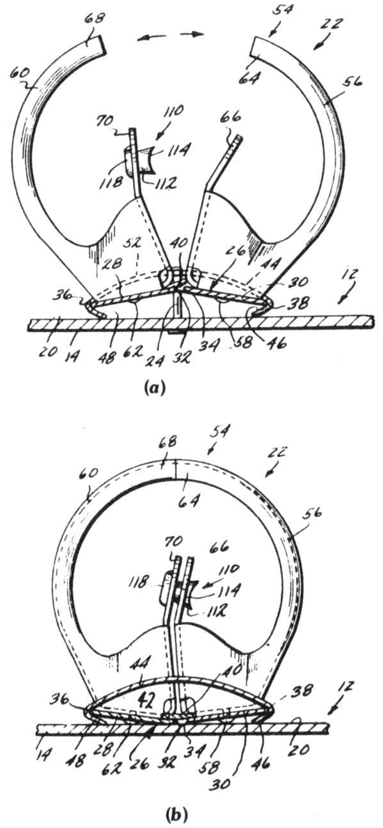

figure 7.11 Two cross sections of the loose-leaf binder-punch combination described in Patent No. 3,349,774 (1967). At left, the combination is open; at right, it is closed. A binder case (12) has a back panel (14) and covers that are made up in the customary way. The back panel includes an inside face (20) to which a loose-leaf binder and paper punch (22) may be affixed by rivet (24). The binder and punch comprise a hinge (26), including a pair of hinge plates (28) and (30) with edges (32) and (34) and outer edges (36) and (38). Tabs (40) on edge (32) engage with edge (34), and tabs (42) on edge (34) engage with edge (32). The plates (28) and (30) may be maintained in the substantially horizontal position shown at left or in the over-center position shown at right by a cover plate (44), which is substantially C-shaped and includes one lip (46) engaged under the edge (38) and a second lip (48) engaged under the edge 36.

Each ring and punch pair (54) is affixed to the plate (30) by rivets 58 and 62. There are paper retainers (64) and (68), a female paper-punch die (66), and a male paper-punch die (70). When rings (56) and (60) are swung from the position shown at left to the position shown at right, the male die engages the female die for punching holes in notebook paper (not shown). This paper may be bound by engaging the punched holes over associated paper-retainer elements (64) or (68) and then closing the ring.

An important feature of the catcher structure is the fact that upon moving the catcher from the closed position to the open position as shown in (*b*), the spring portion thereof tends to move upward past the wall of the jaw so that there is actually very little flexing of the spring. This greatly reduces the chances of fracturing the spring.

The first unitary ring and punch member (56) includes an integral paper-retainer element (64) and a female paper punch die member (66); the second unitary ring and punch member (60) includes an integral paper-retainer element (68) and a male paper-punch die member (70). The members (56) and (60) are aligned and are mounted on the plates (30) and (28) in such a manner that each pair of unitary ring and punch means (54) may be swung from the position shown in (*a*) to the position shown in (*b*), wherein the male paper-punch die member (70) engages the female paper-punch die member (66) for punching holes in notebook paper (not shown). The ring and punch means (54) may then be swung to their (*a*) positions so that the paper may be bound by engaging the punched holes over associated paper-retainer elements (64) or (68) and then closing or swinging the means (54) to their (*b*) positions.

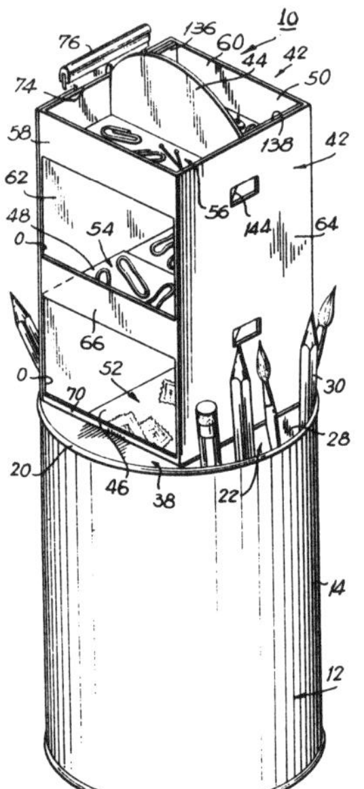

figure 7.12 Front perspective view of the desk organizer described in Patent No. 3,332,688 (1967), with the elevator stack of drawers partially extended. It has three components: a hollow cylinder (12) with a closed bottom and open top; a box, or filler (22) to fit into the can and subject to being folded so as to be shipped flat; and an elevator stack of drawers that is also foldable. The filler provides pockets (28) for such thin, elongated items as pens and pencils and also provides a "shaft" for the drawers. These latter have a handle so that they can be easily lifted. The drawers are designed to hold small items such as paper clips, stamps, and rubber bands.

transform a desktop into a smooth plateau of efficient operation. The first one shows how a comparatively simple idea can be developed to the point where it is indeed patentable (and salable). Careful attention to detail, and a good knowledge of what the average desk generally holds, went into the planning of this invention. Irving Smith of New York had in mind the premium market when he designed his organizer. The idea was that a company whose product was sold in cans could offer to its customers the drawers and filler described, in the proper size and shape to fit inside the can. This is a profitable and useful form of recycling.

Figure 7.13 illustrates how a good product—the rotary desktop file—can be improved upon so that it's possible to market it as a totally new concept. Walter Dulski of Connecticut planned his invention so that the user would be able to take out and replace file cards quickly and easily. The main feature is that a large number of file cards are in sight and accessible without pulling out drawers or going through three big boxes. Even the line illustration gives the impression of smooth efficiency.

Another desktop filing device that will be found on sale in many stationery stores is illustrated in Fig. 7.14. Here the primary consideration for the inventors, Richard McFadyen and James Giltzow of New Jersey, was the overall appearance of the device. They wanted to put a great deal of information—generally, names and phone numbers and perhaps addresses—on cards to be held within a slim container. The names and numbers can be placed in alphabetical order and thus available at the touch of a fingertip. The automatic opening device is kept simple. Cards can be added, removed, or changed without difficulty. This is a popular kind of gadget, one that can be worked on anew with profit.

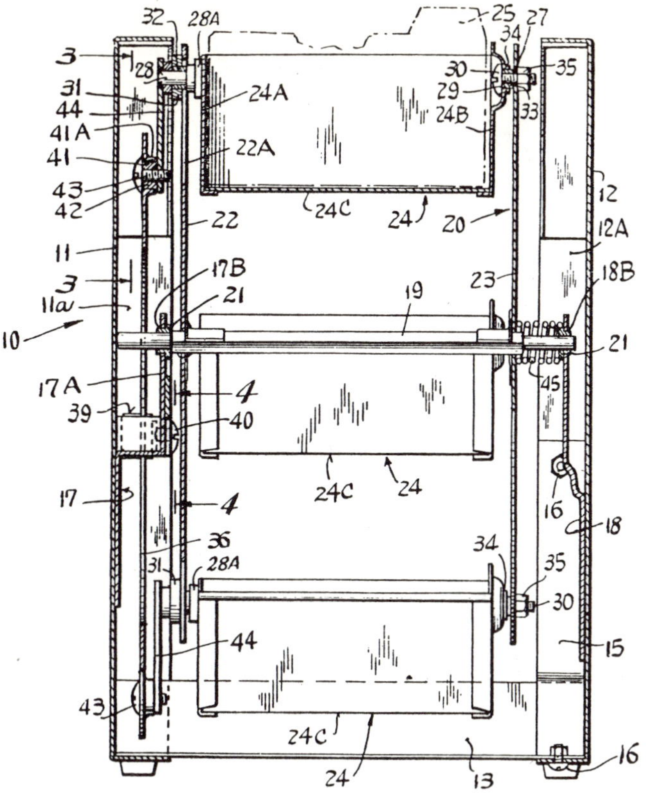

figure 7.13 Sectional view of the rotary card file described in Patent No. 3,231,096 (1966). It is comprised of a pair of substantially similarly constructed walls, members (11) and (12) formed of suitable sheet material, as, for example, metal or plastic or the like, which are maintained in spaced-apart relationship by a pair of front and rear connecting members. They are similar in construction but opposite in hand. Each sidewall (11) and (12) is defined by a panel which tapers upward and inward having inturned flanges (11*a*) and (12*a*) circumscribing the peripheral portions. Accordingly, the width of the base portion of the respective sidewall panels (11) and (12) is greater than that of the top portion.

Connected to the inside surface of the side walls (11) and (12) is Z-shaped bracket support (17) and (18). These supports define the end bearings for the shaft (19) of the rotary (20). As shown, upper leg portions (17*a*) and (18*a*) of the respective Z-shaped brackets are spaced from the adjacent sidewalls and are provided with aligned openings (17*b*) and (18*b*) in which a suitable bearing (21) is journaled for receiving the end portion of the shaft.

The rotary means comprises a pair of similar, spaced apart supporting or drum discs (22) and (23), which are keyed to a shaft (19), the ends of which are rotatably journaled in the bearings (21) of each of the supporting brackets (17) and (18). The shaft is substantially rectangular in cross section with the extremities rounded and received within the end bearings (21) in the supporting brackets. One end of the shaft is sufficiently long so that when it is inserted through its associated bearing, it will thrust against the sidewall (11) of the adjacent housing to locate the rotary means (20) between the sidewalls. The other end of the shaft is supported in its bearing (21), spaced from the adjacent sidewall (12) of the housing. The engagement of one end of the shaft against the adjacent sidewall thereby limits or positively positions the rotary means between the respective sidewalls of the housing.

In accordance with this invention, a plurality of trays (24) adapted to loosely retain therein the card files (25) or other file material, are pivotally connected to and between the spaced-apart rotary drum discs. In the illustrated form of the invention, four such card trays (24) are equidistantly spaced about the circumference of the respective drum disks. However, the number of card trays may be varied as desired.

One sidewall (24*a*) of the respective trays has a pin (28) and a bracket (28*a*) connected there to define a pivotal support for the tray. Directly opposed thereto, the other sidewall (24*b*) is provided with a punched-out aperture (29) adapted to receive a threaded bolt (30) to define the opposed pivotal support for the tray (24). The pivot pin (28) fixed to one side of the respective trays is extended through the elongated slot, whereas the other pivotal connection, that is, the threaded bolt (30), is extended through its respective apertures (27) in disk (23). To provide a bearing for the fixed pivot (28) of the respective trays, a bearing plate (31) is secured to the outer surface of the disk (22), and is provided with an aperture for receiving the fixed pivot pin (28). A bearing sleeve (32) is interposed in the aperture of plate 31 for receiving the pin of the tray.

The other end of the tray is pivoted by the pivot bolt (30), supported in a bearing (33) disposed in the flare (34) defined by the punched out aperture (29), and a nut (35) threaded onto the bolt to secure it to the drum disc. It will thus be noted that each of the trays (24) is pivotally connected between the spaced drum discs (22) and (23) of the rotary means.

The control ring (36) is provided with four equidistantly spaced, punched-out apertures (41) in which the punched-out or peripheral portions of the apertures flare inwardly at (41*a*). By punching out the holes, a simple mechanical expedient is provided whereby such rings can be mass produced. Also, the punched-out or flared portions of the respective apertures provide a support for retaining therein a bearing (42) for the screw fastener (43), which is adapted to pivotally connect the free end of the respective links or cranks (44) to the control ring. The construction is such that the connecting links or crank arms (44) are fixed to pin (28) and are maintained in an upright position throughout the rotational revolutions of the control ring. Also, the axis of rotation of the control ring is offset downward with respect to the axis of rotation of the drum discs (22) and (23), a distance equal to the distance between the center line of the pivots (28) and (43) connected to the respective ends of the crank arms. With the construction described, it will be noted that the disks (22) and (23) are rendered freely rotatable about the shaft (19), and the rotation is such that the opened tops of the trays (24) are maintained in an upright position throughout the rotation of the drum disc. Thus, the cards or card files (25), which are loosely retained in the respective trays, will not be upset therefrom as the respective trays are rotated. The cards, being loosely retained within the respective trays, can be readily replaced or removed without interfering with or affecting the remaining cards.

To resist counter rotation of the drum disks and carrier trays connected therebetween, and to resist the influence of gravity acting thereon, a means is provided for resisting such rotation. This antirotation means comprises a compression spring (45) which is interposed about one end of the shaft (19) so as to be compressed between the drum disk (23) and the bracket (18) supporting the shaft. Accordingly, the spring (45) exerts a bias on the drum that will tend to resist rotation thereof.

An important feature of this invention further resides in the provision that any tray of suitable size may be interspaced between the end walls of the housing without materially affecting the construction of the card file component parts. In other words, the many component parts described can be standardized except for the tray length. Therefore, by merely varying the length of the shaft about which the respective drum discs are supported and positioning a tray of the desired length therebetween, the card file (10) can be constructed and arranged in any of several sizes with a minimum of structural changes.

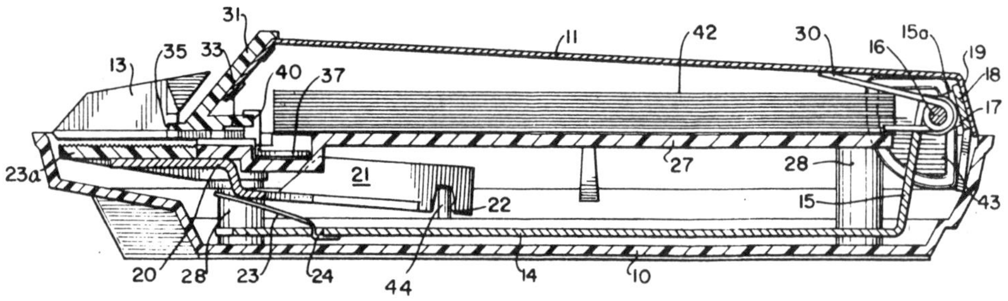

figure 7.14 Sectional view of the desk index described in Patent No. 3,405,466 (1968). The apparatus of the invention comprises a lower housing section (10) and an upper housing section (11). Housing sections may be made of any suitable material, including plastics, metals, or both, and may generally be of any configuration designed aesthetically to please the user of the apparatus. The metal plate (14) is riveted to the bottom interior surface of the housing (10). The rear portion of the base plate (14) has an upstanding wall (15) whose top edge is above the top portion of the housing. Two portions of the upstanding wall are cut back at right angles to form mountings (15*a*) for rod (16), which extends partway across the housing. Hingedly mounted on rod (16) by means of mountings (17) is a narrow, long, flat attaching member (18), which is fixedly attached by means of rivets (18*a*) to the rear wall (19) of housing (11), thereby providing a pivot for the top housing.

Affixed to rod (16) at the back of the assembly is a spring (30) which engages wall (15) of member (14) with one end of the spring and engages the interior of top housing (11) with the opposite end of the spring. In operation, upon release of the latch, spring (30) raises upper housing (11).

Located within the housing is a set of index cards (42), which are mounted on support (43), which is in turn mounted on rod (16) to hingedly support the cards. The front edges of the index cards are progressively cut away or cut back at different locations, so that when cards are piled on top of each other, a small section of each card is visible to finger (37) without interference from preceding or successive cards, different portions of different cards thus being exposed.

In order to operate the apparatus, indexing pointer (13) is slidably moved across the front of member (31) until it points to the designation of the card that is desired to be viewed. Designations may be permanently or temporarily affixed to member (31). At this point, finger (37) will be located under the card above the card desired to be viewed. Depressible portion (23*a*) is then moved downward, disengaging magnet (25) from the metal clip (32). Spring (30) then forces the upper housing (11) upward, carrying with it those cards above the cards to be viewed. When it is desired to view another card, the top is simply closed and the indexing pointer (13) moved to the location of the new card desired to be viewed. Member (23*a*) is then depressed, causing the card now selected to be viewable.

It's been firmly established by generations of office workers that filing is one of the dullest duties known to man. Of course, electronic data-storage devices have provided a partial answer to this problem by putting on tape or drums or cards all kinds of information that used to be put on paper and then stashed in the filing cabinet. With the computer, much of this information is retrievable in seconds. So far, so good, but in many businesses it's still necessary or more economical to process certain kinds of paperwork by hand. Not many labor-saving devices have been developed in this area, though the potential market is surely there and eager to be helped. Charles E. Jones of Illinois patented one new approach to the problem of presorting in alphabetical or numerical order material that will be stored in a file or other area for future reference. His is not a very complicated setup, really, but by making for an easy flow of paper into clearly designated pockets, it makes sorting less tedious and time-consuming—and that's just the kind of thing that inventions are supposed to accomplish.

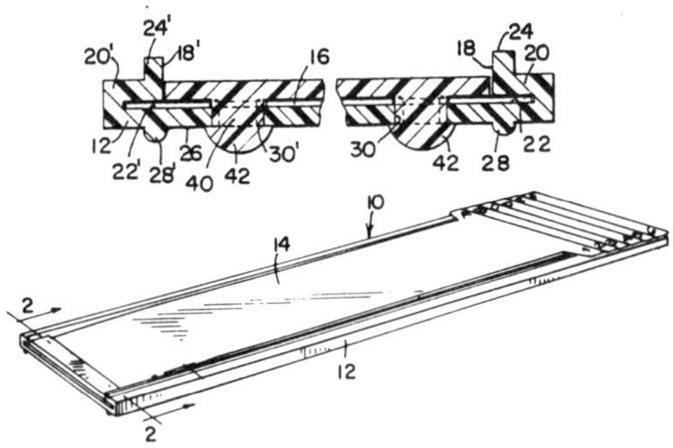

figure 7.15 Perspective view of a file sorter and (inset) a cross section, enlarged, taken on the line 2–2 of the main drawing. Both illustrations are from Patent No. 3,876,077 (1975).

The sorter (10) is in the form of an elongate tray (12) with an assembly (14) of material separating strips, which are mounted thereon in hinged relation for swinging movement, like leaves in a book, so that correspondence or similar sheet material may be interposed between the separating strip or leaf members in the assembly with each of the members having, at its free end, a marginal portion which extends a short distance beyond the corresponding margin of an overlying leaf member for accommodating letters of the alphabet, numerals, or other information or markings constituting the sorting criteria and identifying the material to be placed beneath each leaf member in carrying out the sorting operation. This allows correspondence or similar material to be separated in an orderly fashion according to markings on the separating strip or leaf members. The assembly (14) comprises only six material-separating strips or leaves, which limits the device to separation in six different groups, piles, or categories. It will be understood that the sorter (10) can be made any desired length to accommodate any desired number of different sorting arrangements.

The recess (16) in the top face of the base (12) extends the full length of the base member and has open ends with oppositely disposed vertical side faces (18) and (18′), which are the inner faces of relatively narrow, parallel side portions (20) and (20′) forming side edge guides for the assembly (14) of material separators. At the bottom of the edge guide faces (18) and (18′) slots or slits (22) and (22′) are provided which extend laterally and longitudinally of the recess (16) and which are adapted to receive anchoring portions of the material separators in the assembly. The edge guide faces are extended upward a short distance by ribs (24) and (24′). On the bottom face (26) of the base member (12) longitudinally extending, reinforcing ribs (28) and (28′) are provided adjacent the side margins. The base member (12) has a pair of holes (30) and (30′) in the margin at one end and a pair of like holes (32) and (32′) in the margin at the opposite ends of attaching end stop members. These are of identical construction and serve to close the ends of the recess (16) and also provide a means for clamping in position the separating strip or leaf assembly (14).

The mails play a leading role in everyone's paper handling, both business and personal. A good deal of work has been done on the development of more efficient, more workable mailing pieces, especially those used by large organizations. Here again speed and ease of handling are the characteristics most in demand. The "letter package with breakaway item" in Fig. 7.16 was patented by Wilfred Gendron of Massachusetts. Like many others of its type, it is designed to encourage a quick reply from the recipient. Direct-mail advertisers and fund-raising organizations are particularly interested in inventions such as this to make their mass-mailing operations more economical and, they hope, more productive. You probably have noted an increase in newfangled envelopes in your mail recently, especially from banks. Some of them seem to be rather difficult to open and to figure out. Improvement is certainly called for, and will certainly come. If you can figure out a better way to put these mailers together, do it! The smaller companies that manufacture these products are often looking for new items.

Another type of mail that needs to be improved is the interoffice memo. Many harried paper pushers would like to see it disappear

figure 7.16 Three views of a letter package with a breakaway part, as described in Patent No. 3,912,160 (1975). (*a*) a fragmentary front view of an assembly of series-connected letter packages, a portion of one of the letterheads broken away to show an attachment to the rear surface; (*b*) a rear view of (*a*); (*c*) a sectional view taken along line 3–3 in (*b*). An assembly of continuous form or series-connected letter packages embodying the present invention and indicated generally by the reference numeral (10) comprises a group of individual letter packages indicated generally as (12), (12'), connected together in end-to-end relation. Each package includes a letterhead (14) and an associated attachment (16).

The letterheads are formed from an elongated web of sheet material designated generally by the reference numeral (18). Longitudinally extending lines of weakening (20), (20) formed on the web and spaced inwardly from the side edges thereof define the side edges of the various letterheads and separate a pair of longitudinally extending marginal carrier strips (22), (22'). A longitudinal series of transversely extending lines of weakening (24), (24') formed on the web (18) separate each letterhead from a preceding letterhead in the series. Each letterhead has a generally rectangular form and includes a window (26) opening in the address area. A generally semicircular thumb cut-out opening (28) is preferably formed in each letterhead and opens through one side edge thereof near the window (26), substantially as shown. The assembly (10) is particularly adapted to be automatically fed through a machine. Accordingly, a longitudinal series of configurations or feed holes (30) and (30') is formed in each of the carrier strips (22) and (22') for engaging a sprocket or pin feed mechanism of a machine such as an automatic typewriter, computer-printer, or the like.

The attachment (16) is formed from a single piece of sheet material and includes a plurality of integral parts joined together along at least one line of weakening. A typical attachment comprises a generally rectangular patch (32), a rectangular reply item or postcard (34), and a connecting tab (36). The patch which is somewhat larger than the card, is folded into overlying relation with the card and the tab (36) along a fold line (38). A generally U-shaped line of adhesive (40) secures marginal portions of the patch and a portion of the tab to the rear surface of an associated letterhead. The patch cooperates with the associated letterhead (14) to form a pocket therebetween wherein the card is retained by the tab. A line of weakening (42) coincident with one edge of the card separates it from the tab. In assembly, a portion of the frontal surface of the card is exposed in the window (26) to permit the name and address of a recipient to be imprinted directly on the card as the assembly (10) is fed through a computer-printer or a like printing apparatus. After completion of the printing operation, all the individual packages (12) may, if desired, be separated from each other and from the carrier strips by bursting the assembly along the various lines of weakening. If desired, the assembly may be conveniently fan-folded along the various lines of weakening (24) and (24') for separation at some later time. Each card preferably remains connected to its associated letter package until received by an addressee. The thumb cut-out opening (28) enables the recipient to conveniently grasp the outer or exposed edge of the card to separate it from the letter package. The line of weakening or perforation (42) is so formed to permit the card to readily separate from the tab when pulling force is exerted upon the card.

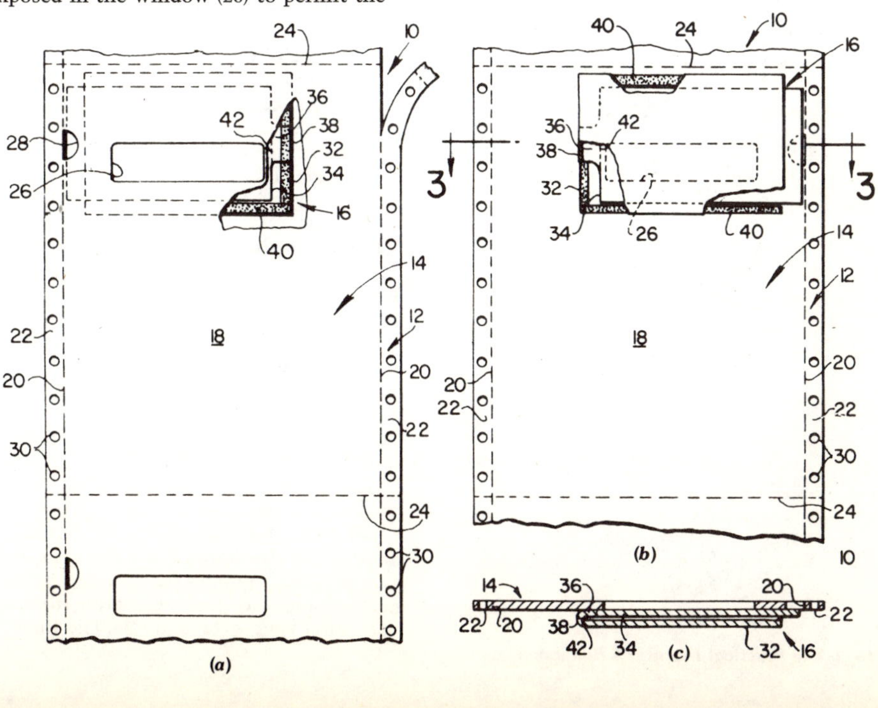

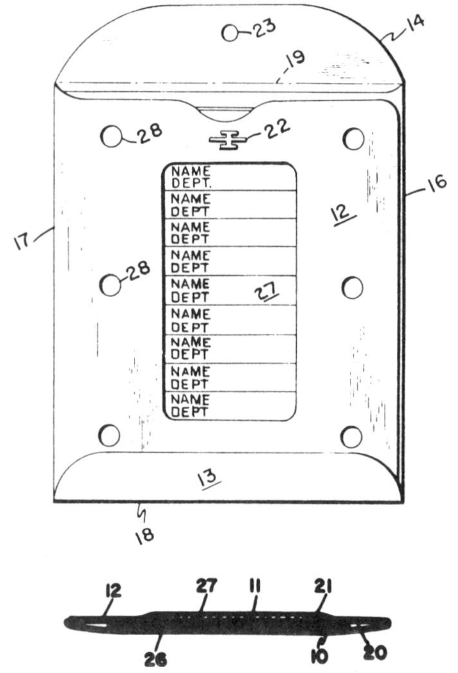

figure 7.17 Perspective view of the interoffice envelope described in Patent No. 3,856,198 (1974), with routing pad in the second pocket. Inset is a cross section of the envelope taken along the lines II–II.

A unitary sheet is scored to provide fold lines which define a rectangular back panel (10), a rectangular intermediate panel (11), a rectangular front panel (12), an end flap (13), and a top flap (14). The front panel has an opening (15). In the assembly of the interoffice envelope, the intermediate panel is folded along score line 16 to overlie the back panel. The front panel is folded along score line (17) to overlie the intermediate panel. The bottom flap (13) is folded along score line (18) to overlie a portion of the front panel and is adhesively secured (25). The top flap (14) is folded along score line (19) to overlie a portion of the front panel.

When the envelope is so folded, two pockets (20) and (21) are formed. The first pocket (20) is bounded by the back panel (10) and the intermediate panel (11) and the second pocket is bounded by the intermediate panel and the front panel. Securing means are provided on the front panel and the top flap so that when the top flap is secured to the front panel, both pockets are closed. The securing means can be in the form of a clip (22) attached to the front panel and a corresponding hole (23) in the top flap. The limits of the second pocket (21) can be controlled by using adhesive to secure that portion of the front panel to the intermediate panel to define the limits of the pocket.

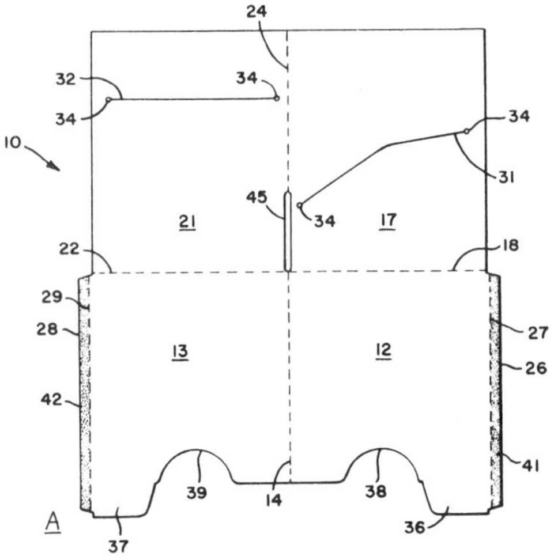

(*a*)

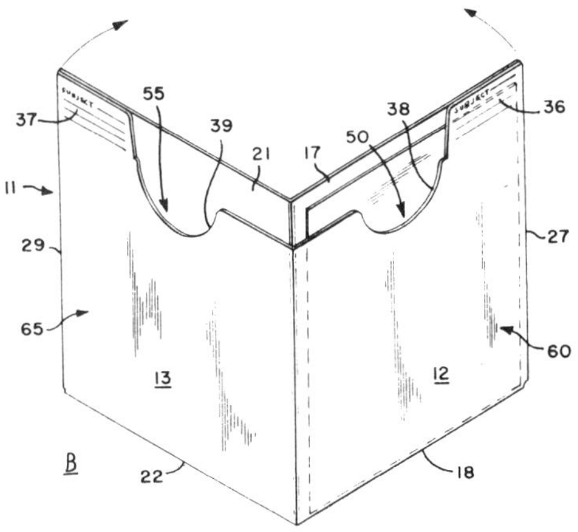

(*b*)

figure 7.18 Two views of the double-wing pocket portfolio and file folder, taken from Patent No, 3,870,223 (1975). (*a*) illustrates the single, unitary blank (10) from which the twin pocket portfolio and file folder (11) in (*B*) is formed. As may be seen, blank (10) is a one-piece blank which includes a front outer panel (12), and a rear outer panel (13) joined by a first center fold line (14). A front inner panel (17) is connected to the front outer panel along a first bottom fold line (18). This line runs along the edge of the front outer panel, which is adjacent and intersects the edge of panel (12), defined by the first center fold line. Similarly, a rear inner panel (21) is connected to the rear outer panel (13) along a second bottom fold line, and to the front inner panel along a second centerfold line (24). The second bottom-fold line forms an extension of the first bottom fold-line (18), and the second center fold line (24) forms a similar extension of the first center fold line (14).

A front flap (26) is joined to the front outer panel (12) on a front flap fold line (27) on the edge of panel 12 opposite the first center fold line. A rear flap (28) is similarly joined by a rear flap fold line (29) to the edge of the rear outer panel (13) opposite the first center fold line.

The front inner panel (17) is partially cut through by a diagonal slit (31), and the rear inner panel (21) is partially cut through by a transverse slit (32). The ends of these slits include round cutouts (34) to resist tearing of the portfolio material at the ends.

The edges of the front and rear outer panels (12) and (13) opposite the bottom fold lines (18) and (22) include respective integral index tabs (36) and (37), as well as generally arcuate recesses or cutouts (38) and (39).

The blank (10) is formed into the twin pocket portfolio and file folder (11) by folding the inner panels (17) and (21) along their respective bottom fold lines (18) and (22) to bring the inner panels into facing relationship with their respective front and rear outer panels (12) and (13). The front and rear flaps (26) and (28) are then folded on lines (27) and (29) around the inner panels (17) and (21) and attached to the inner panels by adhesive lines (41) and (42) on flaps (26) and (28). These flaps and adhesive lines thus join the inner panels to their respective outer panels along the edges opposite the center fold lines (14) and (24).

Next the panels are folded on the first and second center fold lines (14) and (24) to bring the front outer panel (17) adjacent the rear inner panel (21). The center fold lines (14) and (24) serve as hinges for the portfolio, hingedly joining their respective outer panels and inner panels. An inside elongated cutout (45) on the second center fold line (24), adjacent and extending away from the bottom fold lines (18) and (22), promotes the proper and desired folding of the center fold lines (14) and (24) and helps to prevent tenting thereof.

The blank (10) is thus folded twice and adhered at the flaps (26) and (28) to form a front pocket (50) between front outer and inner panels (12) and (17), and a rear pocket (55) between rear outer and inner panels (13) and (21). The slits (31) and (32) then provide access to these pockets from within the portfolio, for reception, filing, storage, and support of papers, notepads, envelopes, and so on.

The front and rear pockets (50) and (55) are open at the top, and the front and rear outer panels (12) and (13) are shorter than the inner panels (17) and (21). That is, the front and rear inner panels are longer along the second center fold line, (24), than the front and rear outer panels along the first center fold line, (14). These panels thus define respective exterior front and exterior rear file folders (60) and (65), with the shorter front and rear outer panels (12) and (13), providing ready access to the file folders (60) and (65). The utility of the file folders is further enhanced by the cutouts 38 and 39, which facilitate access to the folders, and by the integral index table (36) and (37), which index the file folders, if desired.

altogether, but that's a false hope. In the meantime, Robert Daley of Massachusetts has designed a new interoffice envelope that eliminates the old standard format of scratching out the old address and writing the new one in the next available space. The envelope in Fig. 7.17 has two pockets, one for whatever material is being carried, the other for a routing pad on which the name and address of the recipient can be written. Since there is generally no writing on the envelope itself, it should last a great deal longer than other types of interoffice carriers.

Many people carry their desk materials—or part of them—in their pockets or briefcases. Walking offices can be a bit impractical, however, and unless one enjoys groping through a jumble of mixed-up papers, it's important to have some method of organization. Figure 7.18 illustrates the patent granted to John Wyant of Georgia for a pocket portfolio that can be cut and folded from a one-piece blank. The finished twin portfolio has pockets for storage on the inside and easily accessible file folder-type storage on the outside. It's difficult to see how any more use could be gotten out of such a small item. Particularly important here, as well as in some of the other patents described in this chapter, is the attention that has been paid to making the product comparatively easy and inexpensive to manufacture. Unless an item can take a rather high price tag, it must not be a big deal to make. Some patents are in fact based on just this premise—redesigning a product so that it can be produced in a simpler manner. Often an inexperienced person will look at a patent and wonder why in the world anyone would go through the trouble of trying to patent anything so simple. In these cases, a closer examination often reveals that the point has been to improve by simplifying. This is not to suggest that many inventions aren't by their very nature extremely complex and difficult to make—the point is that in a time of increasing costs, simple can be not only cheap, but better, too.

A larger kind of portable office is the well-known attache case, referred to by some as the leather lunchpail. Whatever use they're put to, these cases are really quite handy, and they have sold in the millions. There's always room for improvement, even in such a popular item, as exemplified in an invention patented by Claude Mitchell of California. Many people who carry attache cases do some of their paperwork on planes, trains, and even buses. It helps them to have some kind of convenient work surface. The case illustrated in Fig. 7.19 has a hinged top that can be locked in any one of a number of different positions.

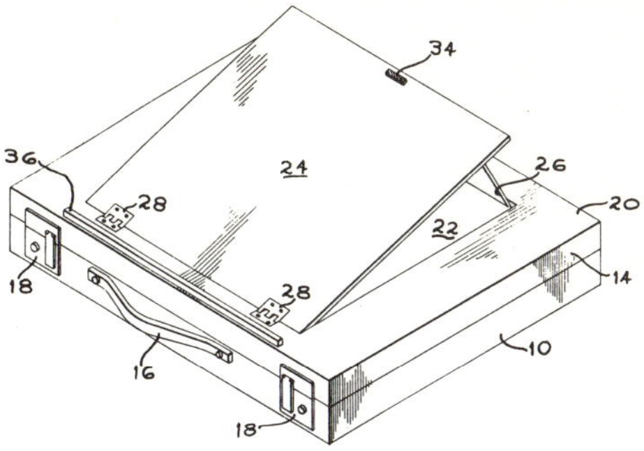

figure 7.19 Briefcase with hinged table plate, as illustrated in (1973) Patent No. 3,773,154. Submerged into a recess (22) is a rectangular plate (24). Its outer surface is substantially flush with the outer surface of the side wall (20) of the case. The plate (24) is hinged to the side wall next to the edge (26) of the recess (22) as shown at (28), for movement to raised positions. Hinged to the plate near its free edge are bars that may be folded out when the plate is raised, to serve as struts that hold the plate in a number of raised positions.

Every year, hundreds of patents are issued for devices and designs that fall into the overall categories discussed in this chapter. Some make it and are successful in the marketplace, but most don't, and are left to languish on the back shelf. Unfortunately, it is almost impossible to predict which will do what. There is no doubt that there are still opportunities for the home inventor, not only in the areas from which we have taken sample patents, but in other categories as well. For instance, the United States is now gradually shifting over to the metric system of measurement. What opportunities for conversion gadgets does this create? Some are already on the market, but there are more to come. Are there measuring devices—rulers, scales, etc.—that can themselves be improved? Are there more efficient ways to store and dispense things like tape and pencils and clips? Staplers have been around for years, but the ordinary desk machines still are not paragons of perfection. Is there a better way? Sure there is—and we're all waiting for someone to find it.

Can the independent inventor compete with Xerox, IBM, and all the others in the race to bring out new and better copying machines? Probably not, but it's a very big market, and there's no reason why the creative thinker in a small workshop can't do just as well as, or better than, the person in the comfortable corporate research center. But the cost and availability of equipment have to be considered. Then it seems most likely that independent inventors will have their greatest success in developing accessories to the machines, rather than entire new machines themselves. Typewriters, for instance, make up a category where major investment is required before major changes can be effected. Beyond that, they represent a kind of item where the curve of change has pretty well flattened out. Here again, accessories offer good possibilities.

The inventor working with a "big-ticket" item will most likely have to look to a large or medium-sized corporation for manufacturing and marketing. The capital required in today's world for such an undertaking is beyond the reach of most individuals. Such companies vary in their receptivity to new ideas, and even the most open-minded among them exhibit traces of the notorious Not-Invented-Here syndrome. That is to say, if the company's well-educated, technically proficient personnel, working with the best equipment, haven't come up with a viable answer to a particular problem, who else could have? And if some individual on the outside did happen to make a lucky discovery, there are many organization men and bureaucrats who would just as soon not hear anything about it. There is no doubt that dealing with some of the large corporations can be a demoralizing experience for the independent inventor. Yet for those who persist, the rewards can be great.

This chapter has only touched on all the categories of equipment useful in the conduct of business—personal, professional, etc. The inventor will undoubtedly think of other categories, other devices, totally new products that would fit into this chapter. That, after all, is the principle of invention, to create something new, original, and useful.

CHAPTER EIGHT
Packaging

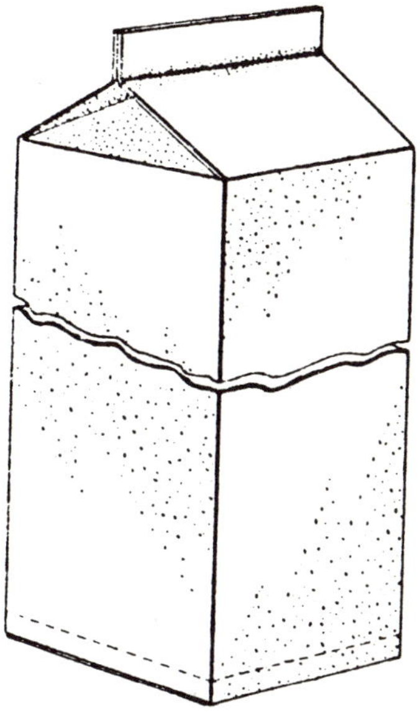

The phenomenon of mass distribution that has spread almost all around the world bears a *Made in U.S.A.* label. The supermarket chains, the consumer-product manufacturers, the advertising business, and many other related businesses, including the great broadcasting networks, are all products of the American system of mass marketing. Another important component of this system is, of course, the packaging industry. You can't have efficient nationwide and worldwide distribution without brand names, and you can't have brand names without packages to print them on. In addition, packages present a clean, efficient, and attractive way of getting a product from manufacturer to consumer.

There are many different kinds of packages, from a utilitarian cardboard carton to a unique coffee bag. This is an important field for many inventors. The designs of a package and, sometimes, of the machinery to produce it can be major components in the success of a product. Even in the nineteenth century and before, some manufacturers and merchants took great pains with the packages for their products. Generally, of course, it was only the more expensive things that were put into the colorful metal and wooden containers of the day. Many of these early packages have become collector's items.

On a more mundane level, the nineteenth century saw the beginning of the widespread use of cardboard cartons. One of the great innovators of the day was Robert Gair, whose company began mass-producing folding boxes around 1880. Cardboard had been in use since the sixteenth century, but not until about 300 years later

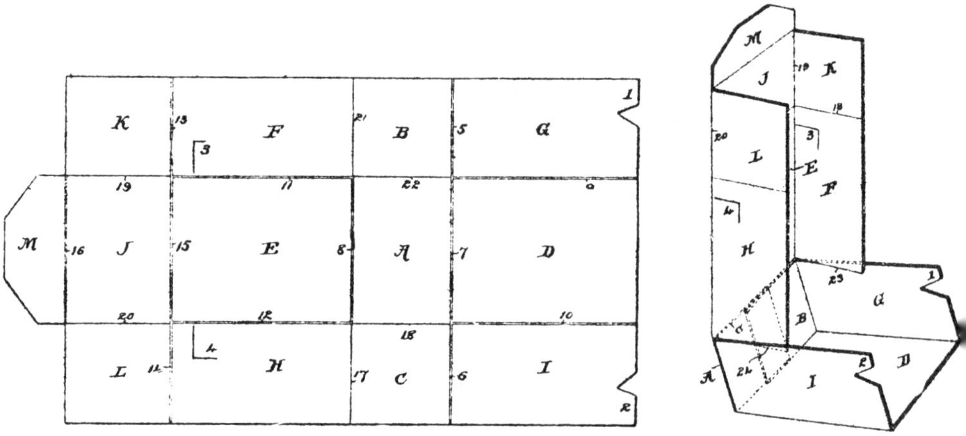

figure 8.1 Two views of a cardboard box, 1884, Patent No. 293,061. Left is a flat unassembled blank, and right is a partially assembled model. The folding lines are (5) through (16) and the incisions are (17) through (22). These latter form the tongues (1) and (2) and the pockets (3) and (4). Other parts are: the bottom (*A*); bottom flaps (*B*), (*C*); front (*D*); back (*E*); sides (*F*), (*G*), (*H*), (*I*); top (*J*); top closing flaps (*K*), (*L*); and tucking flap (*M*).

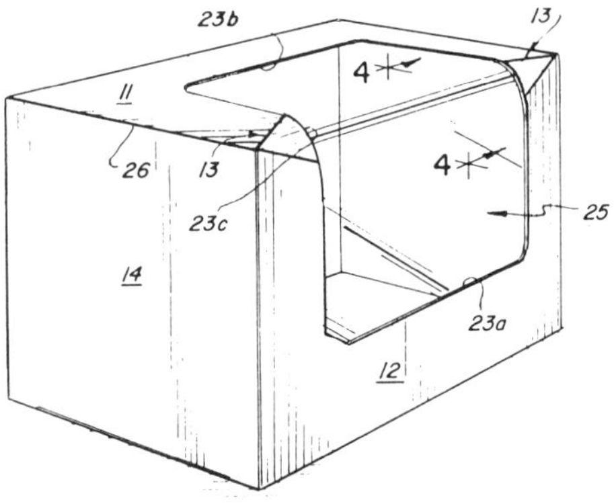

figure 8.2 A carton put together in the fashion specified in Patent No. 3,870,221 (1975). The "window" (25) is formed of a heavy gauge transparent film, and is secured to the carton by adhesive. The top panel (11) and front panel (12) are foldably interconnected by a fold line (13). The ends of the carton are closed by flap (14), and other flaps close the sides. It's a simple design, new, useful, and patentable.

did it become commonly used in the manufacture of cartons, etc. Gair and his associates developed a system whereby multiple dies on a press cut, creased, and scored pieces of cardboard. These "blanks" could be shipped flat and then made into cartons quickly and easily, as needed, at packing stations.

Packaging has changed a great deal since the early days. It is doubtful, though not impossible, that collectors of the future will have any great interest in the packages of the late twentieth century. Although many of the designs are slicker and more colorful than those that have gone before, their manufacture is geared to produce a throwaway item at the least possible cost consistent with the prevailing standard of quality. Modern packages are thus not made to last; durability for the most part is not necessary, and it is generally costly. In mass manufacturing a tenth of a cent, or even less, per item mounts up to a lot of money.

Today's packages must not only be attractive, they must also be easy to load, difficult to damage, simple to pack and unpack for bulk shipment, and light in weight.

Packaging in its many and diverse forms is one of the largest categories of patents in the Crystal City, Virginia, Main Search Room of the U.S. Patent Office. Anyone interested in the subject who is in the Washington, D.C., area should make it a point to look over this mountain of material. The category or class at this writing (sometimes changes are made) is number 229, and it encompasses more than 600 small bins, each containing anywhere from ten to even 100 patents. Things were a lot simpler in the time of Robert Gair.

Another nineteenth-century inventor of paper boxes was Harvey Munson of Connecticut. His 1884 patent, illustrated in Fig. 8.1, is so simple and practical in presentation that it could serve as a model for patents in the same field today. Like most paper cartons, this one starts out as a flat piece of cardboard, which is then cut, scored, and folded to make a box of the size and shape specified. The key point was that the box had to be designed so that it could be manufactured and handled by machine. A modern version of the folding paper carton is shown in Fig. 8.2. Charles Zeitter of Michigan, the inventor, assigned this patent to the Packaging Corporation of America. His box has a kind of window area on the top and front panels. This type of carton enables a bulk purchaser to see that the merchandise is as ordered and in good condition. In addition, this carton can be put out on the retail selling floor to serve as a display bin.

The folding carton has been in widespread use for so long and serves its purpose so well that there would seem to be very few good possibilities for change. But that assumption is incorrect. The time is coming when plastic and similar materials may be used in the construction of boxes. When this happens, it will be necessary to adjust the way some boxes are designed and manufactured. There are other improvements that will probably come along, too. Right now, the best way to open most cartons still involves using a knife. But some people don't know how or don't like to use a blade. Perhaps an inexpensive change could be made in some types of cartons so that they would be easy for everyone to open.

Packaging is important in many areas of American life, but nowhere is it more important, nowhere is more time spent on every single detail, than in food packaging. Several examples are included in this chapter. Procter & Gamble invested millions of dollars in bringing to market a new item where the method of packaging was almost as important as the product itself. Pringle's Old-Fashioned Potato Chips came to the public with all the fanfare of Ford's Edsel—well, perhaps the comparison is not fair, since Pringle's shows no signs of being withdrawn from the market, but there have been some problems with the chips, not the package. The point is that the old way of packing potato chips in cellophane bags resulted in lots of broken chips and also in the retention of small amounts of oxygen and water vapor in the bag, making the product less fresh and less attractive. The bags were also bulky, taking up a fair amount of space for their weight and price.

Two Procter & Gamble employees, Fredric Baur and Harold Hawley of Ohio, worked on and patented a better way of packaging potato chips: "Chips of uniform size and shape are stacked one upon the other in closely fitting relationship . . . and are then placed within a rigid tubular container formed from materials which are substantially impervious to the passage of oxygen and water vapor. The ends are applied to the container to seal it." (Fig. 8.3.) Just as simple as that. The package is indeed a good one; it does a better job than the conventional bags of protecting its product.

Baur and Hawley got their patent in 1970. But the cycle did not end there. Six years later, Nelson Beall of Minnesota had another idea. His uniformly shaped chip snack food products are packed in a circular or loop array (Fig. 8.4). Here again the idea is to keep the chips from chipping.

The quality of newness or novelty is an important factor in packaging consumer goods. The tea bag, for instance, has been around for a good many years, having been invented by Benjamin Hirschhorn early in this century. Since then, many new versions have been brought out. But what about a coffee bag? It should be a good deal cheaper for one person to use a single-serving bag than to brew up a pot. It's less expensive than a spoonful of instant, too. Russell Beck of Pennsylvania patented his coffee bag in 1971 (Fig. 8.5). It seems surprising that it had not been patented in this form before that date. On the other hand, since the invention has not yet taken the country by storm, perhaps the delay is not noteworthy.

The milkman and his early-morning clinking of bottles can still be heard in some parts of the United States, but less with each pass-

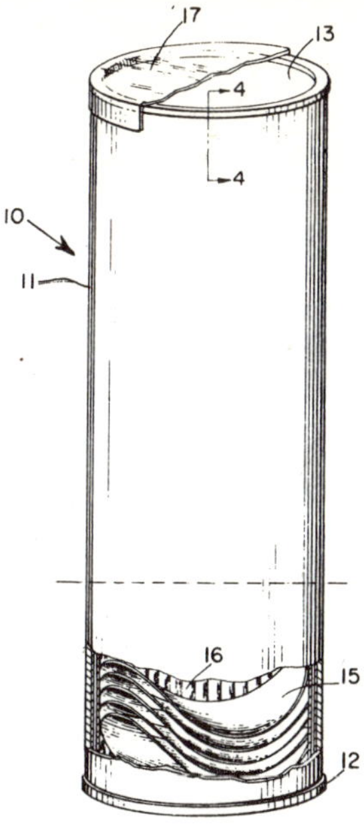

figure 8.3 A potato chip package broken away to show the arrangement of the chips and the structure of the container top, as shown in Patent No. 3,498,798 (1970). The container is rigid and tight, to stand up to the rigors of shipping and to keep the chips fresh. The lids (12) and (13) are hermetically sealed, and the flexible plastic lid (17) is used to seal the package after the initial opening. The chips (15) are manufactured in a uniform shape so that they can be easily stacked. A corrugated sleeve cylinder (16) can be inserted to keep the chips away from the container's walls. The chips themselves are curved so that they fit into each other in a neat stack with more internal strength than would be the case with a stack of flat chips. The interfitting of adjacent chips achieves a laminated structure that helps to prevent the chips' breaking up.

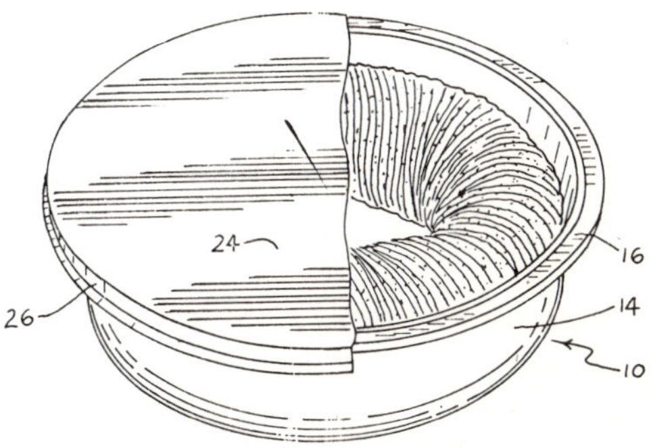

figure 8.4 Another type of package for chips, this one from Patent No. 3,956,510 (1976). The uniformly curved chips are nested into one another to form a closed loop, which rests inside a circular, pie-plate-type package. The plate is shaped so that the chips rest within a track, or depression, on the bottom of the plate that keeps them from moving around. The cover (24) is snugly fit over the flange (16) by means of a lip (26).

figure 8.5 A coffee bag (11), as described in Patent No. 3,607,302 (1971), being dipped into a cup (23) of hot water. An envelope (14) is made either of thin, porous paper or cloth to permit heated water to flow through it. Front and rear panels (15) are formed by folding the paper or cloth material, the edges (16) being secured together to form a pocket (17) within which a premeasured quantity of ground coffee beans (18) has been placed prior to securing the edges either by adhesive, if paper material is used, or by stitching, if cloth is used. One end of the string (12) is secured inexpensively to the bag simply and easily by a staple (20).

The bag is simply dipped up and down into heated water in a cup until a desired brew is obtained, after which the bag is discarded.

ing year. Most people buy their milk at the market in cartons or large plastic containers. The milk bottle was a pretty good package while it lasted—sturdy, convenient to use and store, sanitary—but it proved to be bothersome for the markets to collect deposits, return for refills, in effect to handle each package twice, coming and going. (A few states now require deposits for soda and beer bottles, but many people have doubts about the efficacy of this regulation in reducing litter.) And it became too expensive for most dairies to maintain a staff of milkmen. So it came down to a question of cost, and guess which side won?

The milk carton has gone through a number of different designs. Some of them have leaked, others have been difficult to open, others have been expensive to manufacture. It's a good bet that today's containers will be replaced by new, improved types as time goes on. But there seems to be no possibility of ever going back to the milk bottle. In fact, because of expense and the deposit laws, more liquids will be sold in cardboard and plastic containers; glass and metal may be on the way out. Purveyors of such beverages as juice, soda, and even beer are now exploring means of developing new types of packages. It's worth taking a close look at what the dairy industry has come up with in the way of containers to get some clues as to the possibilities that exist for other liquids.

The patent filed by Kenneth Thompson of North Carolina and Richard Ihde of Ohio, and assigned to the Champion Paper Company, gives a brief rundown on the history of milk cartons. These used to be formed by "setting up a preformed blank with the bottom of the carton closed and dipping the carton in molten paraffin to seal it." This type was replaced in the 1960s by a carton coated with a thin film of polyethelene extruded onto the paperboard before the board is cut, scored, and folded. Leaks still occurred in this type of carton, for various reasons. It was the goal of the invention illustrated in Fig. 8.6 to eliminate these leaks. The idea is uncomplicated and proved to be effective. By reducing the thickness of the film coating except on the bottom and part of the top, where it was increased, less coating was used but to greater effect.

Now, is there still a better way to package milk? Plastic bottles are good for large quantities, half-gallons and more, but they're not cheap. Sacks made of clear, thin plastic are indeed cheap, and easy to fill, but the consumer can't handle them very well without some kind of special holder or by transferring the contents to a container (such as an old milk bottle). Alternatives to conventional packaging are being looked into not only by dairymen but by makers of other beverages, too.

Not all bottles will disappear from store shelves, of course. Most liquors and wines will continue to be sold in glass packages. In fact, some proposals have been made for improvements in these kinds of bottles. In 1975, Mario Mezquita of Spain was issued a U.S. patent for "an artificially aged bottle and method for producing the same." The supply of old bottles being limited, why shouldn't a factory make its own? The idea is, first, to have a label with frayed edges. Then a layer of fluid adhesive is applied to the outer surface of the bottle. Powder—to resemble dust—is sprinkled over the surface at random. Then a cord is wrapped around the

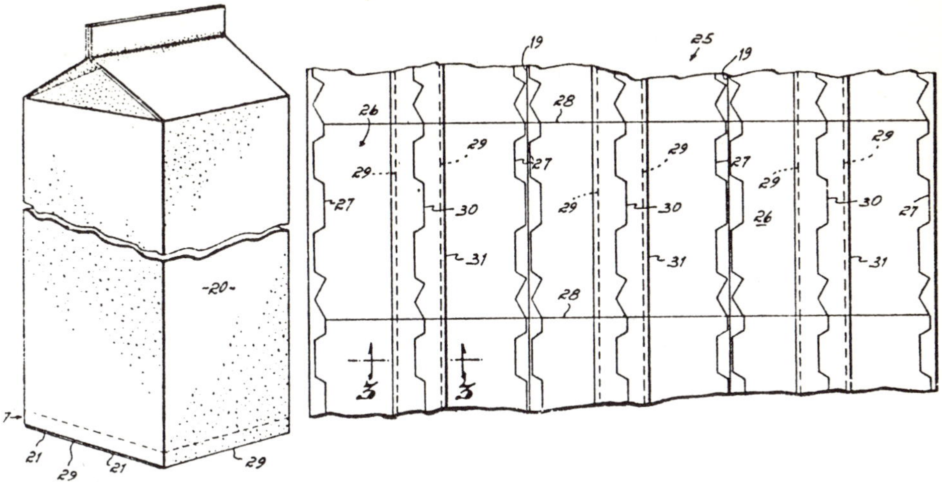

figure 8.6 A milk carton, at left, and at right a blank ready for folding into a carton, according to Patent No. 3,482,278 (1969).

The milk carton (20) is formed of paperboard and coated with polyethylene, inside and out. The carton bottom is formed by folding panels (21) transversely on score lines (29). The structure is completed by applying heat and pressure to the panels, which are folded upon one another. The polyethylene, which has been applied to the paperboard, fuses during the application of heat and pressure to form a sealed bottom structure.

In handling, the carton is subjected to stresses caused by bouncing, vibrating, and the like, which cause the contents to force the side walls to flex. The effect of this stress is most pronounced at the location of the fold lines (29), and it is in this area that the greatest occurrence of leaks is found.

An increased thickness of polyethylene on the inside surface of the carton will cover the score line (29) and increase the resistance of that area. It would, of course, be possible to increase the thickness of the polyethylene film at the score line simply by increasing its thickness uniformly all over the surface of the carton. However, this would markedly increase the cost of manufacture of the carton.

The web (25) is wide enough to space six carton blanks (26) across its width. After being coated, the web will be slit along line (19) and rewound into three rolls. Thereafter, the carton blanks will be cut, scored, formed into cartons, and filled.

The remaining lines shown on the web are merely to illustrate the location of the longitudinal-top-forming cut lines (27), the transerse cut lines (28), major horizontal score lines (29) for the bottom structure, and the bottom forming cut lines (30). As shown, the pattern of each blank is such that transversely adjacent blanks nest with each other along their bottom-forming cut lines. This nesting spaces the major horizontal score lines to be formed in respective carton blanks approximately five inches from each other. Thus, it is possible to apply an extra thickness of plastic film to the score line areas (19) by applying a thicker film of plastic on a longitudinal strip (31), which is approximately six inches wide, the area of the thicker plastic extending approximately half an inch beyond the score lines (29).

---

neck of the bottle, its two ends are joined by a fancy-looking seal, and another rare old bottle comes off the assembly line (Fig. 8.7).

As fewer bottles and tins are used in the packaging of food, other methods are gaining popularity. This development comes at a time when more and more foods are being packaged. Most containers are not complicated. Breakfast foods, cookies, and the like come in paper or plastic bags enclosed in boxes of thin cardboard. There are other packages that take some doing. Figure 8.8 illustrates a can Francis Reid of Minnesota developed for Pillsbury's refrigerated dough. It differs from previous types in that it opens in a single step. All the consumer does is tear off the label and reinforcing strip, and presto! the dough oozes out of the container. There's no need to whack it against the counter or pry the dough out with a spoon. This package is worth studying because it illustrates how one inventor made one package easy to open—convenient to use. Then think of all the packages in the supermarket and department store that are not so easy to open, that are ripe for improvement.

Fast foods are becoming a significant part of the American diet. One rather startling projection has indicated that by 1980 more than 40 percent of the meals in America will be eaten outside the home. Hamburgers, fried chicken, pizza, and even tacos are being served from franchised stands and restaurants at a fantastic pace. A lot of this business is of course take-out, and the fast-food moguls are always on the lookout for packages that will make it easy to pack and carry out their products. One of the most durable of all carry-packs was patented by Frederick Wenzel of Wisconsin in 1961. His object was to develop "paperboard trays which may be

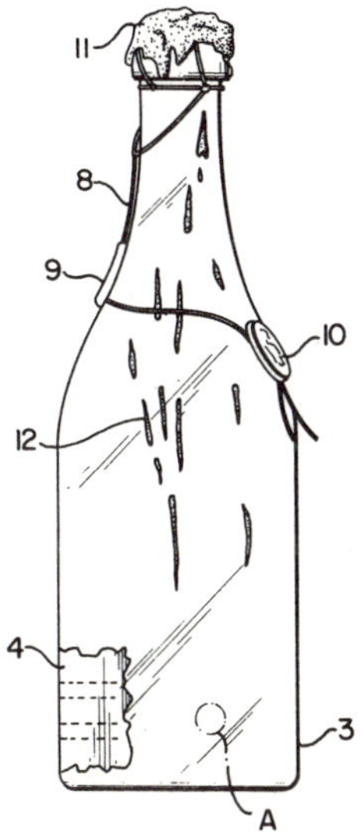

figure 8.7 Artificially aged wine bottle, as described in Patent No. 3,890,449 (1975).

The label (4) may be artificially aged or may in fact be an old label. It is preferred that the edges of the label be frayed.

A fluid adhesive layer (5) is irregularly and randomly treated with a powder, such as dust. A layer of fixer, such as lacquer, is then applied to maintain the position of the dust with respect to the bottle exterior.

Cord (8) may be wrapped several times around the neck of the bottle. The two ends of the cord are then wrapped over the entrance and cork of the bottle and are brought about along the neck and knotted as shown. The two ends of the cord (8) are then further lowered along the neck of the container and joined with a seal (9).

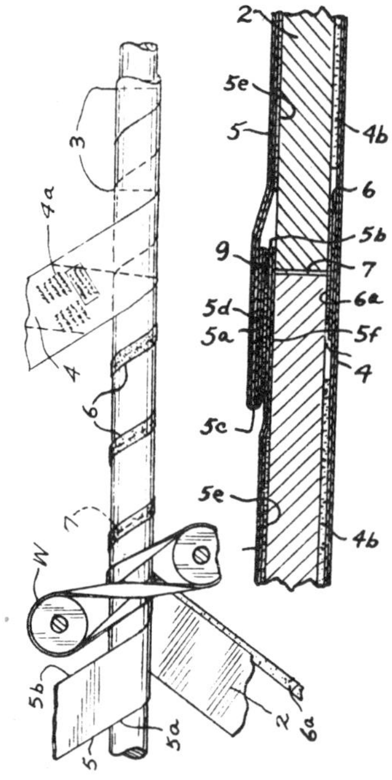

figure 8.8 Diagram of how Pillsbury's easy-opening dough can is made, as illustrated in Patent No. 3,972,468 (1976).

A winding mandrel of standard construction is provided with the usual winder (*W*) having a belt or the like entrained over the mandrel for drawing the various webs including the bodystock layer (2) onto the mandrel as a spiral or helix. The bodystock sheet and liner (5) are wound continuously onto the mandrel and travel from left to right in the figures, thereby forming a helically extending butt joint (7) where the adjacent convolutions of the bodystock contact one another. Over the butt joint is wrapped a reinforcing tape (6), which can be thought of as a tensile member. This is adhered to the bodystock by means of adhesive (6*a*) suitably applied to the strip (6) just before winding. In the winding operation, the strip is important in providing the freshly made tube with sufficient initial strength to prevent what is known as a blow-up, in which all of the coils being wound on the machine suddenly fly apart. The liner (5), which is formed from a relatively tough barrier material such as the composite of kraft paper and aluminum foil, is applied to the mandrel and wound inside the bodystock (2), with adhesive applied to the bodystock to bond the liner in place.

The liner is bonded to itself along its marginal helically extending edges (5*a*) and (5*b*) along a seam underlying and adjacent to the butt joint (7). The seam or joint is formed by folding the edge (5*a*) outwardly at (5*c*) against itself to define a hem or lap (5*d*) that may be on the order of a quarter of an inch wide, and thereafter bonding the overlapping edges (5*b*) and (5*d*) of the liner to each other with a suitable adhesive (9) such as a heat-sealable lacquer, e.g., a polyvinyl chloride lacquer to provide a rupturable joint formed by the lapped edges adjacent to and underlying the butt joint. The peelable joint in the liner provides substantially no strength at the butt joint.

After the tensile member (6) has been bonded over the outside of the butt joint, a label (54), which functions as a protective coating and printing surface, is applied and bonded by adhesive (4*b*) suitably applied to the lower surface of the label. A portion of this adhesive (6*b*) bonds the label to the tape (6) to provide a bond which is stronger than the adhesive layer (6*a*), so that the manual removal of the label also removes the strip (6). Thus, the can is opened in a single manipulative step.

After the tube has been completely formed, it is cut into can lengths as shown by dotted lines (3). The tube is then cut circumferentially only partially through its thickness, specifically through the label (4) and the tensile element (6) along a collar cut no more than about half an inch from the end of the container.

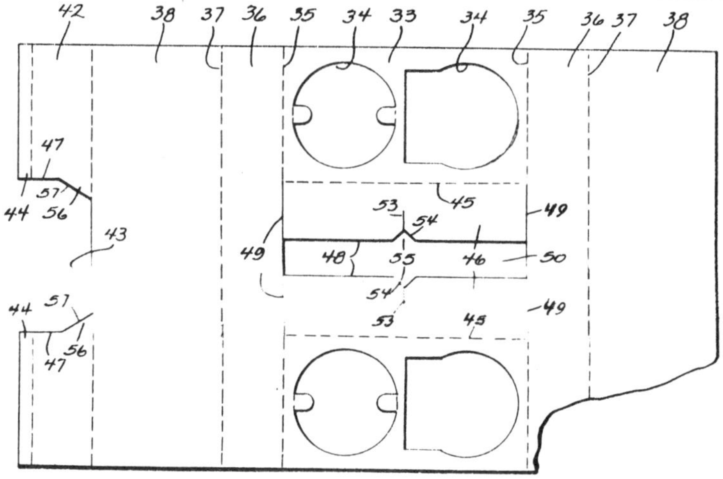

figure 8.9 A blank for food tray, as illustrated in Patent No. 3,001,684 (1963).

The central top panel (33) is connected on score lines (35) to side wall panels (36), which are, in turn, connected on score lines (37) to the bottom wall panels (38), which, when folded, will overlap to provide the glue lap (39).

An additional or intermediate wall panel (42) is provided at the end of one of the bottom panels (38), notched out centrally at (43). The otherwise free marginal portions of notched wall panel (42) are further provided with glue lap flaps (44), which connect with the undersurface of the top panel (33).

Transverse score lines (45) define hinges upon which the bracing flaps (46) will pivot. Note that the cut lines (47) in the partition panel (42) substantially align with the score lines in the top panel.

Flaps are further defined by cut lines (48), and (49) and a waste space (50) between the cut lines (48). Slit the flaps along the cut lines (53) and oblique cut lines (54), which diverge from the end of slit (53) to provide notches (55), which help guide the slit into a frictional interlocking engagement with an obliquely offset strut portion (56) of the partition walls (42). Struts are left in the partition portions (42) between cut lines (57), which are obliquely related to cut lines (47).

Flap portions at opposite sides of the slit (53) straddle the strut. The resilient interlock thus effected will preclude release of the flaps (46) unless they are forcibly disengaged from the struts (56).

Not only do the struts provide an interlock with the flaps, but the margins of the flaps along the cut line (53) laterally abut the strut to further brace the tray.

Partitions (52) intervene between beverage container pockets defined by openings (34) and further help to brace the tray longitudinally.

readily erected from collapsed form and which, when erected, will assume an adequately rigid configuration for the proper support of containers for beverage and food items for the period required for such items to be consumed, whereafter the tray may be discarded" (Fig. 8.9). The illustration clearly shows how this carrier is formed.

The most successful enterprise in the fast-food business has been McDonald's. Packaging is a very important factor in their method of operation. Everything that leaves their counters is put into containers that the customers will find easy and convenient to carry, to use, and to dispose of. Even the French fries are not just put into an ordinary paper bag but into a cardboard container that has been carefully designed to meet certain requirements. First, of course, it must be inexpensive to make, yet durable and attractive. It is cut, folded, scored, and printed, then shipped flat to the indi-

figure 8.10 Top, a blank for the carton that is shown assembled below, as illustrated in Patent No. 3,630,430 (1971).

A pair of sidewall panels (11) and (12) each have a shorter edge attached to opposite sides of a bottom panel (13) along score lines (14) and (15), respectively. The bottom panel is bisected by score line (16), and the edges (17) and (18) of the bottom panel, which are not connected to side panels, are formed as inwardly sloping from the bisecting score line (16) toward, and connected to the ends of, score lines (14) and (15). These lines are curved so as to produce concave edges on the side panels. The opposite edge of sidewall panel (11) is also inwardly curved to form a concave edge (19) to that panel. The form of sidewall panel (12) differs in that its opposite edge is curved outwardly to form a convex edge (21). Sidewall panel (11) has score lines (22) and (23) extending laterally and outwardly, tapering between arcuate score line (14) and concavely curved edge (19). Sidewall panel (12) also has outwardly tapering edges (24) and (25), extending from arcuate score line (15) to the ends of convex edge (21). End wall panels (26) and (27) extend laterally along score lines (22) and (23), respectively, and are thereby connected to sidewall panel (11). The end wall panels are bisected by score lines (28) and (29), respectively. Along their opposite lateral edges at score lines (31) and (32) are attached glue flaps (33) and (34), which have outer free edges (35) and (36), respectively, which are parallel to each other and which have their bottom edges (37) and (38) forming a beveled edge between the outer free edges (35) and (36) and score lines (31) and (32) respectively. This beveled shape is of a sufficient angle to exceed the angle of the arcuate edge (15) so as not to extend below that arcuate edge when the carton is in a glued and either folded or extended position.

Glue lines (41) and (42) are substantially parallel to each other and to outer free edges (35) and (36), respectively, of glue flaps (33) and (34) since parallel application of glue on such a blank is much easier, requiring simpler gluing machinery using the technique and equipment customarily used in the folding carton industry.

End wall panels (26) and (27) extend below the intersections of score lines (22) and (23), respectively, with score line (14), so as to form legs (43) and (44), which are used to support the erected carton in an upright position.

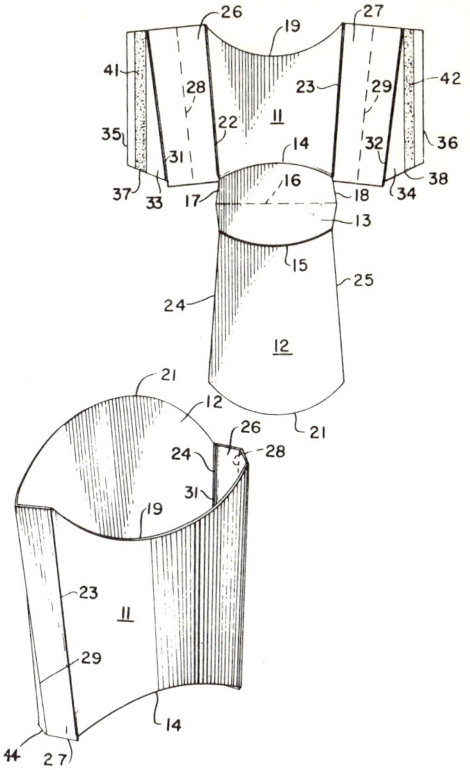

vidual locations. There it must fit into the fast-food flow by being easy to put together quickly and then to fill with exactly the right amount of potatoes. This package, invented by Glenn Struble of Ohio, is illustrated in Fig. 8.10.

Another kind of fast food service is found in hospitals and other institutions. Their kitchens are providing three or more different meals each day, meals that must be nutritious, balanced, and tasty. It must be possible to cook this food in large quantities at reasonable cost and to serve it piping hot at often widely dispersed locations. Packaging can help solve part of this problem. To prevent loss of heat and of fresh taste, the plates of food must be tightly covered while in transit from the kitchen to bedside or wherever the meal is being served. China plates and metal covers formerly did this job, but the expense of handling, breakage, and cleaning up has pushed many administrators to look for new methods. One new product in this area is shown in Fig. 8.11. Invented by Nicholas Commisso of New York, it is a disposable plastic lid that fits tight over a disposable plate. This unit is easy to handle, easy to open.

Packaging is also important in food served on airlines and railroads. Here the standard complaints are poor taste, lack of variety, and food being too cold or too hot. New packages might be able to solve at least part of these problems.

One of the most important elements in modern packaging is speed. This is especially true with products for retail sale. If something can't be packed and sealed in large quantities in a short time, it will not be able to reach a mass market. Cottage cheese, yogurt, sour cream, and similar dairy foods generally come in wide-mouthed, nestable containers made of a thin but rigid plastic. The tops of these containers must be made so that they can be opened

figure 8.11 Part sectional view of the covered food container described in Patent No. 3,794,090 (1974).

A dome-shaped plastic cover (10) fits onto a food serving dish (12). The cover has a sidewall that slopes at a small angle and includes flutes (14), having arcuate upper ends (16) which extend inwardly and merge into upper sidewall portions (18). Generally wedge-shaped indentations separate the flutes from each other and constitute reinforcing ribs (20). The reinforcing ribs also extend up through the upper sidewall portions (18) into the top of the cover.

The top of the cover includes an outer elevated annular rim (22), a wider, generally flat main top section (24) connected to the rim (22) by a curved section (26), a portion (28) that extends downward at an angle of approximately 30 degrees to the plane of the top section and a knob (30), which is a handle.

The cover can be made of a thin-gauge synthetic thermoplastic of about 12 mils in thickness. At its lower end, the cover has a downturned arcuate rim (32) with an outwardly flaring lower edge (34). The cover is attached to the plate by lugs, each of which has an outwardly inclined camming surface and an upper surface inclined inward at a slight angle from the vertical. When the cover is on the plate, the lugs are pressed down into a snap-fit engagement.

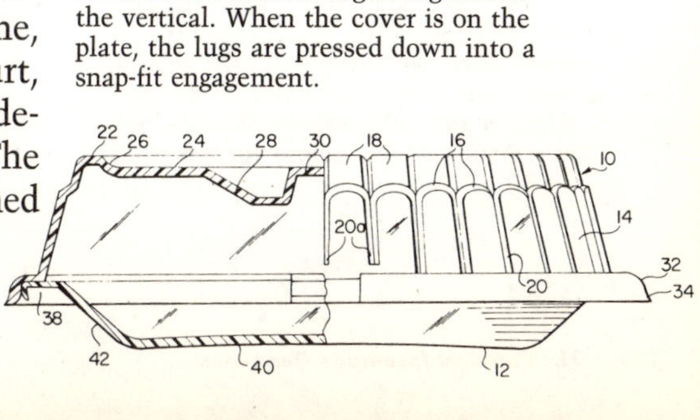

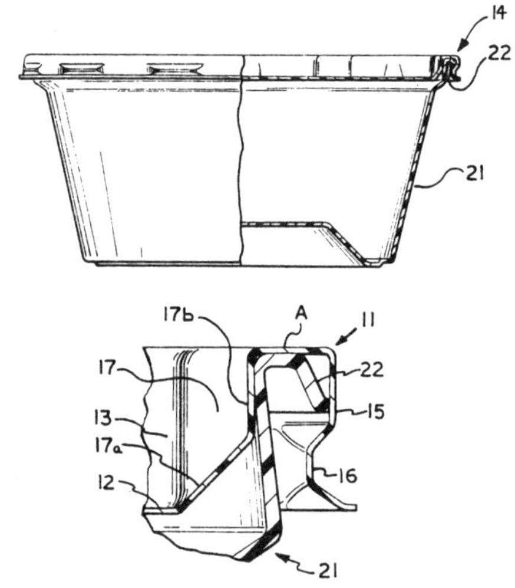

figure 8.12(a) A sectional view of how the top engages the container in the closure shown in Patent No. 3,964,635 (1976), with a partially cutaway view of the container and top.

The central panel (12) of closure (11) is recessed within the head space of the container in order to reduce the volume of air in this space, thus preserving the freshness of the packaged product. To accomplish this, panel (13) is surrounded by a peripheral wall (13) that extends out from the central panel for a distance substantially equal to the distance it is desired to recess the central panel within the mouth of the associated container. The peripheral wall is in turn surrounded by an annular part (14) that is attached above the juncture of panel (12) with wall (13) and that extends laterally outward from that wall. Annular portion (14) is sized and positioned so as to be adapted to rest on the beaded or flanged rim of the container. Attached to it is a peripheral skirt (15) that extends axially downward—in other words, back toward the plane of central panel (12).

To retain closure (11) firmly in place, the peripheral skirt (15) is provided with a series of spaced-apart, inwardly projecting beads (16). These beads are located near the lower or free end of the skirt, below the rim (22). The radial distance from the vertical central axis of the assembled container and closure to the radially innermost portion of the beads is at least slightly less than the radial distance to the outermost portion of the rim (22). This relationship results in interference between the beads and the container rim, thereby making it possible for the beads to securely grip the container rim. The inherent flexible nature of the thin skirt makes it possible to force the closure onto the container mouth by conventional capping equipment.

When using high-speed capping devices, it may be difficult to properly center a closure on the container. To overcome such problems, peripheral wall portion (13) of closure (11) is provided with circumferentially spaced-apart lugs (17) that project into the space between wall (13) and skirt (15), terminating short of the skirt by a distance sufficient to receive the rim of the container. Such lugs, which do not interfere with proper thermoforming of the closure, preferably have at least a lowermost portion of substantial axial extent (17*a*), which tapers at a substantial angle, e.g., 40 to 45 degrees, up and out from the juncture of wall (13) with panel (12), to further aid in the centering of the closure on the container. The lugs can also be provided with an upper portion of substantial axial extent (17*b*), which has a generally vertical orientation for frictional engagement with the inside of the container mouth. This increases the frictional resistance to turning of the cap on the container.

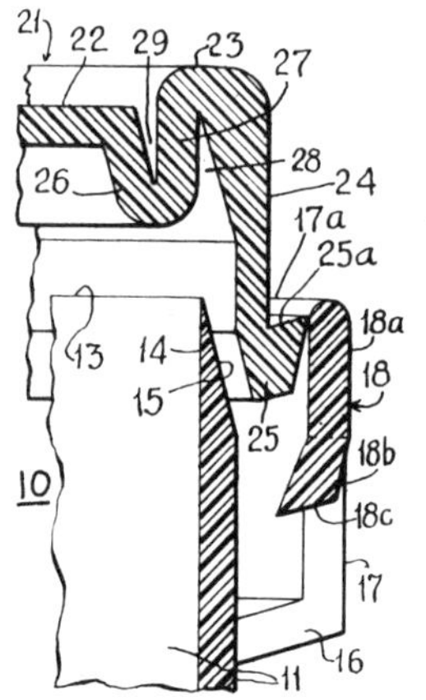

figure 8.12(b) Fragmented view taken on a plane through one of the lock tabs on the body of the container shown in Patent No. 3,840,152 (1974). This illustrates the relative positions of the rib on the lid and the lock tab before locking.

The body (10) of the container comprises a sidewall (11) and a bottom wall (12); there is an open top (13) at its upper end. The upper edge (14) of the sidewall terminates in an inwardly tapered edge (15). Circumferentially spaced, horizontal connectors (16) support a peripheral band (17) adjacent to the upper edge (14) of, and circumferentially spaced from, the sidewall. Preferably, the upper edge (17*a*) of this peripheral band extends no higher than the upper edge of the sidewall. The horizontal connectors are at a slight upward angle with respect to the sidewall to facilitate stacking of empty buckets.

Circumferentially spaced, flexible lock tabs (18) are provided about the peripheral band and are formed intergral with it. The side edges (19) of the lock tabs (18) are spaced from the adjacent wall of the peripheral band by vertical slots. These slots begin at a point about one-fourth the way down from the top of peripheral band; this point defines the upper extremity of the lock tabs and their first flexure point. The upper portion (18*a*) of each lock tab is substantially vertical and in alignment with the peripheral band, while the lower portion (18*b*) of each lock tab is at a slight inward angle of about 10 to 20 degrees toward the sidewall. Preferably, the lower face (18*c*) of each lock tab is inclined downward at a slight angle from the horizontal to insure firm engagement with the lid. Generally, eight lock tabs are provided, depending upon the size and intended use of the container.

The lid (21) of the container generally has a raised shoulder (23) at its circumferential edge. The outer edge of the raised shoulder extends downward, forming a peripheral flange (24) that terminates in a rib (25). The upper face (25*a*) of this rib (25) is inclined upward at a slight angle to insure firm locking with the lower face (18*c*) of the lock tab. A lowered shoulder (26) is provided adjacent to raised shoulder by means of common wall (27). A convergent annular groove (28) is adapted to receive the tapered upper edge (14) of the sidewall. Preferably, a divergent annular groove (29) is also provided between the common wall and the lid surface (22). This groove permits flexing of the common wall when pressure is exerted on the lid in securing it to the body, thereby compensating in part for compression of the upper edge of sidewall (11) when it is inserted in the convergent groove.

In securing the lid to the body, the rib (25) is positioned in the space between the tapered upper edge and upper edge (17*a*). Downward pressure, either manual or by automatic closing apparatus, forces flange (24) between sidewall (11) and peripheral band (17). Tapered edge (14) slides into divergent groove (28) and is slightly deformed by the force. Concurrently, the lower portion (18*b*) of the lock tab is forced outward as a result of its vertical contact with rib (25). At this point the lower face (18*c*) is substantially coextensive with the upper face (25*a*) of rib (25).

Further downward pressure on the lid forces the lower face to clear the upper face (25*a*), permitting the lower portion (18*b*) to snap back to its normal rest position. When downward pressure on the lid is released, it rises as compressive deformation of the tapered edge (14)) is at least partially relaxed, and upper face (25*a*) is permitted to engage lower face (18*c*), thereby firmly securing lid to body.

and closed many times by the consumer without losing their tightness. This is the only way the food can be kept fresh. Added to this requirement is the fact that the capping process must be done by machines that operate at high speeds.

There have been some problems with tops not fitting properly. In some, the surface contact between container and cap has not been sufficient to maintain a frictional hold on the cap. Rodney Ludder of New York invented an improved version of this type of container, a one-piece, recessed, rim-locking container closure, "a wide-mouth container with an externally projecting rim or bead and a female-die-thermoformed plastic one-piece closure in place over the mouth." The cap "contacts the inside of the container mouth at a plurality of circumferentially spaced-apart points for improved frictional resistance to turning" (Fig. 8.12). Another type of cup is discussed in relation to Fig. 5.11 on p. 78. This is the "childproof" top for medicines that should be dispensed only by adults.

Sealable and resealable containers have been popular for many years, even before the use of plastic made them the standard. So many new types have been introduced over the years that this entire chapter could be devoted to a study of them. George Hodge of Pennsylvania invented one with a novel means of locking that is carried on the two essential parts, the container and the cap. On the container there are many flexible tabs, formed into a peripheral band that is attached to the side wall of the body, next to its open top. On the lid there is a flange that ends in a radial, outthrust rib. The upper part of the container wall is tapered to fit snugly into a groove on the lid, firmly sealing it to the body (Fig. 8.12*b*).

Even the old screw-on lid can be adapted to serve new purposes. Caps made from such materials as polystyrene or expanded polyurethane are machined to be exactly the same size—almost. Because of the minor vagaries of the machine and of the material, there are very slight variations in the diameter of the cup's lip and thickness, sometimes making it impossible for lids to fit on tightly. Robert Whiteside of Michigan invented a cup made of soft, cushiony material like polystyrene that threads onto a matching cup even when that cup is slightly larger or smaller than it should be. This design offers just enough additional flexibility over the standard snap-on lids that many spills of coffee and other take-out liquids can be avoided (Fig. 8.13).

The most popular take-out liquid in the country is, of course, canned soda. Here, the major advance—or retreat, depending on your point of view—was the invention of the pop top. There have been problems with this design, particularly with the little tin tails that are torn off the top and then carelessly discarded. Their sometimes sharp edges may cut hands removing them and feet stepping on them. A 1976 answer to these problems was the "nondetachable" tab. Of course, people are able to work these tabs around and get them off the can without too much trouble. Another, more recent idea is the push-botton top, in some cases with one button and in other cases with two buttons that must be pushed in sequence. These arrangements are quite clever—perhaps too clever

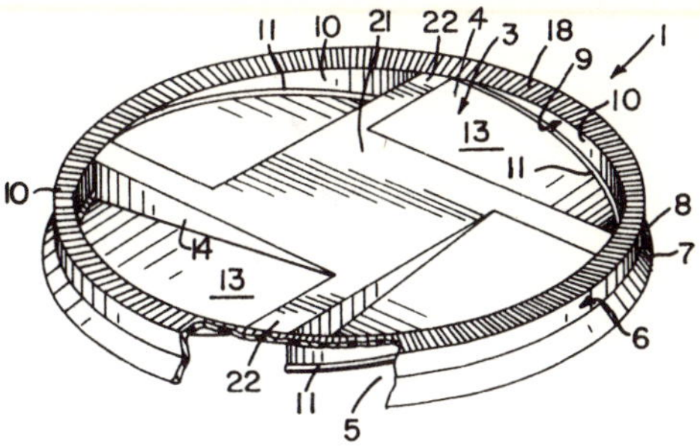

figure 8.13 Perspective view of a screw-on lid, Patent No. 3,904,072 (1975).

The lid has a central section (3) of a generally circular configuration. Extending around the periphery (4) of this central section is a downward-opening, circumferentially extending cavity or channel (5), which has as its outer wall (6) a skirt (7). This wall may actually be shaped as having a lower curvilinear skirt with a circumferential straight sidewall (8) above it, or it may be shaped into a truncated right circular cone or cylinder (not shown). The circumferential straight sidewall connects the external skirt to a top connecting ring of the channel. The inner wall (9) is formed by a series of curved triangular panels (10). Any number of panels may be used, depending on the size of the lid and the thickness of the material of the cup lid. For the ordinary, typical small-size cup and cup cover, four panels are found to be an optimum number.

A screw thread (11) is formed into the surface of each triangular panel. The thread extends at an acute angle to the plane of the top of the lid. The bearing surfaces of the threads are external and project into the channel. A convenient place for the thread is along the bottom of the triangular panel, as shown. In this way, when the lid (1) is turned onto the expanded polystyrene cup, it is threaded into the soft cup wall. The threads on the curved triangular panels and the panels themselves are braced to some extent by an indent bracing panel (13), which extends inwardly from the curved triangular inner wall (10). A bracing panel extends inwardly from the lower edge of each triangular panel of the inner wall until it comes to the indent inner wall (14). The flat section of the bracing panel serves to hold the thread on the inner wall against the soft, cushioned material of the container. As the thread is turned, it threads the container and may even to a slight extent cut the soft inner wall of the cushioned material as it threads down into the mouth of the container. The closure cap is pulled downward, and the container wall is pressed somewhat outward against the skirt (7) and outer wall (8) of the downward-opening channel (5). As the side of the container presses against the outer wall of the channel, it is forced somewhat inward against the threading (11). This process makes the cap more securely tightened onto the inner wall (10) of the container. The upper edge (17) of the container wall finally comes up against the top connecting ring (18), which is made with knurling along its upper surface. When this happens, the operator can feel a sudden resistance to further turning. At this point, the container cap has been turned onto the mouth of the cushioned cup, a tight lock has taken place between the cap and the cup, and no liquid will leak out between the two.

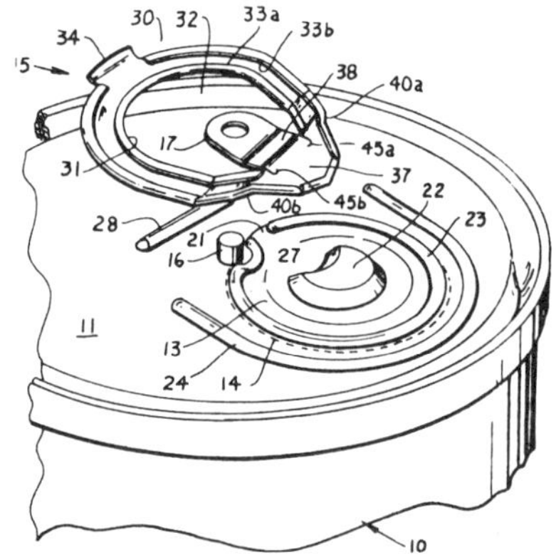

figure 8.14 Top of a can with nondetachable tab, shown in Patent No. 3,977,561 (1976). A container (10) with a top (11) has tear tab area (13) defined by scoring (14). The opening ring (15) is attached to the can top by a rivet (16). The scored area can be pulled off, and the can opened by the application of pressure. However, the scored area has a bendable hinge connection (21) that is nondetachably connected to the can. Thus there will be no scoring in the immediate area of this connection.

The rivet is located on a "peninsula" (27), which juts out and forces the scored line to make a "detour." A stiffening rib (28) may be formed on the opposite side of the connection location (26) from the tear tab (13), to add structural rigidity. The opening (15) includes a handle portion (30) with a generally angular ring (31), which extends generally rearward from the tab portion (17) and which surrounds and defines an opening (32) within the handle. This opening preferably has an outline approximately similar to the outline of the tear tab.

Both the inner and outer peripheries of the annular member (31) may be provided with rolled edges (33*a*) and (33*b*) to add structural rigidity to the handle portion, if desired, and a flange (34) extending rearward from the handle may be provided to facilitate grasping and initial lifting of the handle.

Extending in front of the connection with the rivet, on the opening member (15), is the finger portion (37), which overlies at least a portion of the tear tab (13). Formed at two locations along the outer edge of the opening member are the shoulders (40*a*) and (40*b*), which are positioned to engage locations on the periphery of the opening (49).

The tab portion (17) is connected to the remainder of the opening member along a bend line, which may be defined by any appropriate structural feature, such as the upper bend of the joggle (38). The bend line (43) provides a hingeable interconnection between the tab and the remainder of the opening member. This interconnection is further defined and enhanced by two curls (45*a*) and (45*b*), which are formed at the juncture of the bend with the two sides of the tab.

for some all-thumbers who can't seem to work them properly. Again, the answer will come sometime in the future—and sooner, rather than later. The market is there. For the inventor who comes up with the solution—and gets a royalty on the product—there may be a great deal of money. All that's needed is a top that is easy enough for the average four-year-old to open and that will not create any litter separate from the can. For a recent invention in this area, by Carl Strobe and Robert Wells of Georgia, see Fig. 8.14.

Not many things can be bought today that are not packaged. This means that in most cases the consumers can see what they've bought only after having had the time to get home and unwrap it. Generally, this does not present a problem. One doesn't have to look into a package of cornflakes to make sure that they've been toasted just right. Any standard product such as this is manufactured under pretty stiff quality control that ensures uniformity. Sometimes, the taste is made to appeal to the lowest common denominator, but that's another story. One shipment is like any other, and if you don't like what you get, you can often get a refund or replacement.

Other foods—like meat, fish, and vegetables—are a different matter. They're perishable, made by Nature in a marvelous variety of colors and textures and sizes and shapes, and variable in quality from season to season. So you don't know what you're buying until you've seen it. For years, supermarkets packaged their perishables in cardboard trays covered with clear plastic wrap. All too often, the best-looking parts of the food faced front. The good red meat was right there for everyone to see; the fat, gristle, and less desirable areas were often put face down. Consumer complaints and new laws in some states have changed this system. Many trays are now open to view, top and bottom. Richard Reifers of Connecticut and Henry Lord of Maine combined their talents to develop an open tray that they have patented and assigned to the Diamond International Corporation. Their invention has two parts, actually. The tray itself is shown in Fig 8.15.

There is surely room for improvement here, in packaging not only perishables but other foods as well. There are some products that look so good that it would be an advantage to both seller and maker to have them out on display. And as consumers grow suspicious and want to see what they're getting, there will be a need for inventors to design packages that will let the sunshine in.

In the first part of this chapter, we discussed shipping containers and protective packages. Sometimes it has been possible to economically adapt these packages so that they can serve as display units in stores. Prepacks, dump bins, racks, and cases of all types are being used by manufacturers to make it easy for retailers to display their products. Small packages just aren't put out on counters by themselves very much these days. For one thing, many of the display racks are attractive, eyecatching, and compact. They're easier for the store people to handle and to reorder. And perhaps most important, they can be made so as to discourage shoplifters. Figures 8.16 and 8.17 show two approaches to the display of an increasingly popular item that is easy for the dishonest

figure 8.15 Top plan of an open-bottom packaging tray for meat, illustrated in Patent No. 3,894,679 (1975). Instead of the conventional flat bottom, the tray (10) has marginal ribs (18) enclosing inner ribs (20) between the bottoms of the sidewalls (16).

The ribs form in cross section a pair of vertically extending sidewalls in reinforcing contact, with their apices forming the meat-supporting surface. The apex of each rib is well-rounded so as not to cut or otherwise damage the meat. Each inner rib (20) meets at each end thereof with three other inner ribs in a recess (22). Each marginal rib (18), on the other hand, meets at each end thereof with another marginal rib in a notch (24).

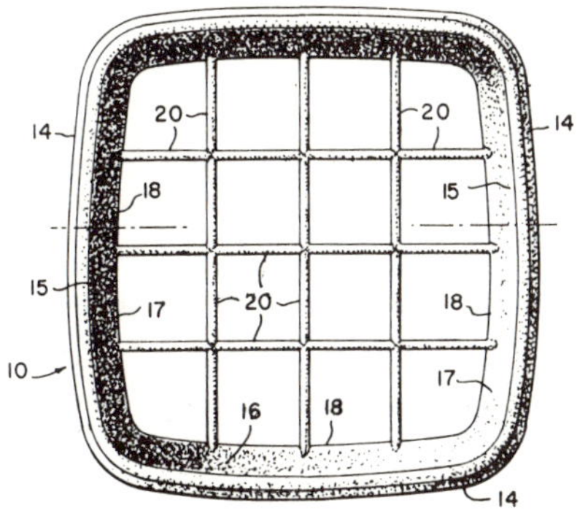

figure 8.16 Cross section of the display case illustrated in Patent No. 3,915,306 (1975).

The structure of the display unit comprises a base (20) that is supported by a four-legged frame (22), sidewalls (24), a sloping front wall (26), and a top (28) that serves as a counter. Leveling screws (23) are threaded into the open ends of the legs of frame (22) to permit the adjustment of the display unit to the surface on which it stands. Top (28) is divided into a rearward part (30) and a forward part (31) that are connected by a hinge (32). A flange (35) is rigidly connected to the forward part and extends downward on the front of display unit (10). A lock (37) is located in the flange.

The back of the unit is provided with several storage shelves (40) on which extra tape cartridges such as 8-track tapes (42) and cassettes (44) are stored. The shelves are supported by sidewalls (24) and a vertical panel (46) that also keeps the tape cartridges from falling into the interior of the display unit. Sliding glass doors (50) permit sales personnel to view the storage area and locate any tape cartridges they may be seeking. A conventional lock (52) is used to prevent unauthorized movement of sliding glass doors (50).

On the front of the unit, cylindrical, generally parallel rods (60) are mounted in brackets (62) and (63) that are secured to front wall (26). A flange or threaded nut (65) at the upper end of each rod (60) keeps the rods from falling through holes in the brackets, and a spacer (68) of suitable length ensures that the total length of the elements on the different rods is approximately the same.

On each of rods, there are housings (70) adapted for the display of tape cartridges. Each housing has a first recess with an open end where there may be inserted the base of a tape cartridge and a second recess with an open end where there may be inserted the top of a tape cartridge.

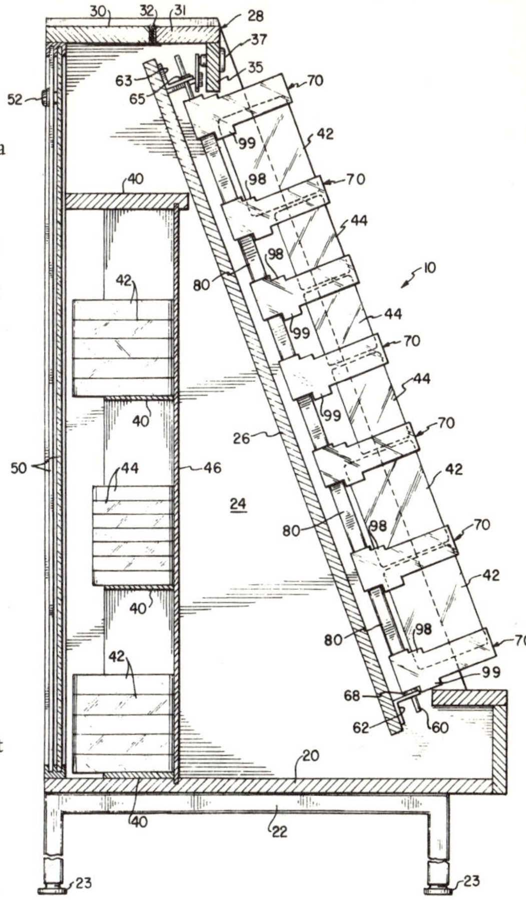

Housings are spaced apart from one another by spacers (80), each of which is a hollow shaft with a square cross section.

figure 8.17 Perspective view of a cassette rack, Patent No. 3,907,116 (1975).

There is a holder (11) suitable for storing and displaying a number of industry-standard, Philips-type magnetic tape cassettes (12), either alone or in protective boxes (13). The holder is constructed to accommodate the greater dimensions of the boxes, and has two spaced, parallel-rack side pieces (14) having integrally formed thereon a number of confronting shelves (16) for holding opposite ends (17) of the boxes. Sidewalls (18) of the side pieces are spaced apart for a dimension slightly greater than the greatest dimension or length of the box, i.e., the dimension between the ends of the box, while the shelves are vertically spaced apart, in the holder position illustrated, for the thickness of the box. Thus the boxes may fit into the shelf recesses with reasonable snugness when the holder is used as a retail sales packaging framework, e.g., in cooperation with a transparent plastic-sheet wrapping, particularly of the well-known "shrink-wrap" type. The shelves slope downward and rearward in the holder orientation illustrated, and each side piece has a back wall (19) defining a backstop for retention of the boxes on the shelves. The depth of each sidewall from the front of the holder to the back wall is approximately the same as the width (depth) dimension of the box so that the box is held with its front side (21) substantially in the plane of the front of the holder for best display of labels. The two side pieces are held together by spacing bars (22) of L-cross section, preferably integrally formed with the side pieces as part of a plastic-molding manufacturing process.

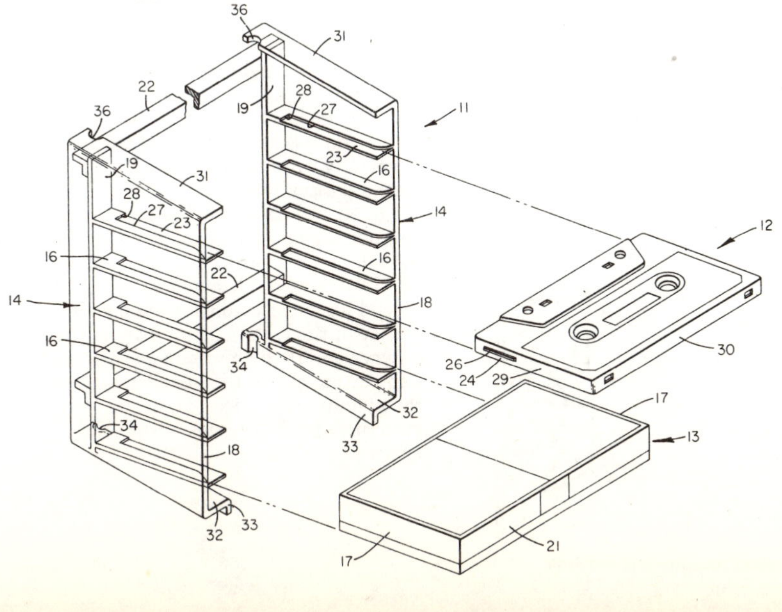

shopper to slip into pocket or purse—stereo tapes and tape cassettes. Many merchants have chosen to lock these items in a glass case or to put empty cartons on display for customers to examine. But inventors have found some better ways. Figure 8.16 shows the plan for a more-or-less permanent rack that can be refilled with various sizes of tapes and cassettes. Four inventors—Frank Desisto of New York, John Madaio of New Jersey, and William House and William Phillips of Massachusetts—assigned their patent for this display case to the J. C. Penney Company. Their design has a

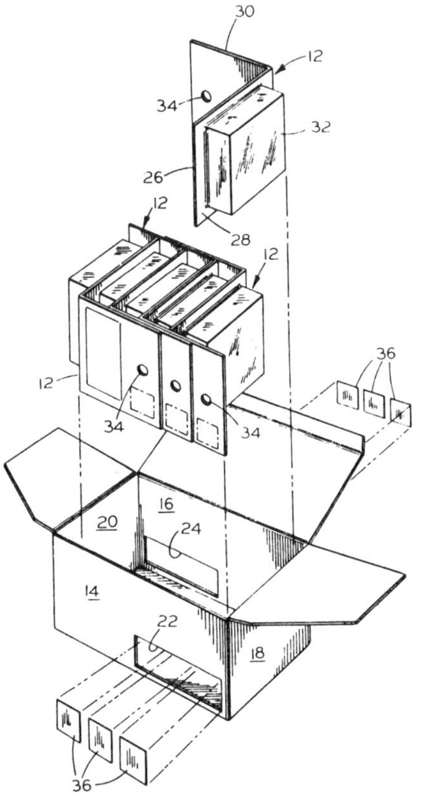

figure 8.18 A display rack that can hold a number of different packages of varying shapes and sizes, illustrated in Patent No. 3,872,966 (1975).

The container (10) is a conventional six-sided container of paperboard or the like and has two parallel sidewalls (14) and (16) of generally elongated dimension extending between the end walls (18) and (20). Each sidewall has an elongated aperture (22) and (24) adjacent the lower longitudinal margin thereof, and the aperture (22) is offset to one side of the centerline of its sidewall (14) and extends closely adjacent the end wall (18). The aperture (24) is offset to the opposite side of the centerline of its sidewall (16) and extends closely adjacent the end wall (20). The two apertures (22) and (24) thus extend over substantially the entire length of the container (10).

Each of the merchandise cards (12) includes a substrate (26) having a body portion (28) and a folded end portion (30). An article of merchandise (not shown) is secured to the body portion by means of the transparent blister (32) which overlies the article and which is adhesively bonded to the substrate. The end portions are provided with apertures (34) by which the merchandise cards may be hung on hooks or other suitable means so as to suspend them for display purposes. Price or other indicia in the form of labels (36) are adhesively affixed to the end portions of the merchandise cards (12) so that they may be viewed through the apertures (22) and (24).

Merchandise cards (12) are stacked in the interior of the container (10) and spaced apart by the body portions (28) of adjacent cards. The end portions (30) are folded with respect to the body portions (28) of the several cards so that they extend away from the blister and article of merchandise supported on the substrate (26), parallel to and along one of the sidewalls (14) and (16) of the container. Half of the merchandise cards are disposed so that their blisters are oriented in the direction of the end wall (20), and the other half are placed in the container so that their blisters are oriented toward the end wall (18).

The end portion (30) of each substrate (26) is of a width greater than the thickness or width of the blister and thickness of the substrate body portion (28) so that it overlaps the adjacent merchandise card, thus providing a support for the folded end portion and a high degree of visibility.

In the method of pricing or applying other indicia, the merchandising cards are first folded to cause the face of the end (30) to extend away from the merchandise and blister so that it will thereby be exposed for viewing through apertures (22) and (24). The folded merchandise cards are then inserted into the container in the fashion heretofore described so as to expose the ends (30).

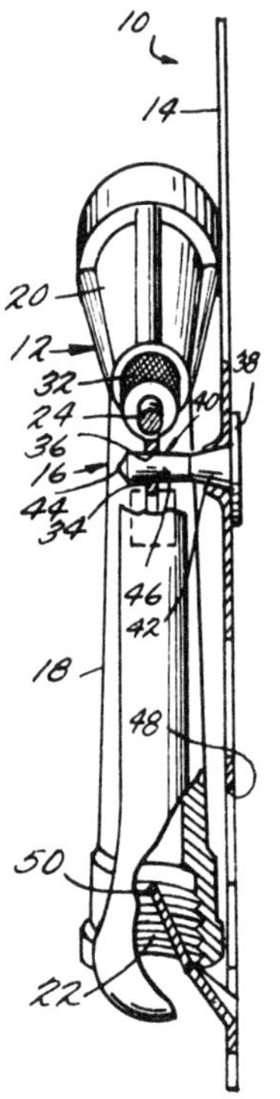

figure 8.19 Side view with parts broken away of the carded nozzle package of Patent No. 3,674,138 (1973). The package is made up of a hose nozzle (12), a card (14), and a fastener for these two (16). This fastener is preferably molded of nylon and includes an enlarged, circular, relatively flat head (38) and a shank structure, generally indicated at (40), extending centrally from one surface of the head. Preferably, the shank structure (14) includes a frustoconical portion (42) adjacent the head (38), a generally conical tip portion (44) at the free end thereof, and a generally cylindrical aperture engaging central portion (46).

The nozzle, card, and fastener may be assembled by positioning the nozzle centrally on one surface of the card and simply driving the fastener with its conical tip end foremost through the opposite surface of the card. The top part of the fastener enters the aperture (36) within the web portion (34) of the nozzle, and finally the cylindrical portion (46) of the shank structure snugly engages within the aperture.

The card is preferably formed with a cutout (48) so as to define a hinged tab (50). The cutout receives the socket (22) of the nozzle, and when the tab is engaged within the socket, this structure not only aids in initially locating the nozzle on the card, but effectively prevents the nozzle from any rotary movement.

The snug fit of the cylindrical portion (46) of the fastener within the web, and the engagement of the relatively large head (38) of the fastener with the opposite surface of the card, provides a strong connection between nozzle and card that will withstand even severe handling. The package can be readily disassembled, however, by utilizing any elongated rigid element such as a screwdriver or the like and prying the fastener out of its engagement within the aperture (36).

number of individual housings mounted on a set of generally parallel rods held in a lockable casing. Two housings join to hold a package in such a way that it can be rotated and examined but not removed. Figure 8.17 shows a rack for both boxed and unboxed tape cassettes designed by Irving Wolf and William Carpenter of California for the Ampex Corporation. This rack can be manufactured inexpensively enough for it to be a giveaway item, though it's sturdy enough to be reused even after all the original shipment of cassettes has been sold. There are no security devices, however, so it would be more useful to some stores than to others. Each shelf has two levels, one on which boxed cassettes can rest, the other for unpackaged cassettes.

There is a discernible trend toward making display racks perform double duty—as advertisement for the small packages they contain, and as pilfer-resistant selling tools for the retailer. There are hundreds of small items that could be handled in this way. Many of them already have been, but some have not. The Stanley Works of Connecticut, known for its line of tools, also produces and sells many smaller items such as nails and screws that are best sold in smaller packages. Two designers, Yale Gordon and John Koley, developed a container assembly to accommodate individually priceable carded packages for the display of Stanley goods in stores (Fig. 8.18).

There would seem to be many items that do not really need packaging at all. Why, then, do manufacturers put just about everything they make for the consumer into some kind of container? Take as one example the hose nozzle. A patent issued to Robert Gilmour of Pennsylvania makes the case for packaging as follows: "The card package permits the article to be commercially shipped in cartons and to be displayed for sale either within the shipping carton or on display peg racks." Packages thus often make things easier for manufacturer, retailer, and consumer—at least that's what the first two groups think, and they're probably correct. So a great many hose nozzles are packaged on cards even though, according to the patent, this "presents particular difficulties because of the shape and weight of the article." A nozzle and its card are soon parted, unless particular care is taken to put them together in the right way. Figure 8.19 shows how Gilmour designed his nozzle package. It works, too; thousands of nozzles have been sold in his packages.

The packaging revolution has been with us for many years, yet has not lost its original thrust. The opportunities for inventors are still great, even though this has become a big business and thus dominated by well-capitalized corporations. There is one significant factor that can help independent inventors: Most of the product manufacturers do not make their own packages. If you had an invention for a new pop-top can, for instance, and wanted to try to license it to a manufacturer, you wouldn't go to a company like Pepsico, you'd go to one like American Can. More important, there are also many smaller companies in the field, always on the lookout for new things. After all, that's the only way they can really compete with the giants.

CHAPTER NINE

Toys and Games

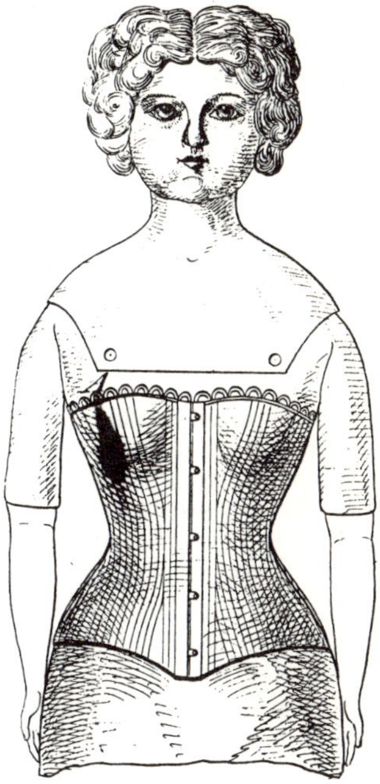

There are some people who, believe it or not, don't think of inventors as modern Thomas Edisons. Some have even expressed the notion that our creative geniuses-in-the-raw are just overgrown children, locked in the workshop, playing with their toys. Sometimes, it must be admitted, that's just what inventors are doing, because toys and games offer a good field for the introduction of new products. There's an almost insatiable market for children's playthings and for games to be used by people of all ages. And despite the lower birth rate, more than 4 million new babies and their parents join this big market each year.

Many purchasers of toys and games want the latest thing, this year's hot new gift. Toy stores often emphasize the novel and the original, especially if the merchandise has had the benefit of TV advertising. Surprisingly often, however, something will catch the public's fancy without benefit of mass-media promotion. Toy stores do, of course, stock some items that have been around for a while, but generally only those classics that have sold well from year to year. Generations of children have rolled along in their little red wagons, and no game closet is complete without a Monopoly set.

Why does one toy become a classic and another one end up on the junk pile? Some manufacturers and sales organizations think they know the answer, since they have the benefit of years of experience, volumes of market research, and all kinds of other data. Even they make mistakes sometimes. The shifting tides of public opinion are difficult to fathom, and sometimes the sure bet to become the toy of the year becomes one of the duds of the decade.

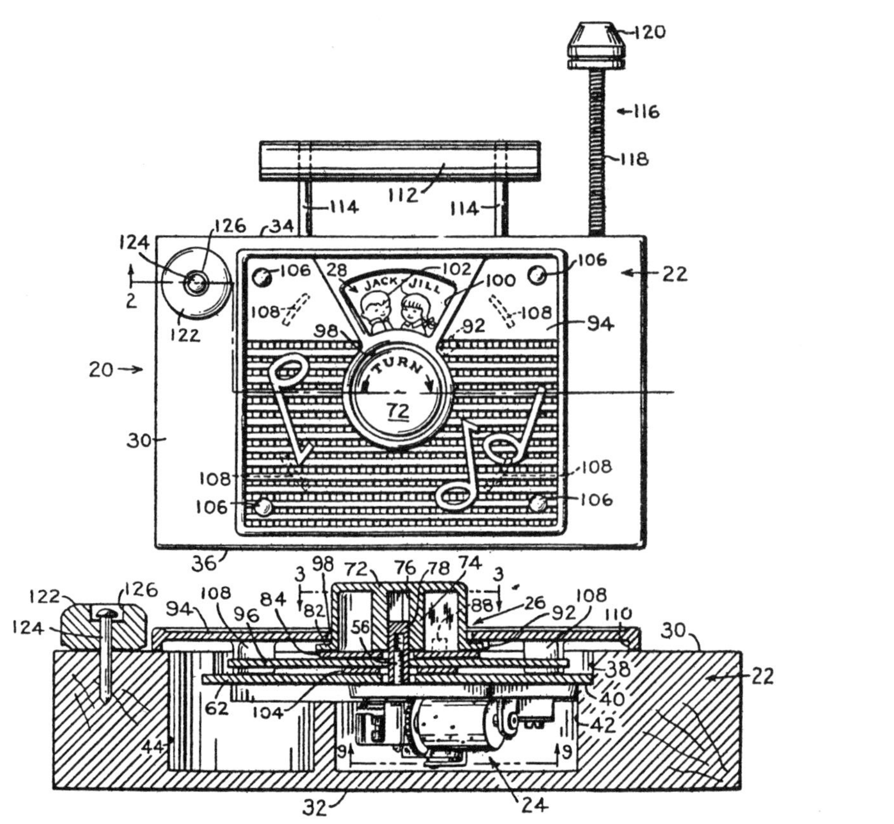

figure 9.1 Front view and enlarged cross section of the best selling play radio described in Patent No. 3,114,216 (1963). The toy (20) is comprised of a case (22), a sound mechanism (24), a winding means (26), and a series of movable pictures (28). The case (22) resembles an adult portable radio. A music box (24) has a rotary drum with radially extending prongs. When the drum is rotated, these prongs engage beneath the ends of a set of tines that produce a tone when vibrated. The drum is rotated by a resilient spring motor (54) aided by the rotation of a threaded shaft (56).

The motor has a knob (72) that is adapted to turn clockwise to increase the energy stored in the motor. This knob is formed with a central boss (74), which in turn is formed with a square recess for receiving an adapter (78), which is substantially square in cross section, but has a threaded opening to receive the end of the shaft (56) as it extends up from the motor through the mounting board (62). When the knob is turned clockwise, energy is imparted to the motor, and when the winding action is over, the motor actuates the music box. There is a novel mechanism for limiting the amount of winding that can be done and at the same time preventing the rapid unwinding of the device. Knob (72) has a single radially extending web (80) that is substantially flush with the bottom of boss (74) and flange (82), which extends around the periphery of the knob. An idler is located directly beneath the knob and is provided with a circular central opening of sufficient diameter to allow the idler to rotate freely relative to the square shaft. The idler has an upstanding projection midway between its center and edge. This lies in the path of a web on the knob when it is rotated relative to the idler. A second projection extends radially and is adapted to engage the opposite sides of a fixed part or stop (92), which is located on the under surface of face plate (94).

Starting with the winding mechanism in its at-rest position, it will be seen that the knob may be wound clockwise for substantially a full revolution until the web strikes against the boss, after which further rotation of the knob will carry the idler with it. This winding mechanism is also well adapted to protect the spring motor from being driven beyond its normal rate of speed, as would occur when the winding knob is forcibly unwound. To this end, there are sufficient threads on adapter (78) so that it requires considerably more than two revolutions to fully thread it on, or unthread it from, shaft (56).

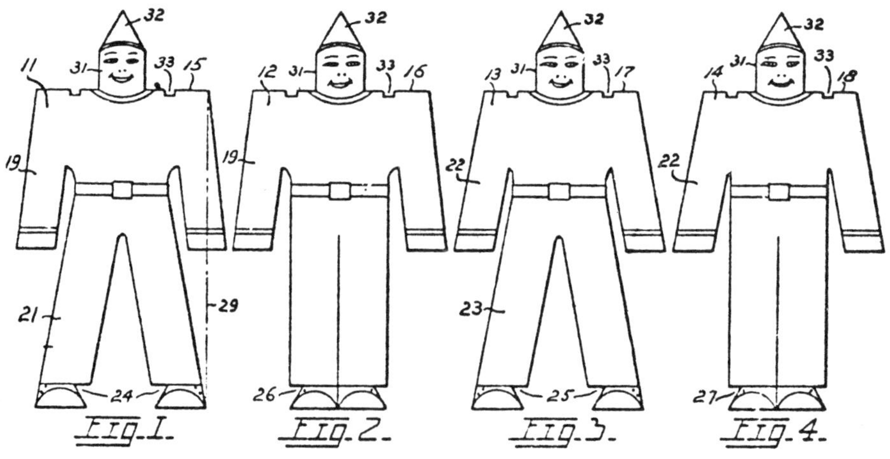

figure 9.2 Four different types of wooden clowns that can be joined to form all kinds of pyramids, towers, and the like, as described in Patent No. 2,078,709 (1937). In the drawings, no. 1 is a typical clown with legs spread; no. 2 is identical with respect to arms and upper portions, but has the legs close together and the feet shaped differently; no. 3 is closely similar, too, but has narrower shoulders for the purpose of making the angle of divergence of the arms the same as the angle of divergence of the legs; no. 4 is identical with no. 3 as far as arms and upper portions go, but has the legs close together and the feet shaped differently. The lower portions of nos. 1 and 2 are identical in dimensions with the lower portions of nos. 3 and 4, respectively. The shoulders (15) and (16) of nos. 1 and 2 are of equal width and are wider than the shoulders (17) and (18) of nos. 3 and 4. The latter are of the same width as each other. The angles of divergence to the arms 19 and 22 and legs (21) and (23) are as shown. Clowns 1 and 3 have similar V notches (24) and (25) on the inside of the legs. Clowns 2 and 4 have similar notches (26) and (27) at the outside of the legs.

In each kind of clown the vertical distance from the shoulders to the lower extremities of the arms is made the same as the distance from these extremities to the bottoms of the feet. The spread of the feet of nos. 1 and 3 is the same as the width of their shoulders. The divergence of arms (19) is made such that a line (29) drawn between a shoulder corner and the corresponding outer corner of the foot bisects the lower extremity of the arm.

Head and neck portions (31) are surmounted by pointed helmets (32). Each shoulder has a transverse rectangular notch (33) at a distance from the neck that is approximately equal to half the width of the neck. Notches are adapted for engagement by the apexes of the helmets, and by many other corners of the clowns. Thus they help to properly position the clowns in groups and to maintain them in equilibrium.

One classic toy is the Fisher-Price music box designed to look like a portable radio. Patented in 1963 by Ralph Crawford and Leslie Sly of New York, it offers some interesting insights into the art of toymaking. First, it imitates an adult (and, perhaps more important, a teenager's) possession, a radio, in a form that is pleasing and useful to young children. It's a clever adaptation of an old favorite, the music box, but is simpler to operate and less susceptible to breakdowns than most music boxes. Taking into consideration the average age of its users, this is not only desirable, it's essential. The radio is relatively inexpensive to manufacture, having a minimum of moving parts, yet it doesn't look or feel cheap. It's safe for children of all ages, who can often wind it up themselves. Like many toys with staying power, it appeals to the adult purchaser and the child who plays with it. (See Fig. 9.1.)

There are many other classic toys that have continued to be produced by their original manufacturers even after the protective patents have expired. Building materials such as Lincoln logs and A. C. Gilbert's Erector sets were popular for many years. Children, especially boys, have always loved to build things. Recently, some successful innovations in the field of light construction have come from abroad. The two best known are the Lego System from Denmark and the Fishertechnik materials from Germany. Their building sets are of high quality. Both of these lines are on the expensive side, and have been sold more in specialty shops than in regular toy outlets. This is changing, however, as they move to fill a partial gap left by a dearth of American-made building sets. Lego, incidentally, was developed by a poor Danish farmer during the Depression. He hand carved playthings for his own and neighboring children and eventually began to sell to others. The company he started is now run by his son.

The time is now ripe for some brand-new entries in this competitive market. Erector sets are enjoying a revival under the sponsorship of the Gabriel Corporation. Some other companies have made modest, often unsuccessful attempts to sell their own building sets. The potential rewards are large. An inexpensive yet well-put-together American product could take over a good share of the market. But it cannot be junk.

Young children like to build, too—witness the durability of blocks of all kinds. One of the most interesting designs for this age group was patented first in 1932. Inventor Clinton Hecht of California called it a "means for exercising ingenuity in marked degree in order to make a large and indefinite number of possible arrangements" (Fig. 9.2). And what could be more fun than building up a structure of these clowns, watching it teeter as you put that last impossible touch to the pile, and then seeing it fall to the floor with a clatter!

New toys are often based on old ones. In one successful invention, Helen Malsed of Massachusetts took a spring—in this case a patented version that was sold under the trade name Slinky—and adapted it for use in a toddler's pull toy. The original invention, by Richard James of Pennsylvania, is shown in Fig. 9.3*a*. In itself it is a clever variation on the spring principle, which not only pulls far

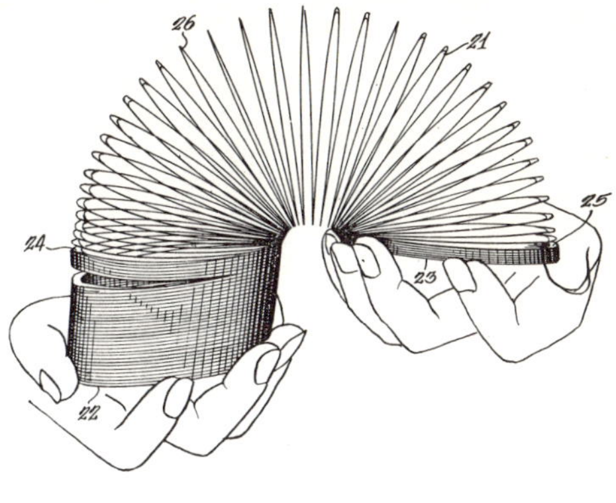

figure 9.3(a) One method of handling the Slinky, Patent No. 2,415,012 (1947). The ends of the coil are grasped in the hands and the coil is bent into an axial semicircle (21) with a lower end (22) and a higher end (23), with the effective top of the solid pile on the low end at (24) and the effective top of the solid pile at the high end at (25), the intermediate coils between (24) and (25) being fanned out at (26).

The turns of the coil are transferred from the solid pile (24) to the solid pile (25), while retaining the fanlike intermediate coils (26). Then, a relative upward movement of the coil end (22) results in a decrease in pile (24) and an increase in pile (25), while turns are transferred across the fanlike portion. The oscillating action can be transferred to a walking motion by the use of a platform such as an inclined plane or a set of steps. The solid pile at the rear end in this case is transferred to the pile at the low end. When the end of the rear pile is reached, the last turn of that pile and the fanlike portion will themselves be transferred across to the lower end. The energy stored in the system will next carry the topmost turns over to the next landing, building up a new solid pile. Assuming a proper incline is used, this action will continue until the end of the platform is reached. The start of this operation may be accomplished either by knocking over the top of the pile downhill, or by bending the coil into an axial semicircle with the lower end placed against a lower landing.

In construction, this helical spring toy has its individual turns made up of spring stock. The coil must be wound without any substantial compression or tension (lateral force) in closed position between the turns when no external force is acting. For best results a suitable spring steel such as S.A.E. 1050 to 1095 is desirable, suitably heat-treated to develop springiness.

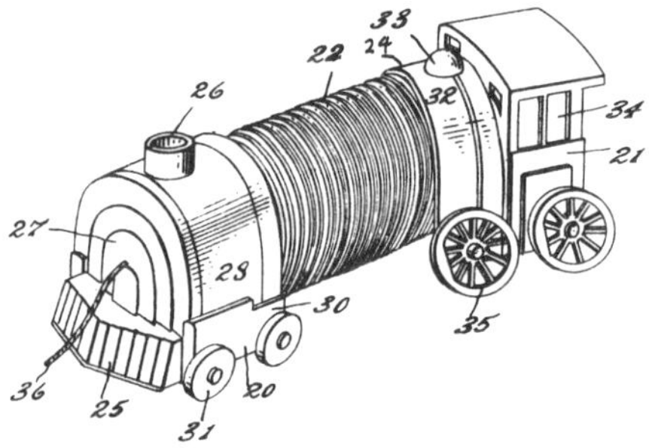

figure 9.3(b) Perspective view of the toy engine shown in Patent No. 2,862,330 (1958). The vehicle front (20) is separated from the vehicle rear (21), and the two are interconnected by a helical spring (22) that is secured at one end (23) and the other end (24) by any suitable means—adhesive, metallic fastening, etc. The locomotive front has a cowcatcher (25), a stack (26), a boiler front (27), and boiler top and side portions (28). A truck is simulated at (30), and mounting wheels (31) on each side support and roll the front of the locomotive. The rear section (21) includes a top and side portion (32) of the rear of the boiler, a steam drum (33), and a cab (34). The wheels (35) support and roll this section. The body of the locomotive boiler is simulated by the helical spring, which is the entire interconnection between the front and rear sections. A pull cord (36) is connected to the front. When the child pulls forward, the front section will come along, but the rear section will remain essentially stationary as the spring elongates. Then, when the spring is under tension, the rear will move forward rapidly and the spring will close up, after which the entire process will be repeated.

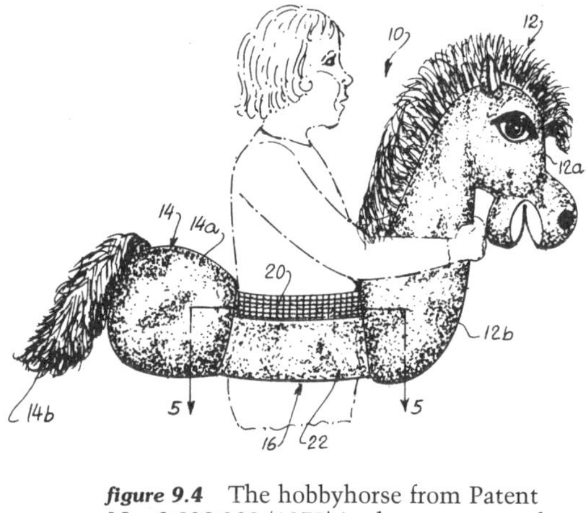

figure 9.4 The hobbyhorse from Patent No. 3,920,239 (1975) is shown as worn by a child. The horse is made up of a fore section (12), a hind section (14), and a pair of connecting side sections (16) and (18). As illustrated, the construction is like that of a stuffed toy, with head (12*a*) and neck (12*b*) and rump (14*a*) and tail (14*b*). Each of the sides consists of an elastic band (20) with a gathered strip (22) sewn onto it. These sides are sewn to the front and rear sections of the horse so that it can be worn around a child's waist. The elastic bands stretch and the side strips expand to accommodate the waist, so that the horse closely conforms to the contours of the body and clings to it.

apart and then comes quickly together, but also "walks" down stairs and can be made to perform other tricks. Ms. Malsed's toy is illustrated in Fig. 9.3*b*. It embodies no great technical breakthroughs—in fact, it's quite simple—yet the idea is such a good one and the execution is done so well that the toy became a success.

The hobby horse is another old favorite that has been updated many times. Rod White of New Jersey developed a modern, inexpensive version of the horse, which can be worn around a child's waist (Fig. 9.4). Not much to it, a critic might say, but fueled by the power of a child's imagination, this hobby horse can become the mightiest steed north of the Pecos. It's a stuffed animal that any cowboy would be proud to own.

Along with their horses, all cowboys—and Indians—need guns. So do soldiers, cops and robbers, spies, and boys of all ages. At one time or another, just about every American boy goes through the "Bang! Bang! You're dead" stage. Most girls seem to escape it. Some boys never grow out of it. Child psychologists who deplore the violence in American culture are just as likely as dedicated NRA members to look out the window and see their kids playing guns. Despite some restrictive legislation and pressure from consumer groups, toy manufacturers continue to make and sell replicas of deadly weapons.

This is another category where you might think that just about everything in the way of invention has been tried—that there are really few opportunities left for the inventors of today. But that is not true. The conflicting desires of two groups have not yet been fully resolved: The youngsters want realism, their elders want improved safety. Most children want a gun that shoots "bullets," a thought that sends many parents up the wall. In 1969, Edward Nielsen of New York patented a double-barreled shotgun that was produced by the Ideal Toy Corporation and is still available in stores (Fig. 9.5*a*). The emphasis here is on realism. Spring-loaded cartridges are provided that can be put into each barrel of the gun. One pull of the trigger releases the latch on first one projectile and then the second, so that the gun can be fired with an authentic double-barreled "bang-bang." Figure 9.5*b* illustrates a more recent invention by Arthur Lange of New Jersey. A safety feature has been introduced. The gun has been fixed so that it will fire only projectiles that have been specifically designed for it. Thus, if a child loads a stone or some other potentially lethal object in the gun, it won't fire. The "bullets" to fit the gun can be designed with a rubber suction cup as illustrated or in some other way so as to be less dangerous.

Safety has also been an important consideration with other toys, especially the equipment for such games as darts and archery. Suction tips were used for many years, but they do not always stick to the target with a high degree of reliability. Jerome Lemelson of New Jersey has been working off and on for almost twenty years with Velcro, a material that has adapted very well to use in target games. Velcro consists of two strips of flexible, wool-like material. One strip has thousands of tiny microfilament hooks protruding from its surface. The other strip has thousands of tiny wirelike fila-

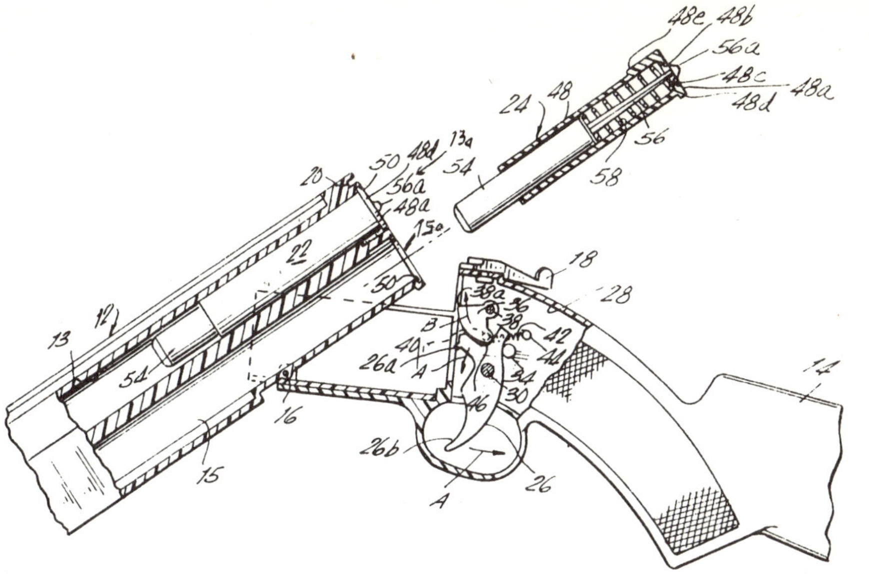

figure 9.5(a) Structural details of the trigger and back end of a toy double-barreled shotgun, Patent No. 3,465,744 (1969). The gun has a front barrel structure (12), consisting of two shotgun barrels (13) and (15) mounted one on top of the other, and a shotgun stock (14) pivoted at (16) to the barrel structure (12). Pivotally mounted on the stock is a locking lever (18), which cooperates with a locking lug (20) to keep barrels and stock together. When this lever is unlocked from the lug, however, the barrel can be pivoted at (16) so that the rear openings (13*a*) and (15*a*) of the barrels are accessible for loading of toy shotgun cartridges (22) and (24). Then, when the shotgun is locked, squeezing the trigger causes the successive discharging of these cartridges.

The forward end of the stock has a chamber (28) where the trigger (26) is housed. The upper portion of the trigger is mounted on a pivot pin (30); the lower portion extends to an accessible position within a guard (32). There is a finger grip surface (26*a*). The drawing shows the trigger cocked, seated against a stop pin (34) and pivotal in a counterclockwise direction. Abutting the upper part of the trigger is a cartridge-discharging device (36), which pivots at point (38). A pin (40) is attached to a lower leg of (36) and is engaged with one end of a return spring (42), which is connected at its other end to a stationary pin (44). In this manner the spring forces the discharger (36) into starting position. A cam surface (46) and a cam projection (46*a*) are in contact with each other. Cartridge-striking edge (38*a*) moves to discharge the upper cartridge, and cartridge-striking projection (26*a*) on the trigger discharges the lower projectile.

The construction of each of the cartridges (22) and (24) is identical. The cylindrical housing has an end wall (48) that includes a large opening (48*a*) and a small opening (48*c*). There is also a radial flange that in practice is accommodated in a seat formed by counterbore (50) in each of the barrel openings (13*a*) and (15*a*). There is also a positioning lug (48*c*) that must be accommodated in a slot (52) formed in the wall bounding the barrel openings. Thus, in order for the cartridges to fit properly into the barrels, (48*c*) must be aligned with (52). The cartridge in the upper barrel has the larger wall opening (48*b*) uppermost; the cartridge in the lower barrel has this larger opening on the bottom.

A projectile (54) slides into cylindrical housing (48). Latch (56) extends from one side of the body of each projectile and has a head (56*a*) that is smaller than the large opening (48*b*) so that it can be inserted through this opening. Between the projectile and the end wall is a compression spring (58), which can force the projectile from the housing when the latch (56) is released from the end wall.

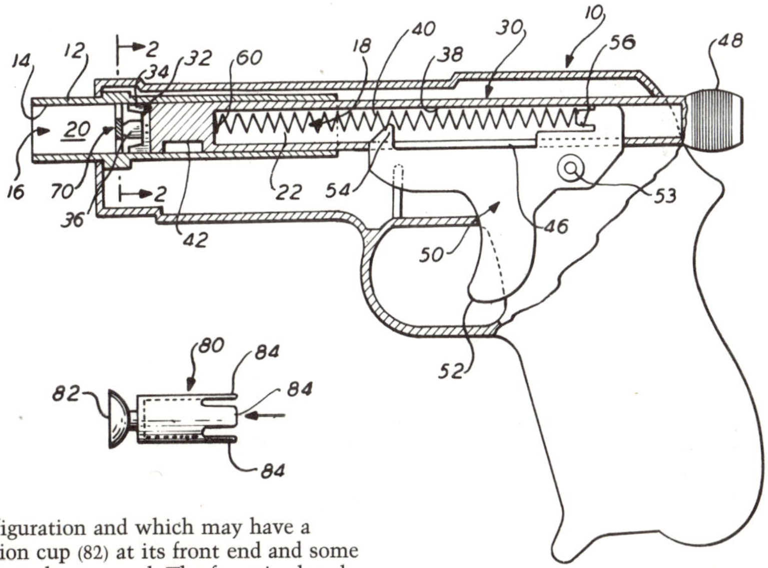

figure 9.5(b) Partial cross section of a toy gun with (inset) a projectile designed for the gun, as shown in Patent No. 3,859,977 (1975). The frame (10) has a cylinder (12) with a bore (14). The latter has an open front end (16), an open rear end (18), a forward portion (20), and a rear portion (22). The spring-operated plunger (30) has a piston (32). The latter has a seal (34) and a stop (36), both secured to the front end of the plunger. A longitudinal bore (38) is formed therein, and received within it is a compression spring (40). There is an indented portion (42) formed on the underside of the plunger, along with a longitudinal slot (46). At the rear of the plunger is a knob (48), which may have knurling, as shown, to facilitate gripping.

There is a trigger (50) with a lower extension (52). This trigger extends upward into the plunger through slot (46) formed therein with the rear portion of the compression spring engaging part (56) of the trigger and with the forward end of the compression spring engaging a wall (60). Mounted in the bore (14) between the forward and rear open ends (16) and (18) are baffles (70), which have a number of openings that are formed in complementary configuration to a number of rearwardly extending parts on the projectile.

Inset shows one of these projectiles (80), which is of generally cylindrical configuration and which may have a suction cup (82) at its front end and some nubs at the rear end. The front is closed, and the rear is open. The projectile is received within the forward part of the bore, with the nubs (84) being inserted first in the forward open end. They extend through the openings formed in the baffle and into the rear part (22) of the bore. Foreign objects such as pencils, stones, marbles, etc., cannot be propelled out of this gun unless they have a configuration like the projectile (80). The forward movement of the piston within the rear part of the bore is halted upon the engagement of the baffle by the piston. A foreign object will not extend through the baffle, and thus will not be engaged by the piston.

ments sewn into loops that also protrude from the surface. The hooks randomly engage the loops, and the surfaces are joined in a bond that can be broken by several ounces of pressure. In other words, if you throw one surface against the other, they will stick together quite well, while at the same time being relatively easy to remove one from the other.

In 1962, Lemelson patented a game using Velcro for target and dart points (Fig. 9.6*a*). Even if the illustrated dart struck a player in the eye, damage would be minimal. But there was one bad feature—the projectile did not always stick to the target surface. In a 1975 patent Lemelson and another inventor, Allan Elfman of New York, described a new version of the dart game that solved the problem by designing a target that was a bit flexible and would "give" when struck (Fig. 9.6*b*). This prevented the Velcro patches from bouncing off each other. Their target game has been sold for use by youngsters and safety-minded adults.

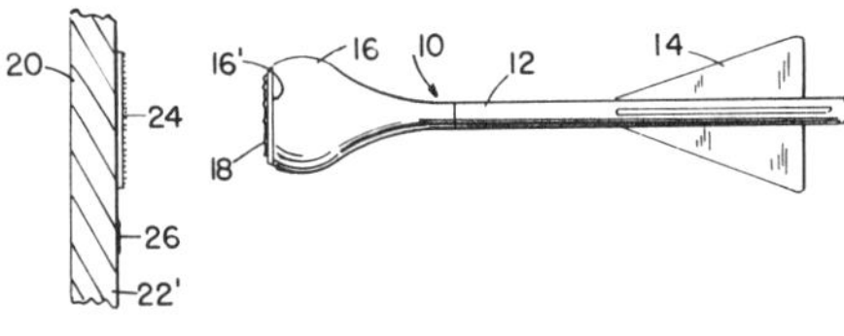

figure 9.6(a) A dart and target as illustrated in Patent No. 3,032,345 (1963). In this game a ball can also be used instead of the dart. The dart (10) is made of plastic and comprises an elongated body (12) having thin, sheetlike guide fins (14), and a head (16) that is weighted to help the dart to ascribe a stabilized path when projected or thrown. Secured to the front face (16′)is a small patch (18) of Velcro closure material.

The target (20) consists of a board (22) with a number of targets (24), which are patches of Velcro. The short, projecting plastic filaments or hooks from dart and board will engage each other and cause the dart to be held against the target, thereby indicating a hit. When the nose of the missile strikes the areas (22′) that do not contain the patches (24), the dart will not be held but will merely bounce off or slide down the surface of the target. The notation 26 refers to numerical markings on the surface of (22′) adjacent the target patches (24) provided to indicate the score.

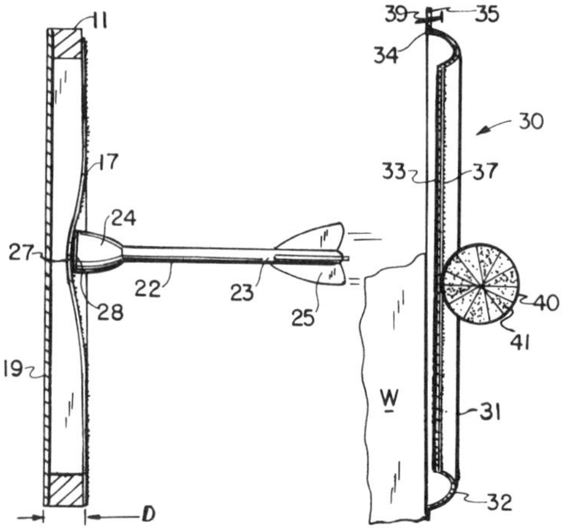

figure 9.6(b) Side view of target with (left) dart and (right) ball, Velcro attached to Velcro, as illustrated in Patent No. 3,857,566 (1975).

Placed across the frame and fastened along its border portion (18) is a rectangular portion of Velcro material (17), which engages with the fastening material secured to the head (24) of a dart (22). The material (17) is flexible and is drawn across the frame a distance (*D*) away from the rear of the frame. The Velcro will be deflected toward the wall when a missile (22) strikes it, as illustrated. This permits the hooklike filamentary formations of the material (17) to engage and retain the broken hook formations defining the patch (27) of fastening material secured to the dart. Thus, the dart is decelerated in such a manner that it and the target engage each other and are not pulled apart, as might be the case if the material were secured to a rigid board or were directly against the surface of the wall.

Action, realistic action, is what most children want from their toys. Miniature automobiles have been a surefire success with them since the days of Henry Ford, and inventors continue to work in this rewarding field. The little car in Fig. 9.7, invented by Chet Wolgamot and Harvey Springer of Ohio, takes its cue from the stunt shows that crash their way across the country. This vehicle has a retractable wheel that upon contact with the surface causes the car to skid realistically. It also has a ejector capable of catapulting a figure out of the driver's seat when the bumper hits an obstruction.

Cars are popular toys; so are guns. Three men from Illinois have taken the logical step and invented a "combination of a small toy vehicle fitted into a simulated ammunition shell, and a simulated rifle-type launching implement." Howard Morrison, Albert Keller, and an invention promoter named Marvin Glass have patented a car that goes roaring off at high (scale) speeds and a "gun" that also includes several safety features (Fig. 9.8). An interlocking mechanism makes it impossible to fire the launcher unless it is flat on the floor or ground. And the mechanism will fire only when it is loaded with the vehicle/shell combination made for it. If it could fire anything from any position, it would of course be used as a kind of gun by some daredevil youngsters.

It's a relatively small step from launching a tiny car to providing the power to lift off a small rocket. Many patents have been granted for many different kinds and sizes of rockets over the years. Long ago, the ideas of America's pioneer in the space field, Robert Goddard, were adapted for use in small-scale toy rockets. And since Sputnik, a great many new kinds of these things have been patented. In general, they have gotten more sophisticated, more powerful, and less cumbersome through the years, following the lead of their large-scale counterparts. In 1976, Charles J. Green of Washington State patented a two-stage rocket with an engine fueled by a self-pressurizing liquid safer to use than some of the other rocket fuels (Fig. 9.9). The stages are designed to separate when the propellant is all used up. As the descent begins, the center of gravity shifts so that a "soft landing" can be achieved and the rocket used again.

figure 9.7 Toy automobile that can skid and that can catapult its driver, as illustrated in Patent No. 3,984,939 (1976). Note that the drive wheel extends slightly farther below the bottom of the body than do the four skids (62), which have the general appearance of wheels. These skids may have rotational capabilities, but this is not necessary, since most of the toy car's weight is supported by the drive wheel. Rotation of this wheel on a surface would impart to the body (22) a linear motion in the direction of the arrow (64). Based on the gyroscopic effect of the rapidly rotating drive wheel (30), the body will tend to continue to move in the direction of the arrow unless something deflects it.

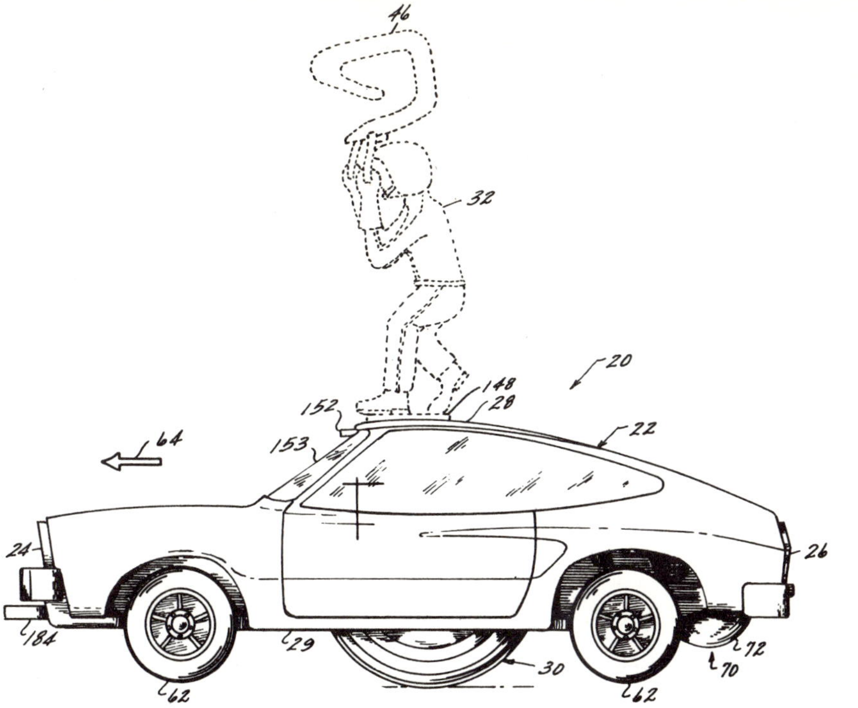

The skid-inducing means includes a rotatable wheel (72) suitable for engaging the surface on which the drive wheel rides. The wheel is journaled near one end of a bracket having a generally elongated shape, the other end of which cooperates with a carriage by which the bracket and wheel are supported. By virtue of a pair of pivot pins at each of its lateral sides, the carriage is supported on the car's body. By means not shown the chassis and the body are held together and the pins are secured. These pins permit the carriage to pivot through a predetermined arc.

Wheel (72) and its supporting bracket are also pivotable by virtue of rotatable engagement between a circular portion of the bracket and an aperture in the carriage surrounded by a generally cylindrical boss. The length of the bracket, the radius of the wheel, and the disposition of the bracket and the carriage with respect to the chassis are such that, in the position shown, the wheel is retracted. Engagement of this wheel with the surface can cause a deflection from motion of the auto from the direction 64 and can result in the appearance of skidding. There is also a manual disabler for overriding the actuator and preventing the rotation of the carriage. This disabler includes a lock arm (152) that projects through a slot in the windshield (153) so that it can be worked easily by hand.

To simulate the catapulting of a person from part of the automobile, there is a special bumper combination at the front of the body. A part of this bumper is an extension (184), formed as part of a pivotable rocker arm that includes an elongated rod. This rod is moved from its normal position when extension (184) engages another object. In operation, a toy figure (32) is mounted, and the automobile is directed so that it crashes into a wall or some other obstruction. This causes the rocker arm to pivot, thereby thrusting the rod into engagement with the lower portion of the figure. When the figure rebounds from the rod, it is catapulted from its position atop the fenders of the car.

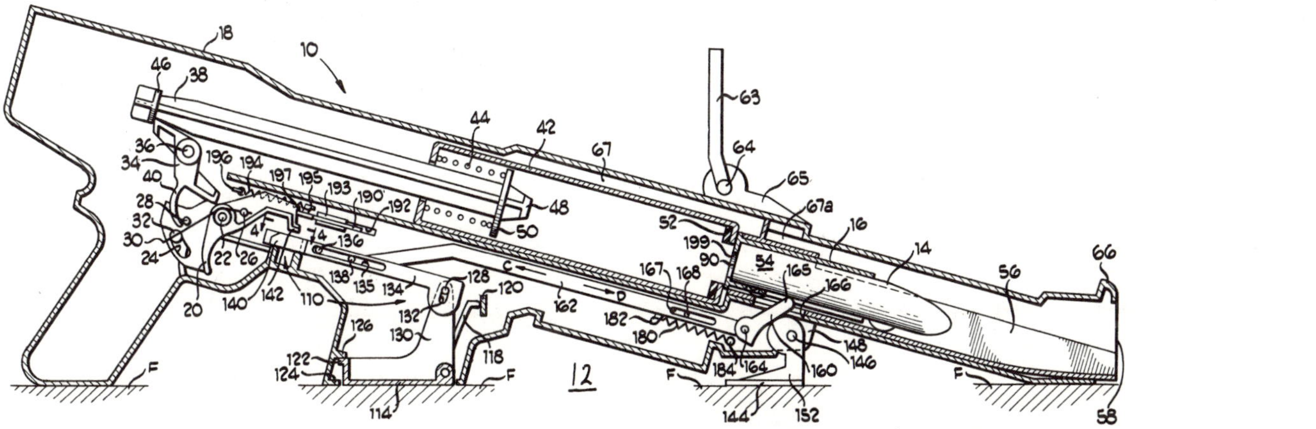

figure 9.8 A launcher that fires a toy racing car, in cocked position, as illustrated in Patent No. 3,895,459 (1975). The launch includes a relatively large, hand-held, generally hollow frame (18) that simulates the shape of a rifle. A trigger (20) is pivotally mounted on a pin (22) that is integrally molded with the frame. A lever (34) is mounted on a pin so that it will release a cross-shaped plunger (38) when the trigger is pulled. A curved flexible part (40) on the release lever returns this lever to its original position by engagement with the post (28) on the trigger. A cylindrical interior bolt housing (42) encloses the plunger (38) and a spring (44). The spring surrounds the plunger so that it is compressed when the rear-headed tip (46) of the plunger engages with lever (34) and the bolt housing is in the forward position. The front tapered tip (48) of the plunger has a disclike flange (50) that contacts a rubber damping gasket (52) at the front of housing when the plunger moves froward under the force of the spring. The front tip of the plunger strikes the rear of the toy vehicle (14), causing it to be impelled down a flanged ramp (56) and out the front open end (58).

A safety interlocking mechanism (110) includes a support pad (114) rotatably mounted on the frame. This pad includes an integrally molded leaf-type spring (118), which abuts against a rib (120) molded with the frame, and a flange (122) that is biased by spring (118) into contact with a rib (124) on the lower portion of the frame. Flange (122) comes into contact with a rib (126) of the frame when the launching implement is placed on a supporting surface, and the pad rotates clockwise against the force of the spring (118). A pin (128) mounted on an upwardly directed ear (130) of the pad (114) engages in a slot (132) on an interlocking arm (134). An elongated slot (135) is provided in the arm to slide it along a pair of pins that are molded as part of the frame. The interlocking arm is provided with an L-shaped rear end (140) such that a blocking pad (142) on the end of the trigger cannot move downward unless the interlocking arm is in its forward position.

figure 9.9 Partial cross section of the rocket, rocket motor, and launcher described in Patent No. 3,943,656 (1976). The rocket (10) includes a first stage (12) and a second stage (14). A motor (16) is positioned within receptacle (18) of the first stage. The motor is held in place by retaining rings (20) that provide an interfering fit between the outer wall of the motor and the inner walls of the receptacle. The forward portion (22) of the motor is surrounded by a sleeve (24) that is connected by pins (26) to the second stage. The sleeve slidably engages the annular space between the wall of the first stage and the outer walls of the forward portion (22) of the motor.

A coil spring (28), shown in compression, surrounds the motor and bears against the upper edge of the upper retaining ring and the bottom edge of the sleeve. The coil spring biases the second stage away from the position shown. A pressure-responsive member (30) protrudes through apertures (32) in the forward part of the motor and bears against the inner walls of the sleeve. Thus the coil spring is prevented from separating the second stage from the first.

The rocket rests on a launcher (36) by engaging the nozzle (38) of the motor. The rocket is vertically supported by launch wire (40), which slidably engages tubes (42) attached by brackets (44) to the outer walls of the first stage. The wire is supported in a base (46). The launcher also rests on this base. A launcher release handle (54) is connected via bowden cable (56) to the launcher for remote control.

Propellant is supplied to the motor from a source (48) via a tube (50). A fueling channel provided in the launcher communicates with the nozzle and thus with a propellant cavity in the motor. A preferred fuel for this motor is a volatile, self-pressurized, halogenated hydrocarbon that is a liquid at atmospheric and superatmospheric temperatures. Freon-12 is suitable, for instance.

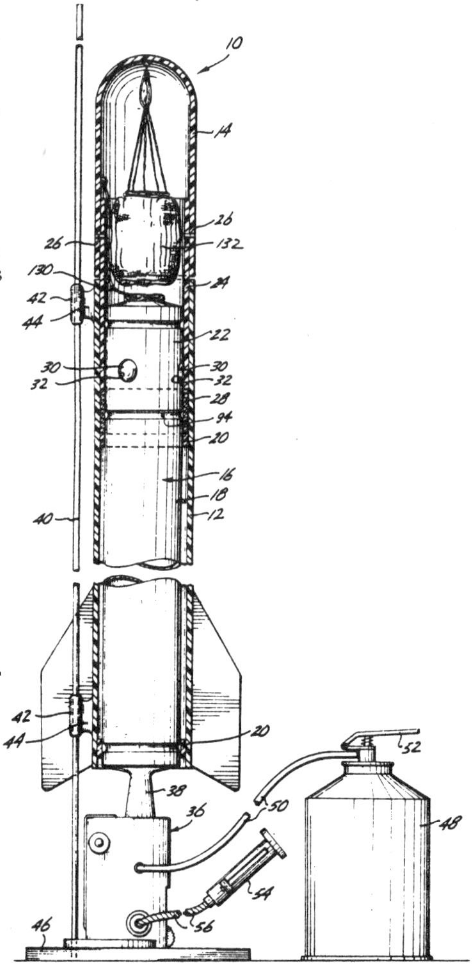

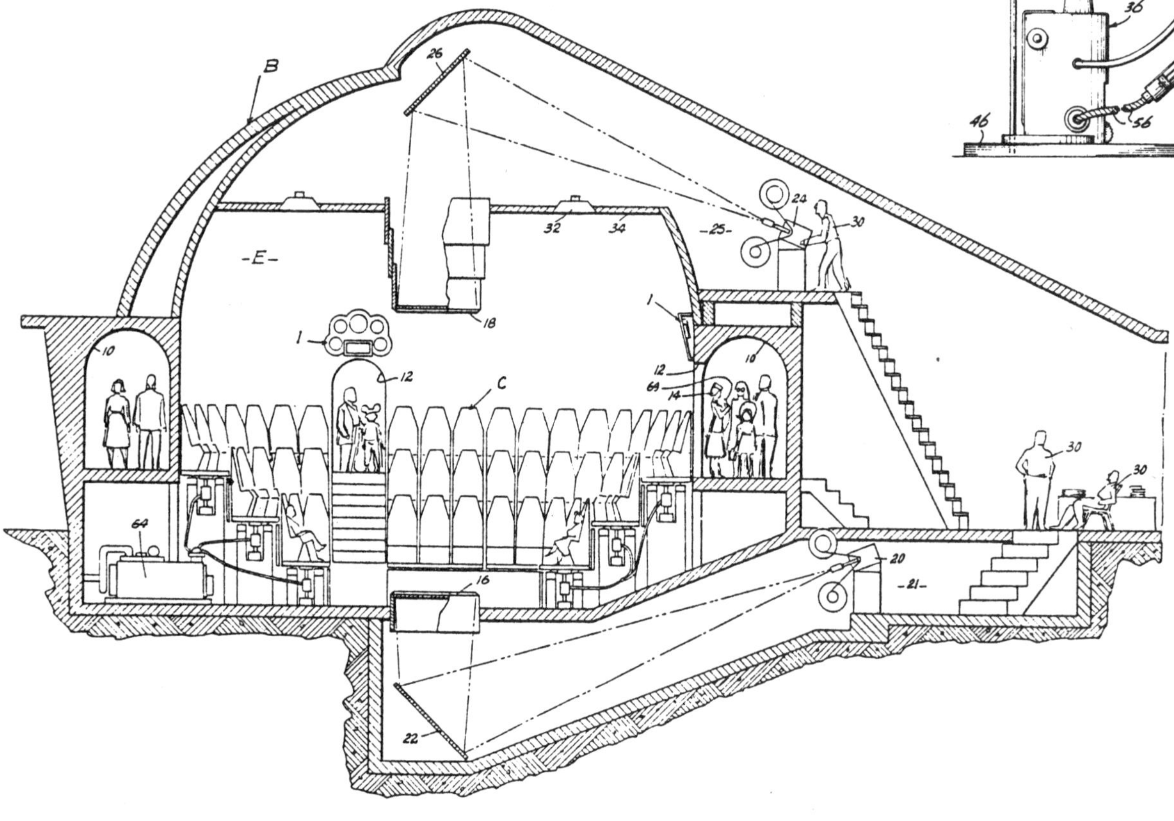

figure 9.10 Side view of Disney's rocket ship ride, Patent No. 2,861,806 (1958). The interior of building (*B*) is formed with a generally annular enclosure (*E*) simulating the inside of a rocket ship and containing chairs (*C*). The "passengers" are conducted to the enclosure (*E*) through passageways (10), entering through any of doorways (12), and are led to their seats by a woman (14) dressed in the uniform of an airline stewardess.

The lower central portion of enclosure (*E*) is provided with a bottom viewing screen (16). The upper central portion of the enclosure is formed with a top viewing screen (18). Both these screens are visible to the occupants of the chairs (*F*). A lower film projector (20) is mounted at one side of the bottom screen in a projection room (21). An inclined mirror (22) reflects light from the projector onto the screen. A similar setup takes care of the upper screen. All projectors and mirrors are hidden from the passengers' view. Speakers (32) are mounted in the ceiling. One or more simulated instrument panels (*I*) are located in the enclosure (*E*). Each will provide simulated indicia useful on an actual rocket trip—there could be an altitude indicator, a direction indicator, a distance indicator, a velocity indicator, a simulated or an actual speaker (48), and a passenger warning screen (50).

Toy rockets offer lots of possibilities, but leave it to Walt Disney to come up with a full-size chamber that simulates a rocket flight for a crew and several dozen passengers. Figure 9.10 shows a drawing from this patent, which is the only one ever filed by Disney himself. Despite this point of interest, it may seem odd at first to have this large, expensive device as part of a chapter on the invention of toys. It does, however, have a good deal of relevance for contemporary inventors. First, Disney was a master toymaker, a man who dared to make reality out of some of his wildest dreams. His careful attention to detail and his skill at creating an illusion are qualities that should be studied by anyone interested in finding out what makes a successful toy. Disney was a genius with a practical nature, a dreamer interested in the art of the possible.

Sometimes inventions can get complicated. The toy satellite system shown in Fig. 9.11 looks as if it would be almost as much fun to play with as a real satellite—and a lot less expensive, too. It's an ambitious project that includes landing pad, launcher, multistage rocket, space station, satellite, and planet. It is probably far too complex for mass production or for use by a large public. However, this patent, filed by Leonard Branneman of Massachusetts, contains elements that might be adapted for use in a simpler layout.

One complicated invention that did make it into production is illustrated in Fig. 9.12. At first glance, the structure of this game looks like one of the rocket-launching monsters at Cape Canaveral. A closer examination reveals that it's not all that involved, though it does present an interesting challenge to the players. The object of the game is to get a self-propelled playing piece, generally a car, through a course by manipulating hazards and obstacles so that they interfere as little as possible with the progress of the car toward the end goal. Burton Meyer of Illinois invented this speedy game.

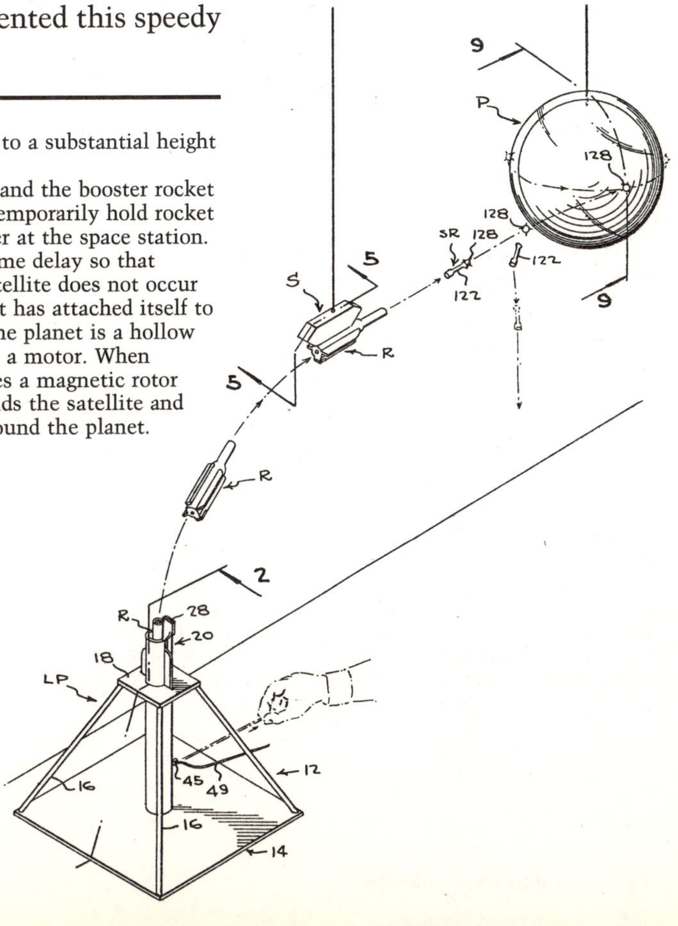

figure 9.11 Schematic view of the satellite system described in Patent No. 3,138,381 (1964), showing a multistage rocket assembly (*R*) in transit between the launcher and a space station (*S*), another multistage rocket assembly in rendezvous with the space station, the second stage of this other rocket assembly being in transit to the satellite. The second stage of the preceding rocket is shown falling away from its satellite, and the satellite of another rocket assembly is adhered to and orbiting the planet (*P*).

The launching pad (*LP*) comprises a frusto-pyramidical stand (12) having a rectangular base plate (14), from whose corners upwardly convergent legs extend to and are fixed to the corners of a rectangular platform (18), the last being smaller in area than, and centered relative to, the base plate. A perpendicular first-stage launching tube (20) is fixed at its lower end to the center of the base plate and extends up through a centered opening (22) in the platform, to a substantial height above the platform.

The space station and the booster rocket have magnets that temporarily hold rocket assembly and booster at the space station. The booster has a time delay so that propulsion of the satellite does not occur until after the rocket has attached itself to the space station. The planet is a hollow sphere that contains a motor. When activated, this rotates a magnetic rotor that attracts and holds the satellite and moves it in orbit around the planet.

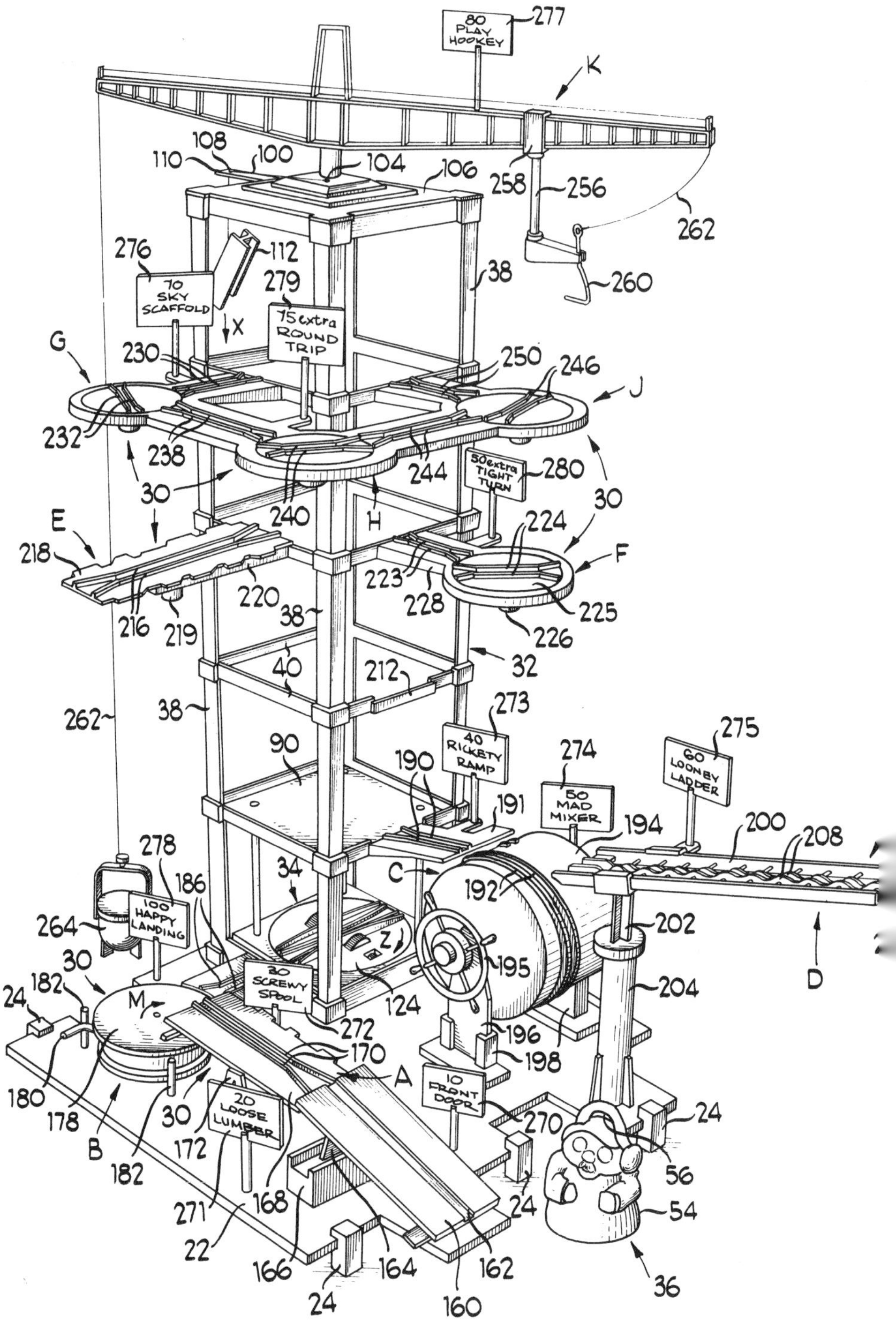

figure 9.12 The game described in Patent No. 3,908,989 (1975) includes a base plate (22) supported by legs (24). There are manually operated stations (*A*) through (*K*). Some of these are mounted on the base plate and some on the skeletonlike structure, as shown, to be reached generally by a vertically movable transport station, or elevator. The game basically constitutes an obstacle course over which a playing piece is caused to traverse in one or more given sequences. This "vehicle" is self-propelled and is formed in the shape of a human figure (54). Its housing includes an opening (56) at the top for use in conjunction with one of the stations, (*K*).

At the beginning of the game, the lower portion of elevator (34) rests on the plate (22), as shown. When it is desired to move the playing piece up to another level, the elevator can be moved within column (32) by means of a cable (100) that is secured to the top of the upper plate (90) by a hook (102). This cable extends through the center of the column through a hole (104) in the roof (106), and then to the left, as shown, to a second hole (108), which is provided in a horizontal arm (110) secured to the roof, and then down to where the cable is connected to a handle in the form of a simulated I-beam (112). This beam can be grasped between a player's fingers and pulled down in the direction of Arrow X to cause the elevator to move up. It can be stopped at any level.

The obstacle course can begin at the lower edge with an inclined ramp upon which the playing piece is positioned. The ramp has a guide groove (162) in the center that is engaged by the guide wheel (86) of the playing piece. This latter proceeds onto station *A*, which includes a horizontally pivotally mounted ramp segment (168) having two guideways (170). These tracks diverge outwardly adjacent to the inclined ramp (160) for the vehicle's easy movement across the connection. This ramp leads to station *B*, which includes a pivotally mounted disc (178). This disc must be rotated approximately 90 degrees in the direction of Arrow M until the handle engages the left stop post (182) in order to direct the playing piece onto a stationary track defined by ribs (186), which then direct the playing piece onto the elevator. The piece begins to rotate the disc (124) in the direction of arrow Z. The player must move the elevator up to the second level at least by the time this disc has completed its rotation. At that time the piece moves off the disc and, if the elevator is in the right position, onto a stationary track defined by its ribs on the top of a platform (191) protruding outwardly from the frame structure (32). The playing piece is then directed by tracks (190) onto a spiral track formed on a rotatable drum (194) of station *C*. The player must rotate the drum in order to align the track ribs with the next station and prevent the playing piece from falling off the drum. And so it goes through stations *D-K*, the latter of which comprises a simulated crane (254) that is pivotally mounted on the roof (106). The crane's arm (256) slides back and forth by means of a bracket (258) and is used to position a hook (260) that engages the eye (56) of the playing piece.

Stations (*F*), (*G*), (*H*), and/or (*J*) are optional, and the player can be awarded extra points for completing them. Different point totals are given for traversing the stations, as, for example, 10 points for (*A*) and 50 points for (*C*).

Races of all types are perennial favorites. Many different kinds of horse races have been patented, some simple and some intricate. The race game illustrated in Fig. 9.13*a* is driven by a motor and gear train. By a careful arrangement of gears and offset bearing surfaces, the forward movement of each horse is varied in an unpredictable manner. Thus the winner of each race is selected in an essentially random manner, impossible to control. It's a game for gamblers, as is the race in Fig. 9.13*b*. Gravity does most of the work here, and this means few moving parts and a minimum of maintenance. In fact, the forces of nature, like gravity, are a common and a desirable way of making games and toys work without the use of electricity or other power sources. Invented by John Carcel of New York, this race involves no apparent skill on the parts of the players—a bit like the flesh-and-blood races it simulates. Provided the "horses" are maintained so that they all have an approximately equal chance in every race, this game can be exciting.

A magician calls upon people to use their imaginations, often trying to figure out how in the world that last trick was done. Believe it or not, many magicians (and manufacturers of magic tricks)

figure 9.13(a) The starting gate of the racetrack described in Patent No. 3,735,984 (1973). The walls (70) form stalls within which the racing figurines, generally horses, are positioned. The barrier (72) is pivotally mounted to the frame (10). The horses are placed on the strips (16) within the stalls, and the barrier is rotated to the horizontal. The motor is activated, and the barrier is rotated up to the position shown, permitting the horses to leave the gates all at once. The finish gate has corresponding stalls. As the horses arrive, they drop into the stalls and cause the bottom shelves to rotate and raise the stall cover. Thus, the winner is designated clearly, as is the remaining order of finish.

The horses "run" on a number of strips mounted for reciprocating movement. An electric motor and a gear train within the frame cause each strip to reciprocate with a continually varying stroke. This causes the horses to move forward. Since the stroke of each strip varies, the speed of each horse varies. Differences in the length of the stroke of each strip are achieved by a series of gears mounted directly below the strips, the gears running at uniform speed. Each of the gears has an offset-bearing surface against which a flange depending from the bottom of each strip bears. As the offset- bearing surface of each gear rotates, the degree of movement of the strip varies. The strips are caused to move toward the finish line by the action of a rotating rod located transversely of the strips, the rod being provided with a series of spokes that contact and push forward a series of other flanges depending from the strips.

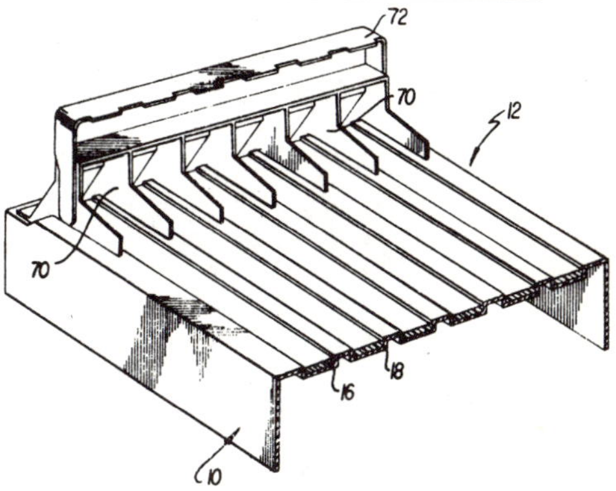

figure 9.13(b) The gravity-operated horseracing game as shown in Patent No. 3,658,333 (1972). The horses are placed in the starting gate at the top of the S-shaped raceway, and a pair of levers open the starting gate and release the horses with an audible sound. Horses (21) include front legs having wire-type rails that form a skate to permit the horse to slide down the track. These rails also have side fenders to maintain the horses in each of the tracks (12) through (17). Similarly, the rear feet include wire-type rails and also side rails that keep the horses between the flanges of the tracks. The toes of the skates, mounted on the front feet, are initially retained in grooves formed in the floors of a removable set of stalls located behind the starting gate.

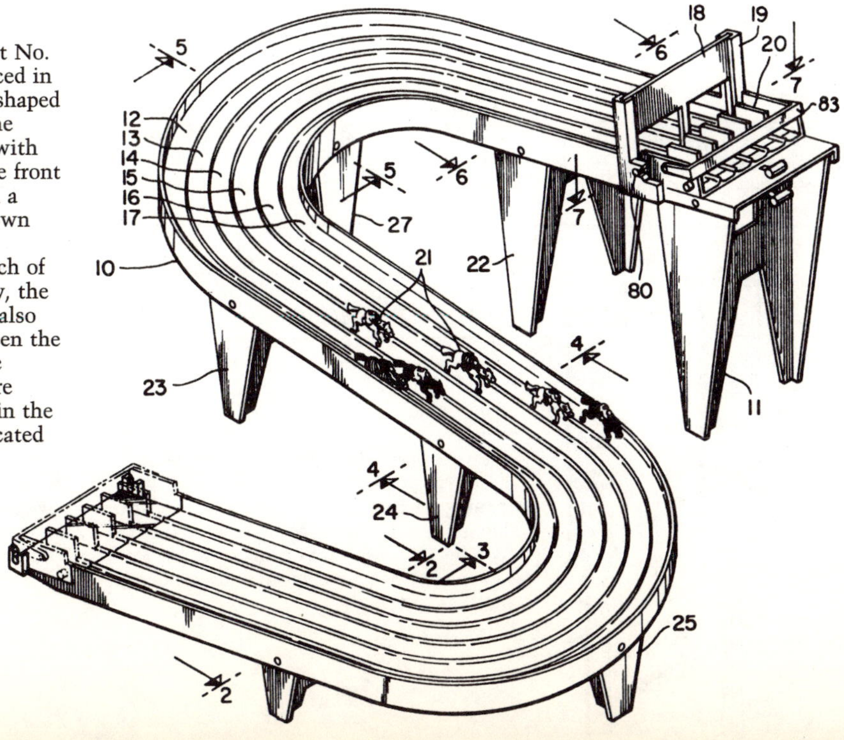

have chosen to reveal their secrets to the public by taking out patents on some of their "illusion devices." Two of these inventions are shown in Figs. 9.14*a* and 9.14*b*. One is suitable only for use on a stage, but the other is readily adaptable for mass manufacture. Both are old, dating from 1907 and 1925, respectively, but could as easily be used in a magic show of today as when they were first patented. Magic seems to be entering a new cycle of popularity. More sets and individual tricks for the young amateur are appearing in the stores and mail-order catalogs. If you're interested in this kind of thing, it would be worth your while to investigate. Remember that card and other tricks that involve manipulation are not patentable. But any device that is new (and useful and practical) is eligible for patent protection.

The most important residents of toyland are, of course, dolls. They've been around for thousands of years, yet are as young today as when they were first cuddled by the children of antiquity. Each

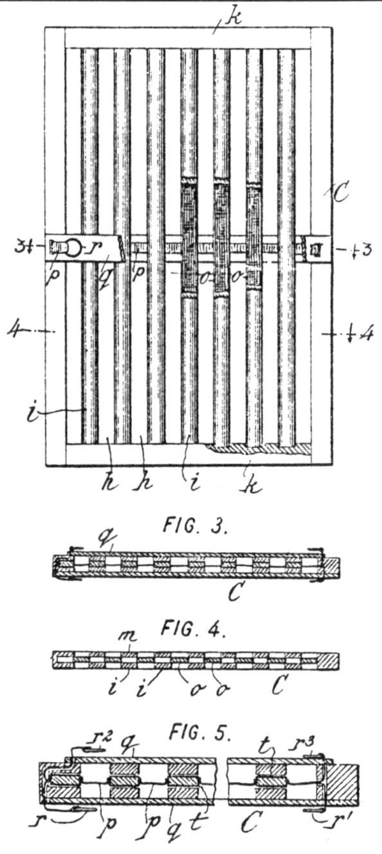

figure 9.14(a) Four views of the illusion device shown in Patent No. 873,447 (1970). Figure 2 is the front door; Figure 3 is a cross section taken along the line 3,3 in Figure 2; Figure 4 is a cross section on the line 4,4; Figure 5 is an enlarged view of Figure 3. In Figure 2, *B* indicates a chamber of sufficient dimensions to accommodate the object that is to appear within the cabinet—generally a person or persons. The chamber is preferably a cabinet with one of its walls of peculiar construction. Such a wall is in the form of a swinging door (*C*) that is mounted on hinges (*g,g*) attached to one of the cabinet's side walls.

The door *C* is provided with a series of openings (*h,h*), shown as elongate slots between a series of strips (*i,i*) extending from top to bottom of the door. At the rear of the (*i*) strips is a second series of strips (*m,m*), the same shape and thickness as (*i*). Front and rear strips are spaced apart sufficiently to accommodate between them a series of movable shutters (*o,o*) in the form of strips that are exactly the same width and thickness as (*i*) and (*m*). The shutters are connected to each other at their middles by a tape (*p*) that is concealed by a crosspiece or bar (*q*) extending across the middle of the door. The ends of the tape (*p*) are passed through opposite ends of the doorframe (*k*) and are provided with rings (*r,r*) on the outside and rings (r^2r^3) on the inside so that the shutters may be moved together to the left to close the slots (*h,h*), or in the reverse direction to open the slots. The cabinet has a concealed door or doors, too, adapted to swing inward.

The entire interior of the cabinet should be patined dull black. The slats of door (*C*) are black edges shaded into a metallic color at the center. This makes it difficult to distinguish between the black areas, and the slats appear thinner than they actually are.

The following procedure is used in a show: Before the curtain is raised, the cabinet is arranged on a table so that a small margin of the tabletop extends in front of the cabinet. The tabletop should be wide enough to leave sufficient room in the rear of the person (or thing) to be produced from the cabinet to hide from the audience. After the curtain is up, the magician does the usual business to "prove" to the audience that the cabinet is not a trick piece. Before the magician turns the table back to front to show nothing is hidden in the rear, the person who is going to pop out of the empty cabinet opens the rear door (*n*) and steps into the interior. This person is made to appear by manipulation of handle (*r'*) to quickly draw back the shutters, whereupon the person instantaneously appears through the bars of the door (*C*).

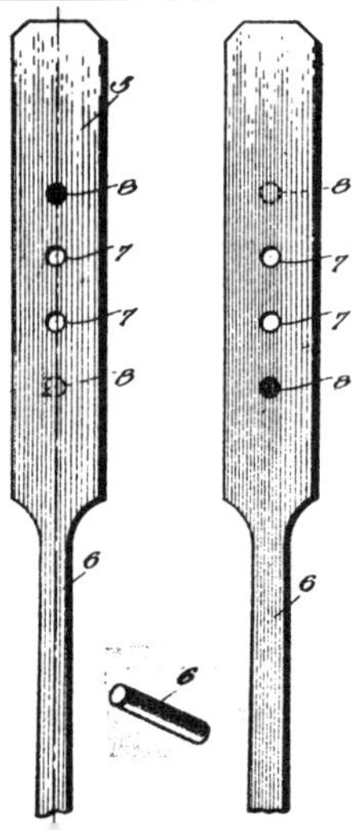

figure 9.14(b) Two views, back and front, of the trick candy holder shown in Patent No. 1,586,137 (1926). A pair of holes (7) is cut through the body of the device. Blank openings (8) are formed next to each of these holes, in alignment with them and of the same size. A short peg (9) is adapted to be inserted in either of the apertures (7); when viewed from one side of the body, these will appear to be in the center of the openings; when viewed from the opposite side, they will appear to be in the end opening. The free end of the stem is tapered and has a piece of candy mounted on it. A groove surrounds the lower end of the stem, so that when the candy has been eaten, the stem may be broken off and the device used for magic.

To use it, the magician holds the stem between the fingers and places the peg in one of the holes. With one side of the device to the view of the audience, the magician asks them to observe the hole in which the peg is placed. Then he places a covering over the stick and by a slight twist of the wrist or fingers turns it over. The cover is removed, and the peg appears to have changed holes.

new generation seems to have its own particular favorites. Over the years, inventors have introduced many new features, and some of them have been well received. Barbie and her friend Ken have been the most celebrated new characters of the last twenty years; there have been many other, somewhat less spectacular success stories, too. And the dolls keep coming—dolls that talk and cough and wet their beds, dolls that laugh and cry and shake their heads. There's even a Real Boy! doll, fully equipped, for those parents who don't want to leave anything to the imagination. Perhaps when Real Boy meets Real Girl and they begin to raise a family, a whole new business will be born.

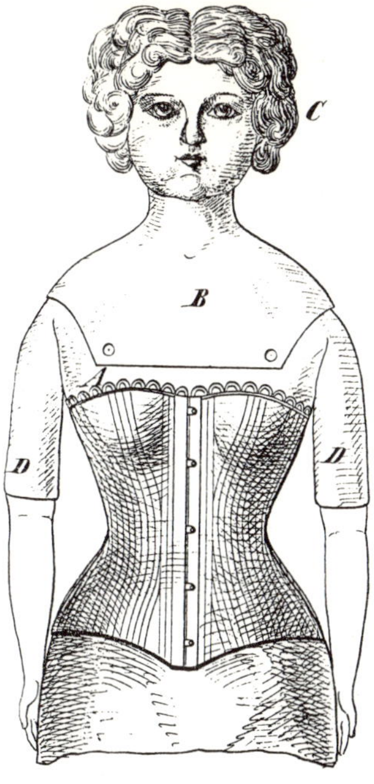

figure 9.15 A doll as illustrated in Patent No. 235,218 (1880). A corset is printed on the body, which is filled with sawdust or other similar material.

Generally speaking, dolls fall into two broad categories, the doers and the cuddlers. The 1880 model in Fig. 9.15 falls somewhere in between. With its corset just like Mama's, this doll is as much a product of its time as, say, the curly Shirley Temple dolls that were popular in the 1930s. Like anything else, dolls are made up of various components, some of which are more important than others. Often, one or more of these components can be improved, lending new interest to a doll that's been around a while. In a patent filed in 1969, Jack Ryan, the inventor of Barbie, and Jack Lemkin, both of California, gave the following background: "Certain dolls are favorites of children today because of the realism which is associated with a particular doll in different play environments. These environments are created by different accessories which may be used with the doll. Some of these play environments are limited, however, by the lack of natural movements and positions for the doll." Their invention, illustrated in Fig. 9.16, was a doll with the "upper and lower torso positions pivotally connected together by coplanar end portions forming a predetermined first angle with a side-to-side horizontal axis and a predetermined second angle with a forward-rear horizontal axis for producing a predetermined twisting action when one torso portion is pivoted on the other." The new doll also had bendable legs swingably connected to the torso; time marches on, and so does Barbie.

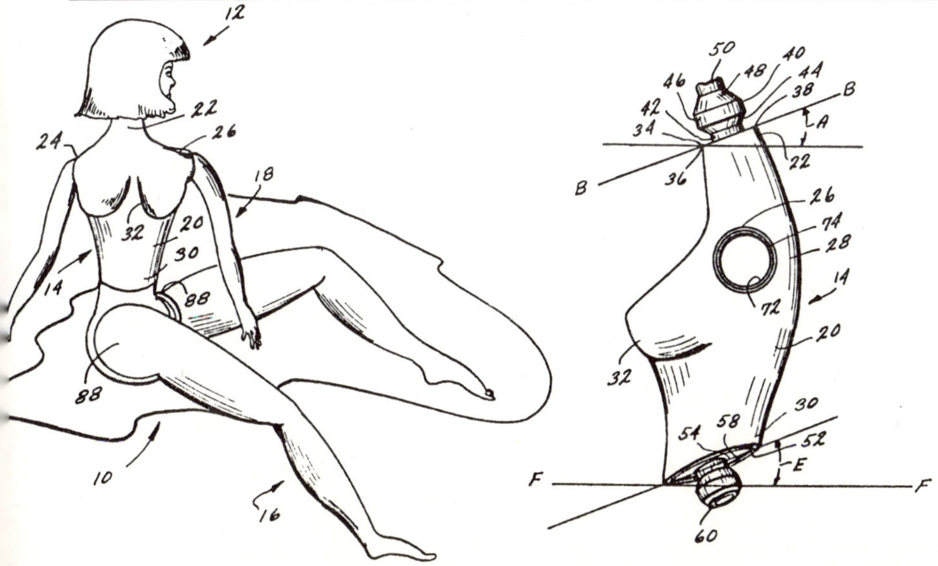

figure 9.16 Doll construction of lifelike movements and positions, as illustrated in Patent No. 3,425,155 (1969). The torso (14) includes an upper part (20) having a neck (22), shoulders (24) and (26), back (28), midriff (30), and bustline (32). The neck includes an upper end (34) forming a shelf for supporting the head (12) on a predetermined angle (*A*) formed between a substantially horizontal plane and a forward-rear axis *B–B* extending from the front edge (36) through the rear edge (38) of the shelf (34). The torso also includes a spool (40) to receive the head. This is formed on the shelf during a molding operation wherein the entire torso is manufactured in a single step. The spool is affixed to the shelf by a cylinder (42) and includes a lower part (44), center (46), and upper part (48). The latter is surmounted by a cylindrical molding piece (50), which may be removed after molding. In use, the doll may assume any one of a number of different positions—through moving of the bendable legs, arms, head, and torso.

The general trend in modern doll design is toward moving parts and lifelike action. Figures 9.17*a-f* show a variety of recent patents in this field. Figure 9.17*a* is a doll that crawls, driven by a power source in its midsection. The parts of the body move relative to one and other in a lissajous motion, causing the doll to crawl in a realistic manner. Jack Ryan (again) and a group of California inventors developed a doll "whose arms move in a manner to simulate hair washing actions, while shampoolike foam is produced in its hair" (Fig. 9.17*b*). The old reliable squeeze motion produces the action.

Frank Johmann of New Jersey has developed a doll that can move its hand in various ways. A motor drives a rather complicated set of gears that enable the doll to play such games as Patty Cake with a child (Fig. 9.17*c*). Another, somewhat similar doll by the same inventor can be adjusted so that "its hand will shake and throw down a pair of dice." With a bit of practice and some imagination on the part of the child, this doll becomes an autorobot opponent for Monopoly and other board games involving the roll of the dice (Fig. 9.17*d*).

From Kosei Kato of Japan comes another innovation, the kissing doll. The torso here is provided with a flexible bellows that can be contracted so that it expels air outward. The air passes through a sound-producing mechanism and then through the open lips, simulating a kiss (Fig. 9.17*e*). If you like your kisses a bit more remote, William Nutting of New York has invented a doll that throws kisses. "When the doll's hand is in the vicinity of the doll's mouth, a spring, which is cocked during the pivoting movement of the arm, is released just about simultaneously with the actuation of a sounding device which produces a simulated kissing sound. . . . The rapid movement of the arm accompanied by this sound simulates the throwing of a kiss by the doll." Nutting's patent was assigned to Horsmann Dolls, an American manufacturer (Fig. 9.17*f*).

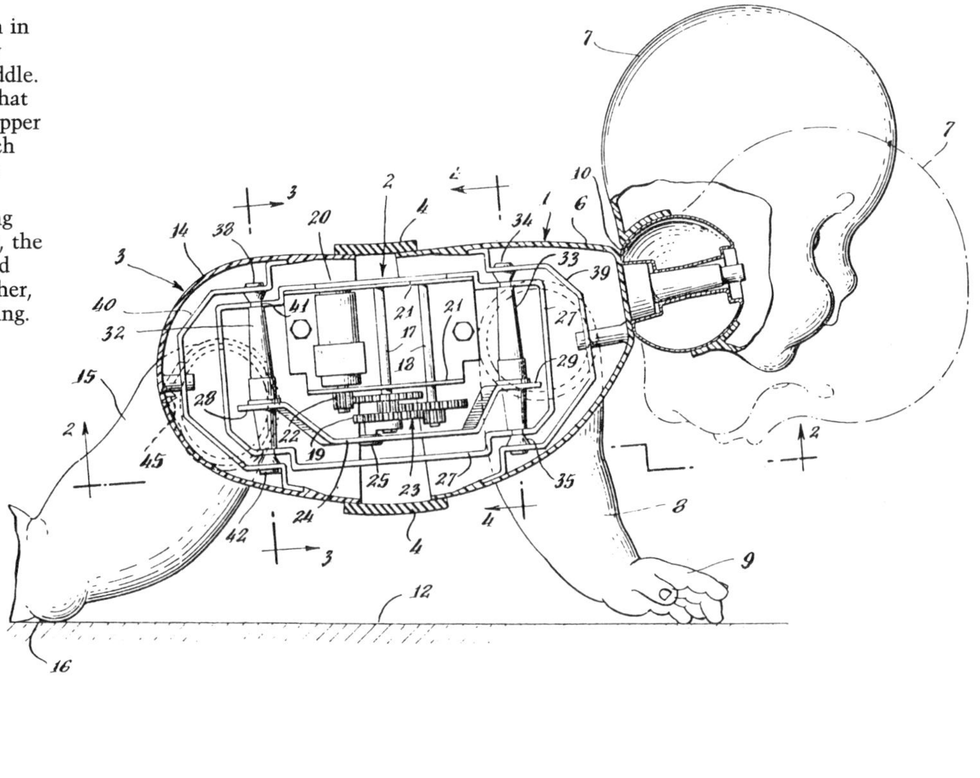

figure 9.17(a) Crawling doll, as shown in Patent No. 3,548, 537 (1970). The body has three parts—lower, upper, and middle. In the last, as indicated, is the motor that powers the articulated motion of the upper and lower parts. The connection to each one is pivotal—middle-to-upper pivots about a vertical axis; middle-to-lower pivots about a longitudinal axis, passing through the upper end of a shaft. Thus, the upper body moves from side to side and the lower twists back and forth. Together, these motions simulate a baby's crawling.

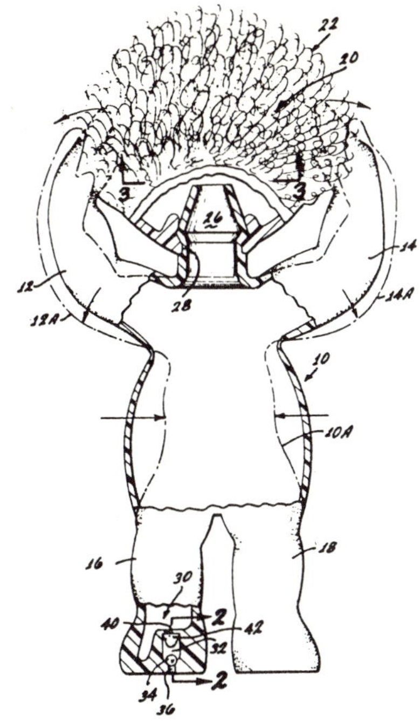

figure 9.17(b) Rear view of the shampoo doll, as shown in Patent No. 3,590,515 (1971). The arms are upraised and can be moved as if shampooing the hair, by squeezing the sides of the doll's hollow body. The check valve (30), which allows air to enter except when the body is squeezed, is located in the left leg (16). Thus, the air being forced out of the body has only one way to go—up, through the head. The threads of hair (22) are inserted with a needle through the surface of the doll's head and anchored. The hole made by the needle is larger than the thread, so that air can pass through when the doll's body is squeezed. When a soapy solution is applied to the hair, the air coming through the hair holes creates bubbles —and a shampoo.

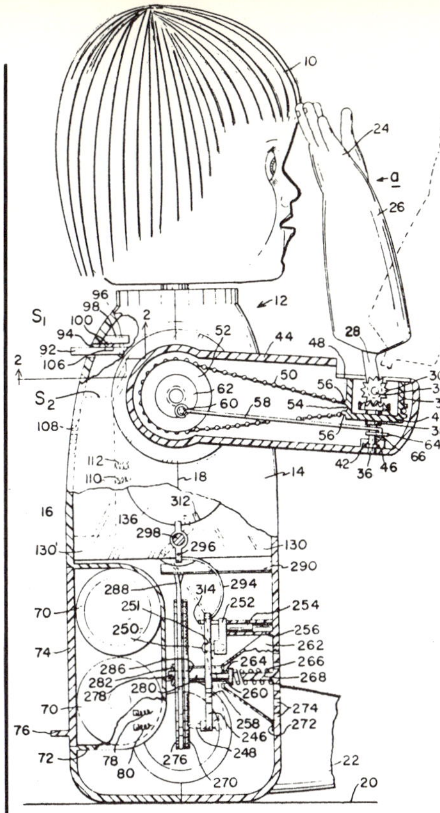

figure 9.17(c) Side view of the game-playing doll described in Patent No. 3,650,068 (1972). The battery-operated motor located in the torso is connected to the hands and can move them through a number of different positions. The doll may also be equipped with a sound-producing unit and may move her mouth as if talking.

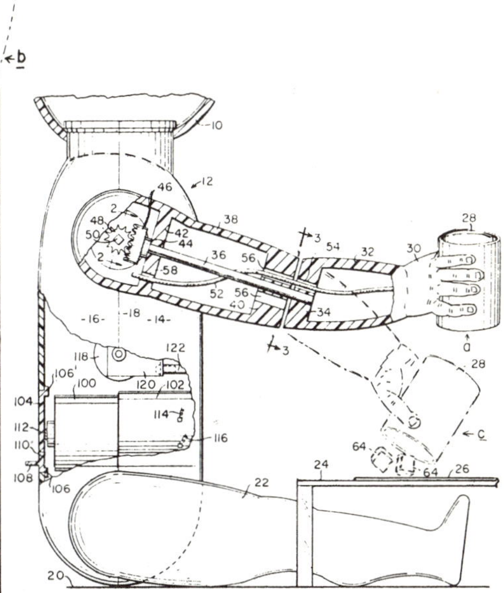

figure 9.17(d) Side view of the dice-throwing doll described in Patent No. 3,684,291 (1972). This doll also has a battery-operated motor located in the torso that is connected to the hands and can make them move. The doll's hands can thus be made to shake and throw down a pair of dice, as illustrated.

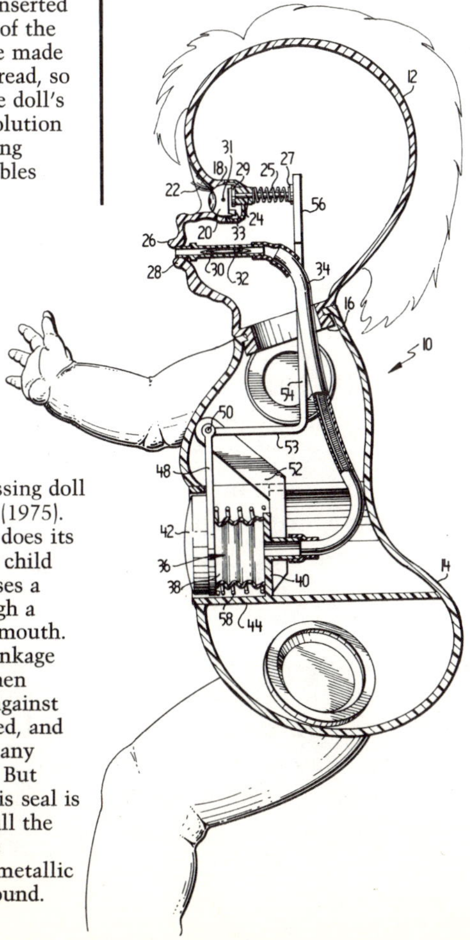

figure 9.17(e) Side view of the kissing doll described in Patent No. 3,916,561 (1975). The doll's eyes open and close, as does its mouth. To operate the device, the child pushes a button (42) that compresses a bellows (36). This expels air through a conduit (34), which ends near the mouth. At the same time, a mechanical linkage closes the doll's eyes. The child then places the doll's lips (26) and (28) against whatever or whoever is to be kissed, and releases the button. This seals off any movement of air through the lips. But when the child moves the doll, this seal is broken, and air passes inward to fill the partial vacuum. This activates the sound-producing mechanism and metallic reeds (32), and there is a kissing sound.

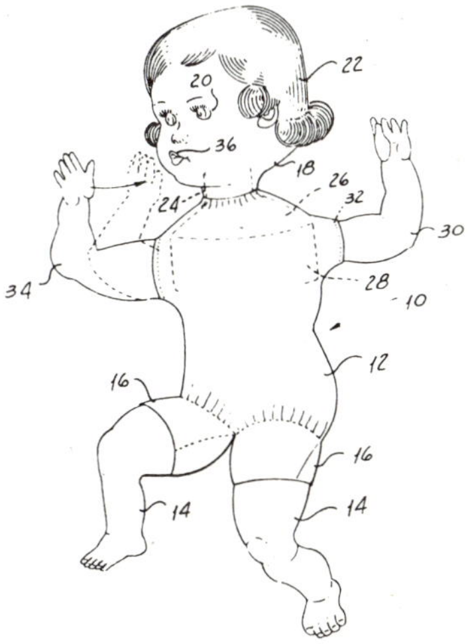

figure 9.17(f) Front view of the kiss-throwing doll described in Patent No. 3,603,029 (1971). Here the child brings the doll into action by pulling a drawstring attached to a driving spool. When the drawstring is released, the driving spool moves back in the other direction, influenced by a coil spring. The pull of the string brings the doll's hand to its lips rather slowly. An attached rotary cam holds the hand to the lips while an upper second cam keyed to the lower first cam compresses a bladder to which a tube containing a sound-producing reed is attached. The bladder is compressed and released at the same instant as the cocked arm is released. So sound and motion involved in throwing a kiss come simultaneously.

A different kind of doll was invented by Richard Bezark of Illinois in 1961. Actually, it doesn't quite belong with this category because it is more a model than a doll, more an aid to education than a toy. The Visible Man and Visible Woman, as they've been named, present a "simulated human body in the form of a transparent skin . . . the whole being split and adapted to receive simulated human integral organs" (Fig. 9.18). Advertised for use in schools, these models sell surpisingly well in the regular retail outlets. The inventor took something from the science lab and adapted it for use outside the school. In this it's reminiscent of the chemistry sets that used to be so popular and now seem to be in the midst of a small revival. Learning and experimentation really can be fun if they're presented in the right way. There would seem to be a great potential in "educational" toys for inventors with the ability to make a game of learning.

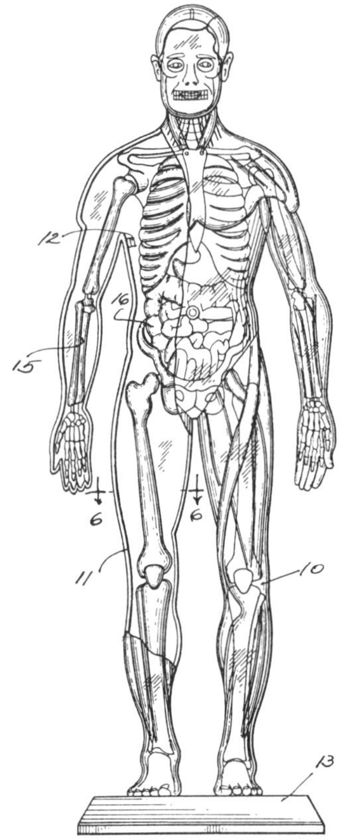

figure 9.18 Front view of the transparent, simulated human body shell, skeleton, and organs, as illustrated in Patent No. 3,009,267 (1963). The front and rear sections (10) and (11) of a hollow shell are composed of transparent plastic. They are molded to show muscles, arteries, veins, and nerves in substantially material bas relief. The shell is split vertically on a somewhat sinuous line that bisects the body on natural contours. As an aid in maintaining alignment, means such as the lugs (12) on component (11) projectable into component (10), can be employed.

Integral with the body shell components are supporting base and foot closure portions (13) and (14), split from each other in a transverse irregular direction so that they fit together to form a platform supporting the shell components in upright position, separately or jointly. The division of the base on a sinuous line also aids in maintaining alignment of the shell halves. The transparent shell is also adapted to receive and support a simulated human skeleton. Also, in contrasting colors, there are simulated human internal organ components (16).

The golden age of mechanical toys is past. As manufacturing costs rose and quality declined, mechanical toys have disappeared from store shelves and have begun to appear at antique auctions and in museums. There's still room for clever ideas, of course, if they're not overly complicated and can be produced at a reasonable price. Arpe Indjian of Massachusetts has invented a mechanical chicken that lays candy eggs (Fig. 9.19). The eggs are placed in a container in the fowl's body and then released one by one through a spring-held sliding door. The chicken's tail feather controls the opening and closing of this door and the operation of a sound box, which emits a cackle each time the door is opened and an egg is laid.

Some of the most popular mechanical toys were automatic banks. Nothing like them has been invented in many years, but replicas of some old models have recently become available through specialty stores and mail-order outlets. Figure 9.20 shows an 1882 patent drawing for a cat-and-mouse-game bank. When a coin is placed in the mouse's paw, the cat lunges forward and gobbles it up as the mouse retreats to its hole. There would seem to be no reason why mechanical banks and similar toys could not be "reinvented"—that is, redesigned so as to reduce the cost of manufacture while maintaining the proper action.

Inventors were coming up with new, improved models of things long before patent offices were established. It's a time-honored procedure that has resulted in many great new products. Recently, Cornelio Fraga of California took the ball and cup and adapted it for modern manufacture and use (Fig. 9.21). A piece of elastic connects ball and cup. The user flips the ball forward, then tries to catch it with the cup as the elastic brings it back. In the present invention the walls of the cup are tapered to a round shape and made just a bit flexible so that the ball will more easily come home to roost.

The bubble pipe is another toy that has been in use for centuries. F. W. Pease of New Jersey patented his version in 1875. With its flexible tube, mouthpiece, and small bowl, it's easy to carry and looks as if it was probably fun to play with. A modern version might garner quite a lot of interest (Fig. 9.22).

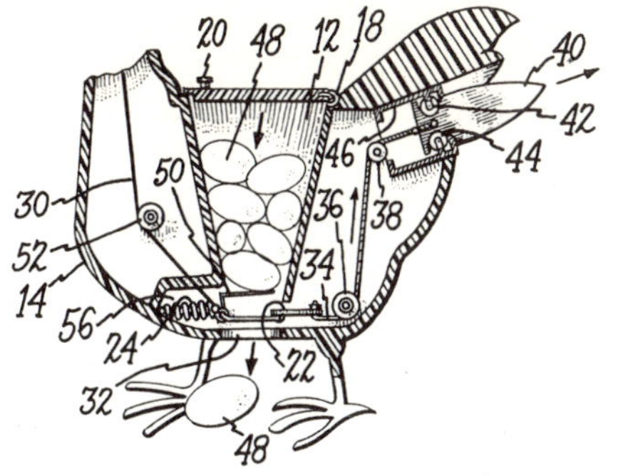

figure 9.19 Enlarged cross section of an egg-laying toy fowl as shown in Patent No. 3,660,930 (1972). The body (14) has an opening (32) below the open bottom of the funnel-shaped container (12). The sliding door (22) is spring-biased to close this opening by a spring (24). One end of line (34) is attached to the end of door (22). One end of this line is trained over rollers in pulleys (36) and (38); the other end is attached to tail feather (40), which is provided with rollers (42) and (44). Tail feathers and rollers roll into a chamber (46) in the tail. When the feather (40) is pulled, it rolls in the chamber on the rollers, pulling on line (34). This pulls the sliding door away from the opening (32). A funnel-shaped container is filled with eggs (48). As this door is pulled away from the opening, one of the eggs drops out of the opening. Spring fingers (50) close the opening so that only one egg at a time can pass through.

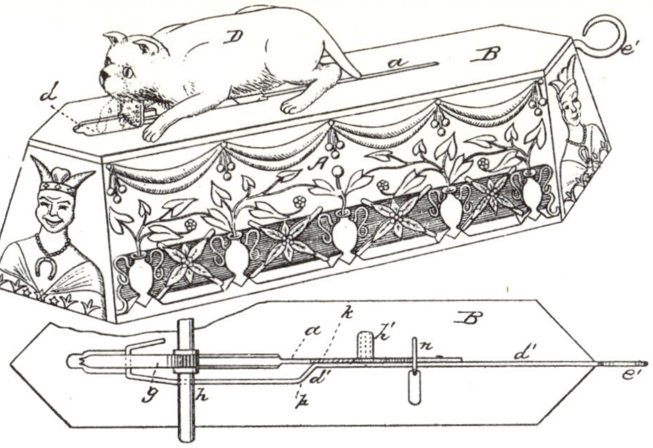

figure 9.20 Cast-iron bank, as illustrated in Patent No. 261,419. There is a slot *a* in the top (*B*), supported by side walls (*A*). To operate the bank, a coin is placed in the bearing (*e*). At the end of the slot is an enlarged opening through which a mouse figure rises and pushes the coin so that it falls through the slot and into the bank. A wire rod (*d*) runs along the underside of the bank's top. One end (*e*) forms the operating handle; the other end is attached to an arm (*g*) that is pivoted on a transverse bearing (*k*) and that carries the mouse figure. When the rod is drawn to the rear, the figure is pulled through the opening (*d*). A springing cat figure (*D*) is pivoted on the upper end of a lever arm, which is in turn pivoted on a fulcrum (*h'*). A strong spring (*S*) is connected to the lower end of the lever and to the rear of the bank. The cat is set in position by drawing it back until notch (*n*) of its lever arm engages the catch (*p*) of the trigger rod (*d*). When the lever is released from the catch, both lever and cat are thrown forward. Thus the cat leaps toward the mouse and appears to seize it.

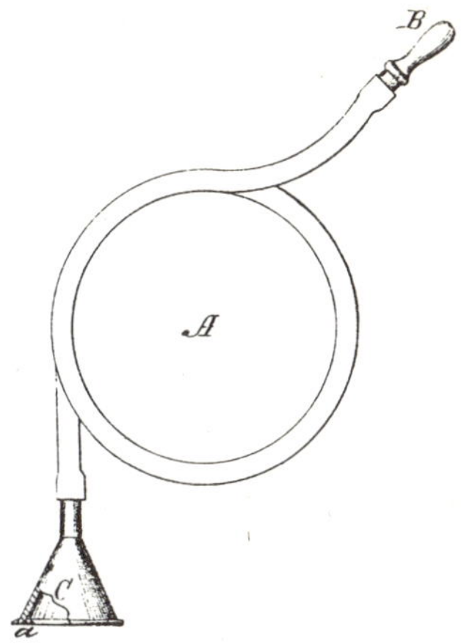

figure 9.22 Flexible, long-stemmed bubble pipe, from Patent No. 158,435. (*A*) is the flexible tube (made of rubber), (*B*) is the mouthpiece (made of wood), and (*C*) is the bubble bowl (made of metal). All parts may be made of rubber (or, today, plastic) with characteristics similar to the above-mentioned materials. At the mouth of the bowl, at (*a*), one or more ledges hold a small quantity of soap solution. This permits blowing a larger bubble than with an ordinary pipe. Separation of bowl from blower by flexible tube permits the user to toss the bubbles into the air as they are being blown.

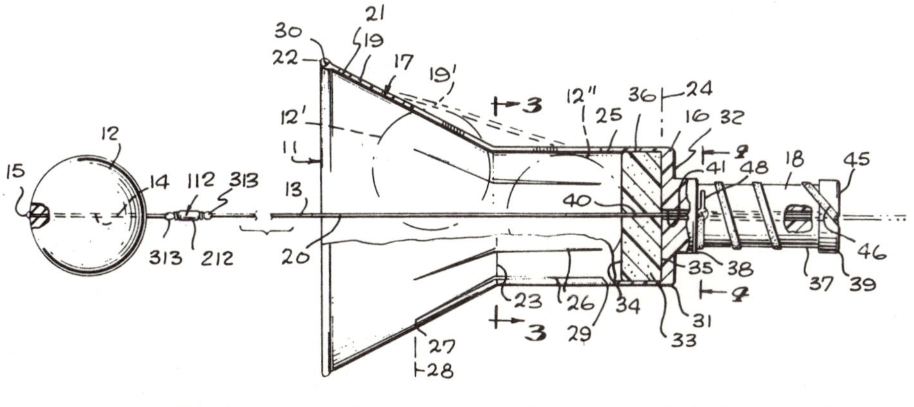

figure 9.21 Ball-and-cup toy, as illustrated in Patent No. 3,687,453 (1972). There is a cup (11), a ball (12), an elastic band (13) connecting ball to cup, and a swivel (112) between ball and elastic band making it possible to turn the ball without twisting the band. The ball may be rubber, containing a diametrical passage (14) through which a short cord (212) connected to the swivel extends, with the cord having a knot (15) at the outer side of the ball of a size too large to slip through passage (14).

The cup includes a main body (16), which may be injection-molded from a suitable plastic and which is shaped to form a tapering cup proper (17) and a handle (18). The cup has a side wall (19) that is annular and is centered about the axis (20) of the cup, and which tapers frustoconically about that axis at 21, from the location of the outer peripheral edge (22) of the cup to the transverse plane (23). From this plane to a second transverse plane (24), the side wall of the cup may extend directly axially, to form a straight cylindrical portion (25) of the wall.

Between peripheral edge (22) and plane (24) the side wall of the cup is thin enough to enable substantial localized lateral deflection of a portion of the side wall, away from the axis (20), if the ball strikes the wall at an off-axis position, such as at 12'. The cup is formed of a substance capable of this sort of flexure, such as polyethylene, and, in addition, the side wall contains elongated slits lying in axial planes and cutting through both the conical portion (21) and cylindrical portion (25) of the wall. And the peripheral edge of the cup is somewhat thickened.

Inwardly, beyond plane (24), the material of the cup body (16) forms a wall (31) having a planar surface (32) disposed transversely of axis (20) and defining the inner end wall of the cup. Next to this wall surface (32) there is an annular cushion (33), having two parallel opposite side surfaces (34) and (35), and having an outer cylindrical surface (36) that is a sufficiently tight fit within the cylindrical portion of the cup side wall to retain the cushion frictionally in position. The cushion is made from a material soft enough to compress readily under the force exerted by the ball when it returns to the position (12"). The internal diameter of the cylindrical portion of the cup should be only slightly greater than the external diameter of the ball, to allow for free movement and a good fit.

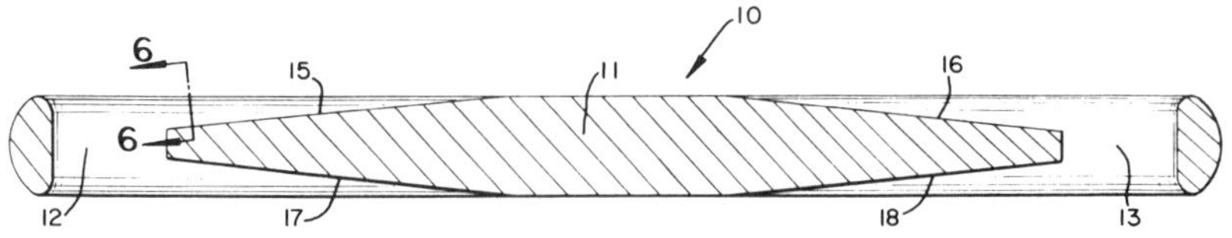

figure 9.23 Flick-a-Stick, as illustrated in Patent No. 3,840,233 (1974). The stick (10) has a long, solid body (11) about twice as wide as it is thick. On opposite ends, a pair of holes (12) and (13) is formed to accommodate a player's fingers. Each side of the body is tapered from the middle toward the ends. The top face of the body has oppositely directed declining surfaces (15) and (16) and the bottom face has oppositely directed inclined surfaces (17) and (18). The surfaces (15) and (16) diverge laterally toward opposite ends of the body. The outer extremity of each surface has a dimension substantially equal to the diameter of each corresponding hole (12) and (13). The tapering surfaces (17) and (18) also diverge laterally to intercept the respective holes (12) and (13) in the same manner as the tapering surfaces (15) and (16). In a game, the stick may be thrown in a spinning, twirling, or end-over-end motion toward an opponent who must try to catch it.

One of the oldest toys in the world is the kite. It probably originated in China more than 3,000 years ago; Egyptian hieroglyphics that are around 2,500 years old describe a form of kite flight. These ancient models were probably made of large leaves that were held by lines made of twisted vines. The early inventors of kites were often thought of as magicians who could make ordinary objects fly toward the heavens. It has always been difficult to exercise strong control over a kite in the air. Lynn Worsham of California sought to solve this problem with the invention of a controllable stunt kite. Though it has no moving parts and can be handled safely and surely by children, this kite can be flown in a horizontal plane, can make vertical dives, can be put through a number of acrobatic turns such as the figure eight, and can land gently. One of the keys to its versatility is the pair of symmetrical bridles used in its construction (Fig. 9.23).

What's in a name? Most inventions are patented under a legalistic moniker, such as Compressed-Air-Propelled Vehicle and Launching System, or under some vague general term such as Game. Norman Patterson of Tennessee departed from this standard practice by giving his simple invention a catchy name that just by itself should have been enough to make the thing a success. Patterson's Flick-a-Stick is an aerodynamic shape that can be thrown by participants in game with all kinds of variations in speed and direction (Fig. 9.24). It's a bit like a Frisbee, but with a different shape. An important point to remember here is that the patent does not protect the name Flick-a-Stick. This would be considered a trademark, and should be registered separately as such.

You don't have to think of yourself as an inventor to believe that you might just be able to strike it rich by creating a clever new board game. Certainly many people have tried, and many manufacturers and distributors have put out many new games. Some have been quite successful, too. But the most popular games are still the ancient ones—chess, checkers, backgammon, and even mah-jong, which is enjoying a resurgence of interest—but there is still room for a good new one. The best games have often been developed by independent individuals, not by corporate research groups. Perhaps only the outsider has the time and the obsessive interest to fiddle with something, to play it again and again with friends and family until it finally comes together as a game that people will really love to play.

It's doubtful that any amount of market research would be able to determine what makes a game popular. There are examples from the past, of course. Monopoly came along during the Depression, when people saw so little real money that they enjoyed handling the hundred-dollar bills of Monopoly money and owning the houses and hotels of the game. Scrabble came to popularity after World War II, a time of the GI Bill and emphasis on higher educa-

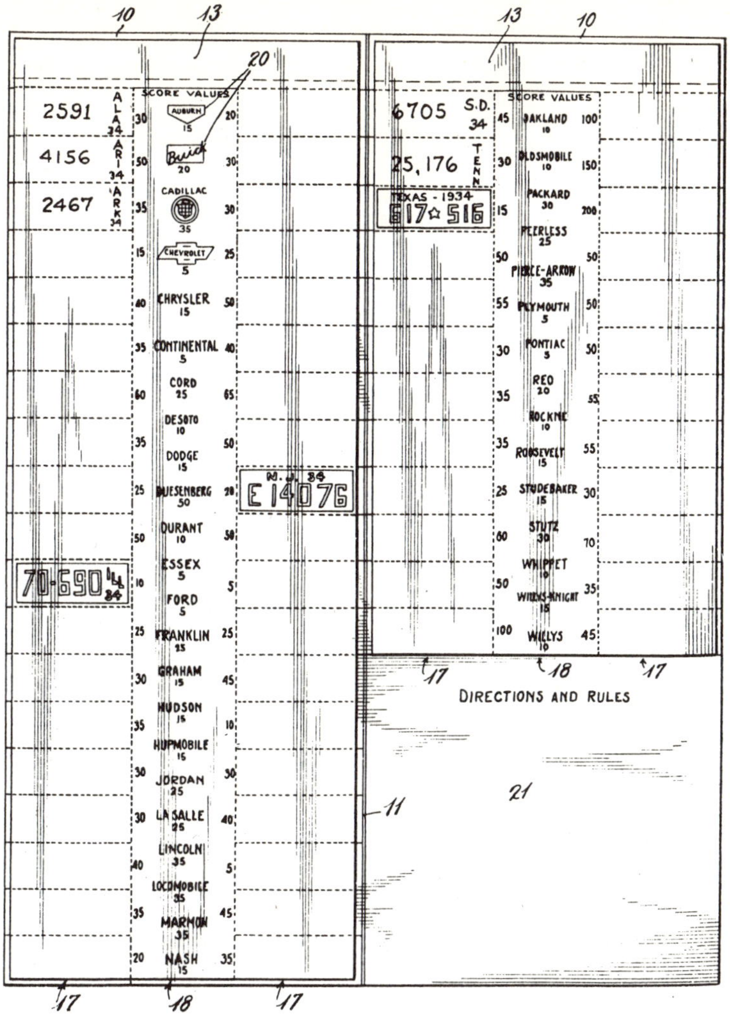

figure 9.24 A sample sticker sheet, showing only part of the printed material for the automobile game described in Patent No. 2,000,369 (1935). These sheets are divided longitudinally into three columns, two of which (17) bear miniature representations of the license tags of all the states and one of which (18) is adapted to carry a facsimile (20) of the name plates of the prominent makes of car. Note how times have changed. Most of these names would not be used in a 1980s version of this game. Each of the license tags in column (17) is a coupon and may be separated from the pack along perforations. The back of each coupon is coated with glue, like a postage stamp. In the lower right-hand corner of the board is a space (21) in which are printed the rules of the game, which are, briefly, as follows: Any number can play. The participants should agree on a time or mileage limit on the length of play. Then the players observe the makes and license plates of passing cars. They select one, detach the miniature of that license, stick it in place on the score board, and write the make of the car in the space alongside. At the same time they write in that space the score value of that license and that make, the sum of the two being the score for that car. For example, an Arkansas Ford is 30 + 5 = 35. The license scores are shown alongside the miniatures, as illustrated, and are based on population. The greater the population, the smaller the score. The values for the different car makes are shown below their name plates and are based on the cost of the cars, the less expensive being generally more numerous and therefore worth a lower score. For any car not listed, a score of 5 is allowed. Only one license of each type is allowed. No licenses except those of private cars count. A player is not required to score a car and a license. For example, if a player sees a Chevrolet with an Illinois license, he may pass it up in the hope that before the game is over, a Cadillac or some other car of higher score will come along with an Illinois plate. This is called a "pass" and must be announced by the player at the time.

tion. It was a good time for a word game that not everyone would have the knowledge to play. More recently, there have not been any blockbusters. There's been a revival of backgammon, and two new analytical games, Mastermind (from England) and Othello (from Japan) have sold well. There have also been lots of novelties trying to capitalize on people's interest in career success, sex, and parlor psychology. War games have made a place for themselves, too. There's a solid core of men and boys for whom battles and strategy and related action hold a strong fascination.

Although it is impossible to predict what kinds of games stand the best chance of success, there is one field that seems ripe for action—games for the car. Energy crises come and go, but the American family continues to move by automobile. Many trips are long and monotonous, with the boredom factor getting higher as the miles of Interstate Highway click off. Surprisingly, there are not many games especially designed for playing in the car. Back in 1935, Carl Zeiss of Illinois put together one of the first license plate games (Fig. 9.25). It's more sophisticated than some, assigning different values to license plates and to makes of cars, depending on their rarity.

One point to keep in mind about inventing toys and games is that lots of people are doing it. The major manufacturers get a great

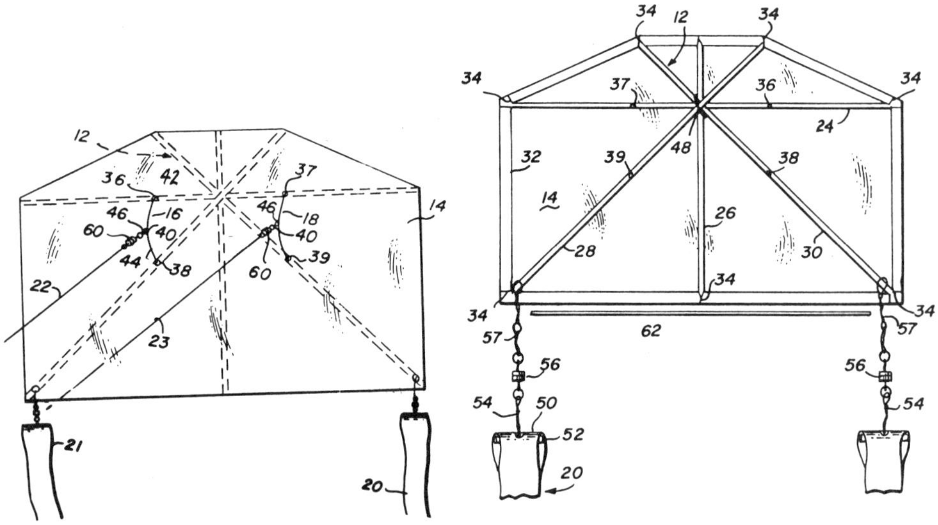

figure 9.25 Front view of stunt kite and rear view with tails attached, Patent No. 3,994,454 (1976). There is a fixed frame 12, a covering, a pair of symmetrical bridles (16) and (18), tails (20) and (21), and control lines (22), (23). The frame is comprised of a horizontal spar (24), a vertical spline (26), and diagonal spars (28), (30). These have grooved ends and are commonly referred to as sticks. The preferred size of these is as follows: horizontal spar, 25¼ inches; diagonal spar, 24 ⅞ inches; vertical spar, 19 inches. Each has a rectangular cross section 11/32 by ⅛ inch.

An outline string is placed around the border of the kite's covering (14), and the border is cut as at (34). An adhesive is applied so that when the borders are folded along the cuts, the outline string is exposed in eight regions in a manner that enables it to be received within the grooves of the sticks (24) through (30). In addition, four holes (36) through (39) are formed. Holes (36) and (37) lie on an imaginary line corresponding to the position on the horizontal spar and are spaced laterally equidistant from a longitudinal axis of the covering that corresponds to the position of the vertical spline. Similarly, holes (38) and (39) lie on an imaginary line parallel to the (36) and (37) line and the same distance from the longitudinal axis. Holes (36) through (39) receive the ends of the bridles (16) and (18) to be disposed outside the outer surface (33) of the kite. The accurate positioning of the holes is essential, so that the bridles can be placed perfectly, thus making it possible to predict how the kite will fly. In relation to the longitudinal axis, if the bridles are too close, the kite tends to spin in flight; if they are too far apart, then the control lines are separated by a distance greater than a human arm span in order to provide the required movement on the kite during flight. The bridles themselves are comparised of a length of string formed into an endless loop that is folded in exactly equal portions and tied with an overhand knot so that an attachment point (40) is formed substantially at the midpoint of the loop. The ends (42) of the bridles are affixed to the horizontal spar with loop knots. The spar is then placed against the back surface of the covering and its grooved ends inserted into the outline string (32). All pieces intersect at a common point (48), where a string is tied to secure them. Next, the covering is made smooth, and the loop knots formed in the ends (42) and (44) of the bridles are tightened, forming the frame and covering into an integral structure.

The tail (20) includes an end (50) formed into a loop. A rod (52) with a string (54) tied at its midpoint is inserted through the end. The string is extended through the hole and attached to an end of a snap swivel (56), connecting tale to frame. The swivel is of a conventional type commonly used by fishermen, and prevents a twist in a line connected to one of its ends from affecting a line tied to another end. A tail string (57) in the form of an endless loop is connected with a loop knot to the bottom end of the diagonal snap (30). When the opposite end of the snap swivel is attached to the string, the tail is connected to the frame. Both tails provide equal stabilizing drag at the bottom outside edges of the covering.

---

deal of mail from people who are trying to sell them everything from half-formed ideas to manufactured products, ready for sale. And some of these people are not quite as bright as me and thee, so their correspondence may be a little around the bend. Unfortunately, some of the companies put all inventors and people with ideas who approach them into one category—not at the top. Beyond this, many of these companies maintain research departments and are not anxious to involve outsiders in the development of new products. There are some success stories, of course, but there are many more quick, thoughtless rejections.

If you think your invention is good, perhaps even great, don't let yourself get discouraged by all the doors that will inevitably slam in your face. Keep on pushing, and don't be afraid to try some new and different approaches. The most important objective is to get your invention on the market, where it can stand or fall on its own merits—or at least on the public's perception of those merits.

CHAPTER TEN

Sporting Goods

The line beteen toys and sporting goods as separate categories in this book is arbitrary and subjective. One of the guidelines used is that toys are for children from the age of a month to about 12 and sporting goods are for children from 12 and up to, say, 95. But this distinction doesn't always hold. Both this chapter and the preceding one could probably have been combined under a single heading such as Recreation. There are so many types of inventions that could be included under this title, however, that it seemed best to split up the category into two fairly coherent chapters.

Not too many years ago, American social planners were worrying about what people were going to do with their increased leisure time. What with the shorter work week, a continuing flow of labor-saving devices, and smaller families to care for, John and Joanna Public were going to have an awful lot of extra time on their hands. The new leisure has indeed come—sometimes simply from unemployment rather than any of the above reasons—and, typically, people have used their free time in many different ways. Some have enjoyed sitting around, watching the grass grow. Others have jumped at the chance to travel. Many have gladdened the hearts of inventors and manufacturers of recreation equipment by participating in sports. Just as the spectator sports have drawn fantastic numbers of new fans, so the do-it-yourself sports have gained increased popularity.

Many of the inventions and manufactured products in the sports area have had as their purpose the improvement of some aspect of well-established games. Very few totally new concepts have been introduced. Dennis Mavrofrides of Massachusetts is one of the exceptional people, a pioneer spirit who has patented a completely new sport. A kind of combination of lacrosse, volleyball, and the 50-yard dash, Mavrofrides' game involves thirty players and a field measuring 90 by 60 yards (Fig. 10.1). The rules of this game are quite complicated—in fact, too long to be explained in full here—and this makes one wonder if the game will ever attain wide popularity. The games that most people play at least start out simply: in basketball, for instance, you score by putting the ball through the hoop, and you cannot move the ball without dribbling (bouncing) it. Even football, with all its complicated plays, is essentially a contest of strength and speed between eleven people trying to get a ball over a goal line and eleven on the other team trying to prevent them from doing so. If you're inventing a game, don't overestimate the intelligence (or the interest) of prospective players.

Many inventions in the sporting goods field have been developed to help players practice a particular skill. Even a sport such as baseball—which has been played for so long by so many different people—has spawned inventions in recent years. No matter how well you do a thing, you are always trying to do it better. There are pitching machines, for instance, that will put the ball right where you want it as many times as you want. They are of course expensive, and thus have a limited potential market. Available at less cost are tetherball devices, which hold the ball in position and return it after it's been hit. More about these in a few paragraphs.

Practice devices often are invented by veteran coaches and players who know their sport and its requirements. An interesting case in point is the "portable projectile return apparatus" patented by Elmo J. Mahoney of Kansas, a former college player. Developed primarily to help basketball players to practice their passing tech-

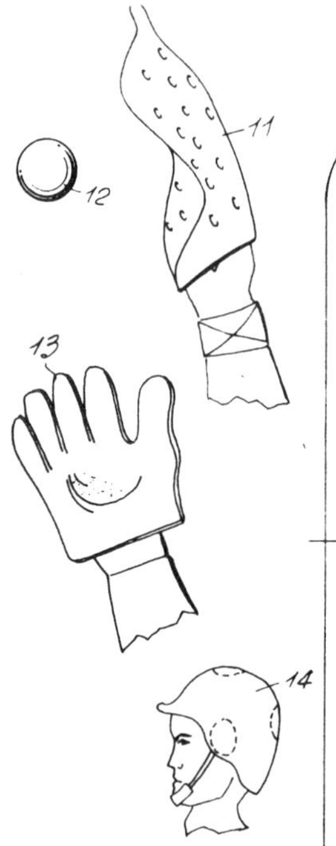

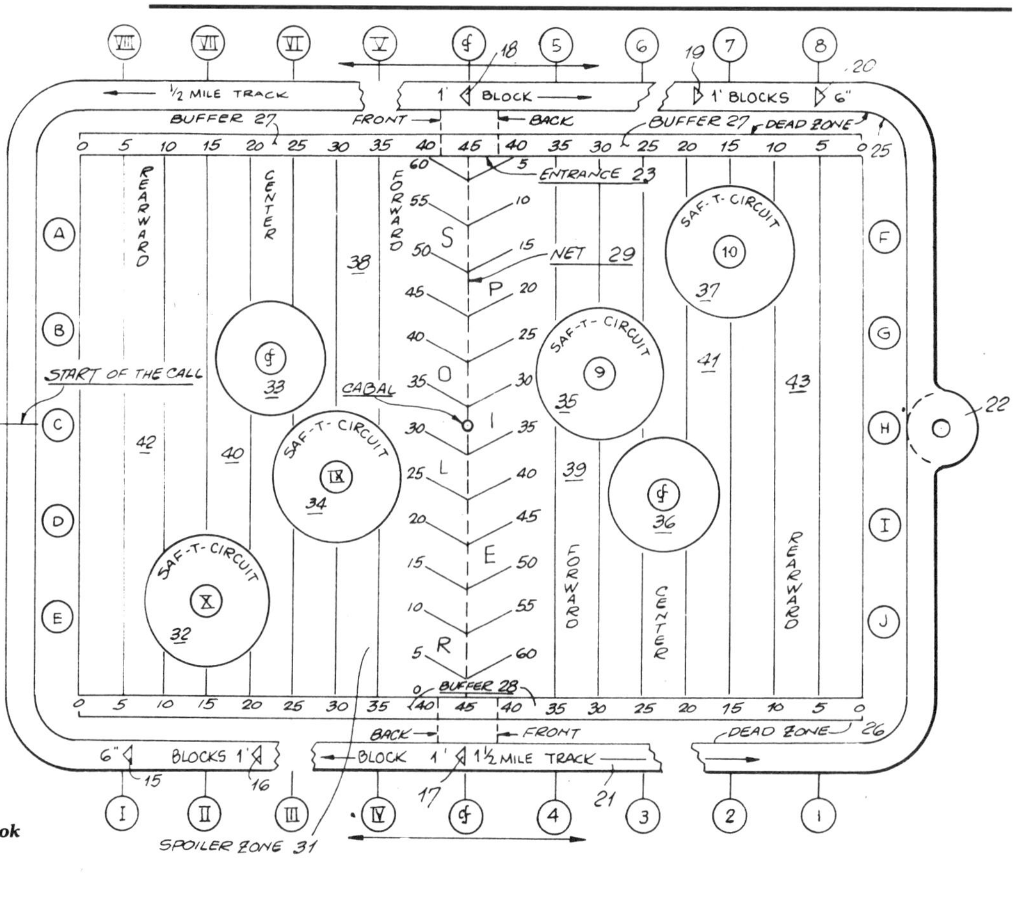

figure 10.1 Schematic diagram of the field for the game that is described in Patent No. 3,841,634 (1974). There are six rope (or wood) blocks (or hurdles). The track is a half-mile around. One block is placed in front of each entrance. Each team's remaining two blocks are placed by the respective captains wherever they wish along the track. At one end of the track is a loop, which is less than or equal to 30 yards and serves to reserve the runners. There are two entrances (23) and (24), 5 yards wide and 50 yards long, which lead runners from buffer to track. The buffers (27) and (28) are themselves 5 yards wide, running along both sides of the grounds. The dead zones (25) and (26) lie between buffer and track. The net (29) is 60 yards long and 5 feet high. The spoilers (30) and (31) are chalked zones, 5 yards wide along each side of the net.

Saf-T-Circuits (32), (33), (34), (35), (36), and (37) are for runners entering from the opposite court. Each court has three circuits—15 yards in diameter, 10 yards in diameter, 15 yards in diameter. The midpoint of the first circuit must be on the center front line, 30 yards from the buffer. The midpoint of the second circuit must lie within the center area, so that the perimeter of the circuit lies between the center and rearward front lines. The midpoint of the third circuit must be on the rearward front line.

The forward area (38) and (39) extends from the center front line to the spoiler. The center area (40) and (41) extends from the rearward front line to the center front line. The rearward area (42) and (43) extends from the dead zone (D.Z.) to the rearward front line. The playable area is between the D.Z.s and includes both courts, the net, and the spoilers. Without the buffer, the grounds measure 90 yards by 60 yards.

niques, this invention has been manufactured and sold in quantity (Fig. 10.2). Success did not come easily. First, the inventor got a lot of help in setting up a small factory at reasonable cost. As he says, "It wasn't difficult to make the first machine, but the next hundred were the rough ones." Mahoney and his brother were able to put together much of the production line by themselves, without having to rely on expensive outside help. The inventor did much of the selling himself, traveling a circuit of basketball clinics and other sessions where he was able to demonstrate his product. As he went farther and farther from home base in order to widen his market, the traveling got pretty expensive. But then he was able to hook up with a large sneaker manufacturer that paid the bills in return for promotion work at the same places where he was showing off the ball return. Mahoney knew and was known in the athletics field. Without this background, even a smart, hard-working man like him might have failed.

There are many other practice machines for other sports, too, some of which will be discussed subsequently. Better equipment has been one of the mainstays of the sporting goods business. Taking baseball as an example again, adjustments have been made in just about everything. To suit changing styles in hitting, weights and shapes of bats have been altered over the years. Leonard McCracken and Thomas Wallace of Illinois took note of another problem in batting: "In order to get a better grip on the bat, especially on days when the perspiration flows, batters have either wrapped the handle with friction tape or some similar material or applied a tacky substance prior to each use. But the tape works loose, and the sticky stuff sticks to the hands and makes them uncomfortable. A specific object of our invention is to provide a bat with a long-lasting friction grip by dipping the handle end into a synthetic rubber base coating containing zinc oxide, rosin, and silica, a blend of slow- and fast-drying solvents permitting buildup of a coating." This substance becomes tackier as the batter's hands heat up and perspire, yet it never comes off the bat to stick to the hands after use. It was designed for use on the Wilson Sporting Goods Company's magnesium bats. (See Fig. 10.3.)

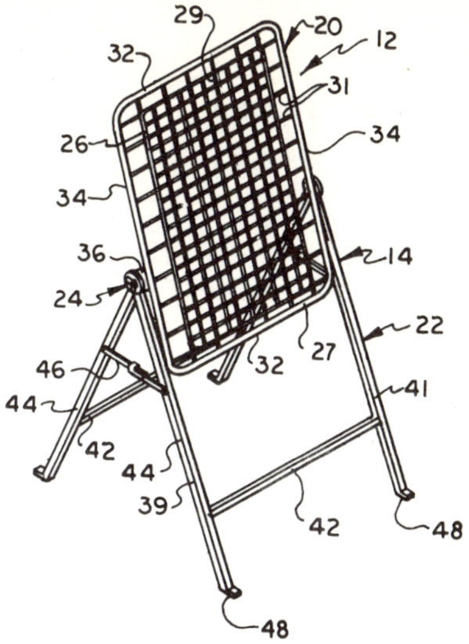

figure 10.2 Ball-return stand, Patent No. 3,836,144 (1974). It includes the overall backstop (20), a frame (22), and a connection between these two (24).

The return panel assembly (26) includes a main support panel (29) constructed of a net fabric material and connected by straps (31) to the panel support assembly (27). The frame (22) includes a pair of A-frame leg assemblies (39) and (41) interconnected by transverse support members (42). The A-frame leg assemblies are substantially identical. Assembly (39) includes a pair of elongated legs having intermediate portions interconnected by assembly (46). The lower portion of each leg has an outwardly turned foot section (48) for vertical support. The upper portion of each leg is secured to a circular disc member having a hole for attachment by the connector means (24).

In use, the frame is extended, and the backstop is anchored. This is the higher position, for practice of volleyball shots, basketball passes, etc. On loosening of the connector (24) through a wing nut, the backstop can be rotated to the lower position, for practice throwing baseballs, etc.

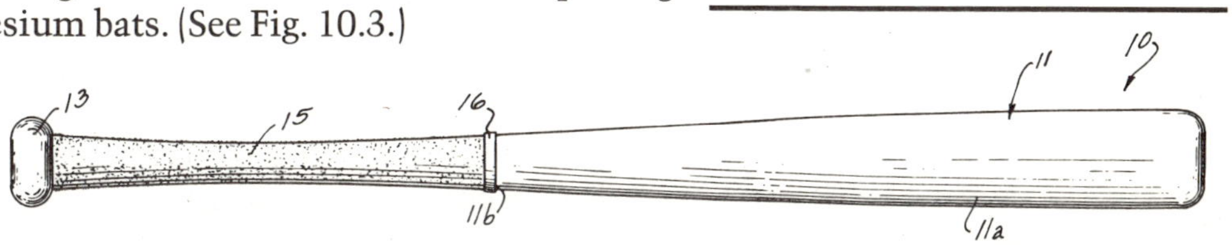

In football and hockey and other hard-contact sports, the emphasis has been more on new protective devices, especially for youngsters. New materials that are harder and tougher, yet often lighter in weight than what has been used, have been developed in recent years. Inventors and manufacturers of sporting goods have been able to use some of these innovative materials in their protective gear. Jhoon Goo Rhee of Virginia has patented a helmet that covers head and face, with openings for eyes, ears, nose, and mouth (Fig. 10.4). It is especially designed for such sports as karate and boxing, and can be adapted for cycle riding protection. It is of simple construction, made from resilient material with a tough outer

figure 10.3 A new baseball bat and grip, illustrated in Patent No. 3,972,528 (1976). The bat (10) is a one-piece tubular magnesium body (11) open at one end and closed by a welded-on cap at the other end, filled with plastic. The body has a large-diameter top (11*a*) tapering down to a smaller-diameter handle (11*b*) with a solid knob on the end of the handle. A plastic cover (13) is cemented over the knob, closing the open end of the tube.

The magnesium body is coated with a polyurethane-base paint. Then the handle is covered with a grip (15) integrated with the coating on the handle. A collar of plastic film (16) is wrapped around the inner end of the grip to provide a decorative line of demarcation between the striking and gripping zones of the bat.

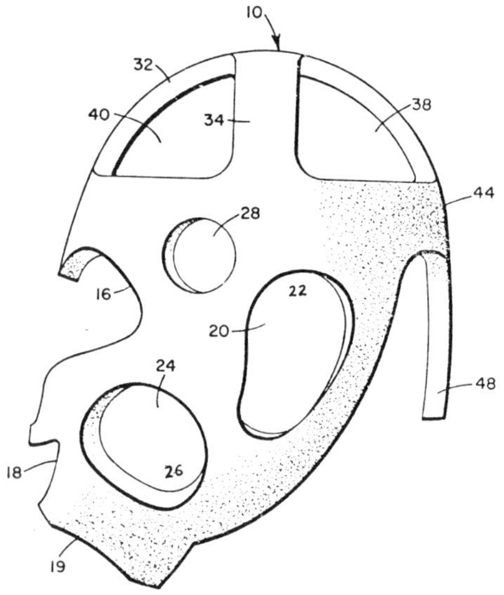

figure 10.4 Side view of the protective helmet, Patent No. 3,992,722 (1976). It is made of a resilient material (14), such as foam, and a smooth, tough outer case. There are various openings for the parts of the head, as follows: nose and eyes (16), mouth (18), chin (19), ears (20) and (22), cheeks (24) and (26). There are also optional vents (28) and (30).

The upper portion of the helmet comprises a pair of cross members (32) and (34) adapted to retain the helmet on the top of the person's head. Openings (38) and (40) provide ventilation and comfort to the wearer of the helmet while the cross members provide protection to the plate. At the rear, a horizontal part is connected to cross member (34) at its midpoint. The rear of the helmet has a large opening, and a downwardly projecting flap (48) extends to, generally, the horizontal center of the helmet, and about to the bottom of the neck of the wearer.

casing. For additional protection, there is a removable transparent shield that can be pulled down in front of the face. The helmet must be easy to put on and take off and be comfortable to wear. Otherwise, people just won't wear it.

In addition to protective helmets, other equipment for ice hockey is in demand. This sport has become surprisingly widespread, with many young players even in the southern states, where a pond will freeze over once in fifty years. The number of rinks with artificial ice has increased, but not enough to meet the demand. In some suburbs, groggy fathers and sons can be seen on the way to the rink at 5 A.M., the only time the rink is available for some school and club games. Practice machines are important here, too. Gilles Deschesnes of Quebec has developed a seemingly complicated gadget to help a player strengthen his shots (Fig. 10.5). It utilizes a common device, attaching an elastic band to the puck so that it will not go far no matter how hard it is hit. It's another version of the tetherball principle. In this case, unless the puck is propelled with a great deal of force, the tension is such that it won't go much of anywhere. Thus the player is encouraged to hit harder and harder. The resiliency can be adjusted to make it tighter for stronger players.

One of the more commercially successful tetherball arrangements is illustrated in Fig. 10.6. Originated in South Africa by Peter Hutt and Ian Gormley, this ball has become popular in many parts of the world. It's not very complicated. A flexible line is secured to a rounded anchoring element inside a hollow ball, probably a tennis ball. The ball never gets very far away from the players, and can be used in many different games. It's not easy to do much better than this for a good, workable tetherball, and most inventors would tend to leave well enough alone. But there are many and varied possible uses for a tetherball, and some of them are patentable. Raymond Lohr and Robert Auer of Pennsylvania invented a support assembly for a tetherball game, and assigned the patent to the Marx Toy Company. It's essentially a setup for two people, one on each side of the bar. The whole thing can be put together quickly and without the use of tools. When not in use, it can be stored in a small space. (See Fig. 10.7.)

An unusual but eminently sensible use of the tetherball is shown in Fig. 10.8. It's a game for astronauts, to be used in the weightless environment of a space station. Invented on his own time by a NASA engineer, Curtis Vetter of Texas, the game has rings and rungs by which the players may pull themselves up or sideways as they pursue the ball. The object of the game is to get the ball through a goal hoop. Ahead of its time? Perhaps, but there is an earthbound version in the works. As of this writing, however, the game has not been produced in quantity. In fact, Mr. Vetter's letters to ball manufacturers and rubber companies inquiring about the specifications for the special balls needed in the game have not been answered. An inventor's life can be frustrating. On the other hand, sporting goods manufacturers generally have been quite businesslike in answering correspondence about the possibility of licensing the production and marketing of the game.

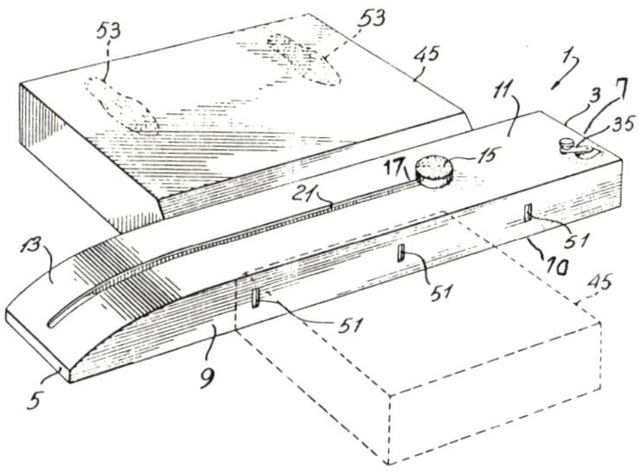

figure 10.5 Hockey training device, Patent No. 3,955,815 (1976). The platform has walls (3), (5), (7), and (9), a bottom (10), and a top playing surface (11), which has a downward curve (13). The puck (15) is connected to a wheeled carriage by a vertical post passing through a longitudinal slot (21). The carriage is H-shaped and is mounted on wheels for linear movement on a pair of tracks running parallel to the slot. These tracks limit the puck's moves to a linear direction.

The carriage (17) is connected to a resilient cord that makes it hard to propel the puck, and that also returns the puck to starting position after each practice shot. The end of the cord is wound about a spool and threaded through a pair of pulleys. The spool has an adjustable spring connected to it, part of the restraining means for the puck. The tighter the tension, the more force it takes to propel the puck.

The platform (3) can be made large enough for a player to stand on it on either side of slot (21). Or a separate stand (45) may be provided. The player can stand on it positioned with his feet on markers (53).

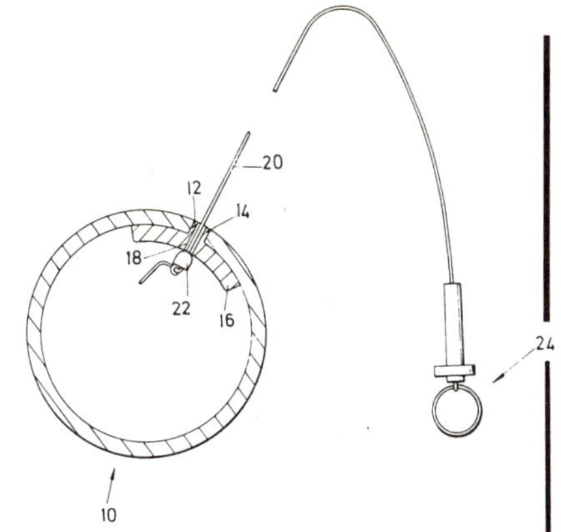

figure 10.6 Hollow ball with a flexible tether line, Patent No. 3,729,195 (1973). Ball (10) has a hole (12) into which fits the nub (14) extending from a base (16). There is an opening (18) in the base. A line (20) extends through this and through a bullet-shaped element, to which it is anchored. In the drawing, this is done with a knot. The other end of the line (24) has a hook for use in attaching it to a sports apparatus or game. In the preferred form, the hook slides along a spiral at the top of a pole. Two opposing players hit the ball with rackets, one trying to whirl it off the top of the pole, the other trying to prevent this.

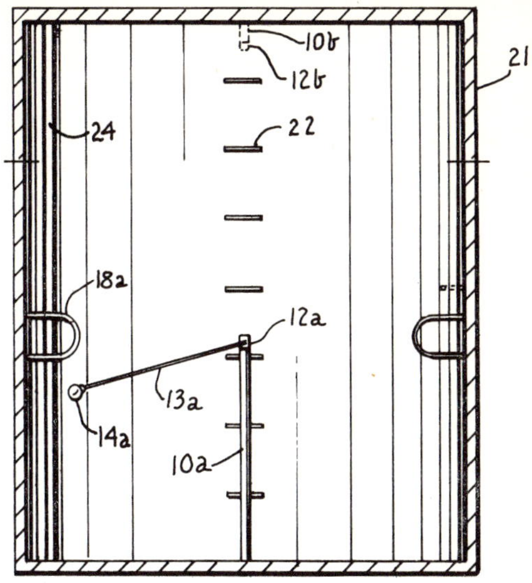

figure 10.8 Tether ball game as it might be arranged in space illustrated in Patent No. 3,856,305 (1974).

In a chamber (21) a post (10*a*) is shown, but since there is substantially no gravity high up in space, the tether (13*a*), shown suspended from the post (10*a*), may as effectively be suspended from a swivel-suspending stud (10*b*), shown in broken lines. The tether has to be swivel-mounted in some way, and in this case a swivel (12*a*) is shown on top of the post (10*a*), and a swivel (12*b*) is shown at the lower end of the stud (10*b*). Since the pull of gravity on the ball is negligible, it will travel in an orbit almost exactly as directed by contact with the bat or racket. As shown, the goals (18*a*) are simple loops.

Vertical hand-lift bars or rails (22) extend slightly, radially inward from the chamber wall. As the players are substantially weightless, they tend to float or rest in space wherever they become disposed, and it is necessary to provide some means whereby they can pull themselves around.

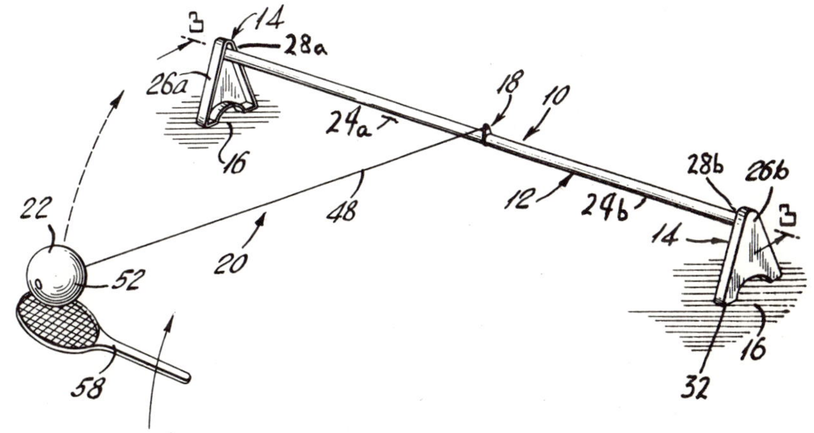

figure 10.7 Exploded view of the disassembled components of the tether-ball support device in Patent No. 3,809,406 (1974). The bar is composed of two hollow pipe sections (24*a*) and (24*b*). Triangular support sections (26*a*) and (26*b*) are joined to the bar (12) by connectors (28*a*) and (28*b*), respectively. All these parts are made of plastic. The attaching elements are made up of a tube (40) with a flange (42) and a hole (44). The ends of tube (40) are tapered so that they fit easily into bars (24). With the free ends of the tube received in the adjoining ends of bar (24), these elements are pushed together so that they directly engage the flange (42). The supports (26) are also pushed toward each other so that the ends of the bars directly engage the inner end of connectors. Thus, assembled parts rest stably on a supporting surface, floor, or outside ground. Components (26) need not be attached in any way to the surface.

On the flange is an opening (44) for receiving a hook (46) at one end of a relatively short ball chain; at its opposite end is a hook to which one end of a flexible narrow strip (48) is attached. One end of this strip is connected to the ball (52), which is made of a suitable plastic and is capable of being inflated. Thus, the lightweight inflated ball (52) is readily attached to the tether means, which in turn is readily attached by way of hook (46) to tube (42).

The ball, while appearing to be rather large, nevertheless is lightweight, and in connection with the elasticity of the tether element (48) provides an erratic movement each time it is propelled toward one side or the other side of the bar.

One of the most popular pieces of sporting equipment to be introduced in the last twenty-five years is also space related, at least in its aerodynamic design. The flying saucer, as it is called in patents relating to its construction, or Frisbee, as it's known to the millions who play with it, is a great success story. The first Frisbees were metal pie plates from the Frisbie Pie Factory in Connecticut. An enterprising fellow named Fred Morrison visited Yale University one day and saw the students tossing Frisbie's pie plates across the campus. He took the idea to California and promoted Frisbees by himself at beaches and other recreation areas, until in 1956 he sold out to the Wham-O Manufacturing Co., makers of Super Balls, Hula Hoops, and other novelties. Since then the Frisbee has been improved in various ways by inventors connected with Wham-O. Figure 10.9*a* shows part of a 1967 patent, filed by Edward Hendrick of California, and Fig. 10.9*b* shows the 1973 changes invented by Richard Gillespie, Sr., also of California. In the '67 version, "spoilers" on the convex side of the saucer were added to interrupt the smooth flow of air around the rim. This reduced drag, especially at high speeds, and increased stability. In the 1973 version, a trough going all around the saucer was added immediately adjacent to the rim. A low profile makes it possible for the Frisbee to fly at a rather high velocity.

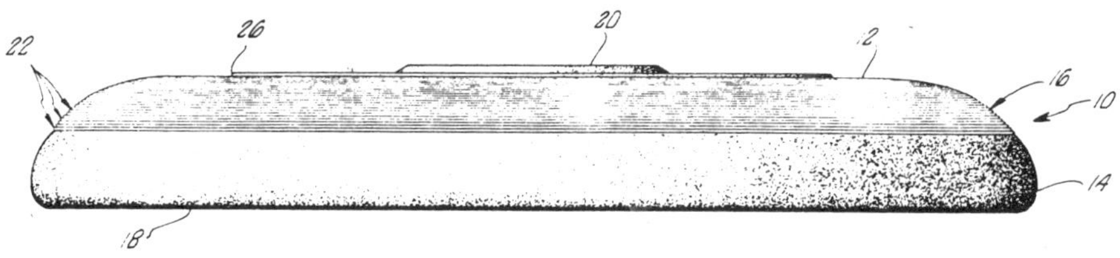

figure 10.9(a) "Flying Saucer," or Frisbee, as illustrated in Patent No. 3,359,678 (1967). A rim (14) is joined to a center (12) by a curved surface (16). The bottom edge (18) of the saucer, with its convex side, resembles an air foil. The crown (20) on this convex side probably contributes to the saucer's stable flight. A number of ridges (22) are superimposed on the curved surface (16). These ridges also influence the saucer's flight.

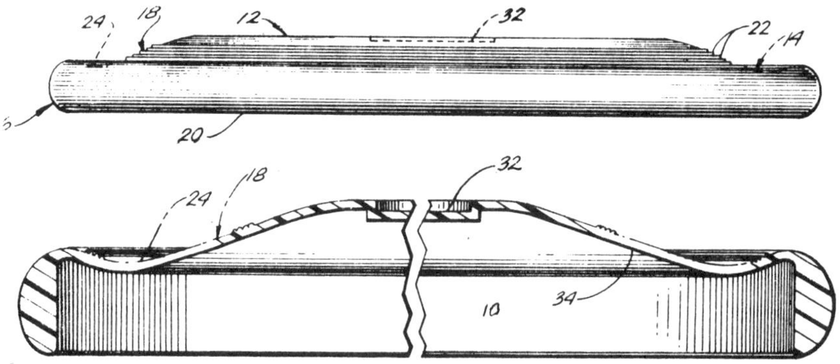

figure 10.9(b) Two side views of an improved Frisbee, Patent No. 3,724,122 (1973). The saucer (10) has a bottom edge (20), a crown (12), an inner circumference (14), a curved surface (18), an outer circumference (14), and a rim (16). A number of ridges (22) are superimposed on (14) and part of (18).

Surface (14) is curved so that a reverse or "S" curve is produced in the transition from rim to intermediate surface to crown. This results in a trough (24) extending around the saucer next to the rim. In addition to making a lower profile for the saucer, the trough is a means whereby the saucer may be easily held. The fingers of the user grip the under or convex side of the saucer and rim portion while the thumb rests in and grips the trough on the concave side.

A circular, shallow recess (32) is located inside of spoilers (22) in the center of the crown portion, forming a closed circle concentric with the center (28). The central region of crown portion (12) in such an embodiment is thereby essentially flat. The concave side of the saucer has an essentially smooth surface (34) extending from the rim to recess (32). This recess appears as a circular disc raised from the surface of the concave side.

Another successful invention of the last twenty-five years is the Wiffleball, which was first patented in 1957 (Fig. 10.10). It was, and is, a kind of simulated baseball, to be used in small spaces or by people who don't want to play with a standard hard ball. The Wiffle can be hit very hard with a bat, but it will never go very far. Its slatlike construction pattern sets up an air resistance that slows the ball down. It is also flexible enough to absorb a strong impact without breaking.

The purpose of including the Frisbee and the Wiffle here has not been to suggest to inventors that they work on making improvements to these old favorites. This might be done, of course, but the odds are against it. The point is to show how creative minds can use a knowledge of the principles of physics together with a sense of the different properties of new materials, especially plastics, to develop something new and useful and profitable. Inventions do not come sailing out of left field on the wings of inspiration. They come out of a broad knowledge of the laws of nature.

The invention illustrated in Fig. 10.11 looks like a Frisbee but is actually quite a bit older and with a much different use. It's the first really practical clay pigeon, invented in 1885 by George Ligowsky. He was an accomplished trap and skeet shooter who understood some of the principles of aerodynamics and applied them to solving the problem of getting something to shoot at other than live birds, glass balls, and other targets that had become outmoded. He developed a "cup-shaped flying target formed as a thin shell of clay. . . without slot or tongue or extraneous handle for propulsion which might constitute an impediment to the axial rotation of the target." A more modern clay "bird," patented in 1974 by Allen Ludwig and John Dale of Texas, is a combination of elemental sulphur and an "inert filler such as limestone dust. . . . By the inclusion of materials that are dimensionally unstable in water, such as bentonite clay, it is possible to produce an environmentally degradable target." Different ages and times demand different techniques. In the 1970s, the call is for an inexpensive target that will break down fairly rapidly so as not to litter the landscape. Both inventions have one important feature in common—they can be mass produced economically.

Another important consideration in today's world, especially as related to firearms, is safety. Joseph E. Smith of California invented a unique device that prevents a trigger from being pulled all the way by anyone except the owner of the gun. He uses a "magnetically attractable bar pivotally mounted on the back of the trigger or in the gun stock housing and pointed in an upright position. The bar is balanced about the pivot point so as to maintain a center position," inhibiting the trigger pull. The authorized user of the firearm wears a magnetic ring that pulls the bar out of the way, giving free movement to the trigger. (See Fig. 10.12.)

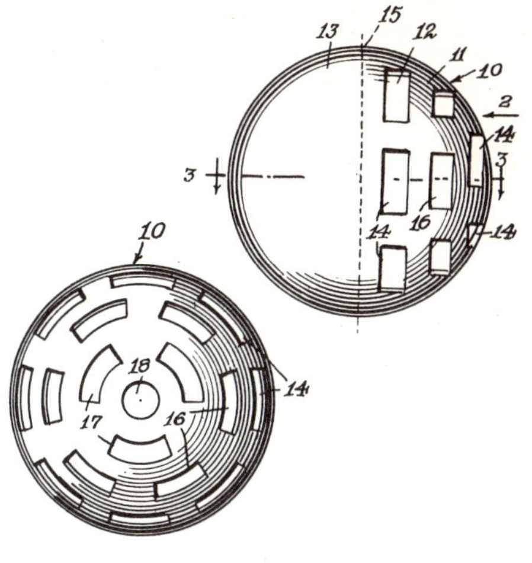

figure 10.10 Side and top views of the Wiffle ball, Patent No. 2,776,139 (1957). The Wiffle is a hollow sphere about the size of a baseball. The shell (11) is made of plastic and has a number of holes (12) in one hemisphere and a smooth, unbroken surface elsewhere. This arrangement sets up a pattern of varying resistance to air when the ball is thrown. It also slows up the ball's flight. When hit or thrown, the ball curves and wobbles in a fairly unpredictable manner.

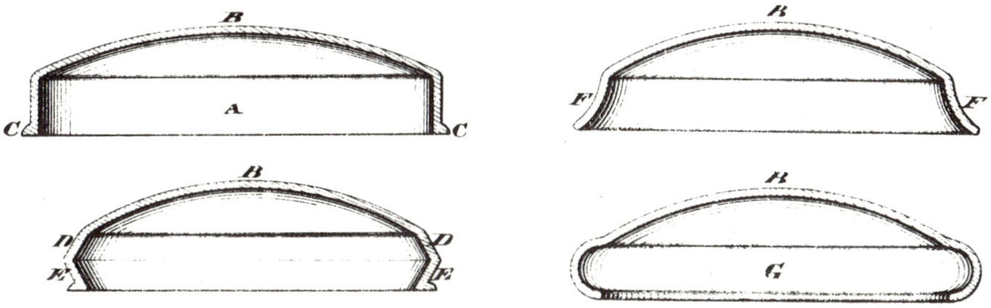

figure 10.11 Axial sections of four different forms of the saucer-shaped flying target shown in Patent No. 311,768 (1885). Top left, the shell (*A*) has a crowning top (*B*) and, if desired, a circumferential bead or molding (*C*) at the margin of the cylinder. The target is made of earthenware or other fragile material capable of being readily punctured or cracked when struck with shot. The target is light and thin, constructed without a slot or a lug, tongue, or other lateral projection of any kind whatever.

Bottom left, the crowning top (*B*) is used; but the body, instead of being exactly cylindrical, inclines outward at *D*, and then slopes inward at (*E*), causing the peripheral rim of the target to have the shape of an obtuse angle in cross section.

Top right, the crown is joined to an outwardly flaring rim (*F*) that may be either curved, as shown, or may be simply a straight incline or pitch, as represented at *D*, bottom left.

Bottom right, the crown merges into a concave rim (*G*) of any appropriate degree of curvature.

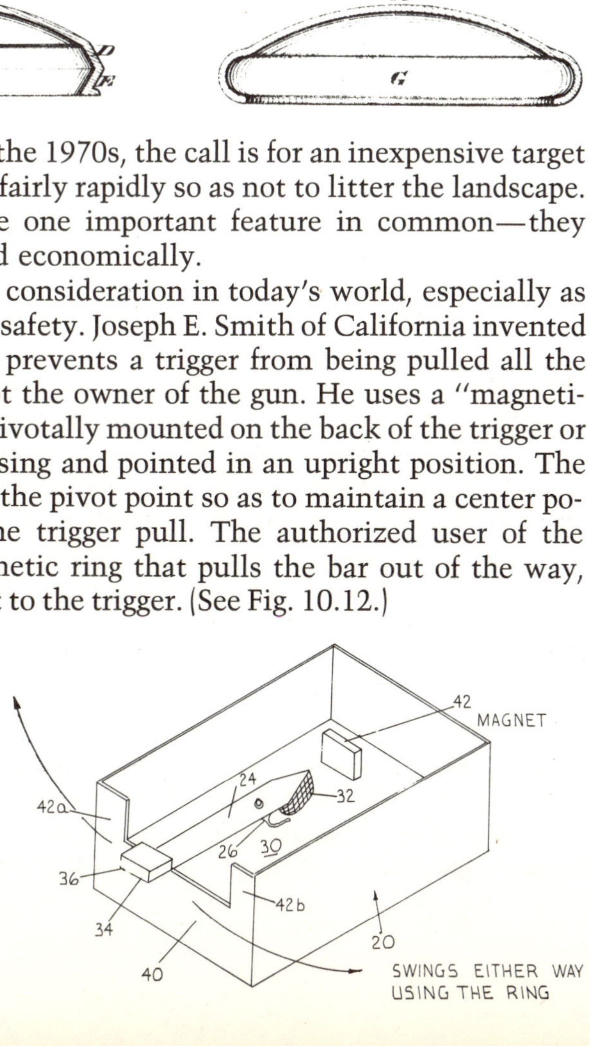

figure 10.12 An enlarged cutaway view of the trigger safety device illustrated in Patent No. 3,978,604 (1976). The device can be attached on the handle to the rear of the trigger guard. A nonmagnetic bar (24) is pivotally mounted on post (26) attached to floor (30). Weight (32) balances the bar and keeps it in the center position of the rear magnet (42). At the opposite end (34) of the bar is a block (36) that is attracted to magnetic material. The front wall (40) of the housing (20) has an aperture with side walls (42*a*) and (42*b*), which serve as stops to prevent bar (24) from moving too far to either side. In this way, the bar remains centered in relation to the trigger or handle.

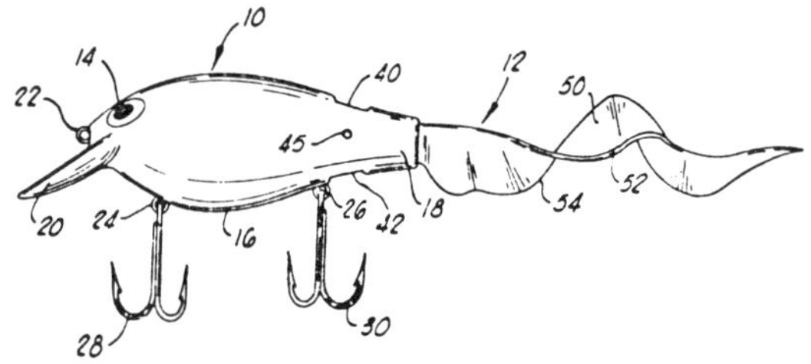

figure 10.13 Side view of the fishing lure described in Patent No. 3,979,853 (1976). The streamer (12) is extended, as it would be when pulled through water.

The rigid flotation body (10) is made to resemble a minnow or small fish, with an eye (14), an underbelly (16), and a tail (18). Just above the diving plane (20) in the nose of the body is an eyelet (22) that is glued into the body and constitutes an anchor point for the fishing line (not shown). A pair of hook hanger eyelets (24) and (26) is glued to the underbelly, and holds a pair of treble hooks (28) and (30).

The body is hollow and includes an interior air chamber containing some small lead shot or weights. The streamer (12) is an elongated strip of flexible material, such as rubber or polyvinyl chloride, and is relatively thin over its length. The curved section (50) includes one side edge (52) that is substantially longer than the opposed side edge (54), with the side edges extending parallel to each other. At the end of the streamer, the side edges converge in a pointed tip.

Outdoor sportsmen are inveterate inventors. Although the flow of patents related to guns and hunting has slowed somewhat, the volume of inventions related to fishing has been on the increase. Fishing lures are especially popular. William and Gary Storm of Oklahoma designed the lure shown in Fig. 10.13 and then assigned the patent to their own company, Storm Plastics. Their invention is described as "a rigid body of generally fish-shaped configuration tapering from midsection to tail, and an elongated flexible resilient streamer element connected to the tail of the rigid flotation body." Fishermen being what they are—absolutely sure that their own methods are right and the other follow's are just a bit strange—one might question the need for going through with the procedure of getting a patent for a lure. Of course, a rival manufacturer might choose to copy your idea, but that could not be done overnight, and meantime your lure (if it's any good) would have made a place for itself in the market. Caution: The Patent Office files hold a large number of lures, most of them commercially unsuccessful. Proceed at your own risk.

Inventive types will be found among the ranks of hikers, too. They can always find a better way to do something, even the most minor chore. Martin Saunders of Nevada took on a major project when he developed a new design for a sportsman's saw. One edge of his saw has coarse teeth, the other has fine teeth. The saw thus can be used for all kinds of jobs in the outdoors, and is made to be carried in a scabbard on the belt or in some other convenient place. The handle can be used as a hammer, too (Fig. 10.14).

The biggest tinkerers of all are, of course, the boatmen—or, if the craft is more than 25 feet and they have a fancy cap, the yachtsmen. Almost 140 years ago, William Henry Porter of Great Britain invented a better anchor. He assigned U.S. rights in the invention to Jonathan Osborne of New York, and they were granted patent no. 2,497 in 1842. Porter's object was to make an anchor that would be less liable to be fouled when the vessel was swinging. When one fluke is holding to the ground, the other fluke lies on the shank so that the vessel's bottom will not be damaged (Fig. 10.15*a*). Another mooring device was patented by Arthur Kosmatka of Wisconsin in 1976. As can be seen from Fig. 10.15*b*, his object was to tie a boat to a dock but keep it a fair distance away so that there is less chance of damage in rough weather, when a closely moored boat would hit the dock again and again. One mooring line is attached to the bow, and another line goes from the stern to the dock through an arrangement of pulleys. There is enough slack to keep the boat from breaking these lines, but they are taut enough to keep the boat away from the dock.

There's always a certain amount of bailing to be done around a boat, but it's not always easy to get someone to do it. Rex Ridgeway of Iowa has patented a simple, portable device powered by direct current, which is the type of electricity generally available aboard a boat. The bailer is automatic, going into operation when the water in the bottom of the boat gets to a certain level. No one needs to be aboard during bailing operations. Thus the inventor has taken an age-old device and brought it up to date. (See Fig. 10.16.)

figure 10.14 Sportsman's saw, Patent No. 3,837,024 (1974). It includes a blade (12) having a fine saw edge (14) and a coarse saw edge (16). The coarse-edge teeth extend all the way from top to bottom. The fine-edge teeth extend from the top of the edge down to a point (18) near the tip (20) of the saw blade. The portion of the blade extending between point 18 and tip 20 is sharpened to form a convex cutting edge (22).

The blade is 12 inches long and about 0.038 inches thick. The coarse saw edge is provided with six teeth per inch, each set 90 degrees to the cutting edge with an 88-degree hook. The fine saw edge is provided with eleven teeth per inch, each set with a 90-degree pitch to the cutting edge with an 88-degree hook. Cutting edge (22) located between point (18) on fine saw edge (14) and tip (20) is about 1½ inches long.

Attached to saw blade (12) by means of rivets (24) is a handle (26), which takes the form of a planar extension (28) attached to a body (30) in such a way that the longitudinal center line of the body is roughly perpendicular to the longitudinal center line of blade (12). Extension 28 defines a hole (32) so that the user can wrap his fingers all the way around the body and thereby tightly grasp the handle (26).

The saw is preferably housed in a scabbard long enough to completely house the saw blade, and with a strap for keeping the saw in place.

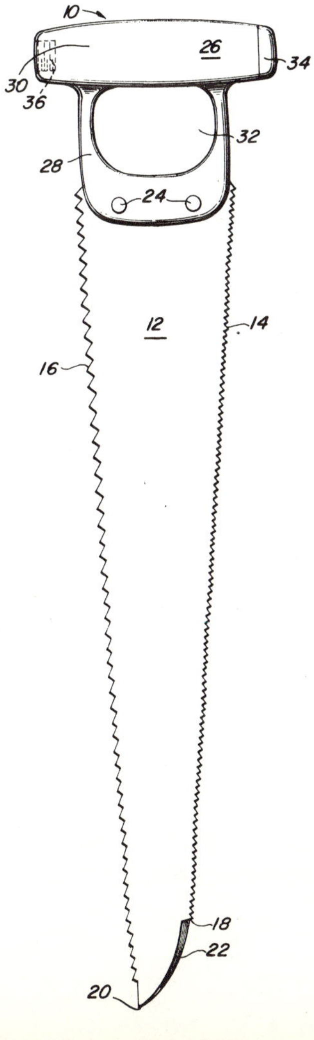

figure 10.15(a) Anchor patented in 1842, No. 2,497. The shank, or stem, is at (*A*), with two checks (*B,B*) that the arms (*C*) may be received in the space (*D*) between them, the checks and arms being of sufficient strength to receive a pin or axis (*E*) through them, such a pin or bolt being secured in its place by a split collar or other fastening, and washers or otherwise as needed; the washers have a buoy chain (*F*) attached to them.

The stock (*G*) is represented to be of iron, but a wooden stock may be applied to anchors constructed according to this invention in the manner which is usual to any of the known forms.

The projections (*H,H*) are for the purpose of opening out the underarm and fluke as the anchor is drawn along the ground when it has fallen with the under arm close to the shank.

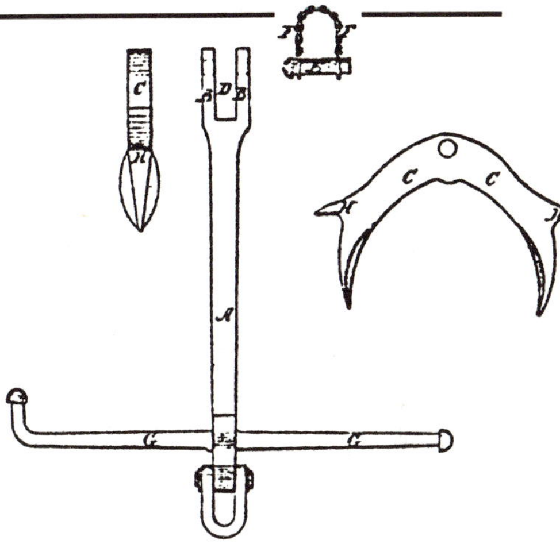

figure 10.15(b) Mooring device, as shown in Patent No. 3,971,329 (1976). A boat (10) is provided with a mooring eye (12) at its rear end and a mooring bracket (15) at its forward end. A line (16) has one end attached to the bracket and its opposite end tied to a mooring eye (17) that is rigidly secured to the pier (11).

When the hook (20) is disengaged from the eye, it is kept above the waterline (21) by a float (22). A line (25) is attached to the float and extends about a pair of pulleys (26) and (27). The pulley (26) is journaled in a bracket (28) that is mounted on a bar (29). Pulley (27) is journaled in a bracket (30) mounted on pier support (31).

Most of line (25) rests on the pier beyond its knot on the eye (17). To pull the boat to the pier, the operator merely unties line (25) and the eye and pulls on line (16). This will cause the line to draw away from the pier; sufficient extra line is provided so that it remains extending through the eye. The person using the boat then disengages hook (20) from the eye and pulls line (25) through the eye onto the pier so that float (22) and its associated hook (20) float directly above pulleys (26). The person then unties line (16) from the eye, and the boat is free to be used.

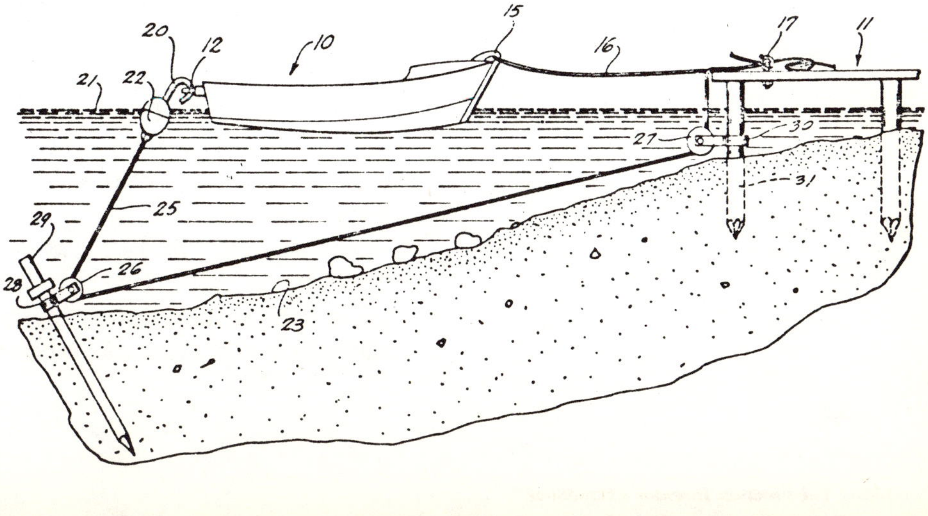

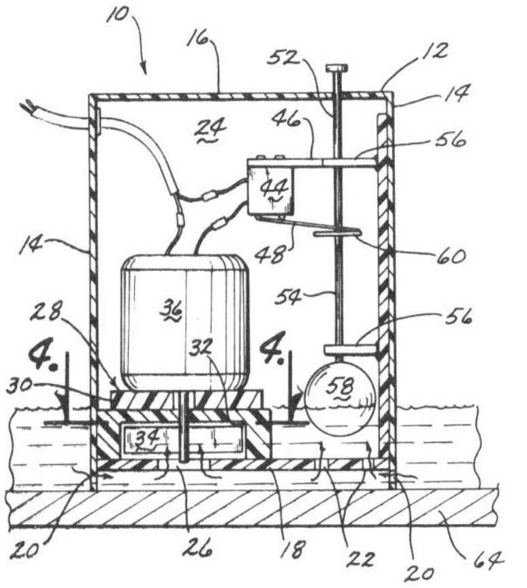

figure 10.16 Cross section of a portable bailing device, Patent No. 3,941,073 (1976). The bottom wall (18) includes a number of apertures (22) for permitting water to enter compartment (24). There is also a pump inlet opening (26) in this bottom wall.

Within compartment (24) is a pump (28) with a pump housing (30) and a cylindrical chamber (32). Within this chamber is a rotor blade (34) connected to a direct-current motor (36). The inlet opening (38) is in registered alignment with pump inlet opening (26). Chamber (32) also includes a tangentially extending outlet opening (40) that is connected to a tube (42).

Also within the compartment is a limit switch (44) held in position by a bracket (46). Switch (44) is normally open, but switch arm (48) may be moved upward to a closed position. The switch and DC motor are connected in series with a battery.

Float device (52) includes a vertical rod (54) slidably mounted to a pair of brackets (56). The lower end of this rod includes a float (58), and the upper end extends up through the top wall (16). The switch can be moved by this rod.

In operation, the device is placed in the bottom of a boat, positioned so that the bottom edges of its walls are in any water that may be in the bottom of the boat. Because the switch is normally open, the device is not actuated unless the water rises sufficiently high to lift float (58) to its upper position. This causes the switch to move to its closed position, thereby actuating motor (36) and pump (28). As the pump lowers the water within the boat, the water lowers correspondingly within compartment (24) thereby causing the float to lower and return switch to its open position.

As winter sports increase in popularity, they are also drawing the interest of inventors. Arnold Thompson of Illinois has patented an ice skate sharpener designed for use with a portable electric drill (see Chapter 13, Tools, for a discussion of other attachments to the hand drill). According to the inventor, standard ice skate sharpeners are not able to produce straight, parallel, and uniformly sharp cutting edges on both sides of the skate runner. These sharpeners generally consist of an abrasive that is enclosed except for a longitudinal slot through which the blades are drawn. But since blades are of varying widths, the slot has to be made to accommodate the widest. Given this arrangement, with its room for slippage, most people cannot keep the skate perfectly straight and centered. So it is seldom possible to do a full-scale sharpening with these devices, which are used mostly for touchup work. One of the features of inventor Thompson's sharpener is a blade-positioning unit that can be clamped into place and adjusted to control blades of different widths. (See Fig. 10.17.)

Winter sports people seem to be fussier about their equipment than their warm-weather counterparts. The array of products and materials in ski and skate shops grows with each passing year. Ski boots are a case in point. A large number of patents for different kinds of boots have been issued in the past ten years, to both American and foreign inventors. Figure 10.18 shows two different approaches to the patent process, one a standard utility patent and one a design patent. The latter, from Willy Kaufmann of Switzerland, contains no written explanation because there are no apparent moving parts. The patent is the drawing. In Fig. 10.18*a,* a patent by Georg Turnheim and Axel Kubelka of Austria, there is a recess in the sole of the boot to receive parts of the ski binding. In other boots of this type, this recess opens only downward, making it difficult to remove snow and ice that tend to accumulate there. In the present invention, the recess is a longitudinal groove that is open at both ends. Also, the sole is curved both toward the heel and toward the tip, which makes walking easier for most people. So this patent, too, is concerned mostly with the design of the boot, but because it is complex and needs to be explained, a utility patent offers the best protection.

All the equipment skiers have is often difficult to tote, especially if the parking lot is far from the lift. Many people like to tie everything together so that it can be carried in one hand. But according to Allen Rosenthal of California, "conventional ties are only suitable for holding skis together, are bulky and awkward to use, and, because of angular metal parts, are difficult and, indeed, dangerous to carry while skiing." So he invented a tie that made use of Velcro, the binding material discussed in connection with Fig. 9.6 in the chapter on toys. This tie readily straps any size skis and poles together, can be quickly released, and is easily carried when not in use (Fig. 10.19).

Golfers are another group generally quite interested in new equipment. Here the emphasis is on improving the game, shaving a stroke off the score wherever possible. Many training devices have been developed, including an aid to putting patented by

figure 10.17 Side view of an ice skate sharpener, Patent No. 3,812,626 (1974). It is designed to fit into a portable drill chuck (12). The tubular casing (14) contains a grinding wheel (16) supported on a drive shaft (18). The skates are inserted in slot (20) for sharpening. There is an assembly (22) which automatically positions the skate properly; thus the sharpening can be done by both skilled and unskilled operators with equal success. This assembly can handle skate blades of different widths. It consists of a rigid, one-piece, framelike support having two substantially identical arched side wings (62) and (64) connected at their ends by transverse base struts. Side wing (62) has clamped to its inside face a shoelike wear plate (74). Side wing (64) has a socket within which is the outer end of a helical compression spring. An adjusting screw (82) holds an adjustable shoelike wear plate (86) in place. The head of the screw is located adjacent to the outer side surface of side wing (64) and limits inward sliding movement of the screw and the wear plate.

The two wear plates (74) and (86) are between the two arched side wings (62) and (64), with plate (74) fitting against the inside face of wing (62) and plate (86) in a position slightly spaced from the inside face of wing (64). These two wear plates thus define a skate blade-receiving channel or slot (88) through which the blade is drawn in a longitudinal direction during sharpening, while the edge of the blade contacts the periphery of the rotating grinding wheel (16).

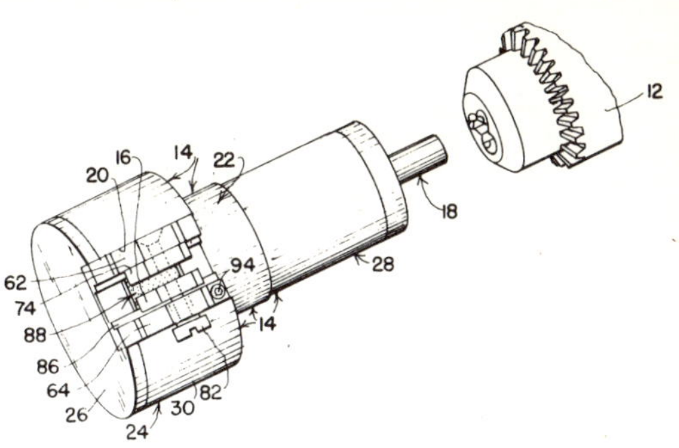

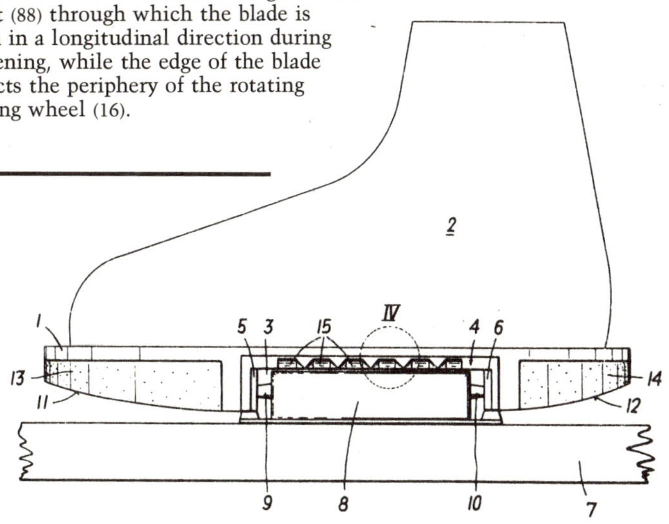

figure 10.18(a) Side view of a boot on a ski, from Patent No. 3,957,280 (1976). A groove (3) is provided in the sole (1) of the ski boot (2), extending transversely to the longitudinal direction of the boot and being open at its ends. In this groove the ski binding part (4) is supported and carries two parts (5) and (6), forming locking receptacles. The binding part (8) provided on the ski (7) has two spring-loaded locking elements (9) and (10), which engage the locking receptacles formed by the parts (5), and (6).

During an overload, for example, during a fall, the ski boot (2) can be released both laterally and also upwardly by pressing the spring-loaded locking elements from the parts forming the locking receptacles. For stepping into the binding system, the ski boot is pressed in a simple manner to place the recess (3) onto the binding part (8) arranged on the ski.

The stepping surface of the sole has both toward the tip and also toward the heel, respectively, appropriately curved surfaces (11) and (12) by means of which the ski boot can properly roll during walking. In order to make walking even easier, the stepping surface can be formed in the area of the curvatures of elastic material.

figure 10.18(b) Two illustrations from design Patent No. 218,794, (1973), for a ski boot.

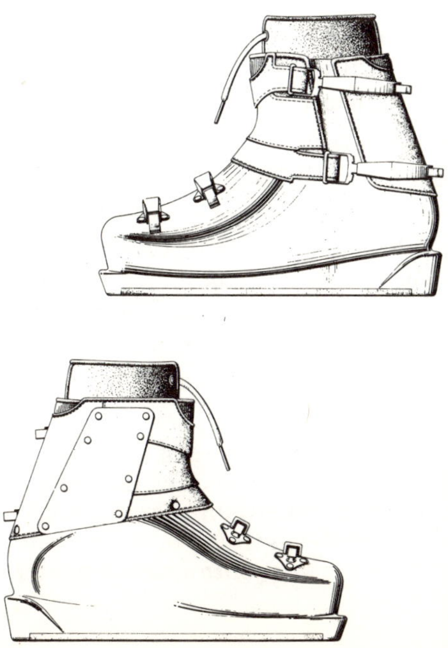

figure 10.19 Ski tie and ski strap from Patent No. 3,947,927 (1976). Made of a flexible tape material, the strip includes a center (12) with two free ends (14) and (16) connected. The upper surface of each end and of the center is made of Velcro, so that they can be joined. These surfaces where the hooks of one surface become enmeshed in the loops of the other surface, may be readily separated from each other by being peeled apart, but when fastened, they strongly resist longitudinal movement. There is also a pair of eyes (18) and (20), each formed of metal or plastic loops connected to the central segment. These eyes serve as loops so the end can be doubled back, cinched tightly, and coupled to the center to form the loops of a figure eight. The eyes are located between the end segments, close to one another on the same half of the central segment and at approximately one-third of the total length of the tie. By such configuration, loops of different sizes can be formed, for skis and poles.

The eyes are secured to the tie in various ways, such as bypassing the tie through and forming retaining loops by stitching. Additionally, a reinforcing strip of the same type of surface can be secured by stitching to the central segment, bridging the location of the eyes to add strength at the eye junctures.

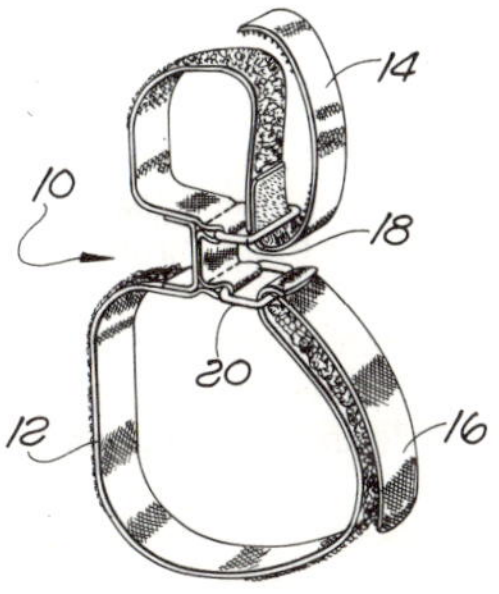

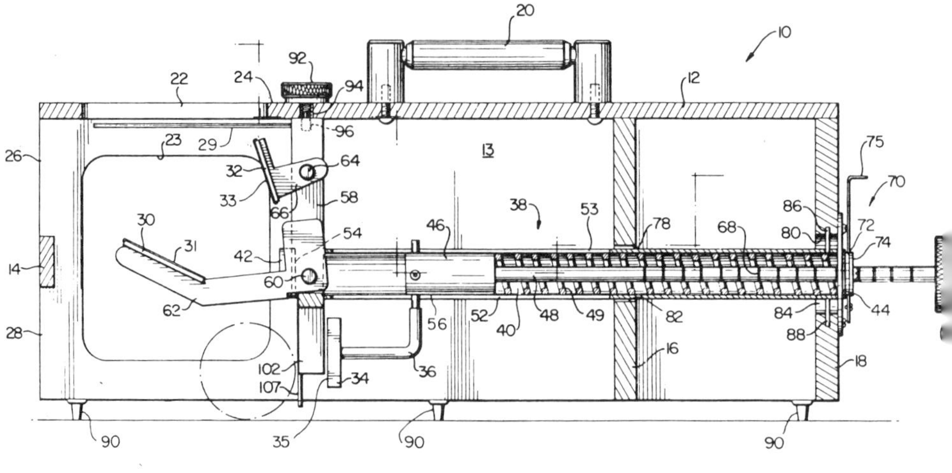

figure 10.20 Side cross section of a golf putting trainer, Patent No. 4,005,870 (1977). The device has walls (12), (13), (14), (16), and (18) with a handle (20) and a center line (24). Openings (22) and (23) allow for adjustment of the optical system.

The front wall (14) has two apertures (26) and (28). Aperture (26) is used in conjunction with the optical system; aperture (28) allows the ball projected by the putter head to leave the housing. Extending horizontally from the putter head assembly is a sight rod (29) corresponding to the center of the putter head. Pivotally mounted to the putter head assembly are two adjustable mirrors (30) and (32), which are used in conjunction with the housing center line (24) and the sighting rod to align the housing with the cup or flag.

Putter head (34) has a center line (35) and is attached by an arm (36) to the force-inducing mechanism (38). There is a spring-biased plunger having a head (46), a shaft (48), and a handle (50). The spring (49) exerts force between the rear cap (44) of the cylinder and the head (46) of the plunger. The putter head arm is attached to the head and rides within grooves 52 and 53 of the spring housing (40). Mounted on the housing at cap (42) is a shock-absorbing elastomeric material (54) to stop the head of the plunger at the end of its forward motion.

Mounted on cylindrical housing (40) is optical assembly support (58), to which sight rod (29) and adjustable mirrors (30) and (32) are mounted. By adjustment of these mirrors, the golfer may view targets at different distances from the machine through the optical system.

The plunger shaft (48) has ratchet surfaces (68) that coact with a trigger device (70) having a pawl or locking member (72), which engages the ratchet surface (68). Encompassing shaft and pawl is a guide (74), which limits the pawl to vertical movement into and out of engagement with ratchet surface (68). Handle (75) is connected to pawl (72) to release the trigger device.

The force-inducing mechanism (38) transverses the center wall (16) through an aperture (78) and the rear wall (18) through an aperture (80); it rests on bearing surfaces (82) and (84), respectively. The plunger is pivotally mounted to the rear wall by pins (86) and (88). This allows for adjustment of mechanism (38) within the horizontal plane of the housing. The housing is detachably secured to the putting surface by golf spikes (90) secured to the bottom of the housing.

Henry Grace of Maryland and Robert Conrad of Pennsylvania. Most of the previous similar machines were based on the assumption that the golfer will stroke the ball perfectly every time. The present putting machine and method enables golfers to calibrate their own putting strokes, to adjust the force they apply according to distance and condition of the green. The apparatus includes a transparent viewing plate, a spring-loaded putter head assembly having an adjustable optical system with a center line, and a traversing mechanism that enables the user to line up putter and cup. The ball can be propelled again and again down the same path with varying amounts of force until it finally goes in the cup. This kind of practice sharpens a golfer's judgment of just what is required in different situations (Fig. 10.20).

In tennis, too, equipment has become more important as more players model themselves after the professionals and want nothing but the best. One of the phenomena of recent years has been the sneaker boom. Not only in tennis but in squash, basketball, running, jogging, sailing, etc., shoes have become the badges of expertise and the symbols of status. Two brothers in Germany have seen their rival companies grow large and prosperous supplying athletic footwear to a sports-minded world. Both Adolf Dassler of Adidas and Rudolf Dassler of Puma have continued to improve their products through the use of patented inventions and new designs.

Tennis racquets have undergone a number of changes in recent years, too. Materials have been an important factor, with steel and aluminum being used instead of wood for the frames and new types of synthetics being used for the stringing. During the last week of 1976, Howard Head of Maryland was granted a patent for a radically new kind of racquet that has since been purchased and used by many players. Instead of the normal string surface of approximately 70 square inches, this one has closer to 100 square inches of hitting area. In particular, the distance between handle and tip has been increased. (See Fig. 10.21.) According to the inventor this results in the following:

> Circumscription of the center of percussion, a larger zone of high coefficient of restitution, an average more accurate return of shot by the player, an increased polar moment of inertia about the longitudinal axis, a greater ease of applying spin to the ball, a longer "dwell time" in which a returned ball is in contact with the strings for greater control, and a greater strength to the racquet.

In the foregoing paragraphs, we have cited one or two inventions from many of the major sports played by Americans. The purpose has been to show inventors some of what is going on in this area, and how things can be changed. New ideas should spawn other new ideas. If one person can invent a new kind of tennis racquet, another person might be able to develop a new kind of golf club. If a method of practice works well for one sport, it might have the potential to be adapted for use in other sports.

Like many sports, bicycling is enjoying a boom—it's almost as popular now as it was in the late Nineteenth century, its heyday in America. Inventions continue to be patented in this area, both for all kinds of accessories and for new kinds of bikes, too. The first example to be cited here is essentially a modification of an existing product to meet changed conditions. Robert Boudreau of Pennsylvania, an employee of the Hedstrom Company, was given the problem of developing a handlebar that would not turn relative to the bike's front fork when a torque of at least 35 pounds is applied. This was a new safety requirement of the federal government, and was not easy to meet without using relatively expensive machined parts. In the economically priced sidewalk bikes made by Hedstrom, this would have put costs out of line. Other solutions would have changed the method of assembling the bikes. Again, cost factors were important. It is most economical to ship these bikes knocked down, to be assembled by the purchaser. Naturally in this case, it would have to be fairly easy to put the thing together without the use of complex tools or procedures such as, for instance, welding. Boudreau's solution to this situation is shown in Fig. 10.22.

Many inventors have worked many years to develop a good collapsible bike, which would, of course, be ideal for travelers and people living in small apartments. According to Eric Jaulmes, an inventor living in France, the models that existed before his own patent was granted had many disadvantages. They were too heavy, mechanically complex, difficult to put up and take down, and uncomfortable to ride. In addition, their tires were a different size from the norm. Jaulmes set out to correct all these faults. Part

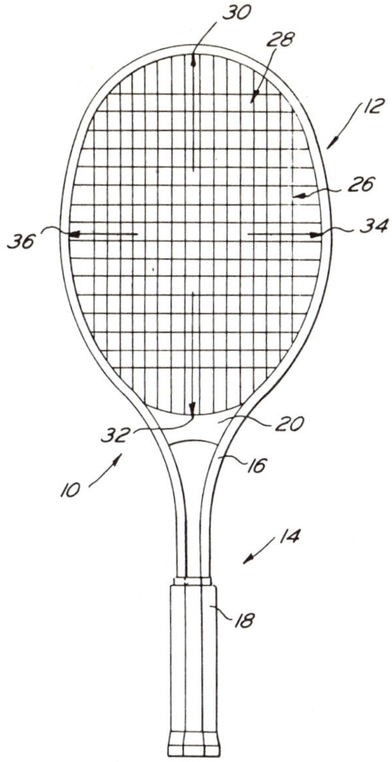

figure 10.21 Tennis racket described in Patent No. 3,999,756 (1976). It includes a head (10), a handle (12), and a frame (16). The frame is aluminum, and the grip (18) is plastic or leather. The length of the strung surface (28) is defined by the line (30–32), the width by line (34–36). The length measures between 12 and 15 inches—45 to 58 percent of the racket's total length—and the width measures between $9^{1}/_{2}$ and $11^{1}/_{2}$ inches. The string pattern is nonuniform, there being few strings near the frame and more toward the center of the impact area.

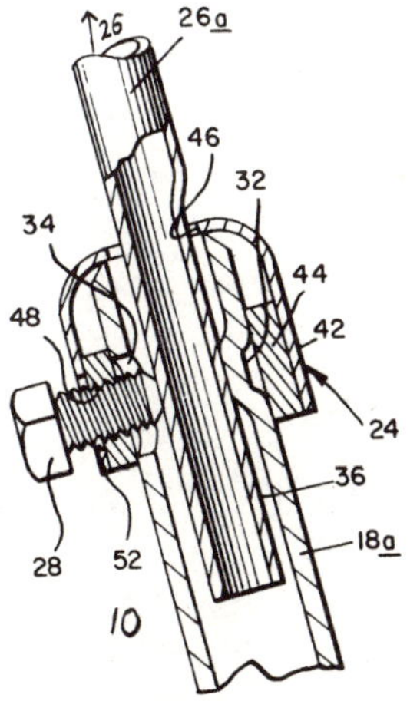

figure 10.22 Fragmentary view, with parts cut away, of a bicycle handlebar assembly, illustrated in Patent No. 3,955,828 (1976).

A collar (24) comprises a generally cylindrical exterior shell (42) and an internal circular ring (44) formed of relatively heavy bar stock press-fit into the shell. The lower end of the shell is fully open and the inside diameter of the ring is slightly larger than the outside diameter of stem (18*a*) so that the collar (42) can be slid over the end of the stem projecting up through head tube (10). The upper end of the shell is formed with a reduced diameter opening (46), which is slightly larger in diameter than the outside diameter of the handlebar post (26*a*), permitting post to fit into stem.

A threaded passage (48) is formed in ring (44). A coextensive opening (52) is provided in the shell to accommodate the set screw (28). In use, the collar is engaged over the projecting end of the stem and oriented so that screw is positioned opposite opening (34). Then the handlebar post is slid down through opening (46) and aligned so that its flat area (36) is positioned opposite flat area (32) on the fork stem. This permits further downward travel of the handlebar post, yet prevents relative rotation between the post and stem.

When the handlebar (26) is positioned at the correct height above the head tube assembly (10) for the child, screw is turned down into the threaded passage (48) until its inner end bears against the handlebar post inside the ring. Further tightening of the screw tightly presses the flat area (36) on post (26*a*) against the opposing area (32) on stem (18*a*) so that those two parts are securely retained against relative rotational or axial movement.

Even when the wheel assembly (16) is trapped in a relatively deep crack or rut during normal use of the bike, the rider can turn the handlebars (26) to escape and the torque will be transmitted to the wheel assembly without any rotational slippage between the handlebars and the wheel assembly.

Aside from the set screw, which is an off-the-shelf item made in extremely high volume and thus at low cost, and the threaded passage in the ring, the assembly involves no machining of parts. Only conventional stamping and forming techniques are required to make the various components of the assembly.

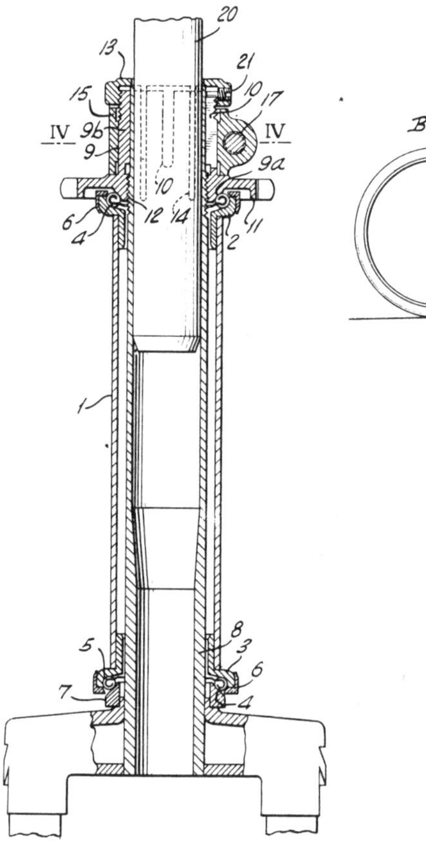

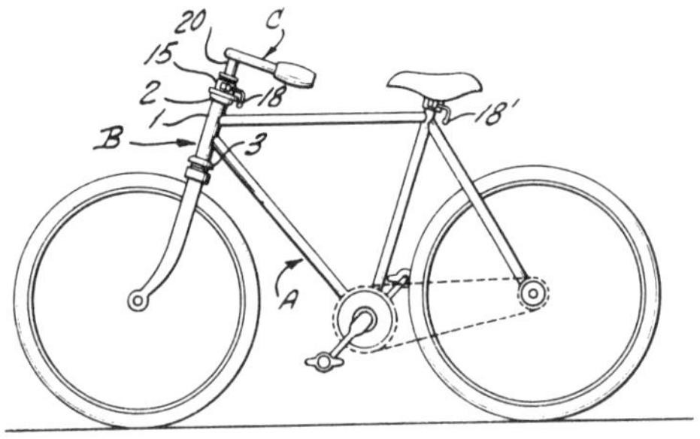

figure 10.23 Cross section of the steering, pivot, and plunger tube assembly and locking mechanism of a collapsible bicycle, Patent No. 3,260,535 (1966). The bike itself is shown in the inset.

The steering tube (1) forming part of the frame (*A*) is fitted at both ends with flanges (2) and (3) forming the races of ball bearings adapted to receive a number of spherical balls (4), retained or mounted within cages (5).

Secured to the lower end of the pivot tube (8) forming part of the fork (*B*) of the bicycle is a lower cone (7), forming a cooperating race in conjunction with the dish or race (3) in the assembled position of the bicycle. The upper cone (*9a*) forms the cooperating race in conjunction with the race (2), part of a removable locking head or mount (9). Cylindrical sleeve (*9b*) has longitudinal slits (10) and is fitted with internal threads adapted to join with external threads (12) of the pivot tube (8), to longitudinally adjust the head and to establish a proper relation between the balls (4) and associated races (2), (*9a*), and (3) and (7), respectively, by the aid of a manual locking ring or wheel (11) also forming part of the mount (9).

In order to preset the bearing pressure, suitable stop or abutment means are permanently fixed during manufacture. A perforated screw cap (13), set upon the end of the sleeve (*9b*), acts as a stop in conjunction with the upper edge of the pivot tube (8), to limit the extent of relative adjustment of setting of the head (9) and, in turn, to positively predefine the thrust or bearing pressure between the balls and races. Proper function of the bearings is thus ensured independently of the assembly and disassembly operations.

In order to mount and remove the handlebar (*C*) having a plunger tube (20), there is a split mounting or clamping ring or collar (15) adapted to engage the outside of the sleeve (*9b*) of the locking head between the adjusting wheel (11) and the limiting cap (13). One of the radial extensions of the collar (15) has a smooth perforation (16), while the other extension has a tapped perforation (17) to receive an angular clamping lever (18) having an abutment or limiting shoulder (19) corresponding to a predetermined clamping pressure. As a consequence, with the plunger tube (20) of the handlebar engaging the inside of the pivot tube (8), the latter being advantageously longitudinally slitted as at (14), and secured by the clamp (15), slitting of the tube plunger (20) and the use of a special wrench is dispensed with.

The set screws (21) serving to lock the cap (13) against rotation engage one of the slits (10) of the sleeve (*9b*) of the locking head. With the bicycle being assembled and the main component parts being in position, disassembly is effected in the following manner:

At first, the lever 18 is unscrewed, loosening the split collar and the elastic expansion in the radial outward direction of both the sleeve of the locking head and of the end portion of the pivot tube (8). As a consequence, the plunger tube (20) will be loosened to slide within the pivot tube (8), to allow easy withdrawal of the handlebar (*C*).

The locking head is then unscrewed and separated from the pivot tube, while the cap remains in its previously set or adjust position. As a consequence, the pivot tube is allowed to descend through the steering tube (1) to result in the separation of the parts. Due to the presence of the retaining rings (6), the sets of balls (4) are confined within the races (2) and (3) attached to the tube (1). The separated components, parts (*A*) (*B*) and (*C*), may be stored and/or transported in collapsed position or separately.

---

of his patent, which was assigned to the French manufacturer of high-quality bicycles, Motobecane, is shown in Fig. 10.23. A somewhat unusual bike patent is illustrated in Fig. 10.24. It's a design patent, granted to Lee Carpenter and James Cotter of Ohio. These men did not come up with anything new in the way of the bike's moving parts. But their design was new enough and sufficiently unique to be worth a design patent, which prevents anyone else from selling a machine that looks exactly like the model illustrated.

figure 10.24 One figure from a design patent for a bicycle, No. 237,478 (1975).

Bike riding is just one aspect of a trend in modern American life—the pursuit of physical fitness. All kinds of health clubs and exercise classes have sprung up over the years, many of them relying on gimmicks without much staying power. This is a good field for inventions, especially those machines that enable a person to keep in shape and, perhaps, to lose weight without expending very much effort. This is of course a paradox that most people would reject—though it certainly is compatible with human nature.

Another approach was taken by Robert Genin of Florida. He invented a physical training robot programmed to demonstrate any

figure 10.25 A physical training robot, illustrated in Patent No. 3,888,023 (1975). The robot's hollow legs and feet (11) are anchored on top of a pedestal (12). Then there are a head (13), a torso (14), and left- and right-hand assemblies (15) and (15'). Each of these assemblies has a hand (*A*), swiveled on a forearm (*B*), joined by a wrist (*C*). Forearm (*B*) is pivotally connected to an arm (*D*) to define an elbow (*E*); the arm is pivotally joined to the shoulder of torso (14) to define shoulder joint (*F*). The torso element is mounted above pelvis (16). The torso may be bent or rocked from side to side or oscillated back and forth in a twisting motion. The pelvis is trunnioned on (11) so as to bend forward.

The robot is linked to a motorized cam mechanism within pedestal (12). There is a cam cylinder having a series of endless cam tracks producing a distinct program of calisthenic movements, the program being repeated with each full revolution of the cylinder. Selection of a particular program is effected by a selector (18), which is manually shiftable with respect to a row of track-positioning notches (19) formed in an opening in the front wall of the pedestal.

The anchored legs remain stiff, for the legs are not articulated and no means are provided to manipulate the leg positions.

Because the articulated left- and right-hand and arm assemblies are manually adjustable, the PT robot may be set by the user to assume any desired calisthenic posture.

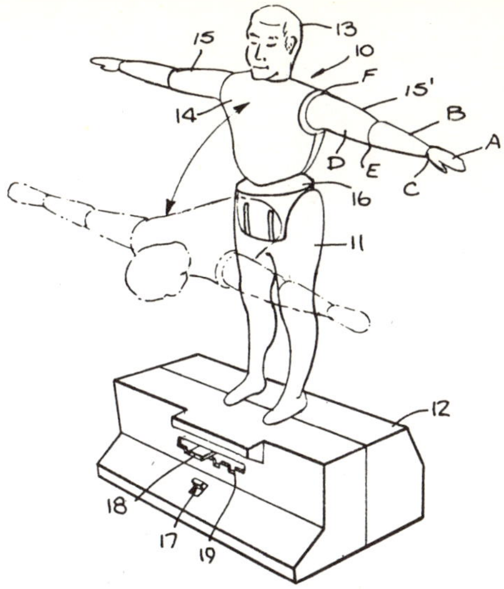

figure 10.26 Self-contained push-pull exerciser, Patent No. 3,999,752 (1976), which is supported entirely by the body. There is a flexible belt (14) and a buckle (16). The rear end of the belt is looped around the intermediate bar (22) and attached to the buckle. The opposite end of the belt is passed through the opening between the front end (18) of the belt and intermediate bar (21) and then is slipped through in between the intermediate bar (22) and end (20) so the buckle is adjustable. The buckle has an extension (24) at the front end.

Secured to the belt (14) is a pair of pulleys (26) and (28). Pulley element (26) has a yoke (30) which supports a pin (31) on which a pulley (32) is rotatably supported. Extending from the opposite end of the yoke is a stem (34) which terminates in a ring (35). The ring is looped around a metal band (36) which, in turn, is anchored to the belt. The band can be slid manually if desired. The other pulley (28) is similarly constructed and secured to the band (36). It supports a pulley (33). Each of the pulley elements (26) and (28) is therefore independently secured to the band (36) attached to the belt; they are maneuverable independently of each other.

Exercising element (38) comprises a flexible cord (42), preferably of nylon or the like, which passes around the pulley with each of the opposite ends of the cord passing through spaced openings in the triangular-shaped plate generally indicated at (44). This plate is thus connected to the cord and passes outwardly thereof so that a loop portion (46) is formed. The loop is covered by a nylon sleeve (47). The two loops and their sleeves are normally used as hand grips.

All of the triangular plates (44) are of identical construction, whether they be used for forming the hand grips of cord (42) or for forming the loops (48) of the other exerciser (40), and serve as stirrups for the feet. The plates are molded of plastic, one side of which is substantially a flat surface (50) with a raised border (52) extending from the underside so that a recess (53) is formed on the underside. A hole is formed in each of the triangle's three corners (54), (56), and (58). Extending from the underside is a pointed projection (60), which is adapted to be engaged by the nylon cord when the cord is in a locked position. The upper triangular plate, which holds the loop for the hand grip, is secured as shown with the flat surface facing upwardly, whereas the other triangular plate is shown inverted for purposes of illustration so that the underside is visible. Normally, both triangular plates would be positioned similarly to the upper triangular plate. Cord (42) is shorter than cord (40), which is used to form the leg stirrups.

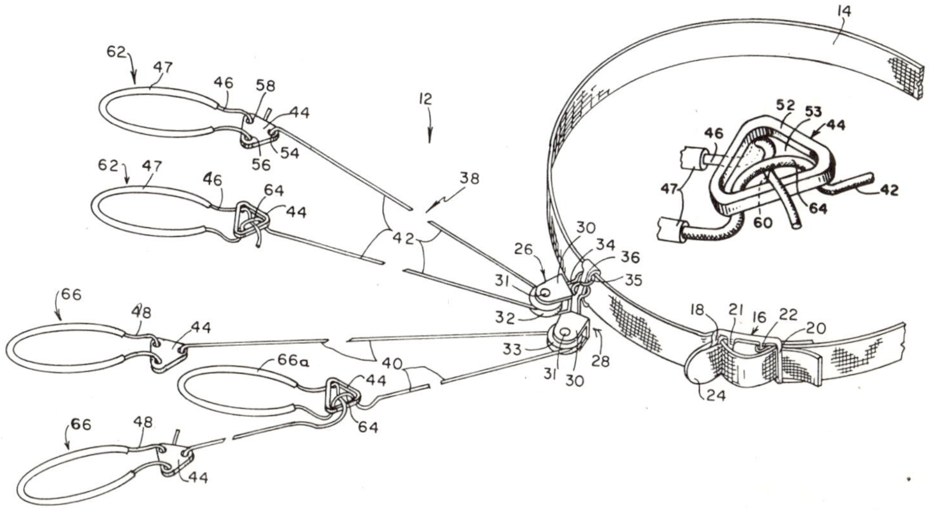

one of a series of exercises (Fig. 10.25). The robot is designed to replace PT instructors in certain instances, since they cannot be on call whenever you need them. All the student has to do is follow the robot's motions, which have been planned by experts in the field. The robot's hands, arms, torso, and pelvis are all movable, controlled by the interior machinery as described in the illustration caption.

Another type of exercise apparatus is represented by the push-pull machine illustrated in Fig. 10.26. Most of the machines that had been produced before this one had to be secured to some fixed

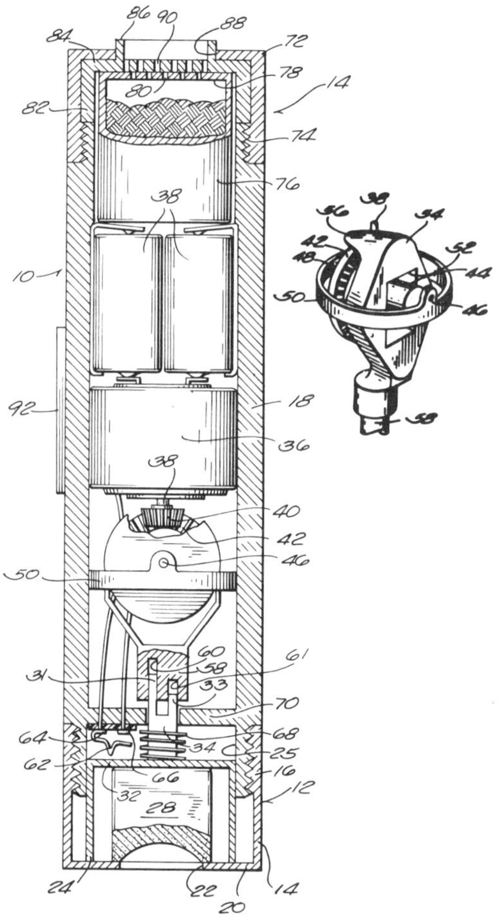

figure 10.27 Cross section of a cue stick chalker and talcum powder dispenser, Patent No. 3,963,237 (1976). There is a cue chalk end (12) and a talcum powder dispenser end (14). A cue chalk holder (24) is located within the cap. It is generally circular in transverse cross section and includes four equally spaced notched retainer ribs (26) on its interior surface. The retainer ribs extend substantially the entire length of holder (24). The holder has a loose running fit with wall (25) of body (18) so as to be held generally in axial alignment but to be capable of oscillating movement.

A conventional cue chalk (28) is in the form of a cube, and notched retainer ribs are spaced on the holder so as to receive the corners (30) of the cue chalk cube. The holder includes an inner wall (32) and a drive projection (34). The cue chalk holder is held within cap but is free to move.

A motor (36) is located within central body portion and is powered by batteries (38) in a conventional manner. The motor includes an output shaft (38) which carries a drive pinion (40) that engages with a circular bevel gear (42). A stub shaft (44) is connected to bevel gear, the shaft and bevel gear being supported for rotation on pins (46) and (48), which are mounted in a retaining ring (50). Shaft (44) is mounted on pins (46) and (48) and projects into a rectangular opening (52) in yoke (54). This yoke includes a projection (56), which is mounted on, but not driven by, shaft (38). The connection between projection (56) and shaft (38) is such that the projection is free to move relative to the shaft and will not rotate as the shaft rotates. The cross section of rectangular opening (52) is generally equal to (slight clearance being provided for smooth running) the diameter of shaft (44). The shaft being eccentrically mounted, when it is rotated by bevel gear (42), in response to driven movement of pinion (40), it will cause yoke (54) to oscillate back and forth a distance equal to twice the difference between shaft (46) and that point on the diameter of the shaft farthest from pins (46) and (48). This in turn imparts an oscillating motion to drive member (58), which is part of the yoke, being either cast integrally with or attached to the yoke.

support such as a doorknob. This unit, on the other hand, invented and patented by Sam and Dennis Kupperman of Illinois, has a belt that is attached to the user's waist. The hand grips and foot stirrups are formed of a flexible cord that is adjustable in length (Fig. 10.26).

Not all participants in sports are concerned with physical fitness. Some of the people who frequent pool halls and billiard parlors, for instance, are not known for staying in shape. For them, Todd Bushberger of Wisconsin has patented something to make their lives even easier—a hand-held, battery-powered cue stick chalker and talcum powder dispenser. As shown in Fig. 10.27, this unit has the chalk cube at one end and the talcum holder at the other end. When the cube is engaged with the stick, it automatically begins to oscillate. There's even a file attached for shaping the chalk.

A summary of this chapter can do little more than repeat the obvious. There are a great many different sports played in the U.S.A., and more people are getting involved with them each year. Americans being generally buyers of equipment, there are some good opportunities for inventors and manufacturers. If you're active in a sport and have developed or can see a way to make a better piece of equipment, an aid to practice, a safety device, a needed accessory, a clever novelty, or even an entirely new sport, do it! The market is there, and if your invention has any merit, it stands a fair chance of success—provided you are willing to promote it patiently and persistently after you have invented it. It also wouldn't hurt to have the secret ingredient known to all people who have made their marks—luck.

CHAPTER ELEVEN

Solar Energy

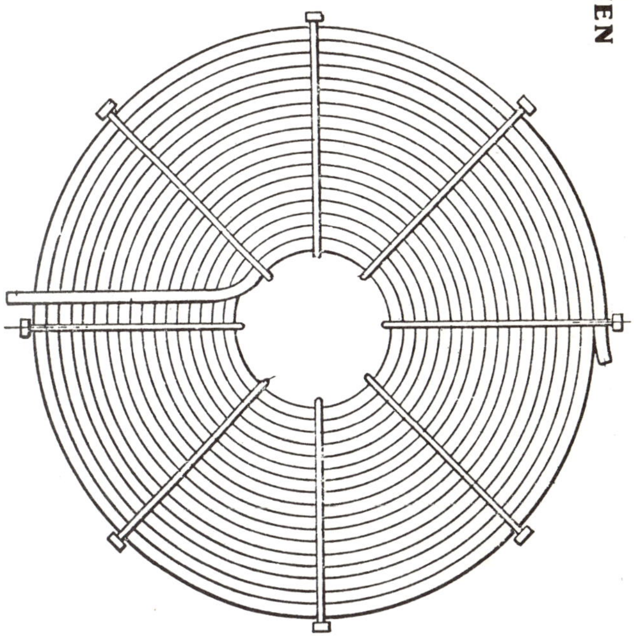

In the good old days, primitive people didn't have to worry about energy crises, oil shortages, or the high cost of electricity. They lived a simple life, providing themselves with the food, clothing, and shelter that were necessary to sustain life, and protecting themselves from predators. Some lucky folk in benevolent climates lived off the fat of the land. Most lived at a subsistence level, unable to save anything from year to year, vulnerable to the droughts and floods, famines and gluts that are nature's way in most parts of the world.

The best things in life are indeed free, but in many cases they were in short supply, and people had to compete in order to get their share. Increasing population brought this situation to a crisis in some areas. The land could not support all the people who lived on it. There was great poverty and even famine.

In the Eighteenth century, scientific and industrial progress began to change man's life. It was found that machines could do the work of many hands, could increase production, as long as there was a good source of energy. Electricity, steam, and waterpower turned the wheels of the new society with increasing speed. Conditions in the early factories and slum cities were generally horrendous; nevertheless, they were like magnets for the laborers leaving the fields. The people wanted a chance at a better life. Power was, and is, power. It opened a gap between what we now call the developed and the underdeveloped nations that in many cases is growing wider even today.

The cost of power, of energy, is growing year by year. Even now, certain people, certain industries, and certain countries are being priced out of the market. They simply can't afford to keep up anymore. That's the problem this chapter is all about. People are looking to inventors for help.

The greatest source of energy in the world, the sun, is available to everyone. Without it, we'd be nothing—there would be no plants, no water, no fuel, no life. Even the fossil fuels like coal and oil drew their energy from the sun, in effect storing it for future use. These fuels stayed in the ground for millions of years, not coming into widespread use until the 1700s and 1800s. Primitive man knew a lot more about the sun than he did about converting fuel to energy. The sun's rays have been used for drying food and for distilling water since antiquity. There is reason to believe that in 212 B.C. Archimedes used carefully placed mirrors to direct the sun's rays against an invading Roman fleet. The concentrated heat set some of the ships on fire, and the Greeks won the day.

A great deal of experimentation on solar energy was carried out during the Seventeenth and Eighteenth centuries. Much of it had to do with ways of melting metals and other hard materials. Furnaces were constructed with mirrors to direct and intensify the sun's rays so as to create very high temperatures. France was the leader among nations in this work. But as coal came into wider use—first in England, where Abraham Darby patented his blast furnace in 1735, and then worldwide—interest in solar energy waned. Lots of inventors took a crack at it, but other sources of energy were much easier to tap. So the Industrial Revolution was fueled by coal. Later, in the Twentieth century, petroleum took on the major role.

In certain areas of the world, there has always been a great interest in solar energy. In the American Southwest, for example, there was a shortage of fuel in the early days, and of course a surplus of sunlight, along with wide differences in temperature between night and day and among the seasons. Given these conditions, many experts thought that this was the place where the breakthrough in solar energy research was bound to happen. And indeed a great deal of work has been done and many successes have been achieved. Some promoters have taken advantage of this situation and sold the public on a number of get-rich-quick-schemes that never worked out. These operators can be found today, too, as solar energy once again begins to capture the public's imagination.

The whole history of solar energy has been one of fluctuating expectations. Bright hopes have been dimmed by poor results. And then new discoveries have rekindled imaginations. It's not commonly realized today that the 1950s was a period of great interest in solar potential. In 1952, a study group sponsored by the federal government took a careful look at the situation, liked what they saw, and concluded that by 1975 approximately 13 million American homes would have some solar heat. Their prediction was approximately 12,999,817 off the mark. In 1954, Bell Labs engineers were right on the mark with their invention of a solar battery using silicon cells, or chips. This development has found its greatest application so far in the space program. Well over a thousand satellites have been powered by solar cells. Mounted in arrays of several thousand to a panel, these photovoltaic energy converters have been a key element in the success of space flights. Their future in earthbound usage looks bright, too, even though their efficiency ratio in converting solar energy into usable energy is only about 15 percent, compared with 35 to 40 percent for modern steam turbines. With this kind of loss, a great many panels are necessary to produce a significant amount of power.

The 1970s and 1980s would seem to be the right time, and the United States the right place, for significant progress to take place in solar technology. The federal government is interested, and has begun to put significant money into research and to give tax incentives for private individuals and corporations engaged in solar work. Perhaps more significantly, coal, oil, and natural gas have become a lot more expensive than they used to be, in terms of both dollars and environmental costs. As these conventional fuels rise in price, the cost of solar energy looks more reasonable. There's still a big difference, but it's getting narrower all the time. This is especially true as more inventors and manufacturers turn their attention to making solar energy more practical and less expensive. The capital, the creativity, and the manufacturing know-how are all there. Inventors, especially, are active as never before. It's the hottest subject the U.S. Patent Office has had in a long time. In the Search Room just outside Washington, D.C., each day several hundred people consult the file of solar patents to see what has gone before.

There are many possibilities for the practical application of solar energy. Some of them have already come to fruition, others are on the threshold, and others hold great promise for the future. It is useful at this point to look at the opportunities in each of these areas.

First, we know that the sun's heat has been used for centuries to dry food and to extract salt from seawater. The sun has also been used in the distillation of freshwater from saltwater, but this process has not succeeded on any large scale. In the 1950s, President Eisenhower predicted that new prosperity would be brought to arid regions of the world through the harnessing of the sun's energy to distill water. So far this has not happened, but the oil money flowing into the desert kingdoms of Arabia may change this. It takes enormous capital to construct a distillation facility anywhere near the size necessary for it to be practical. The conventional wisdom would be that the independent inventor does not belong in this field. However, there may be some possibilities for the development of small stills. Two inventors from Taiwan, Yaw Jenn Tsay and Kuo-Chuan Lin, have been granted an American patent for an automatic distilling device that can be installed on the roof of a house or other small building or even on a boat. It's comparatively low in cost and is not difficult to build. An insulating layer and a heat-absorbing plate are layered together to form a vaporizing chamber, which also serves as the reservoir. A lens focuses the sun's rays on various points in, on, and above the water to accomplish the distillation (Fig. 11.1).

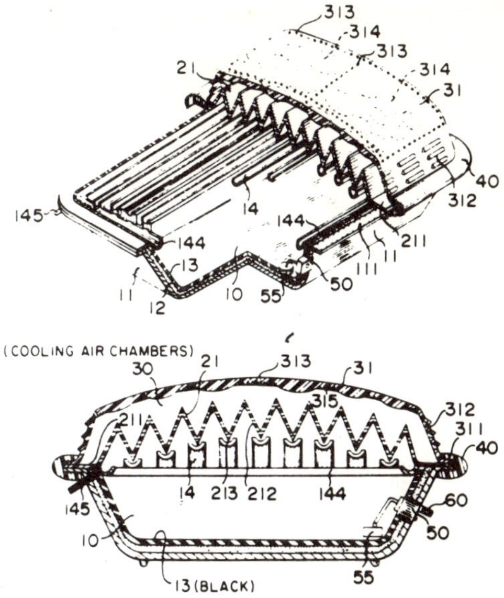

figure 11.1 Cross sections of a solar-powered distiller, Patent No. 4,010,080 (1977). The bottom plate (11) has sides inclined outwardly and is made of metal. An insulating layer (12) and a heat-absorbing plate (13) are layered together to form a vaporizing chamber (10), which also serves as the reservoir. The upper portion of the vaporizing chamber is layered over with condensate collection troughs (14-14) that end in a main water-collecting trough (144). A transparent lens cover (31) is supported over the condensation plate (21) by virtue of end support sections (311), which are supported on top of the end support sections, (111) and (211). In this way, a cooling air chamber (30) is formed between the condensation plate and the lens cover. The end sections are sealed together by a rubber or plastic strip (40).

Each downward-sloping wall of the lens cover is provided with vents (312-312); the top portion is provided with small holes (313-313). On the inner surface are convex lenses (314-314) of various diameters and thickness. The spaces between them are formed in alternately concave and convex annular rings (315-315) in a ripplelike design.

Water enters the vaporization chamber by means of a feed hose (60) and an automatic valve (50). A float (55) opens the valve when the water level falls below a specified level and closes the valve when the level attains a second specified level.

When the vaporization chamber is filled with water, the water is heated as the lens cover (31) causes rays of sunlight to converge at various focus points in, on, and above the water. As the water is vaporized, the vapor rises between the troughs (14) and comes into contact with condensation plate (21). Cool air enters chamber (30) by means of the vents. As cool air passes between the condensation plate and lens cover, the temperature of the plate is maintained at a much lower level than that of the water vapor, so that when vapor comes into contact with the inclined sides (212) of the plate, it will be condensed to water droplets. As the droplets accumulate and increase in size, they will flow down from the inclined sides to sharp bottom edges (213) of the plate and into the troughs. The water then flows down and into the trough (144) through a hose (145) and into a receptacle.

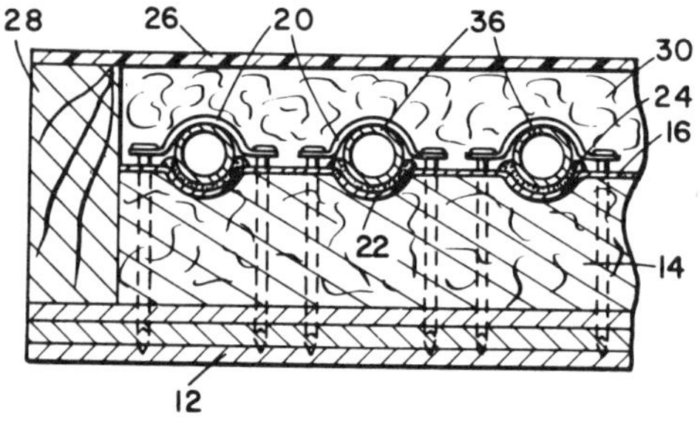

figure 11.2 Cross section of the solar panel illustrated in Patent No. 3,952,725 (1976). Overlying the backing panel (12) is a layer of resilient insulation material (14), such as fiberglass wool. The other type of sealant is black paint with dispersed particulate graphite. It has a relatively thinner consistency and can be applied over the entire surface of the foil and tubing after the tubing is installed. The paint will seep into the troughs to provide an effective thermal bond. It is preferred to use the graphite paint even if the cement sealant is also used, inasmuch as it accomplishes the desired blackening of the upper surface of the foil and tubing, thus increasing conductivity. Either or both of the sealants provide an increased thermal conductivity between tubing and foil that is as effective as welding or soldering.

Above and parallel to the foil sheet is a pane of light-transmitting material (26), preferably a two- or four-mil panel of Tedlar, a material that can withstand the elements and that is opaque to long-wave infrared rays. The Tedlar may be reinforced with wire mesh or supported on a transparent fiberglass panel.

Each unit has a peripheral border that provides a mounting for the pane (26) and seals the unit from outside air currents. In order to minimize convection currents an open-fibered or loose-knit layer (30) of glass wool may be provided between the foil and the light-transmitting pane. Especially when the panels are used in a nonhorizontal position conducive to the circulation of convection currents, the baffling effect of the wool more than compensates for the slight shadowing it produces on the foil.

Sunlight enters through the pane and impinges on the blackened upper surface of the foil (16), where it is delivered to the fluid circulating in the tubing (generally water). The black-body radiation from the upper surface of the foil is in the infrared zone and is partially reflected by the Tedlar pane and the filter media and reabsorbed, then partially absorbed by the Tedlar, causing a reduced temperature gradient in the heated area of the unit. The tight construction minimizes heat escape, and fairly high temperatures are developed in the circulating water.

An area that would seem to offer good possibilities for investigation by independent inventors is the heating of water by the sun. In Australia, which basks in the sun for a good part of the year and where more than half the land is in fact desert, many houses have solar water heaters. In Japan and North Africa, too, the sun's rays have been utilized for water heating. Some manufacturers in the United States have begun to approach this area cautiously. The breakthrough has not yet come. The fall 1978 edition of the Sears catalog did not offer any solar heating equipment. When it does, you'll know the field has really "arrived." That time may be close.

An independent inventor who has developed a domestic hot water heater is William Edmondson of California, who is also editor and publisher of *Solar Energy Digest.* He calls his invention SolarSan.™ It was designed generally to heat water for domestic use, though there are other possible applications, such as running an expansion engine. According to the patent, "the sun penetrates one or more panes of light-transmitting material (e.g., glass) and impinges on a blackened layer of soft-tempered conductive foil on which are disposed conductive water-conveying tubes. Loosely packed glass wool between the pane and the foil inhibits convection currents, and resilient, dense glass wool . . . supports the foil." (Fig. 11.2.)

The inventor claims that anyone can build the SolarSan "with simple tools and from readily available materials." He has published a booklet of instructions that he sells to do-it-yourselfers, and also has some commerical licensees for the heater. So here is one inventor who has been able to profit from his patent, not by going into manufacture, but by selling plans and instructions. He invented a water heater that is comparatively low in cost and that can be built without great difficulty—and people liked the idea and bought it.

The heating of swimming pools by natural gas or other fuels is one American custom that may become quite rare within a decade. Costs will force most people out, and good sense will convince even the wealthy that there are better ways of achieving the same thing. Since most pools are located in sunny areas, there seems to be every reason why the sun's rays should be used to supply heat. Richard H. Smith of California has patented a floating solar pool heating unit (Fig. 11.3). Any number of them can be used, depend-

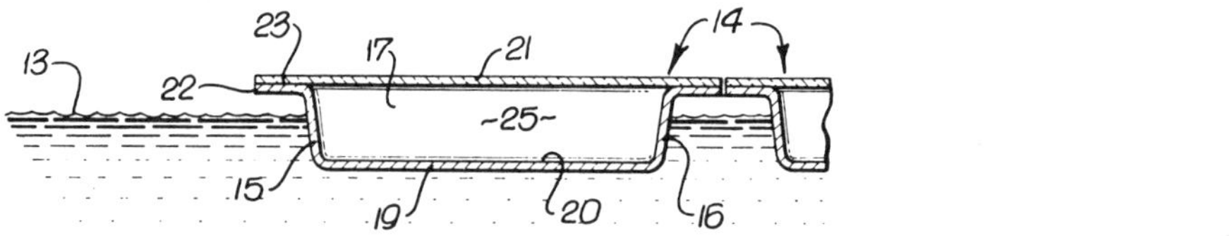

figure 11.3 A floating solar pool heater as illustrated in Patent No. 3,893,443 (1975). Floating on the surface (13) of a pool are any number of units (14) consisting of floating pans. The units are rectangular, but any shape that will allow them to fit together is acceptable. Extending over the top edges of the side and end walls is a transparent or translucent cover (21) resting on flat areas (23). There is a chamber (25) of relatively thin dead air space; the height of the side and end walls is about 1¼ inches and the aggregate area of the cover is about 10 square feet. It is advantageous also to have the cover sealed to the flat areas so that water won't get into the dead-air space (25). The pan can be made of aluminum or fiberglass. The material cover should be either of glass or translucent plastic of a type that lets the short rays of the sun pass through it and be absorbed by the blackened surface. The long rays emitted by the blackened surface cannot pass back out through the transparent cover, and the short rays are readily converted to heat that is then conducted through the bottom of the pan directly to the pool water surface.

ing on the size of the pool and the temperature of the water. They can be removed easily when the pool is in use, or if the day is chilly, they can be placed at one end of the pool, out of the way but still transmitting their heat to the water. These units are basically shallow heating pans with a translucent cover made of aluminum or plastic. The bottom of each pan is painted an absorbent black.

There are other methods of heating pool water with the sun's rays. Arrangements of coils or tubing on the top of a building can capture and hold a good deal of hot water, but then it must be circulated to and from the pool. The same holds true for shallow reservoirs in the sun. Flexible covers of heat-absorbing material can be stretched over the pool when it is not in use, but they're cumbersome to take away and then replace.

Home heating—space heating—is a more complicated, yet equally fascinating subject for inventors. The first real solar house in the United States was constructed in the 1930s by F. W. Hutchinson of Purdue University. Since then a few others have been built, but so far there has not been the kind of success that has given manufacturers and builders a prototype they have been able to translate into use for the mass market. Certainly, the solar energy that annually falls on the roofs of most houses far exceeds the energy used to heat and cool those houses each year. A home of about 1,500 square feet in a temperate climate could at the present state of the art receive most of its heating and cooling energy through a collector of 600-800 square feet. That is, of course, quite a large installation, too large and too expensive for most people.

Nor can solar heat do the full job in most locations. A backup unit fired by oil or some other fuel is necessary to help out at periods of temperature extremes and when the sun is obscured by clouds for more than a few days. A dual system like this means increased costs—not double, to be sure, but added to the already rather high costs of a solar collector, costs have been enough to discourage most nonexperimental installations. The systems as they exist tend to be cumbersome. Sunlight must be collected and then either transmitted directly to the areas to be heated or stored for future use. Each of the two different types of collectors—hot air and hot water—offers advantages and disadvantages. Hot water is more efficient, but it must be protected from freezing at night. A hot-air system is simpler to build and to maintain, but the controls that regulate the flow of heat are quite expensive.

One of the first patents for an "apparatus for collecting, storing and utilizing heat" was granted in 1887 to John Hittell and George Deitzler of California. Like many of the solar inventors of that day, they did it with mirrors (Fig. 11.4). They positioned lenses "so as to intensify in a complete or partial focus . . . all the forces conveyed in solar rays." Their storage reservoir was a mass of iron, which heated air that was conducted to various rooms through pipes or flues. Another device patented in the past that has some interest for the present is the solar-heating apparatus invented in 1913 by Marcos de la Garza of Texas. His particular interest was the lens that focused the sun's rays and thus concentrated the heat. Figure 11.5 illustrates his layout, which is quite different from the approach illustrated in Fig. 11.4.

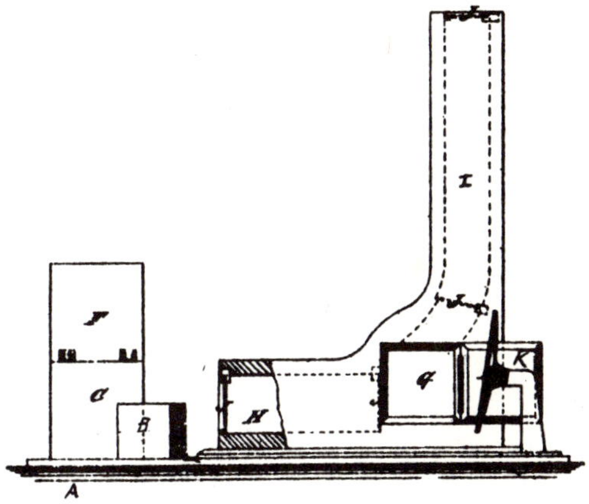

figure 11.4 Cross section from the side of the solar apparatus described in Patent No. 188,517 (1877). (*A*) is a concave mirror for concentrating the solar rays upon the heat reservoir (*B*), which is a mass of iron or other suitable material. (*C*) is the heat box for continuing the heat until needed, and also for serving as package for transporting the heat reservoir when hot. The heat box may have different forms, some movable and others fixed—the latter for keeping the heat without transporting it—and these last may be either under or above ground. The movable heat box may also be used in keeping the reservoir when the heat is being utilized for some purpose.

(*G*) is the heat reservoir chamber, in which the heat is communicated from the hot reservoir to the air. Under certain circumstances the heat reservoir may be heated in the heated reservoir chamber. (*H*) is a devaporizing chamber for extracting the moisture from the air by means of a deliquescent substance or other material or treatment. A vertical stack or flue (*I*), communicates with the heat reservoir chamber, for conveying the heated air away for use.

(The device for concentrating the solar rays may be either stationary or movable, and if movable may be moved by hand or automatically to follow the sun.)

The various chambers mentioned will have valves (*J*) at the ends to regulate the passage of the air, and there will be a door or doors (*K*) at the side or the bottom.

The air passing to the heat reservoir chamber may be required to pass through the devaporizing chamber, so as to deprive it of its moisture and thus increase its capacity to absorb other moisture.

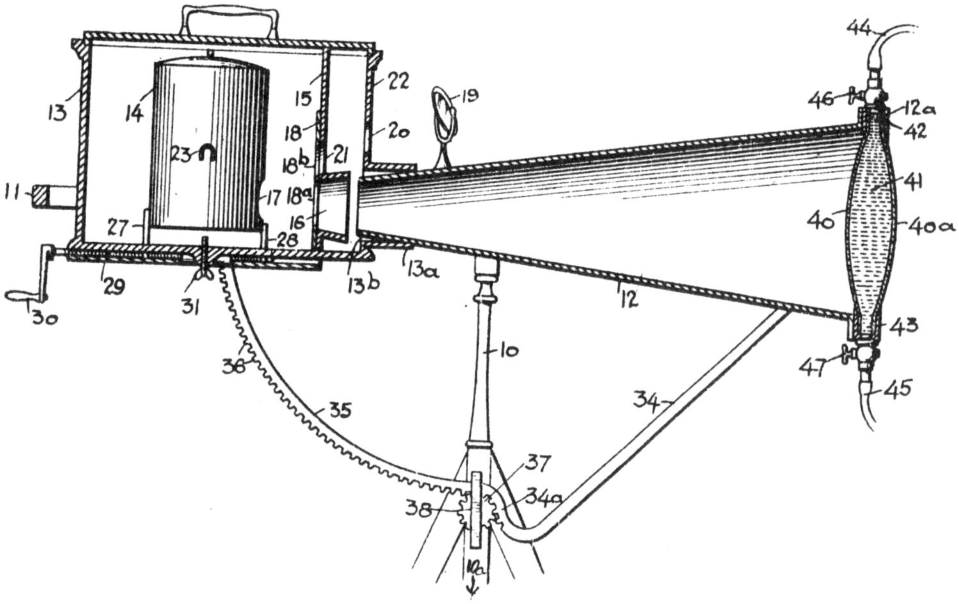

figure 11.5 Solar-heating apparatus shown in Patent No. 1,081,098 (1913). A base frame (10) is provided with standards (10*a*). The opposite end of the lens tube is received in a nipple or socket (13*a*) at an opening (13*b*) in the casing (13), which may contain a heater (14) adapted to receive a cooking utensil or any other article or substance to be heated. The casing has a transverse partition (15) with an opening defined by a conical flange (16) in axial alignment with the lens tube (12), and an opening (17) also in axial alignment with the lens tube is formed in the heater.

A mirror (19) is mounted on the lens tube (12), and openings (20) and (21), are formed in the adjacent side (22) of the casing and in the partition, so that the mirror will reflect the inner portion of the heater at the opening, that the user may observe the direction of the light rays and adjust the apparatus accordingly. The cut-off slide (18) has an opening (18*b*), which registers with the opening in the partition.

At each side of the heater is a lateral, arched hanger or trunnion (23), resting on the end of a journal (24). One of these has a crank handle (25), there being sleeves (26) on the sides of the casing through which the journals (24) pass. The arrangement is such that the vessel rocks on the journals and thus remains upright when the lens tube is tilted. Posts (27 and 28) are fitted in the bottom of the casing, their lower ends connected to a shaft (29) so that it will turn relatively to said stops but will move the latter longitudinally, the shaft being provided in practice with a suitable crank handle (30). A set screw (31) may be provided in the bottom of the casing to engage the bottom of the heater. To connect the casing with the vertically rocking frame, any suitable means may be employed, as, for instance, brackets (32), there being similar brackets (33) for holding the lens tube (12) in the frame (11).

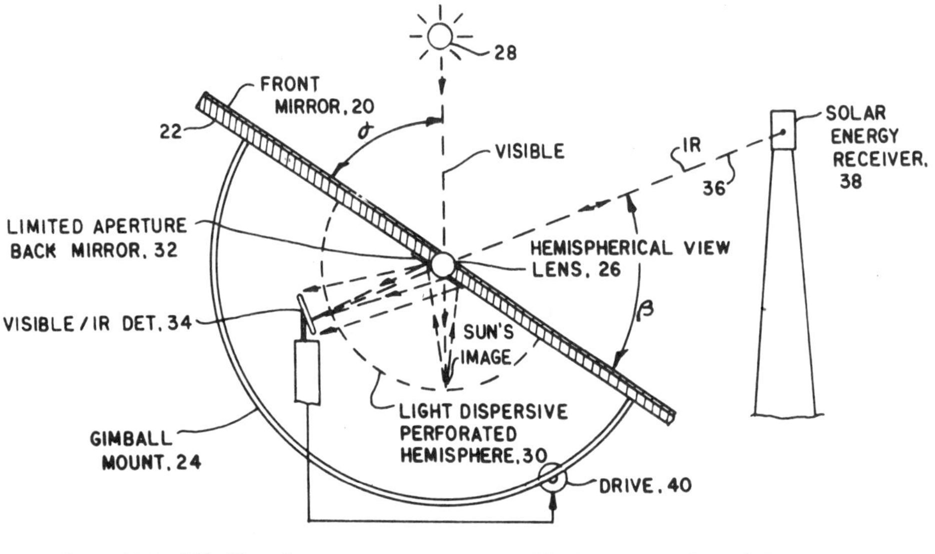

figure 11.6 Tiltable-mirror arrangement as described in Patent No. 4,013,885 (1977). A large front mirror (20) is gimbal mounted on a support (22) and (24) so that it may be rotated about its center. The center of the mirror contains an aperture at which a hemispherical view lens (26) is located. One of the purposes of this lens is to form an image of the sun (28) onto a hemispherical imaging surface (30) attached to the back side of support (22). This imaging surface forms a projection screen and is perforated so that it is semitransparent or transparent at a large number of points. The surface may be removed slightly from the image plane of the lens so that the image of the sun is spread out.

The imaging surface of this screen may be light dispersive, or it may be made of tiny corner, spherical, or other narrow-beam retroreflectors when an extensive back-mirror reflector is utilized. When an image of the sun is formed on a light-dispersive surface, it is reflected by a limited-aperture back mirror (32), which is concentric with lens (26), to a solar/receiver image detector (34) located adjacent to hemisphere (30). This detector, which contains a central IR detector surrounded by a visible-radiation detector, is centered on a line (36) that passes from a solar energy receiver (38) through the center of the aperture in the front mirror. If tower sway or earth motion is not a problem, only a solar-image detector need be utilized.

The output of the detector (34) actuates the gimbal drive (40) to rotate the mirror about axes in the X and Y directions so that sunlight impinging on the surface of the front mirror is always reflected along line (36) toward the solar energy receiver.

The mirror must be tilted so that the angle between the sun and the mirror, here α, is equal to the angle between the solar energy receiver and the mirror, here illustrated as β. When this has been accomplished, the image of the sun, as reflected by the limited-aperture back mirror, will be concentric about line (36) and also concentric to the infrared image of the solar energy receiver at detector (34). As the sun moves across the sky, its image on surface (30) will shift, causing a shift in the position of the reflected image so that it is no longer concentric with either line (36) or the smaller infrared image of the solar energy receiver. Since the back mirror has a limited aperture, when the image of the sun has shifted far enough, none of the light from the image will fall on the detector. Before this happens there will be less incident light on one side of the detector than on a diametrically opposite side, and a servo system generates signals to gimbal the mirror to the point where equal light is detected on diametrically opposite sides of the detector. Alternatively, the operation is the same when retroreflectors are used on the solar-imaging surface in conjunction with back reflectors that are not aperture-limited.

Reflective devices still hold great promise, and work on them is continuing in the present day. Daniel Blitz of Massachusetts has patented a solar energy receiver designed to be mounted atop a tower, perhaps as high as 300 feet. Automatically steerable mirrors at the base of the tower redirect the sunlight toward the receiver. The mirrors are tiltable and are provided with a centrally located aperture and lens, a hemispheric, semitransparent imaging surface attached to the back, a limited aperture back mirror, and a detector located on the line from the receiver running through the mirror's central aperture. (See Fig. 11.6.)

Another modern solar heating device is the Skylid™, a variation on Benjamin Franklin's Pulse Glass. It is an insulated, louvered device that opens during sunny weather and closes itself at night. Thus it maximizes transfer into the interior of solar energy when it is available, minimizes heat loss when it is not, and provides light. No electricity is needed for its operation. The tubes and reservoirs of the heater are filled with a partially vaporized fluid, such as Freon, so that a higher temperature on the outside raises the relative vapor pressure on the inside (Fig. 11.7). The Skylid was invented by Stephen Baer of New Mexico. He sells it and many other products related to solar energy through his own company, Zomeworks.

John G. Johnston of California has invented a solar heater that utilizes the "greenhouse effect" of glass, whereby solar heat is passed into a structure but not radiated from it. According to the patent for this invention, a fluid such as air is placed within a housing that forms a flow path. Solar rays pass through a porous barrier to heat the air. This barrier is constructed in layers from a fibrous material or from other material such as mesh. A heat check valve is arranged in the housing along the flow path to take in heat but to prevent it from getting out (Fig. 11.8).

A complete solar heating system was patented in 1977 by Max Taylor of Wisconsin. It is not a particularly sophisticated device; it can be built by homeowners who know their way around tools. The solar heat enters the collector unit, which is a top and a bottom plate supported by a frame. The heat is then circulated into the storage unit, which in this case is a bed of small rocks. They retain their warmth for a long time after the sun goes down, and can thus warm up the air that is blown into the house or other area to be heated. To provide hot water, pipes are run through the bed of hot rocks. This invention is fairly basic, without much refinement, and has not met with any commercial success. Yet the principles are good, and the arrangement is a logical one. Study Fig. 11.9 and its caption, and then consider how the inventor arrived at his final product.

Many corporations are interested in solar energy, especially in developing their own products for the market. A research associate of PPG Industries, Pandit Patil of Pennsylvania, has patented a solar heat collector, as illustrated in Fig. 11.10. It has "an outer cover plate, an intermediate cover plate, and a solar radiation absorber mounted in spaced relation to each other by edge packing. This (1) removes moisture from the air space between the cover

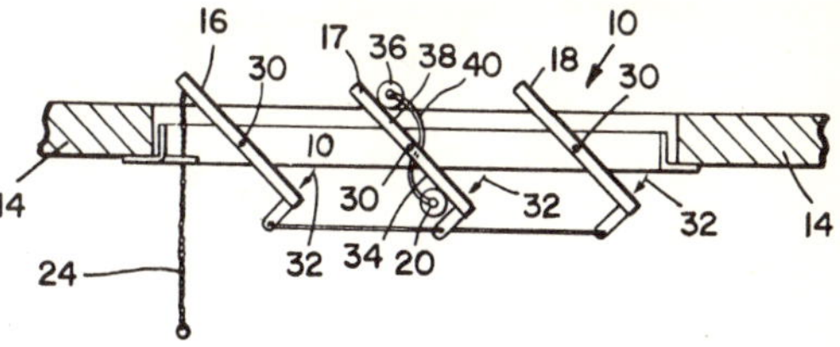

figure 11.7 Side view of the apparatus (10) of the solar-heating device described in Patent No. 3,884,414 (1975). Louver panels (16, 17, and 18) are pivotally attached to frame (12) along axis (30) passing through their respective centers of gravity. A skylight (not shown) is ordinarily used to cover the panels. Panels (16), (17) and (18) are rotatable first in a direction as illustrated by arrows (32) to open the panels. Rotation of the panels in the opposite direction closes them. Interior canister (20) is mounted on the interior face (34) of louver panel (17) at one side of the pivotal axis of the panel. A corresponding exterior canister (36) is mounted on the exterior face (38) of panel (17). A tube (40) provides fluid communication between canisters (20) and (36). A volatile fluid such as Freon is provided in the canisters and occupies the tube as well. At ambient temperatures the Freon will be partially vaporized and consist of both liquid and vapor.

When exterior canister (31) is exposed to direct sunlight or when the exterior temperature of structure ((14) is greater than that of the interior, the internal temperature of the exterior canister will be greater than that of interior canister (20). As a result, the relative vapor pressure of canister (36) will increase, forcing liquid Freon through tube (40) until the vapor pressures are stabilized. The weight of canister (36) will decrease and that of canister (20) will increase, causing louver panel (17) and slaved louver panels (16) and (18) to open as illustrated by arrows (32). In this configuration the solar energy is transmitted to the interior of the structure. Conversely, when the internal temperature of exterior canister (36) is less than that of interior canister (20), the vapor pressure of the canister will be greater than that of canister (36), causing liquid Freon to flow into exterior canister (36), and the window will close. Hence, the operation of apparatus (10) is automatic and is responsive to the ability of solar energy to heat the structure at any given point in time. Freon is sufficiently volatile so that the apparatus responds to a one-degree variation in temperature. In order to prevent overheating of the structure on extremely warm days, chain (24) can be manipulated to close louver panels (16), (17), and (18) and override the action of the canister.

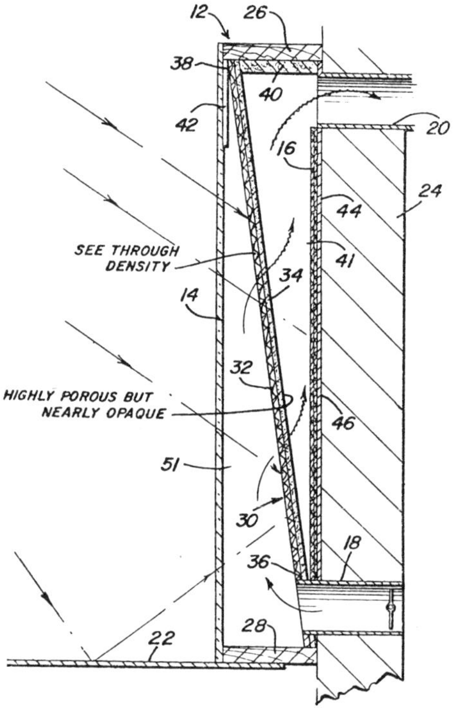

figure 11.8 Vertical cross section of the solar heater described in Patent No. 3,875,925 (1975). A housing (12) forms a flow path for air or other suitable gas to be heated. This housing includes a face (14) made of material that is clear and readily transmits solar radiation, especially short waves.

Inlet (18) and outlet (20) are arranged as part of the flow path. A reflector (22) may be connected to the housing to increase the amount of solar radiation reflected. A wall (24) of a building to be heated is provided with openings for receiving the conduits forming the inlet and outlet. A heat valve is arranged across the flow path for accepting light received from the face, and blocking the passage of heat back toward the face. This valve is constructed from a porous material such as fiberglass and is formed from two layers (32) and (34). Layer (32) is directed toward the face and has a very low density, while layer (34), arranged toward back wall (16), has a higher density. Both layers should generally be colored black.

The mat (30) extends from a point (36) on the inner end of inlet (18), across the flow path to a point (38) on the face opposite the outlet. This outlet is arranged at a higher level than the inlet for creating a natural circulation along the flow path defined by the housing. Some insulation (40) along the inner surface of the top wall cooperates with mat (30) to form a chamber (41) in which heat is collected. A double facing (42) is arranged along the upper edge of the face from the top wall to the depth of outlet. Heat dissipated through back wall (16) goes into the enclosure to be heated. A thin layer of insulation (44) covers the back wall, and a black fabric (46) covers the insulation. This covering absorbs heat or, more specifically, solar radiation, passed through mat (30), and completes the heat-collecting chamber (41).

---

figure 11.9 Side view, partially cut away, showing the major portions of the solar-heating system described in Patent No. 4,016,861 (1977). The solar-heat collector is mounted on the front of a porch (12) so as to be exposed to the sun in winter and effectively absorb heat. A reflecting cover (14), which has a reflecting surface on its inner side and is covered with siding on its outer side, is hinged at its bottom edge to this same collector. When the system is not in use, this cover is swung to its closed position, indicated by the dashed lines, and provides a decorative cover for the solar-heat collector. In its open position, indicated by the solid lines, the cover is supported by a block (16) at such an angle as to reflect sunlight onto the collector, as indicated by arrow (18), to increase the intensity of illumination thereof.

A heat-storage unit (20) is formed underneath the porch by a bin (22) full of relatively small rocks (24). A blower motor (26) is coupled to the top of the collector and to the top of the bin by conduit (28). The bottom of the collector is coupled to the bottom of the bin (22) by conduit (30). The blower mower causes air to circulate, transferring heat from the collector to storage. A second blower (32) is coupled between the heat-storage bin and the various rooms of the house to circulate heated air. Air returns through air-return duct (33). The heat-storage capacity of the rocks (24) will tend to retain the temperature of the air circulating through the house at a constant level even when the sun goes under the clouds.

To provide hot water, a water conduit (36) is mounted in the bin on staggered supporting panels (38), (40), and (42), which can be plywood panels. This conduit is arranged in loops on the sides of panels 38–42, facing the heat collector to intercept the hot air stream entering through air conduit 30 from the collector. The staggered location of the panels causes them to act as baffles to maximize airstream contact with the looped water conduit (36) and thus maximize heat transfer to the water. Fins can be added to the looped portion of the conduit to enhance heat absorption, and the surface of panels (38-42) facing the conduit can be lined with a reflective coating for the same purpose.

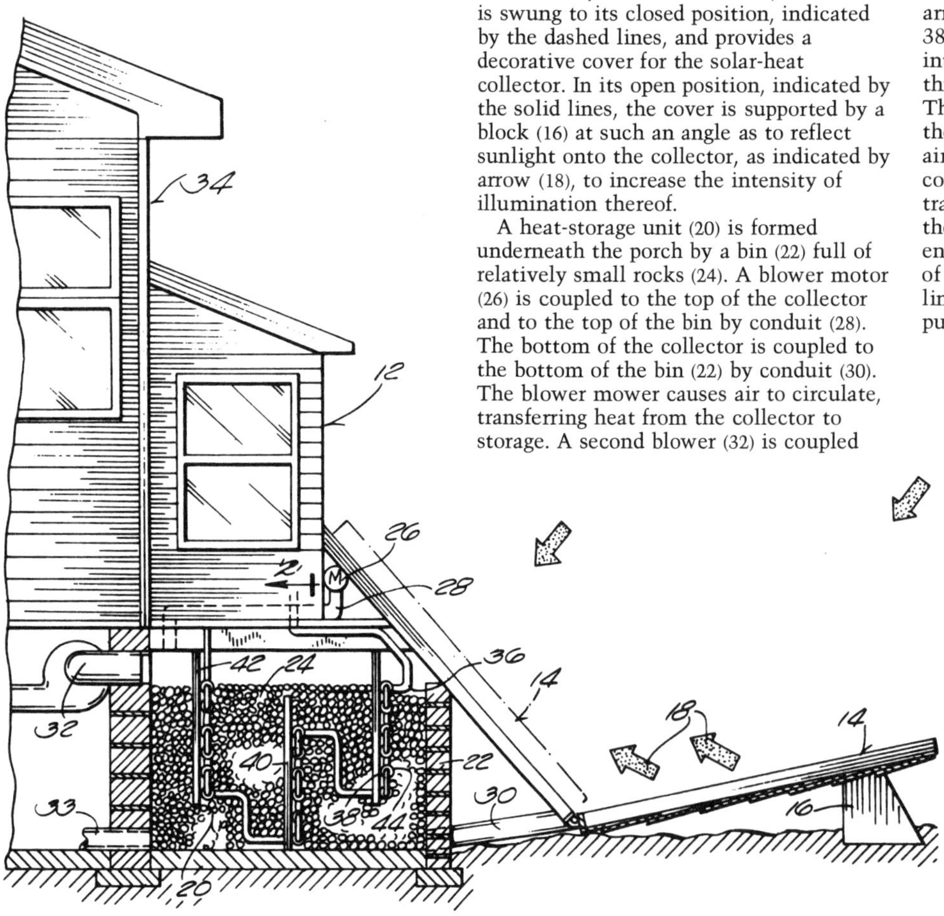

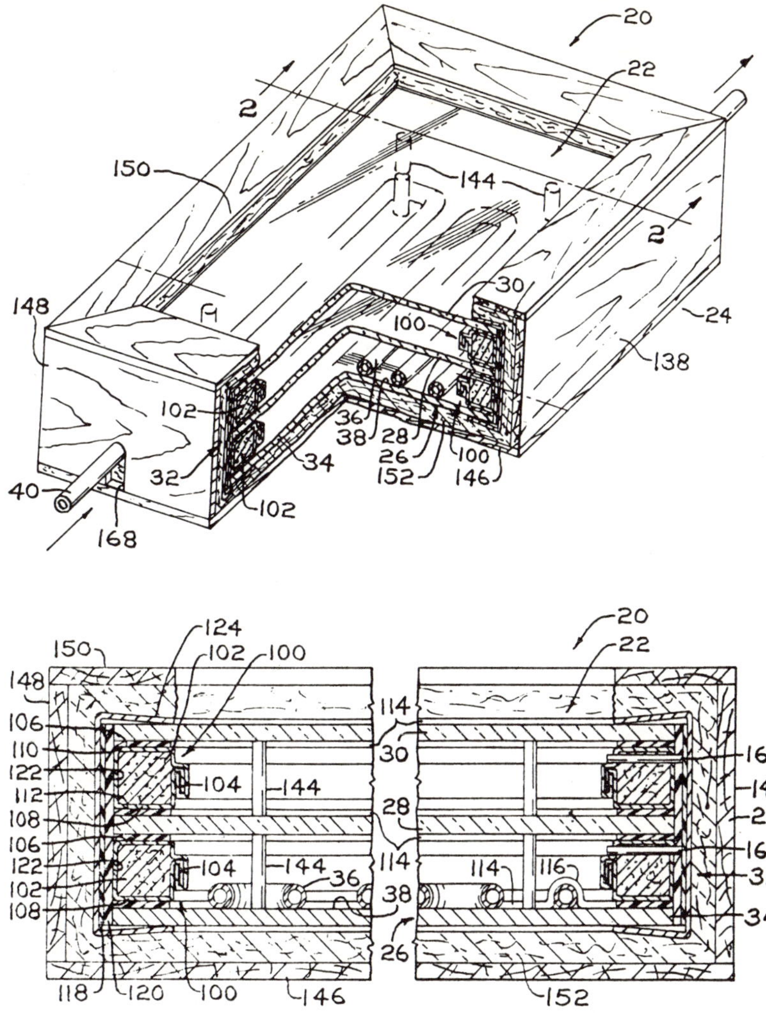

figure 11.10 Isometric view (top) of a solar heat collector, as shown in Patent No. 3,974,822 (1976). At bottom is a cross section of the collector unit taken along lines 2-2 of the top figure. The collector, mounted in an insulated box (24), includes a solar-radiation and infrared-energy absorber (26) and intermediate and outer cover plates (28) and (30). These plates (1) pass solar radiation to the absorber; (2) reflect infrared energy back toward the absorber; (3) trap infrared energy between the cover plate and absorber; and (4) reduce convection heat losses. The absorber and cover plates are held by edge packing (32), which (1) absorbs moisture trapped between the cover plates and the absorber during fabrication of the collector, and (2) prevents moisture from moving into the space between the cover plates and the absorber during use.

The absorber (26) includes a plate (34) having a conduit (36) mounted on surface (38) of the plate. Both plate and conduit are made of a heat-conductive material such as aluminum, steel, or copper. The conduit is mounted on the surface of the plate so that ends (40) and (42) extend beyond the insulated box (24) to provide external access to the conduit for moving a heat-absorbing medium such as water or ethylene glycol mixed with water. As the absorber is heated by solar radiation and infrared energy, the heat-absorbing medium moving through the conduit is heated by conduction and convection.

The surface of the plate (34) and of the conduit (36) should absorb solar energy in the wavelength range of 0.3 to 2.1 microns, and infrared energy in the wavelength range of 2 to 15 microns. The amount of solar radiation and infrared energy absorbed and reflected is a function of the absorptivity coefficient and emissivity coefficient, respectively, of the absorber surface.

In general, the amount of time required to heat the heat-absorbing medium in the conduit depends on (1) the length of the conduit; (2) the inside diameter of the conduit; (3) the heat-absorbing quality of the medium moving through the conduit; and (4) the flow rate of the medium moving through the conduit. For example, as the length of the conduit increases, the time required to heat the medium to the predetermined temperature decreases.

The cover plates (28) and (30) are selected to pass solar radiation and to reflect and/or trap infrared energy. The plates may be made of commercial-soda-lime glass or borosilicate glass. To improve the reflectance coefficient of the plates while passing sufficient solar energy to the absorber, one or both may be coated.

By selecting (1) the absorptivity coefficient and emissivity coefficient of the absorber (26) and (2) the transmittance coefficient and reflectance coefficient of the cover plates, the efficiency of the solar-heat collector can be improved.

A spacer element (100) includes a tubular portion of generally rectangular cross section that is filled with a desiccant material (102), such as granular or powdered silica gel. Communication between the air space between cover plates (30) and (28) and the desiccant (102) is provided through channels (104). The spacer (100) extends completely around and between the marginal edges of the cover plates.

Opposed surfaces (106) and (108) of the spacer element (100) are adhered to marginal surfaces (110) and (112) of the cover plates, respectively, by a thin continuous film of a moisture-resistant adhesive (114).

A portion of the spacer at (116) is removed to receive outlet end (42). Similarly, a portion of the spacer (not shown) is removed to receive the outlet end of the conduit (36), which ends extend beyond the solar-heat collector. A resilient, moisture-resistant strip (118) with a layer of mastic (120) adhered is bonded to the flat side (122) of the metal spacer and the peripheral edges of the absorber and cover plates. A channel (124) of essentially U-shaped cross section also extends completely around the perimeter of the windows to protect the edges. Several sections of channeling are joined or butted together at their ends. This invention is not limited to any particular edge packing, only to those edge packings that absorb moisture from between the cover plates and absorber and prevent moisture from moving between the cover plates and absorber.

figure 11.11 Solar energy collector panel shown in Patent No. 4,014,314 (1977). It is formed by coiling an elongated piece of cylindrical tubing (12) into a toroidal shape stiffened by radial clamps (20) and endpieces (22).

The tubing ends outside the circumference of the panel in a discharge stub (14) and an inlet (16) that extends, in substantially radial alignment, from the inner diameter of the toroid to the outer periphery. The material of the tubing may be chosen from several plastic and metal compositions readily available; a darkened finish is preferred to increase the collection efficiency.

The radial clamps are rigid and elongated, bent into a U-shape, whose internal spacing corresponds to the outer diameter of the tubing. The ends of the tines of the clamps are engaged by endpieces (22), which prevent the discharge of the tubing from the outermost winding of the collector. The tubing is closely wound to form the collector panel. Adjacent turns of the spiral winding are closely packed to present a continuous absorbing surface to incident radiation.

An interconnecting frame may be provided to rigidly interconnect the several clamps into a self-supporting assembly and to provide support for inclining the collector panel at the angle to the sun best for its geographical location. When the panel is laid directly on a surface (such as a roof) which is at the appropriate inclination with the mean solar vector, and where one or more of the clamps may be attached to the surface, a support structure is not necessary.

The panel (10) is assembled at the use site, the clamps and the tubing being delivered as separate items and wound into the toroidal collector immediately prior to final connection.

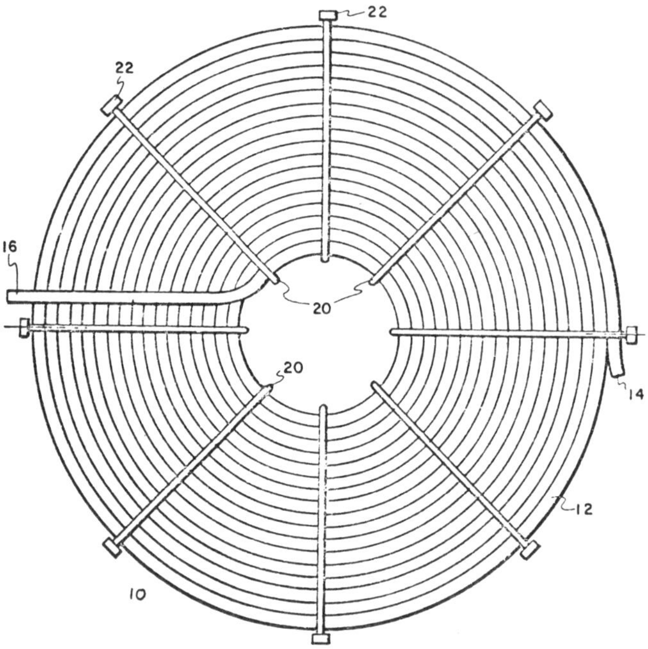

plates and the absorber and (2) prevents moisture from entering the air space between them." Collectors are useful for heating a fluid and/or generating electric power. According to Mr. Patil's patent, the design of prior collectors had certain disadvantages that he tried to correct. Moisture was occasionally trapped in some of the air spaces. When the temperature went down, this moisture condensed. Even when the water evaporated as the temperature went up, there was still the probability that a stain would be left that would impair the collector's overall efficiency. This collector solved this problem by the construction cited above, and was also designed to be less expensive than previous models.

Another collector panel was patented in 1977 by T. Lawrence of California. Its primary purpose is to heat water, and to this end, its collecting surface is composed of a continuous length of metal or plastic tubing coiled into a kind of doughnut shape, ten square feet or more in area. This collector is rather simple in construction (and thus comparatively inexpensive) and does not have to be continually adjusted as the sun moves across the sky and hits it from different angles. The circular shallow spiral shape of the tubing arrangement creates a collection surface that is not dependent on the angle of incident solar radiation (Fig. 11.11). The collector simply has to be aligned with the local mean solar radiation vector in order to do its job.

Frank Carroll of Illinois found that a diffuse-reflecting surface can significantly increase the efficiency of a tubular collector behind which it is placed. He also found efficiency increased if the tubes are spaced no farther apart than four tube diameters on centers and no more than four tube diameters away from the diffuse-reflecting surface. In his invention, a continuous tube is joined to a manifold so that liquid can flow from the manifold into the tube, where it is heated, then out the other end of the tube and back into the manifold (Fig. 11.12).

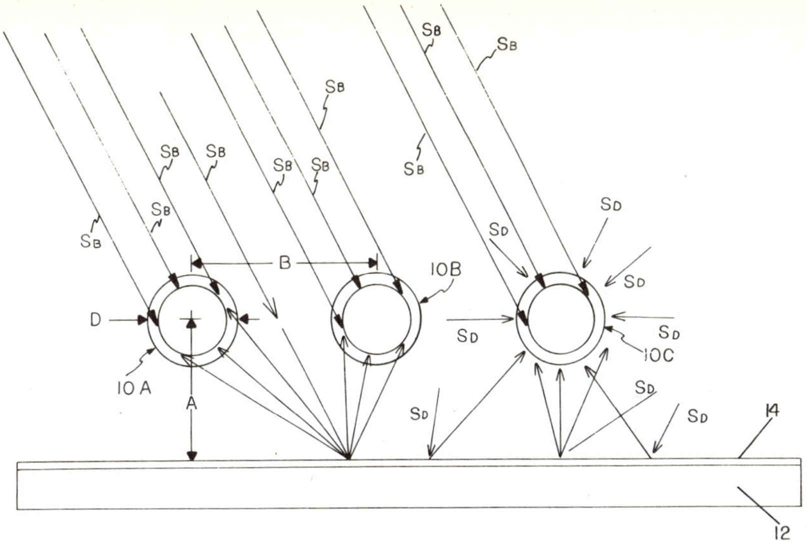

figure 11.12 Schematic side view illustrating the operating principles of the solar energy collection system described in Patent No. 4,002,160 (1977). As is well known, solar radiation is made up of two primary components. First there is (S_B), which is incident upon the surface of the earth from the position of the sun at any given time. It is a collimated beam of light shown at an angle of about 60 degrees from the horizontal. Second, there is a diffuse-radiation component designated as (S_D). This component is not collimated and the rays are not parallel; it is available from many directions. In the case of a flat plate collector the major energy absorption takes place when the beam (S_B) is overhead or at solar noon. When the sun's position is lower in the sky, the angle at which the beam (S_B) strikes a horizontal flat surface is lower, thereby causing loss in energy available to a horizontal flat surface. Despite this deficiency, the flat plate allows the maximum availability of energy for absorption as versus an array of tubular collectors. However, the latter offer significant advantages in that they are capable of operating at higher output temperatures than flat-plate-type collectors and in that the weight and expense of a tubular array may be somewhat less than a flat-plate collector. In addition, a tubular array is somewhat independent of the position of the sun, since as the sun moves across the sky the sides and backs of the array may collect energy at one time of the day, while the upper portion of it may collect energy during another part of the day. At all times, the direct beam (S_B), which impinges upon a cylindrical collector, sees a constant intercept area, thus allowing maximum transfer of the solary energy. Shadowing effects can occur from adjacent collectors when the sun is low on the horizon, thus reducing the unit's efficiency. Tubular collectors (10*A*) and (10*C*) can absorb solar energy with the use of heat-transfer media and transmit this energy for use in space heating, etc.

In order to increase the efficiency of the tubular array, a back plate (12) is placed behind the collectors. The plate's surface faces the tubular collectors and is coated with a thin layer of diffuse-reflecting material, such as white paint. Reflecting surfaces behind solar collectors have normally been polished and even more frequently have attempted to focus back to the collector those direct beams (S_B) that missed a collector. Polished surfaces and focusing surfaces are not easy to make, which raises the expense of a solar collector array. In addition, exposure to the elements for a period of time causes such surfaces generally to lose their polished sheen and would, therefore, cause focused-type surfaces to lose precision of focus. It is well known to place a single tubular collector at the focal point of a parabolic reflector. The tubular array of solar collectors (10*A*) and (10*C*) does not need to be moved in order to absorb maximum solar energy. A diffuse-reflecting layer (14) on the back plate (12) reflects back any collimated beams (S_B) or diffuse beams (S_D) which impact upon it in a diffuse pattern. There is no attempt made to focus the reflection of any beams of light which are incident upon the layer (14) to any particular one of the tubular collectors. Rather, there is a random back reflection of light incident upon the layer (14) to the collectors (10*A*) and (10*C*), which are positioned above the back plate (12). The collectors are positioned a distance (*A*) above the diffuse layer (14). This distance may usually be defined in terms of the outside or overall diameter (*D*) of the tubular collectors (10). Likewise, the tubular collectors themselves have a center-line spacing, generally designated as B. There are relationships between the distances A and B and the diameter of the tubes (*D*) to achieve maximum energy absorption from a tubular collector array.

One of the most influential people in the field of solar energy research is Abraham Mlavsky of Massachusetts. In 1976, he patented a tubular cell device that can be coupled together electrically in series and in parallel arrays to form an integrated structure. Solar energy concentrators are combined with these solar cells to maximize their power output (Fig. 11.13). According to Mlavsky's patent, which he assigned to the Mobil-Tyco Solar Energy Corporation,

> More specific objects of the invention are to provide solar cells and arrays thereof which have a modular form, can be easily cooled, have structural integrity, can be made by existing techniques, and are capable of withstanding changes in dimensions due to thermal cycling. . . . Leakage is minimized by the use of a geometry which minimizes the ratio of exposed active surface area to exposed junction region area.

Another invention in the area of solar generation of electrical power was patented by Bill K. Davis, a NASA employee from Alabama. He attempted to solve some of the problems that have confronted inventors in this field by developing a solar boiler of relatively high efficiency. A fluid with a low boiling point, such as Freon, is vaporized in the boiler so that it powers a turbine. To

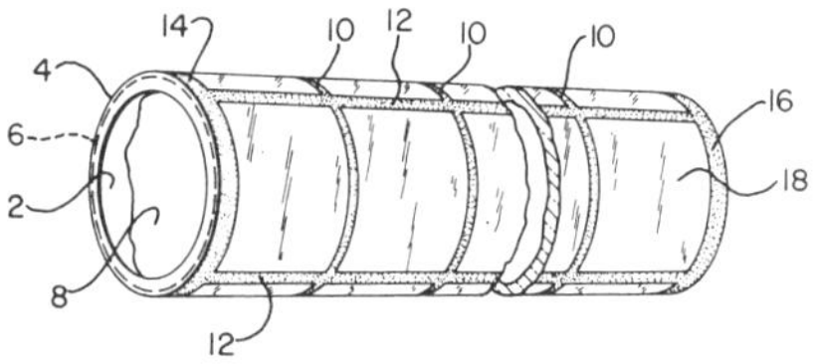

figure 11.13 Tubular solar cell, partially broken away, as illustrated in Patent No. 3,976,508 (1976). The cylindrical silicon tube (2) of N-type conductivity has been subjected to diffusion of boron into its outer surface to form an outer P-type conductivity region (4) and a P-N junction (6). The inner surface of the cylindrical tube is provided with a first electrode in the form of an adherent metal-conductive film (8), which forms an ohmic contact with the tube. The film (8) covers the entire inner surface of the tube and consists of a selected metal or metal alloy having relatively high conductivity. The outer surface is provided with a second electrode in the form of a grid consisting of conductors connected together by other conductors (12). The opposite ends of the outer surface of the hollow tube are provided with two circumferentially extending terminal conductors (14) and (16). The arrangement leaves relatively large areas (18) of the outer surface of the tube exposed to solar radiation. Preferably, but not necessarily, the conductors (12), (14), and (16) are made wider than the conductors (10) since they carry a greater current than any of the latter. These conductors are made of an adherent metal film like the inner electrode (8) and form ohmic contacts with the outer surface of the tube. When the unit is connected by its inner and outer electrodes into an exterior circuit and the exposed portions (18) of the outer surface of the unit are exposed to solar radiation, electron-hole pairs are generated in the tube with the result that current will flow through the exterior circuit via the inner and outer electrodes. The open-circuit potential of the unit is approximately 0.52 volt. The same results will be produced if the solar cell unit is made by providing a tube made of P-type conductivity and treating its outer surface to provide regions of N-type conductivity with an intervening P-N junction.

A plurality of P on N or N on P units may be combined to form a solar cell array with the individual solar cells being interconnected electrically either in series or in parallel. Preferably, but not necessarily, the several units are mechanically connected end-to-end to form an integrated structure.

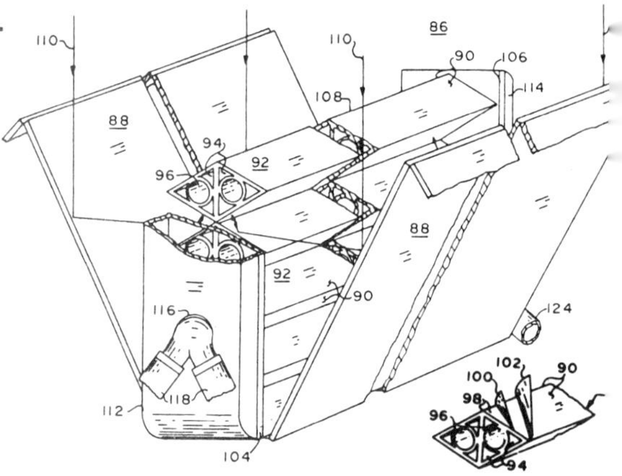

figure 11.14 Part of a solar boiler as illustrated in Patent No. 3,957,030 (1976). Solar energy is directed into a collector assembly (86) by pivotally interconnected reflecting panels (88), which form a V-shaped channel to direct the sun's rays onto each opposing surface (90) of tubes (92).

These tubes are typically formed from a length of heat-conductive material such as aluminum, etc., and are diamond-shaped in cross section. Each includes six longitudinal, essentially triangular passages (94) and two cylindrical passages (96) through which the working fluid is circulated. The outer faces (98) are then coated with a thin layer of copper (100) or other suitable material, which is adapted to receive as a bonded surface an outer-skin member (102) such as a titanium alloy.

Exposed surface (90) of the titanium is then anodized so as to produce a desired shade of blue. This shade occurs when the absorptivity (α) ranges from 0.70 to 0.74.

Tubes (92) are supported by end plates (104) and (106), and the apices (108) of obtuse angles of each of diamond-shaped tubes (92) are joined. When thus mounted, the exposed surfaces are displaced at an angle of 60 degrees with respect to each other, so that any ray (110) of the sun, reflected from any reflector, impinges on one surface from reflector (88) and then to the adjoining surface, wherefrom it is directed along the same path back to the original surface and to the reflector. Thus, there is achieved a three-point contact with the surfaces of the tubes. Once the tubes are assembled to end plates (104) and (106), end covers (112) and (114) are sealed to support plates (104) and (106), respectively, through which working fluid is circulated by way of Y-shaped fittings (116), only one being shown. When combined, collector assemblies (86) are interconnected by lengths of flexible hose.

maintain operation of this turbine in the absence of sunlight, the system includes a thermal storage unit that contains a meltable substance such as Pyrone. This chemical is melted to a hot liquid during the periods of sunlight, thus accumulating heat. When the sun's rays are not available, then, vapor is generated by this stored heat to keep the turbine running. The boiler has a blue anodized outer surface that enhances the effectiveness of translating the sun's rays into useful heat. When polished, it provides an improved ratio of heat absorption to heat emission (See Fig. 11.14).

Different inventors approach the solar field from different points of view. The solar cell is a comparatively small device that is designed to fit into different situations. When thinking about a solar house, however, it's sometimes useful to think of the construction as a whole. It's often difficult and unsightly to tack something onto a house that was built and designed without any future attachments in mind. Way back in 1893, William van der Heyden, a Dutchman living in Japan, was issued U.S. Patent no. 504,544 for a "sanitary house." This inventor-architect thought that people should live in glass houses. As his basic unit of construction, he

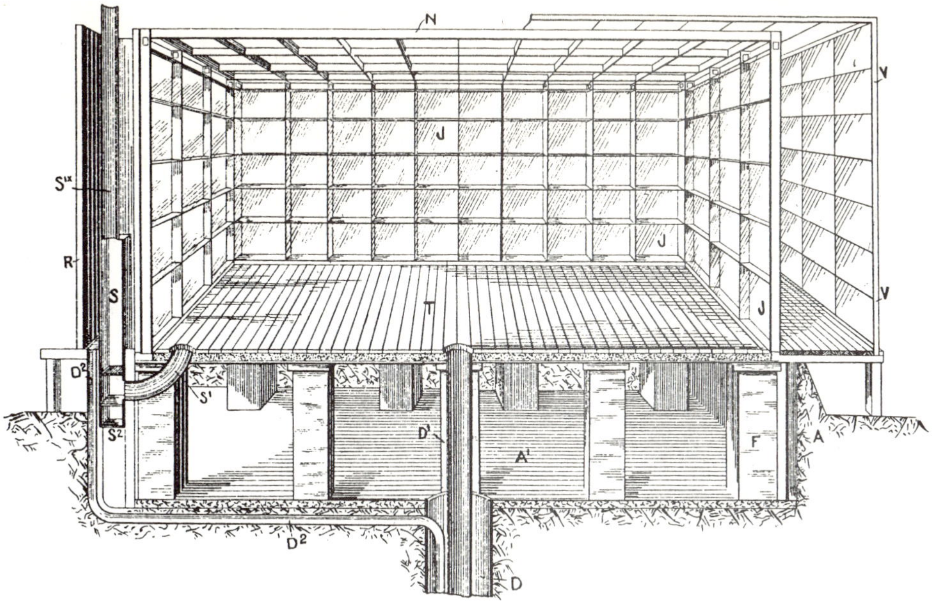

figure 11.15 Cutaway view of the "sanitary house" described in Patent No. 504,544 (1893). It is built of double walls, that is, double glass cases that contain some kind of liquid. The double walls of heating tower (*R*) may in hot or moderate climates be filled with water or a solution of iodine, and the chimney may be coated outside with carbon to absorb heat, and thus keep up air circulation. The stove (*S*) may in warm climates be replaced by a simple iron tube coated outside with a layer of carbon. The heat of the sun warms the air inside the iron tube. This air will rise and must then necessarily be replaced by the air from the room. The carbon coating of the tube absorbs heat and will give this off during the night, continuing circulation of air. When the weather is cold and the stove (*S*) is in action, air passes to the stove from the upper room by the tube (*S′*), and the upper room is supplied from the tower (*R*), which obtains filtered air by the tube (D^2) from the basement or well (*D*). In warm weather, the heat of the sun acting on the iron chimney draws air through the stove from the dwelling, or the stove may be put in action with the same effect. The stove then receives its supply of air from the pipe (*S′*), through the tube (*D′*) from the well (*D*), where the temperature is nearly constant.

designed a kind of brick made of two layers of glass or other transparent material separated by water and albumen in solution and held in place by metal frames (Fig. 11.15). Heat is not introduced to the house through these "boxes," but through a ventilation system connected with a heating unit in a tower just outside the house:

> The double walls of the tower may in hot or moderate climes be filled with water or a solution of iodine, and in such circumstances the chimney may be coated outside with carbon to absorb heat, and thus keep up air circulation. . . . A simple iron tube may be coated outside with a layer of carbon, and the heat of the sun warms the air inside the iron tube, which air will rise and then necessarily be replaced by air from the room. The carbon coating of the tube absorbs heat and will give this off during the night.

A contemporary version of a solar-heated house was patented by Volkmar Heilemann of New Jersey (Fig. 11.16). The system is based on continuous flow of a fluid from solar cells to heat reservoir and back to cells again. The cells are, of course, on the roof, where they get maximum exposure to the sun's warming rays. Fluid from these cells flows down to the interior floor, where it circulates through pipes that pass through a layer of rocks that serve as a kind of heat sink. A concrete slab is above the rocks, between them and the flooring, and a layer of peripheral insulation is below them, to prevent significant heat loss. As the water cools, it is pumped back up to the roof, where it flows through the cells to be heated once again. This is a somewhat oversimplified description; it can be mitigated by reference to the figure and its caption.

As solar energy gains the interest of more people, more aspects of home construction will have to change or be modified to accommodate the new technology. This is perhaps one of the greatest areas of opportunity for the independent inventor who might not have the lab facilities to develop a new kind of solar cell but who would certainly be able to see the ways in which important new inventions can be fitted into practical applications. George Mather

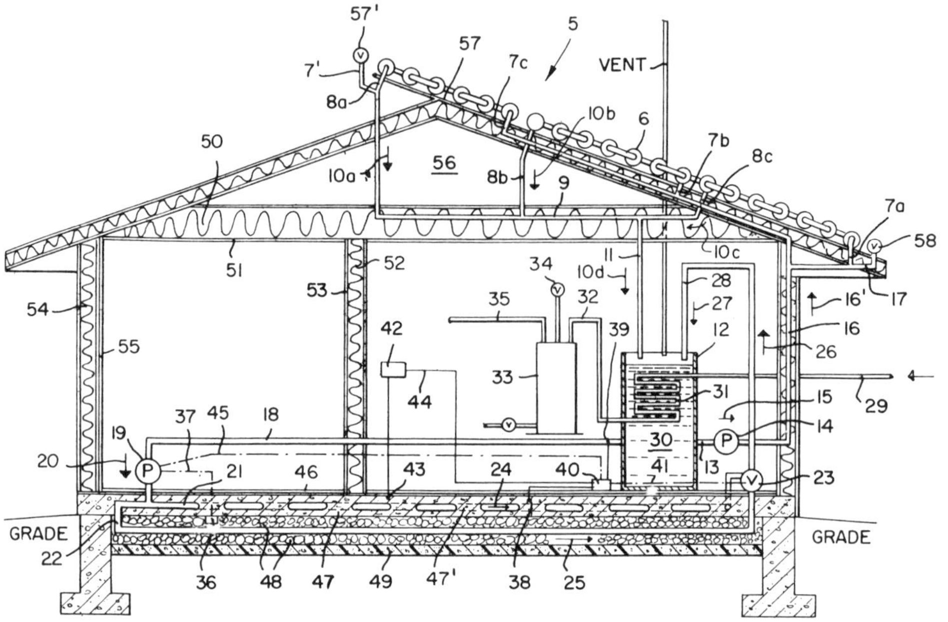

figure 11.16 Side view of a solar-heated house, as illustrated in Patent No. 4,000,851 (1977). A system of solar cells (5) is arranged along a roof (57), lying horizontally along the surface. Cell inlet pipes (*7a*), (*7b*), and (*7c*) and cell outlet pipes (*8a*), (*8b*), and (*8c*) are collectively channeled through collector pipe (9) and down pipe (11) into the reservoir tank (12). Coils heat cold water entering through source-water pipe (29) with an outlet to the tank (33) by the outlet pipe (32). Outlet (13) leads cooler water from the tank to pump (13), delivering water by up pipe (16) in direction (16′) to the safety valve mechanism (17) and to the inlets (*7a*), (*7b*), etc. Also from the tank space (30) outlet pipe 18 leads the water to pump (19) in direction (20) through one or both alternate flow-path pipes (21) and (22). The extent of flow through these pipes is regulated by the butterfly valve (23), which is controlled by thermostat (40) as dependent upon the temperature of the concrete slab (47) as read by sensor (38) as well as the temperature of the water in pipe (18) as read by sensor (39). The valve (23) directs a flow through pipe (21) when the upper surface of the concrete slab (47) drops below a set temperature, and directs a flow through pipe (22) when the upper surface temperature sensor (38) exceeds a predetermined value. Heat then becomes stored in the rocks (48) (preferably of large gravel size), and the peripheral insulation (49) prevents any significant heat loss downward, as water flows in direction (25) until valve (23) again shifts flow to pipe (21). The differential thermostat (40) also in part controls the pump (19), turning it on whenever sensor (38) senses a predetermined low. Sensor (36) turns on the pump whenever the temperature of the rocks (48) is below a predetermined minimum. Pump (19) is off only when both the sensors (38) and (36) are off. It remains running when flow is shunted from pipe (21) to pipe (22). Safety sensor (43) causes switch (42) to break circuit if the temperature of the concrete upper portion exceeds a danger level, thereby turning off power to the system and shutting down the pump (19). Downflow through the down pipe (11) in direction (*10d*) results in a heating of water in the tank to a predetermined temperature, at which the pump (14) shuts down. This is controlled by the differential thermostat.

The inventor utilized the concept of the "combination" of low-temperature utilizable water to adequately heat the living spaces of the building by virtue of solar energy, supplying only low temperatures during most of the day and/or evening hours, in conjunction with the floor-heating unit. Other solarheat units have relied solely upon conventional radiators, baseboard heat, and the like. They also kept water in the reservoir tank at too low a temperature to heat the house after heat loss from pipes during transit to radiators and the like. Thus large amounts of electricity for pumping, as well as supplemental heating by conventional routes, have heretofore been necessary.

---

figure 11.17 Roof deck construction as described in Patent No. 4,006,731 (1977). The sheet metal shape utilized in box section (20) has a lower horizontal base (21) resting upon purlin structure (30), vertical sides (22) and (23), and upper horizontal flanges (24) and (25). Slot (26) between the terminal ends of the flanges is continuous to permit the most flexible use of the sheet metal box section. It may be discontinuous, with the flanges joining to enclose the top of the box section, where it is not necessary to utilize the sheet metal clips.

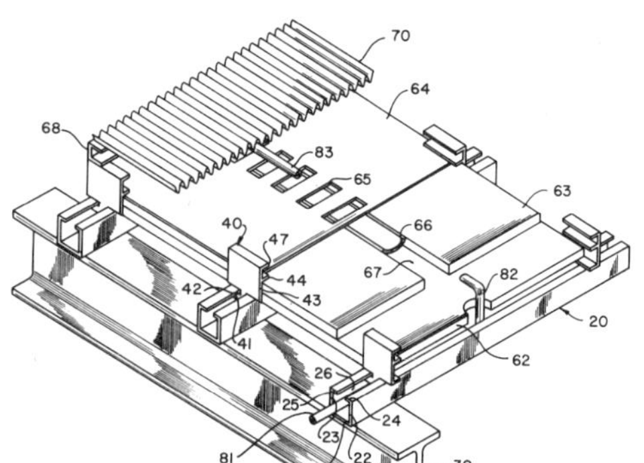

The sheet metal clip (40) has opposing slots (42) between vertical portions (41) and (43). These engage the flanges of the sheet metal box section so that when the axis of the clip is 90 degrees with the axis of the box section, the clip is anchored with the opposing flanges of the top of the box section engaging slots (42) with vertical portion (41). Vertical unit (43) extends upward a suitable height to provide horizontal portion (44), providing snug fitting of desired rigid board such as form-board (62), polymeric insulation (63), and solar energy absorber (64). Vertical portion (43) may extend substantially vertically farther upward for the desired distance to provide a support for structure (68) and polymeric sheet (70).

Clip (40) is readily inserted into slot (26) simply by turning it so that the axis of the clip and the box section are parallel and inserting the clip for the distance so that slot (42) engages flanges (24) and (25), then turning the clip so that its axis is approximately 90 degrees to the axis of the sub-purlin. The vertical portion of the clip extending above the box section may be about 1½ to about 4½ inches to accommodate the thickness of various form-boards and insulation.

A metallic absorber plate (64) providing a "black body" is placed directly above the insulation (63). This plate may be a solid absorbing surface having heat-transfer pipes (83) in contact with its upper surface so that the solar energy passes through polymeric sheet (70), is absorbed by the surface of sheet (64) acting as a "black body," and transferred by conduction to liquid, flowing through pipe (83) and riser (82) to pipe (81). The liquid is then carried within the box section for distribution to the desired volume to be heated or to supply thermal energy for air-conditioning cooling. The box section acts to transfer thermal energy from the solar absorber to the volume desired to be heated or cooled. Pipes (81), (82), and (83) may be eliminated, and air from the room or volume to be heated may be circulated directly over absorber plate (64) and passed through box sections (20) and provided directly to the volume to be heated by having slots in the bottom of box section (20). When such airflow system is utilized, blowers (not shown) must be located at the entrance to or the exit from the roof deck structure.

of Ohio patented a roof deck structure that is readily adapted for the installation of solar energy collectors. The deck's structure, or framework, is formed of sheet metal, and its top is a plastic sheet. The spacing of the deck's joints is flexible enough to accommodate solar collectors of various sizes (Fig. 11.17). This is a neat, straightforward idea of how to accommodate a new technology into a time-tested design. More complicated still would be devices and systems whereby solar heat could be channeled into existing home-heating systems. Could this be done? It is certainly worth a try.

In addition to space heating and water heating, solar rays can also supply strong heat for cooking. Solar ovens have been known since the later half of the Nineteenth century, for instance. They have never attained great popularity because of various problems, including the rather obvious fact that the stove remains stationary while the sun moves across the sky, thus striking the stove from different angles according to the time of day. In addition, the sun can disappear behind the clouds for hours (even days) at a time. Stella Andrassy of New Jersey saw these problems and invented a solar oven so designed as to compensate for them. First, Ms. Andrassy's oven is mounted so that it can be moved or tilted so as to receive maximum exposure to the sun's rays. The food that is being cooked, of course, must remain horizontal. A transparent shield is available to cover the oven and to prevent the escape of heat when the sun is obscured by clouds (Fig. 11.18).

Another version of the solar stove was developed by Merton Clevett of California. This unit has foldable, petal-shaped panels that are designed to reflect and concentrate solar rays through a clear plastic window and into a foil-lined oven. The stove is capable of preheating up to 450–500 degrees Fahrenheit. (See Fig. 11.19.)

Solar radiation isn't concerned strictly with heat. It's also an unparalleled source of energy for all sorts of uses—even air conditioning. The first step in such an invention, patented by Gershon Meckler of Georgia, is to dehumidify the air by passing it through a

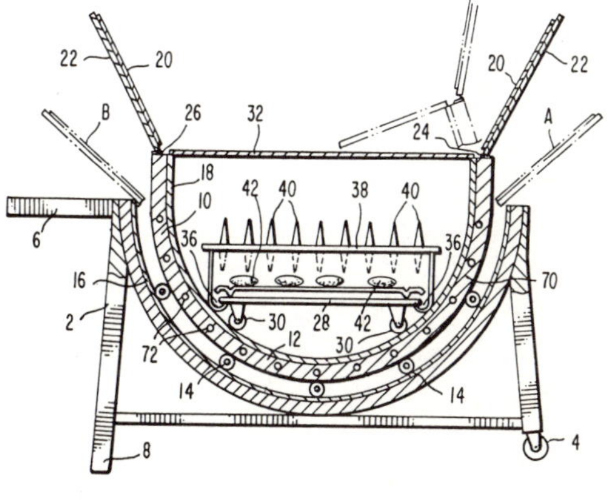

figure 11.18 Cross section of solar cookstove as illustrated in Patent No. 3,938,497 (1976). The frame (2) has wheels (4) and a handle (6) to enable the apparatus to be readily moved about so it will be effectively exposed to the sun. A leg (8) is secured to the frame to hold it in the fixed position to which it has been moved.

The oven (10) is supported on the frame and is formed with a lower portion (12) designed to rest upon rollers. The oven presents a concave inner reflecting surface (18) adapted to direct reflected rays of the sun inward toward the central portion of the oven. Additional reflecting surfaces (20) on the inner surface of members (22) are movable to concentrate the reflected rays on the articles to be heated within the oven.

The upper edges (24) and (26) of the oven project above the adjacent sides of the frame (2) when the oven is in its central position as shown in full lines, but is tiltable as shown in dotted lines (*A*) and (*B*) in the most effective locations for directing the sun's rays into the oven. The oven is provided with rollers (30) and is thus movable so as to maintain a horizontal position while the reflecting surfaces are tilted or moved.

A transparent cover (32) extends across the oven above the article support and may be hinged to the edge (25) of the oven. The cover may be formed of glass, transparent plastic, or other material which will effectively transmit the sun's rays, particularly the infrared rays which serve to develop heat upon contact with articles located on the support (28).

The support may serve as a cooking utensil and, as shown, may be in the form of a pan within which a rack is located. A marginal trough (36) in the bottom of the pan drains liquids away from the food being heated. A cover (38) has heat-transmitting elements (40) in the form of pins or fins by which additional heat may be transmitted to or from the cover. Thus, as shown, the cover may be formed with upwardly extending fins (40) for conducting heat most effectively from the space above the support to the cover. The cover may also be inverted to conduct heat directly to the food enclosed between the support (28) and the cover (38).

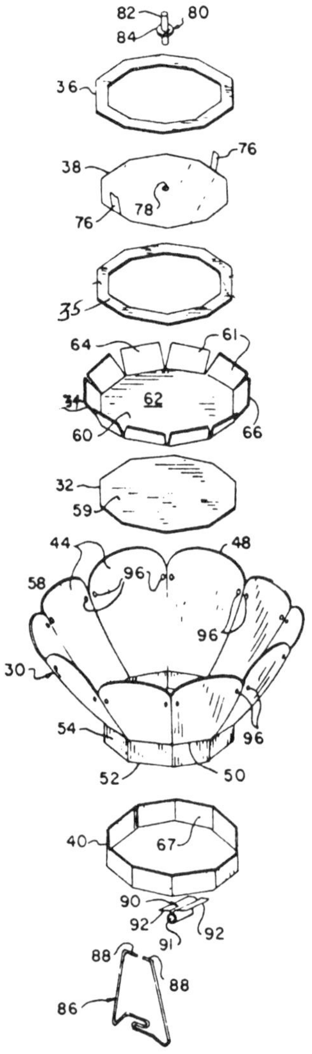

figure 11.19 Solar stove as illustrated in Patent No. 3,896,786 (1975). The stove includes a petaloid panel base assembly (30), a spacer insert (32), an oven liner (34), an oven window (35), and a retention band (40) adapted to fit externally around the base (30) to keep the stove together.

The petaloid panel base assembly (30) is made of a blank piece of nonconductive material such as corrugated paper, having a symmetrical decagon bottom (42) with integral petaloid panels (44) radiating from each side. Each petaloid panel is identical, with sides that diverge slightly outwardly and a semicircular outer end (48). At the connection (52) of each petaloid panel with the decagon bottom, the blank corrugated paper is bent, establishing a pivotal connection of each petaloid panel to the bottom. The blank from which the petaloid panel base assembly is made is, therefore, adapted to be folded into the configuration shown, wherein the inner section (54) is folded to form a right angle with the decagon bottom and the remainder of the panel forms approximately a 60-degree angle with the decagon bottom. The inner section of each panel defines one side of a continuous ten-sided wall around the bottom to form the oven. The inner face of the paper blank (30) has a bright aluminum foil lining (58), for reflecting the sun's radiation. Each of the panels (44) has a slightly concave, substantially parabolic, longitudinal cross section so that parallel sun rays that are normal to the bottom of the petaloid panel base assembly will be reflected by the panels into the center of the decagon bottom.

The spacer insert (32) is a disc of nonconductive material with a bright aluminum foil lining (59) on top and with a decagon like that of the base assembly. The insert is placed in the base (30) and doubles the thickness of the bottom, adding a second reflective layer of aluminum foil to the bottom of the stove to prevent heat from passing downward through the assembly.

The oven liner (34) is also made of nonconductive material with an aluminum foil lining (60) on its upper face (61). Rectangular tabs (64) radiate outwardly from the bottom portion (62). The tabs (64) fold so that the liner (34) can be inserted into the base (30), overlying the spacer insert (32) and with the tabs (64) engaging the inner-foil-lined surface of the inner portions (54) of each petaloid panel. The top edges (66) of the tabs (64) form a surface on which the window (35) is seated. Thus the bottom thickness of the device is tripled, with each layer having a foil lining to help prevent heat loss through the bottom layers of the stove. The sides of the oven portion (56) of the device are doubled in thickness when the oven liner (34) is inserted.

The retention band (40) can be slid upward to rigidify the stove. The band also has an aluminum foil lining (67) so that the oven (10) has a triple thickness, with three layers of reflective aluminum foil to retain the heat in the oven.

To align the stove with the sun, the window pane (38) is provided with a centrally located vent aperture (78) into which a direction finder (80) fits. By moving the stove until the pin (82) casts no shadow on the disc portion (84) of the direction finder, the stove can be aligned so that the sun's rays will be reflected by the petaloid panels (44) into the oven portion (56). After the stove has been properly aligned, the direction finder can be removed and the hole (78) will thereafter serve as a vent to prevent pressure buildup. A stand (86) may be utilized to brace the stove so that it remains aligned with the sun. The stand is made up of a U-shaped metal frame having inwardly turned end portions (88), which are adapted to be inserted into opposite ends of an open tubular connector (90) anchored to the bottom of the base (30).

---

contractor, where it comes into contact with a hygroscopic solution of glycol. (This dehumidification makes it a lot easier to cool the air.) The hygroscopic solution is periodically exposed to solar energy in a regenerator, where it is concentrated and then returned to the contractor, ready for use again (Fig. 11.20). The solar energy is also used to operate the generator, the condenser, and the evaporator to cool the air after it has been dehumidified.

A patent granted to Robert Neidigh of Indiana describes an apparatus for converting solar energy into mechanical energy. His invention utilizes lenses to concentrate solar rays on a number of heat-actuated devices that pump fluid into a high-pressure accumulator. This pressurized fluid can be used to perform such mechanical functions as operating hydraulic motors. The advantages of this invention include its comparative simplicity of construction, its ability to operate continuously, and the fact that it relies on an external power source (electricity) only during the first few minutes of its operation.

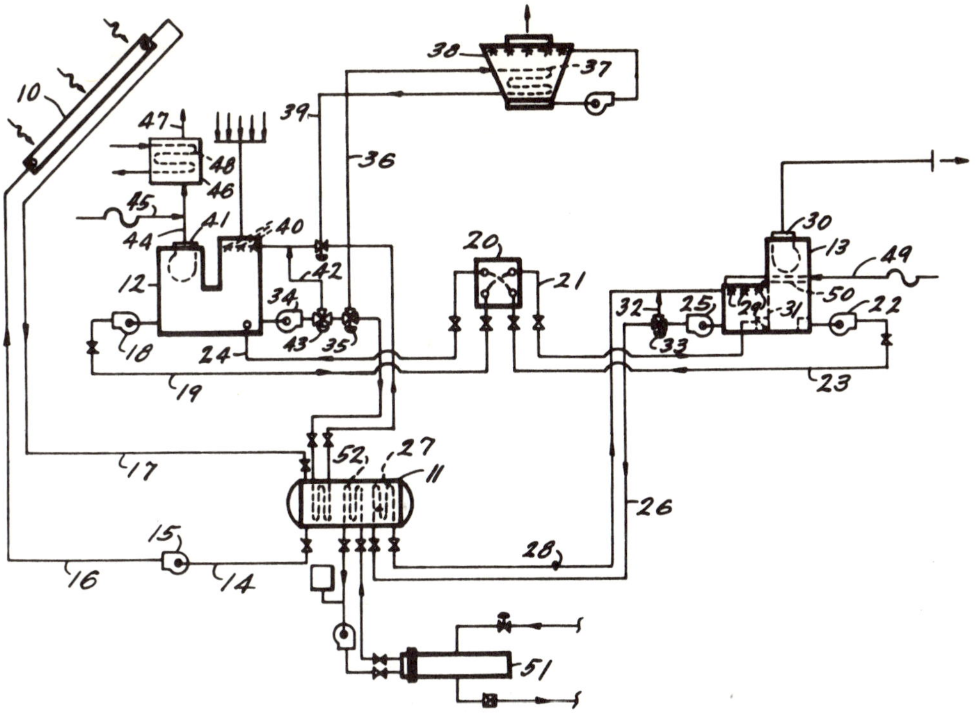

figure 11.20 Schematic diagram of a solar air conditioner, as illustrated in Patent No. 4,011,731 (1977). It is made up of a solar collector (10), heated-water-storage tank (11), and dehumidification apparatus including a contactor (12) and a regenerator (13). A heat-transfer fluid, usually treated water, is circulated from the tank through a line (14) to a pump (15), then through a line (16) to the collector. Heated fluid returns from the collector through a line (17) to the tank. The pump (15) is controlled to maintain a predetermined fluid temperature, say 140 degrees F., within the tank.

Fresh air is dehumidified by being drawn through a spray of an aqueous hygroscopic solution of a glycol in the contactor (12). The hygroscopic solution is recirculated through the contactor. A portion of the solution in the reservoir is also circulated through the regenerator (13), where it is concentrated by evaporating water from the solution. Concentrated solution is returned from the regenerator to the contactor.

The solution is circulated by a pump (18) through a line (19), an indirect-heat exchanger (20), and a line (21) to the regenerator. The concentrated solution is then returned by a pump (22) through a line (23), the indirect-heat exchanger (20), and a line (24) back into a reservoir in the contactor. The glycol solution is also circulated by a pump (25) from the regenerator through a line (26) to a heat-exchange coil (27) positioned within the tank (11) where the solution is heated, and thence through a line (28) to spray nozzles (29) within the regenerator. Preheated air enters the regenerator at the upper left, travels downward with the heated glycol solution sprayed from the nozzles, past a baffle (31), and then upward through the blower (30) to be discharged from the system along with water evaporated from the heated glycol solution. The regenerator can be controlled conveniently by utilizing a bypass (32) under the control of a valve (33) to maintain the temperature of the glycol solution, leaving the nozzles at a predetermined control level. Hygroscopic glycol solutions are available, for example, for which a predetermined controlled temperature on the order of 120 to 130 degrees F. is sufficient to evaporate the water from the solution.

The glycol solution is also circulated from the reservoir in the contactor by a pump (34) through a three-way valve (35) and a line (36) to a coil (37) of an evaporative cooler or cooling tower (38). The cooling tower transfers to a heat sink—the atmosphere—heat of sorption from the dehumidification process. ("Heat of sorption" may be defined as a change from latent heat to sensible heat, including the latent heat of condensation of water vapor and any heat of solution resulting from the mixing of the water removed from the dehumidified air with the glycol solution, or other hygroscopic material.) The cooled glycol solution from the coil flows through a line (39) and is sprayed from nozzles (40) within the contactor. Fresh air is drawn by a blower (41), is dehumidified and, usually, cooled sensibly, by contact with the glycol solution sprayed from the nozzles (40). The contactor can be controlled by using a bypass (42), under the control of a valve (43) to maintain a predetermined dry bulb temperature at the inlet to the blower (41).

The contactor requires a cooled glycol solution, while the regenerator requires a heated solution. The warm concentrated solution pumped from the regenerator to the contactor is therefore passed through the heat exchanger, where some of the unwanted heat is transferred to the cool dilute solution being pumped from the contactor to the regenerator. Thus, the heat exchanger increases the efficiency of the dehumidification apparatus.

Conditioned air leaves the contactor in a duct (44), where it is mixed with return air in a duct (45) of a conventional air-distribution system. The mixture of dehumidified fresh air and return air flows from the duct (44) through an indirect heat exchanger (46) and into a duct (47), from which it is delivered to the air-distribution system. In the indirect heat exchanger (46), the mixture of dehumidified fresh air and return air is cooled by contact with an indirect-heat-exchange coil (48) through which chilled water from a conventional source (not illustrated) is circulated as required to maintain a desired dry bulb temperature in the duct (47).

There are other alternative sources of energy, too, and we should touch on one of them here—the wind. Windmills used to be employed in many ways, including the grinding of grain, the operation of pumps and other machinery, and even the generation of electricity. The old-fashioned windmills were relatively inefficient, however, and when centrally generated electric power became widely available, they became for the most part obsolete. Recent advances in aerodynamics may change this. In 1977, Thomas Sweeney and Walter Nixon of New Jersey patented a windmill blade that is twistable when subjected to high winds. It is thus capable of reducing the undesirable effects of wind drag without using a complicated and expensive rotor hub. Nor is it necessary to vary the blade pitch angle at the root of the blade. The windmill blade illustrated in Fig. 11.22 is semi-rigid, and the degree of twist is readily controllable. It can generate a dependable supply of electrical power for use by small installations such as the family farm.

There is no doubt that the ancient science devoted to the study of solar energy has undergone a spectacular renaissance in recent years. Indeed, the movement is growing. New ideas and inventions are announced almost every week. The changes are fast and dra-

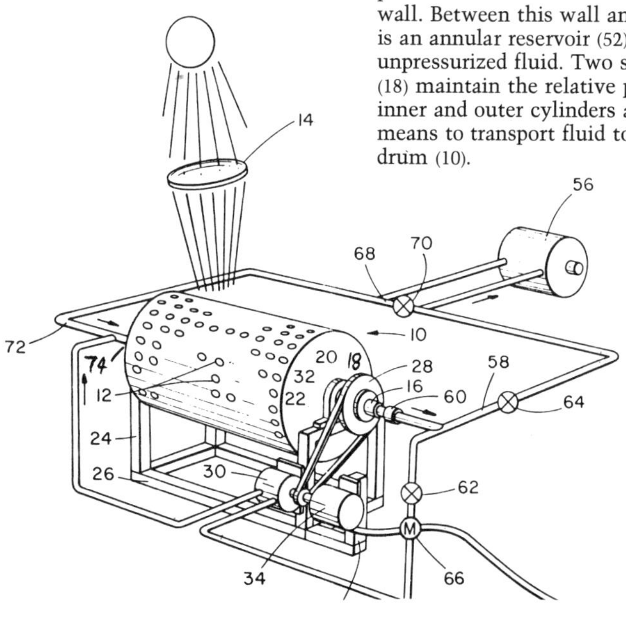

figure 11.21 Apparatus for converting solar energy into mechanical energy, as described in Patent No. 3,989,417 (1976). Cylindrical drum (10) has a number of thermally actuated pumps (12) on its periphery. The rays of the sun are concentrated on these pumps by one or more lenses (14).

Tubular shafts (16) and (18), extending from the drum along its longitudinal axis, are journaled in bearings (20) mounted on upper support (22), which are connected by vertical legs (24) to a base (26). Sheave (28) is fixed on shaft 16, enabling the drum to be rotated on its longitudinal axis at a constant angular velocity by means of a hydraulic motor (30) and drive belt (32). The drive shaft of the motor is connected to an auxiliary electric motor (34), mounted on a vertical unit (36) connected to an extension of base (26); this enables the drum to be rotated during the initial minutes of operation.

The drum's core is an accumulator for a pressurized fluid surrounded by an inner wall. Between this wall and the outer wall is an annular reservoir (52) for unpressurized fluid. Two shafts (16) and (18) maintain the relative positions of the inner and outer cylinders and serve as a means to transport fluid to and from drum (10).

Pressurized fluid from the accumulator is transferred through shaft (16) to motor (30) and device (56), which can be any type of hydraulic device such as a hydraulic motor or hydraulic cylinder, by inlet lines (58). These lines have pressure-actuated valves (62) and (64) which enable operation to be started or stopped manually at any time. But if the pressure in the accumulator is below a minimum value, as, for instance, during the first minutes of operation, the valves will not allow fluid to flow to the motor or the device. The inlet line to the motor has a metering device (66) associated with valve 62 such that, if the pressure in the accumulator is low, the absence of fluid flow will cause electric motor (34) to become operational.

If the device (56) is not operated for a period of time, the pressure in the accumulator will increase as long as the drum continues to rotate. To prevent possible damage to the pump units (12) by excessively high pressures, relief line (68) is connected to the inlet and outlet lines of device (56). This line has a pressure-actuated valve (70), similar to valves (62) and (64), such that excessive pressure in the accumulator will cause valve (70) to open, permitting fluid to pass through the relief line, bypassing device (56), thus preventing the pressure within the accumulator from increasing beyond a maximum value. Low-pressure fluid returns to reservoir (52) for reuse through the outlet lines (72), connected to hydraulic connector (74) which is threaded onto one end of tubular shaft (18). Hydraulic connectors (60) and (74) are constructed so as to permit drum (10), shafts (16) and (18), and hydraulic connectors (60) and (74) to be rotated as a unit, at the same time maintaining sealing engagement with inlet lines (58) and outlet lines (72), which remain stationary.

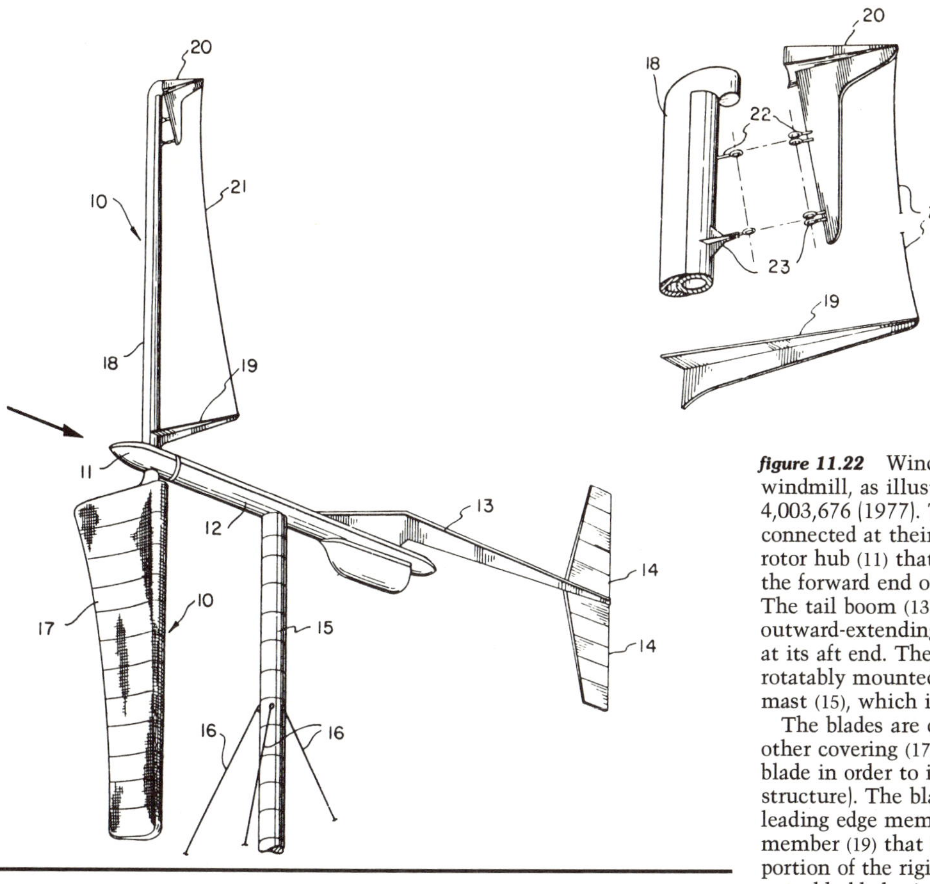

figure 11.22 Windmill blades and windmill, as illustrated in Patent No. 4,003,676 (1977). The blades (10) are connected at their root ends to a blade or rotor hub (11) that is rotatably mounted on the forward end of a windmill body (12). The tail boom (13) has two outward-extending stabilizing tail fins (14) at its aft end. The windmill body is in turn rotatably mounted on the upper end of the mast (15), which is supported by wires (16).

The blades are covered by a sailcloth or other covering (17) (omitted from the upper blade in order to illustrate the internal structure). The blade has a rigid elongated leading edge member (18), a rigid root end member (19) that is connected to the lower portion of the rigid leading edge, and a movable blade tip (20) that is connected to the outer end portion of the rigid leading edge.

The blade also has a flexible elongated trailing edge (a cable) (21) whose ends are connected to the outer end of the movable blade tip and the outer end of the root end. Hinges (22) and (23) connect the blade tip to the upper portion of the rigid leading edge.

The pitch angle of the end (19) can be set to give optimum performance for given wind conditions and power-output requirements by fixing the location of the blade root with respect to the rotor hub (11) to which it is connected.

In high winds, the cable (21) can be curved to the right side of the center line of the blade root end and hence away from the plane of rotation for the blades. In moderate winds, the center line of the blade tip is substantially in line with the center line of the root end, when viewed from the end of the blade, and the blade is in an untwisted state. In winds of comparatively low velocity, the blade tip is rotated toward the oncoming wind and the cable (21) is curved toward the oncoming wind.

matic. Although many thousands of people are working in the field, so far the big breakthrough has not come. Perhaps it never will—in those terms. It may be that the state of the art will simply advance steadily over the next years, making step-by-step progress that will fall into place only when solar energy occupies a significant, perhaps the preeminent, place among the sources of power available to us.

No one knows what the future holds. For the present, however, the field right now is one in which no large corporate or governmental entity holds a commanding position. Thus there is room for the person with a useful invention and the energy and enthusiasm and persistence necessary to get it before the public. Several of the inventions described in this chapter have been marketed by their inventors with a satisfactory degree of success. The mail-order route has proven to be a good one for many entrepreneurs in this field. The overhead is fairly low, you can do just about everything yourself, and you can reach a particular audience all over the country by using mailing lists and by advertising in special-interest magazines. As solar energy gains in popularity, more and more retail outlets are getting interested, too.

If you do not want to market your own invention, you can try to interest a corporation in a licensing agreement or some kind of similar arrangement (see Chapter 2). Many different kinds of companies are coming into the field with varying degrees of interest. Many of them are looking for new products. You can get an idea which ones are in the field by consulting the advertisements in the trade magazines.

CHAPTER TWELVE

The Automobile

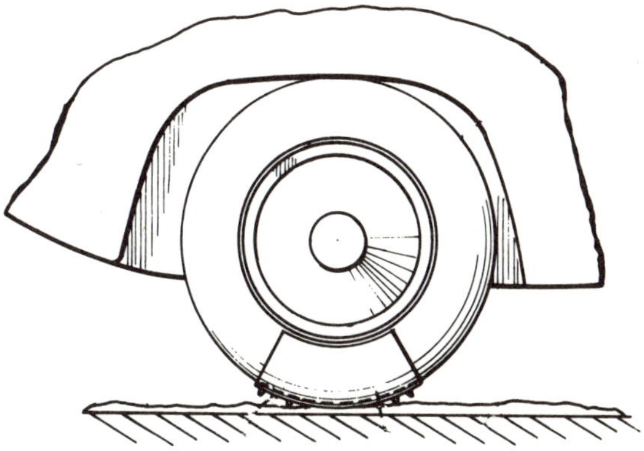

Inventors and designers have been tinkering with the automobile ever since the mid-Eighteenth century, when the Frenchman Nicolas Cugnot first built and ran his steampowered carriage. Progress was not rapid. Not until the 1880s was the first practical internal combustion engine developed, and manufacturing did not really get going until the turn of the century. According to the figures compiled by the American Automobile Association, about 4,200 cars were produced in the United States in 1900. The number went up to 550,000 by 1914, to 4.5 million in 1929 (before the Crash), back down to 1.1 million in the depths of 1932, up to 5.1 million in 1949, to a high of 11.4 million in 1973. That figure will almost surely be surpassed in the 1980s.

Patents have played a most important role in the development of the auto industry and its many satellites and offshoots. In the early days, especially, there were some bitterly fought infringement cases. The powerful Association of Licensed Automobile Manufacturers sued Henry Ford for infringement of the landmark Selden patent of 1895, which had attempted to encompass all horseless carriages (see Fig. 12.1). The case was a honeypot for lawyers, with litigation stretching out over almost eight years. The final decision found Ford not guilty.

This case and some others did not deter the workers and dreamers along America's gasoline alleys from sending their inventions to Washington for protection. There were probably as many outlandish accessories developed as there were attempts to change one part or another of the basic engine and structure. The motorcar

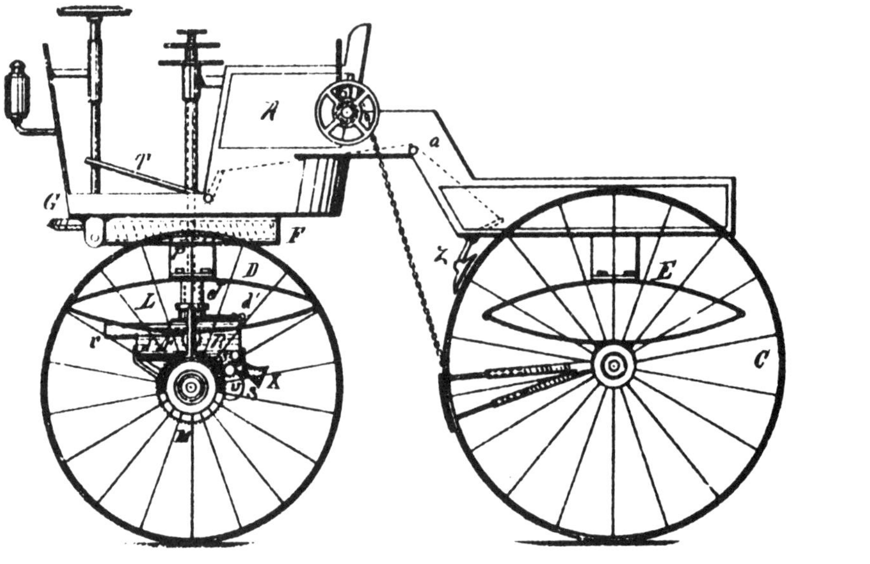

figure 12.1 Side view of George Selden's famous Improved Road Engine, Patent No. 549,160 (1895).

(*AA*) is the body, (*BB*) the driving wheels, and (*C*) the trailing wheels. (*D*) and (*E*) are, respectively, the springs of the two pairs of wheels. (*F*) is the fifth wheel, to which the springs (*D*) are connected. (*G*) is the steering device, and (*H*) the handwheel for operating it. (*Z*) is a brake, which is controlled by a cord running forward to the foot lever, (*T*).

The liquid-hydrocarbon gas engine (*L*) may be connected with either the steering or trailing wheels. The most convenient way of arranging the cylinders of the liquid-hydrocarbon engine is transversely to the driving shaft. The gas engine is provided with a power shaft carrying a pinion (*N*), meshing into the gear (*M*), fixed to the driving axle. By attaching the cylinders, which may be of any convenient number, to the air reservoir (*O*), which is a shell closed at each end and arranged parallel to the driving shaft, a compact arrangement is obtained. If the driving shaft passes through the air reservoir, it should pass through a tube secured to the heads at each end in order to avoid stuffing boxes. A portion of the upper side of the air reservoir is removed and the cylinders of the gas engine are cast in one piece with a curved flange, which is riveted over the top of the air reservoir. On their upper sides, the springs (*D*) are connected with a frame (*P*), which is fastened to the worm gear constituting the fifth wheel.

(*X*) is the exhaust pipe through which the products of combustion are discharged underneath the body.

The cam shaft (*S*) is driven by the gear (M) on the axle of the driving wheels. The gears are boxed in to exclude dust. A clutch (*Y*) may be interposed between the engine and the gearing (*M-N*) to disconnect the two, in which case the camshaft (*S*) should be driven positively from the engine shaft.

The traction wheels (*B*) are attached to the axle by clutches (*j j'*), splined on the driving shaft and held in mesh by springs in order to permit the wheels to rotate independently of each other to facilitate progress over rough roads and the turning of corners. These clutches may also be used for the purpose of disconnecting the engine from the traction wheels. This is accomplished by connecting the clutches with the rotating plate (*m*) by means of the rods (*i i'*). This plate is rotated from one of the hand wheels (*I*) by the flexible connection (*J*).

was a great new toy, and people made the most of it. It's just about impossible to get statistics on this sort of thing, but surely automobile-related patents comprise one of the largest continuously active categories in the patents on file in Washington. In fact, auto patents are found in many of the more than 400 main classes there. The Big Three and the other automakers, including the foreign companies, have large research departments working on all kinds of improvements—from something as small as a new type of hood latch to a revolutionary innovation like a new kind of engine. The major changes, naturally, have come slowly. Automatic transmission, for instance, was patented in various forms during the 1930s, but it did not come into widespread use until after World War II.

We're long past the stage when it was possible for an enterprising fellow to go out and start his own auto company. The market is dominated by the giants, but there is certainly still room for the independent operator. There'd better be. All over the United States, there are garages where the lights often stay on far into the night, where people with an idea are working to prove it out. The love affair between Americans and their cars has gone beyond the ga-ga stage, but divorce is far in the future. Billions of dollars are spent each year on automotive products, and there's plenty of room in that pool for anyone, big or small, who's got a better idea.

In general, of course, the little guy can't take on big projects. It's a paradox. Precisely the people who have the unfettered, independent outlook needed to come up with revolutionary changes in the family car can't buy the chips that would get them into the game. So we will probably have to be content with the evolutionary changes that come out of the big corporate research facilities. There may be exceptions. One of the most basic ideas of recent times, the Wankel rotary piston engine, came out of the brain of one man, Gerhard Wankel. This German inventor had a difficult time convincing his fellow countrymen to take him seriously. Perhaps the culmination of his career came when General Motors signed a licensing agreement that gave them the right to manufacture the engine. Unfortunately, there are currently many doubts about the viability of this engine, especially as fuel conservation becomes a more important issue. Whatever the future holds, Wankel's contribution to automotive research is substantial. He pursued an idea that had been rejected or downgraded by most of the experts in the field. He persisted, and eventually proved his point.

Another important consideration for automobiles of the 1970s and 1980s is safety. Prodded by consumer groups and strong-armed by the federal and state governments, the auto industry has been looking for ways to reduce the number of deaths and injuries caused by car crashes. Even after discoveries are made, it can be a long time before improvements are accepted by the public. For instance, the safety belt had been known for many years before Ford gave it and some other safety features a big push with its 1956 models. The public was more interested in fins and horsepower at the time, however, and safety equipment became standard only as required by government regulation.

A great deal remains to be done, and there are many opportunities. One of America's greatest inventors, George Westinghouse, made his reputation with the safety devices he developed for railroading, including the air brake and signal. Before he came along, rail travel was risky. Many travelers and more employees lost their lives in those early days. Now, with road deaths and injuries a major public health problem, the time is ripe for some giant strides to be made in automobile safety.

There are problems, of course. For one, even some of the safety features need safety features. Jesse Hollins of New York discerned that many persons fear they will not be able to release their safety belts after an accident and thus will be trapped in the vehicle. His invention, illustrated in Fig. 12.2, automatically unhooks the belt whenever the engine stalls—which would certainly be the case in almost all serious accidents. Naturally, there are times the engine would stall, too, mostly in very slow-moving traffic, when belt protection is not particularly needed.

Another safety device in which people seem to have less than full confidence is the air bag. Insurance companies and others have conducted research that has proved the efficacy of these bags. If every auto were equipped with devices that automatically and instantaneously inflated upon any sharp impact, they say, lives would be saved and injuries prevented because a vehicle's occupants would be provided with a cushion against the crash. The automakers have gone ahead with the development of the air bag and related devices, but fear that their customers will be unhappy at the extra cost and be wary of the effects of accidental inflation. In fact, many people feel exactly this way. They are afraid of something new, even though the old ways mean over 50,000 deaths per year. Publicity is a manipulating factor here. If there were 100 aircraft crashes (two per week!) in each of which 500 persons were killed, investigating committees would proliferate like ants at a picnic, and the cries of outrage would be heard from coast to coast. In 1977, the government did finally impose a deadline early in the 1980s for the installation of air bags in all new cars. As with air pollution standards imposed in the 70s, this will probably be subject to negotiation and delay. Again, the public's attitude is the important factor. Most people remember the law that required new

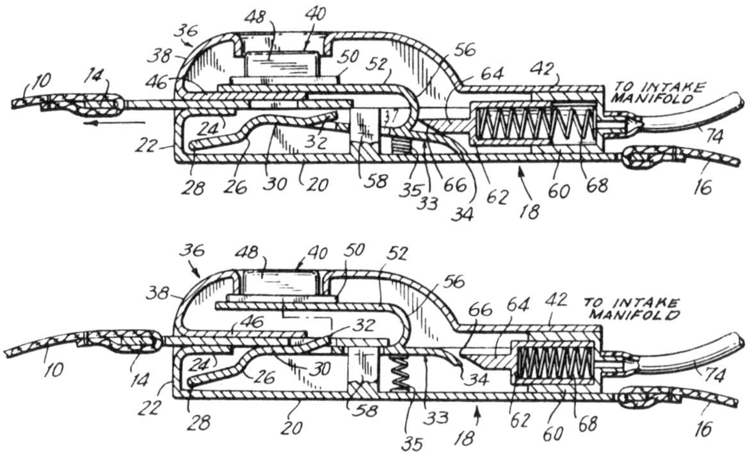

figure 12.2 Two views of a seat belt unlatching system, as shown in Patent No. 3,963,090 (1976). At top is a cross section where the tongue is secured by the latch. At bottom is a cross section where the tongue is freed from the latch. A seat belt strap (10) has two ends, one attached to an auto chassis and the other holding a tongue. Another seat belt strap (16) also has two ends, one (not shown) attached to the chassis and the other holding a tongue-latching mechanism (18).

More specifically, seat belt latching mechanism (18) includes a flat plate (20) that is secured to the end of seat belt strap shown in the drawings. Located at the forward end of the plate (the end farthest from seat belt strap) is a vertical section (22), which is the same width as the plate and perpendicular to it. Extending from the upper portion of the vertical section toward the portion of the plate secured to the seat belt strap, while being parallel to the plate, is a guide lip (24).

A rigid, metallic latch strip (26) includes a forward end (28) that is pivotally connected to the plate so that the strip can pivot relative to the plate. The strip includes an upper, substantially flat section (30), which for the orientation shown in bottom figure is substantially parallel to plate (20). Projecting from section (30) is a latch (32). There is also a rear section (33) and a slightly depending, downwardly curved tail (34). A spring (35) rests on plate (20) and urges flat section (30) to the position shown in top figure.

A push-button housing (36) is provided including a frame (38) and a well (40). The rear of the frame includes a cylinder-securing section (42) that is parallel to the plate at all times. Extending inward from the forward position of the frame is a guide (46), which is spaced from and parallel to lip (24). Located in central opening (40) is a button (48) that rests on a button plate (50) affixed to button support (52). A curved section (56) extends from the button support to rear section (33).

A bumper (58) extends through an opening (37) in latch strip (26). The top of the bumper prevents the tip of latching tongue (14) from dipping downward.

Affixed to the portion of plate (20) adjacent to the end of seat belt strap (16) (shown in top figure) and to cylinder-securing section (42) is a cylinder (60). Located in this cylinder is a piston (62), which is adapted to slideably move within the cylinder. The piston is cylindrically shaped and includes a closed end; projecting outward therefrom is a shaft (64). The free end of the shaft includes a cam surface (66) for a reason that will soon be readily apparent.

A compression spring (68) is located within cylinder (60) and biases piston (62) in a leftward direction. An external force acting on piston spring causes the piston to be in a position such that cam surface (66) abuts tail (34), causing depression of the tail and latch strip (26) bottom. A conduit (74) connects the right end of cylinder (60) to the engine intake manifold.

When the motor vehicle engine is not running, there will be no vacuum created by the engine intake manifold. Consequently, spring (66) will move piston (62) to its left position (bottom figure) so that cam surface (66) will depress tail (34) and strip (26), overcoming the bias of spring (35). As a result, latch (32) will be depressed, and it will be impossible to latch seat belt latching tongue (16). When the engine is started, the vacuum created by the engine intake manifold will overcome the bias of spring (66) maintaining piston (62) in the position shown at top. Spring (35) will maintain latch (32) in the position shown at top. Seat belt latching tongue (14) is inserted between guide lip (24) and guide (46) and instantaneously slightly depresses latch (32) against the bias of spring (35) until the latch is received within the opening in latching tongue (14). Strip (30) slides downward along bumper (58) when the latch is momentarily depressed.

If the motor vehicle engine should stall, as occurs due to an accident, there will no longer be a vacuum created by the intake manifold. Consequently, spring (66) will expand, moving piston (62) to the position shown at bottom with cam surface (66) depressing tail (34). Latch strip (26) will be depressed, overcoming the bias of spring (35). As a result, latch (32) will no longer be received in latching tongue (14)—thus the latching tongue (14) will be free of the latching mechanism, thereby freeing the seat belt and the occupant.

If it is desired to release the latching tongue from the latching mechanism while the motor vehicle engine is operating, push button (48) is manually depressed, thus depressing latch strip (26) against the bias of spring (35) so that the latch (32) is no longer received in the opening in the latching tongue, which can be removed from the latching mechanism.

cars to be equipped with a buzzer that activated when driver and passengers did not buckle up their seat belts. It wasn't long before ingenious mechanics figured out how to disconnect this gadget, and most Americans went back to their beltless driving habits. The buzzer regulations went the way of Prohibition.

What about the state of the technology in the air bag field? Figures 12.3*a* and 12.3*b* illustrate two recent patents, both assigned to General Motors by the inventors, Charles Hay and Robert K. Brown, Jr., respectively, both of Ohio. There are basically three types of inflator systems for air bags—the pure gas inflator, which involves a storage unit to hold pressurized gas; the gas generator, where a propellant can be ignited to create the gas; and a hybrid, or augmented type, which involves the use of both a generator and stored gas. Both of the inventions illustrated in Fig. 12.3 are of the latter type.

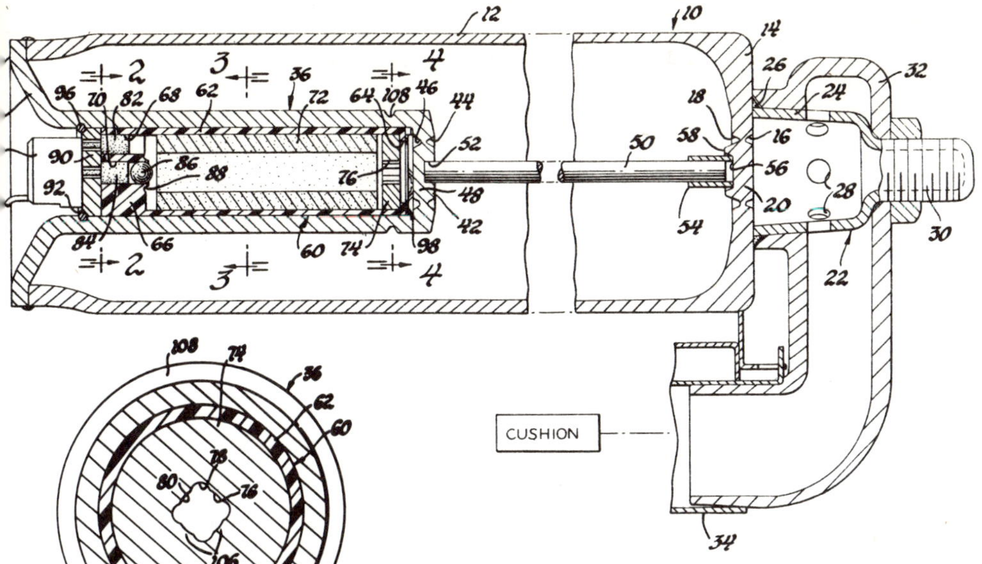

figure 12.3(a) Longitudinal section view of an inflator for an air bag, or occupant-restraint system as it's called in Patent No. 3,968,980 (1976). The inflator (10) includes a cylindrical pressure vessel (12) of metal. An integral end wall (14) of the pressure vessel includes coaxial, opposed, generally V-shaped cross-section grooves (16) and (18), which define an integral, severable plug or diaphragm (20) of wall (14). An outlet (22) of tapered configuration has its base wall (24) welded at (26) to wall (14) around the diaphragm (20). The outlet includes a number of spaced radial openings (28) and an integral, externally threaded cylindrical lug (30). The outlet is received within and bolted to a manifold (32), which in turn communicates with a diffuser (34) of a conventional occupant-restraint cushion. When the diaphragm is severed, the stored gas within the vessel (12), as well as the resultant gases from the gas generator, flow into the outlet (22) and thence to the cushion, after passing through the openings, the manifold, and the diffuser.

A cylindrical combustion chamber (36) of metal includes an outwardly flaring skirt or end wall (38), which terminates in a peripheral flange (40). This abuts against, and is welded to, the open end of the pressure vessel (12) to close it. The pressure vessel is filled with any suitable pressure fluid or gas—in this case with argon at 2400 psi.

An end wall (42) of the combustion chamber includes a pair of opposed V-shaped cross-section grooves (44) and (46), which define a severable plug or diaphragm (48) that is coaxial with the diaphragm (20) of the pressure vessel. One end of rod 50 is received within a bore (52) of the diaphragm (48); its other end is frictionally fitted within a sleeve (54). This sleeve is received within a recess (56) integrally formed within the diaphragm (20) by providing a rib (58) on the inner surface thereof.

Received within the combustion chamber is a cartridge assembly (60), which includes a cylindrical housing (62) of plastic. One end of the housing includes a partial end flange which defines a cylindrical opening (64). The other end of the housing is closed by an integral end wall (66) of substantial thickness, which is provided with a pair of cylindrical bores (68) and (70), the latter being coaxial with the housing.

Contained within the housing is a hollow cylindrical mass (72) of a suitable propellant which produces high-temperature, high-pressure gases when burned.

A metal orifice plate (74) seats against the end flange of housing (62). This plate includes a central outlet or orifice (76), which is defined by flutes (78) that open to a central circular aperture (80). Although four flutes are shown, other numbers of flutes can be used with equal success. The orifice (76) is of predetermined cross-sectional area and is coaxial with the bore (70) and the diaphragm (48).

Contained within the bore (68) is a conventional electrically fired squib or igniter (82), and contained within the bore (70) is a conventional electrically fired squib (84). A ballistic element or ball (86) of metal is received within the bore slightly ahead of the squib. The ball may be a conventional ball bearing and may either be frictionally fitted within the bore or may be retained therein by a flange (88) heat deformed out of the material of wall (66) after the ball has been first inserted in the bore. The ball (86) is coaxial with the bore and the orifice (76).

A seal plate (90) seats against the wall (66) within the combustion chamber and is provided with a beveled edge or wall (92). The seal plate may be integral with or separately formed and secured to an electrical plug (94). The leads from the plug are connected to the squibs (82) and (84). A split ring (96) of circular cross section is received within a similar cross-section groove of the combustion chamber (36) and engages the wall (92) to retain the cartridge assembly (60) against movement outwardly of the combustion chamber. The cartridge assembly is seated against the split ring by a spring.

Although the connection is not shown in the drawing, it will be understood that the electrical plug (94) is connected to a known sensing and actuating system. Such a system may, for example, include velocity-responsive sensors mounted on the vehicle bumper and acceleration- or inertia-responsive sensors mounted on a part of the vehicle's occupant compartment, such as the fire wall. These sensors are set so as to sense impact of the vehicle with an obstacle and to sense the level of that impact. The inflator (10) is shown in unactuated position, and it will be understood that the inflatable occupant-restraint cushion will be located in a folded position and mounted in a conventional manner within the occupant compartment, such as within the instrument panel. The recess (56) and the rib (58) cooperate with the sleeve (54) in always locating the rod (50) coaxial of diaphragms (20) and (48), despite possible vibrational movement of the rod during vehicle operation.

Assume now that the sensing and actuating system senses low-level impact conditions due to impact of the vehicle with an obstacle. The squib (82) will be fired or ignited and ignite the mass (72) of propellant. As the propellant burns, it will produce high-temperature, high-pressure gas that flows through the orifice into the space between the plate (74) and wall (42). As the pressure within the combustion chamber (36) increases and reaches a predetermined level, the diaphragm (48) will be severed from the combustion chamber along the grooves (44) and (46) to provide a combustion chamber outlet. When the diaphragm (48) is severed, the rod (50) will be instantaneously shifted to the right within the sleeve (54) and in turn impact the diaphragm (20) to sever this diaphragm from the wall (14) and provide an inflator outlet. Thus both diaphragms (20) and (48) are severed to provide outlets from the combustion chamber to the pressure vessel and from the pressure vessel to the outlet (22) and in turn to the occupant-restraint cushion. The resultant gas from the combustion chamber will mix with the stored gas in the pressure vessel, and then both will flow through the openings (28), the manifold (32), and the diffuser (34) to inflate the cushion.

Assume that the sensing and actuating system either initially senses high-level impact conditions or senses such conditions subsequent to sensing low-level impact conditions. When this occurs, the squib (84) will be ignited and will explosively fire the ball (86) out of the bore (70) and through the combustion chamber (36) into engagement with the aperture (80) of orifice plate (74). The ball deforms as it engages the ribs (106) so as to be tightly wedged against these ribs. The OD of the ball is larger than the diameter of the aperture. The ball thus reduces the area of the orifice (76) so that the resultant gas now can only flow through the flutes (78). When the orifice area is reduced, the pressure within the combustion chamber (36) will increase. This in turn increases the rate of burn of the mass (72) of the propellant, so that the resultant gas is now at a much higher pressure and will flow at a much higher rate through the flutes into the pressure vessel (12) and thence to the cushion.

Thus the inflator of this invention utilizes a single charge or mass of propellant and provides for inflation of the occupant-restraint cushion at a relatively low rate under low-level impact conditions and at a relatively high rate under high-level impact conditions by controlling the rate of burn of the propellant material.

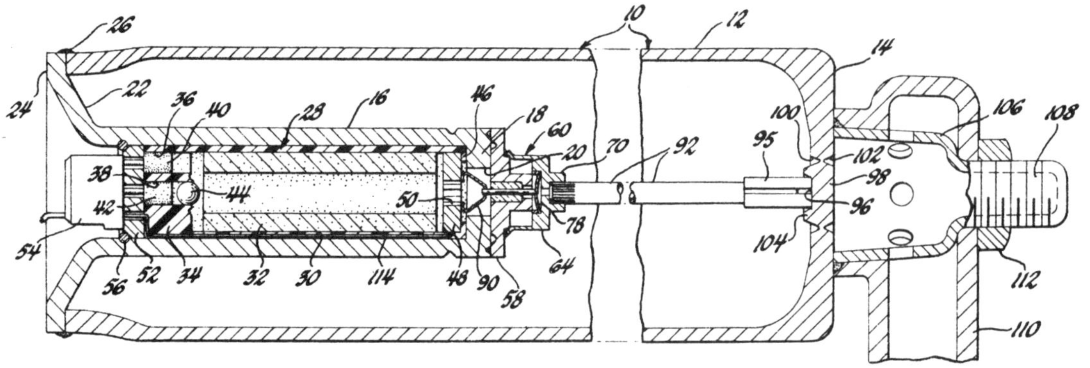

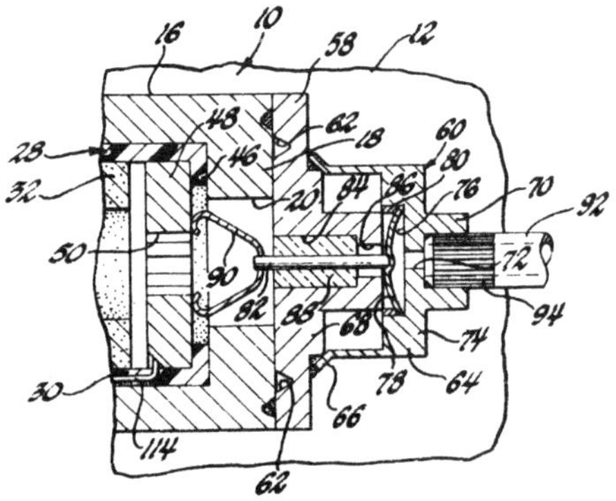

figure 12.3(b) Sectional view of the inflator for the air bag described in Patent No. 3,877,721 (1975) and (inset) an enlarged view of the critical part of the inflator, which includes a cylindrical pressure vessel (12) with an integral wall (14) closing one end thereof. A cylindrical combustion chamber (16) fits within the pressure vessel and includes an inner wall (18) provided with a cylindrical outlet (20) coaxial with the combustion chamber. The other end of the combustion chamber includes an outwardly flaring (conical) wall (22) which terminates in a radial flange (24) that seats against the open end of the pressure vessel and is welded thereto at (26) in a conventional manner in order to close the pressure vessel. The pressure vessel is filled with any suitable gas under pressure, such as argon at 2400 psi.

The combustion chamber (16) includes a cartridge assembly (28), which includes a cylindrical plastic housing (30) that fits within the combustion chamber and contains a cylindrical charge (32) of propellant for generating high-pressure, high-temperature gases when the charge is initiated. The outer end wall (34) of the housing (30) is substantially thick and is provided with a pair of bores (36) and (38). Bore 36 contains an igniter or initiating element (40), and bore 38 likewise contains a squib (42). Retained within the bore 38, to the right of the squib, is a conventional ball (44).

The inner end of the housing includes a partial radial flange (46), which seats and retains a metal closure plate (48), having a central fluted orifice (50). The outlet (20), the orifice (50), and the ball (44) are coaxial. A sealed plate (52) is secured to the wall (34) and mounts an electrical plug (54). This plug is connected to the igniter and the squib by electrical wires extending through the sealed plate. The cartridge assembly is retained within the combustion chamber by a circular split ring (56) engaged in a semicircular groove in the combustion chamber wall and also engaged by the chamfered outer edge of the seal plate.

The plug (54) connects the igniter and squib with a known multilevel actuation system for a vehicle occupant-restraint system. Such actuation systems can include a velocity-responsive sensor and an inertia-responsive sensor. Should the velocity-responsive sensor or the inertia-sensor sense low-level impact conditions, the igniter (40) will be connected with the vehicle battery or other source of power to thereby initiate the charge (32) of propellant to generate high-temperature, high-pressure gases. Should the impact conditions sensed be of a higher level, then the squib (42) is connected to the same source of power to fire the ball (44) into the orifice (50). This restricts the orifice area and increases the pressure within the combustion chamber to increase the burn rate.

A seal plate (58) of a pressure sensor (60) fits against the wall (18) of the combustion chamber and is welded thereto adjacent the outer periphery thereof and radially outwardly of a continuous-V cross-section groove (62) thereof. A cup-shaped housing (64) has a flared flange (66) thereof welded to a boss (68) of the plate (58). A circular boss (70) of housing (64) opens to an inlet aperture (72) in the base wall (74) of housing (64). The aperture opens to a bore (76) in the wall.

A dish-shaped, pressure-responsive diaphragm (78) fits within the bore and is retained therein by a press-fitted ring (80). A switch contact (82) is received within bores (84) and (86) of the plate (58) and is glass welded at (88) to the bore (84). The outer end of the switch contact engages and seats against the diaphragm (78). A U-shaped spring finger contact (90) is secured to the inner end of the switch contact (82) and slidably bears on the plate (48).

A hollow tubular impact member (92) has one roughened or lightly splined end (94) thereof force-fitted within the boss (70) in order to secure the impact member to the pressure sensor (60). The other end of the member (92) is frictionally fitted with a split sleeve (95), which seats within a counterbore (96) of a severable or rupturable seal (98) formed integrally with wall (14) of the pressure vessel. The seal (98) is defined by opposing V-shaped circular grooves (100) and (102) formed in the wall (14), with the material from the groove (100) providing the rib (104), which defines the counterbore (96).

A hollow, tapered outlet (106) is welded or otherwise secured to the wall (14) around the seal (98) and includes an integal threaded extension (108). A manifold (110) is received over the outlet member (106) and secured to extension (108) by a nut (112). Although not shown herein, the manifold (110) communicates with a conventional diffuser of an occupant-restraint system, and the diffuser in turn communicates with a conventional inflatable occupant-restraint cushion.

After the plate (58) has been assembled to the housing (64), the space therebetween to the left of the diaphragm (78) is filled with a suitable gas at a predetermined reference pressure, such as 1800 psi. The pressure sensor can then be assembled to the combustion chamber by welding the plate (58) to the wall (18) to close the outlet (20) of the combustion chamber.

A wire (114) extends from the plate (48) through a groove in the housing (30) and through wall (34) and plate (52) to the plug (54). From this plug the wire is connected into an external indicator circuit, which gives a signal when the pressure of the fluid within the pressure vessel (12) drops below the reference pressure, such as 1800 psi, within the pressure sensor, and the diaphragm (78) moves out of contact with the switch contact (82).

The fluid within the pressure vessel is admitted to the bore (76) to the right of the diaphragm (78) by passing through the slot of sleeve (95), through the bore of the tubular member (92), and then through the inlet aperture (72).

When the high-pressure gases generated by initiation of the charge (32) of propellant reach a predetermined level, the plate (58) will rupture at the groove (62) and the gases will then blow or move the pressure sensor (60) to the right to open the outlet (20) and permit these gases to flow into the pressure vessel. Simultaneously, the sensor (60) will shift the rod (92) to the right and the right end of the rod will move within the split sleeve (95) and impact seal (98) to remove the seal and permit the combined generated gases and pressure-vessel gas to flow to the occupant-restraint cushion and thereby inflate the cushion.

It will be noted from the foregoing description that both the pressure sensor and the switch contact (90) are easily and effectively removed to open the outlet (20) when the generated gases within the combustion chamber reach a predetermined level.

If at any time the pressure within the pressure vessel (12) drops below the reference pressure level, 1800 psi, the diaphragm (78) will move to the right to actuate the indicator circuit and indicate to the vehicle operator that the pressure within the pressure vessel has dropped below such minimum.

One of the good features of an air bag is that it can be inflated at a relatively low rate under low-level impact conditions and at a relatively high rate in a heavy impact. This can be accomplished under any of the three systems of inflation noted above.

As a study of these two patents and others in the field indicates, the technical state of air bag development has reached a fairly high level. Yet the controversy over air bag use will continue, and this presents inventors with the chance to come up with a better solution—or at least one that will be more acceptable to the general public. Only a few hardheads would dispute the need for some changes in automotive engineering to cut down on deaths and injuries. It's easier to change a vehicle's design and equipment than to change the driving habits of more than 100 million people. The two important elements that cannot be ignored are cost and public confidence. The ideal invention would be a simple one, cheap to manufacture and install, and easy to understand and maintain. Can it be found? Nothing's ever perfect, of course, but that doesn't prevent people from trying. What is most likely is that an evolutionary change, not a revolutionary one, will take place here, as is the pattern in many other areas.

There are several different approachcs to the problem of safety. One is to forget about occupant-restraint systems and to concentrate on improving the "package," the vehicle itself. Many automotive designers have worked in this area; in fact, many structural changes have been made in recent years by the big companies. An independent inventor in Florida, Cheston Lee Eshelman, has been attacking the problem of vehicle design for some time. He has secured a dozen or so patents thus far, and is still working at finding a device that would be suitable for mass production and sale. His approach is simple and straightforward: "The modern bumper," he says, "is an ornamental feature rather than what its name implies."—and would be of some use only in straight-on collisions. One of his suggested improved models is shown in Fig. 12.4. At first glance, his invention seems obvious, perhaps even ineffective. But is it? So far, Eshelman's ideas have not met with much commercial success. Nevertheless, he insists, "Tests have shown that a car equipped as described [in the patent] can drive head on into a concrete wall at 40 miles per hour and not even break the headlights."

Safety considerations, when you think about it, are responsible for a good many components of the modern automobile—windshield wipers, defrosters, headlights, directional signals, disc brakes, shatterproof glass, even the tailpipes that keep the engine's poisonous vapors from seeping into the passenger area. There are other places, too, where safety could become more important. In recent years, engineers have begun to pay more attention to the rear window, which often gets fogged up in rainy weather and packed with snow in the winter. Various devices have been patented that would improve a driver's visibility to the rear, and more are coming into use.

Another contributing factor to overall safety is the comfort of the driver. Figure 12.5 illustrates a design patent for a ventilated

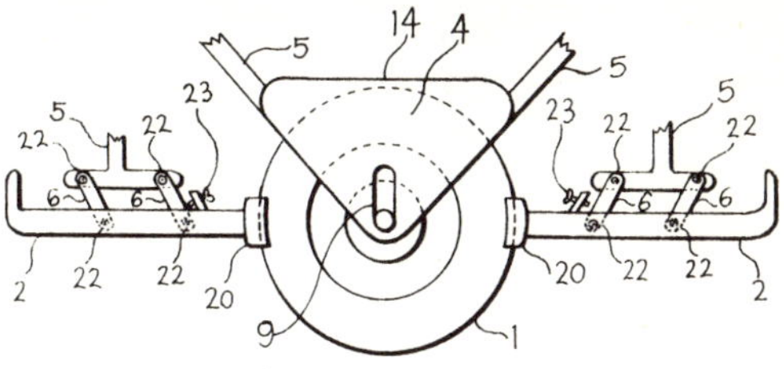

figure 12.4 Top view of a pneumatic bumper system, as illustrated in Patent No. 3,845,977 (1974). The spare tire can be utilized as an energy absorber in collisions since it is mounted in front of the car about standard bumper height. It is held in position by a pressed steel housing (4), a king pin, and a slotted hole (9). Tire and wheel (1) fit inside the housing. Bumper bars (2) are held in place by fulcrum-hinged, mounting arms (6) or links attached to frame (5), which extends from the frame (3) to the bumper bars. These arms rotate on pivot bolts (22) and are set on an angle so as to force the bumper bars inward upon impact with a barrier (19). The small foot (20) on the inside ends of the bumper bars compresses into the tire and thus moves inward, and this tends to dissipate the forces of energy into the tire. If the collision is directly in front of the car, the front profile of the tire will absorb the crash and then push it in the slotted hole (9). The profile of the tire flattens out against the back stop (14) as it travels aft in the slot (9).

If the collision is severe enough, it will make contact with the bumper bars, and they will also move aft and inward together. They are mounted on fulcrum-hinged mounting arms (6) or links that rotate on pivot bolts (22), which thrust the bumper bars into the tire. This squeezes the tire from all four sides on a hard frontal impact or on each side on a corner impact. An adjustment retainer (23) prevents the bars moving outward and also adjusts them for clearance from the tire to install a cover.

The tire and wheel can be removed by hand without tools, for the purpose of using it for a spare. A lock at the top of the king pin and/or retainer cable prevents theft or loss.

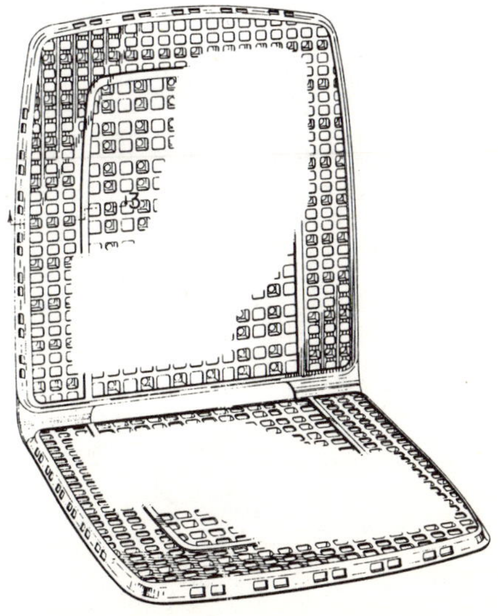

figure 12.5 Ventilated air seat, Design Patent No. 233,868 (1974).

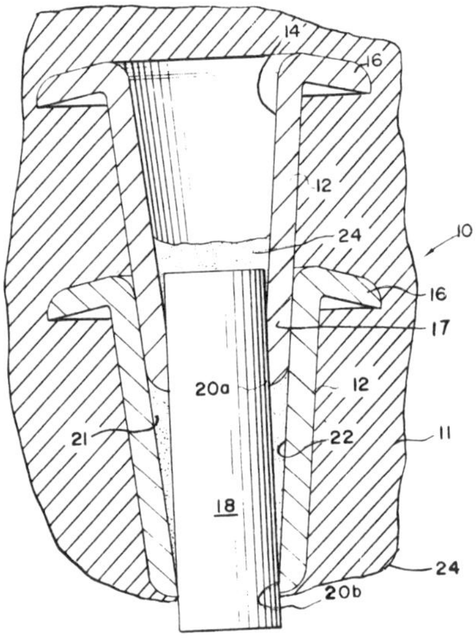

figure 12.6(a) Cross section of a stud inserted in a tire, as illustrated in Patent No. 3,396,772 (1968).

The stud (10) comprises body means defined by a plurality of like flanged tubular body members (12), which are made from hard metal and include a bore (14) extending entirely through the body member. At one end of each body member (12) there is an annular, outwardly extending flange portion (16). The flange or head is preferably disklike and circular in cross section, although other configurations may be used if desired. The bores are tapered back from the flanged end.

The walls of each body member are substantially uniform in cross section, and therefore the exterior surfaces taper generally in the same manner as the walls defining the bore (14). The body members are sized so that the end (17) of one body member remote from the flanged end (16) is adapted to be received in the bore of the other body member adjacent to the flanged end.

A wear-resistant, rod-like insert (18), fabricated from a hardened material such as cemented tungsten carbide, is secured within the bores extending through the body portions. The wear-resistant insert is cylindrical and is of an external diameter slightly larger than the internal diameter in the portion of the bore in each body portion remote from the flanged end. Thus, there is a forcefit between each body portion and the wear-resistant insert in the areas indicated generally at (20*a*) and (20*b*).

In addition, a bonding agent (21) such as a low-melting-point silver solder or copper or a thermo-curing adhesive is permitted to flow into the space (22) between the insert (18) and the body (12). The bonding material will also be deposited in the area (24) above the insert within the bore to further enhance the bonding.

A cylindrical insert is generally less expensive than a tapered insert. The insert is securely locked in place within the bore extending through the body by the bonding agent in the areas indicated at (22) and (24) and also by the interference fit of the insert with the body in the areas (20*a*)) and (20*b*). In addition, there is a collet effect between the end (17) of the upper tapered body and the insert for further locking the insert in place. The walls of the outer body adjacent to the lower end (17) of the upper body reinforce the walls of the upper body next to the area of joinder to the insert, and the walls of the end (17) are compressed between the insert and the upper end of the lower body (12) so as to define a wedge for more securely retaining the insert in place in the body.

The end of the insert (18) extends beyond the end of the lower body portion a predetermined amount. When the stud is in place in the tread of the tire, its tip extends beyond the surface (24) of the tread and engages the road surface. The stud is formed with two flanges (16) for stabilizing it in the bore within the tire tread and for preventing it from being moved inward, toward the fabric of the tire.

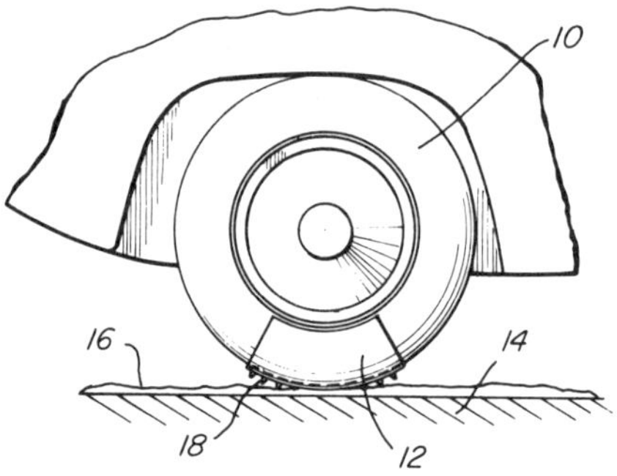

figure 12.6(b) An aid to traction attached to an automobile tire, as illustrated in Patent No. 4,036,272 (1977). The tire (10) is on a surface (14) that has a cover of ice or snow (16). A series of studs extends from the device (12) to provide additional gripping as the tire rotates. The "Sure Grip" can be constructed from a discarded tire and should be slightly larger than the vehicle tire. To make this device, the studs are pierced through the tread, and then a layer of rubber or synthetic rubber is capped onto the tire remnant.

seat. This seat keeps the driver's back on a cushion of air, rather than on the sometimes-sticky upholstery.

Tires are another important element contributing to the overall safety (or dangerousness) of the automobile. Most of the work in this area has been carried on at corporate research centers and has been quite productive. As tire reliability and durability continue to improve, many new cars will not even be equipped with spares. Independent inventors have come up with many new elements for tires, too. In 1968, Raymond McCarroll of Michigan patented a tire stud that gives added traction to snow tires (Fig. 12.6*a*). Poy Lee of Massachusetts has invented a simple gadget that incorporates tire studs. His Sure Grip Winter Tire Traction Device can be carried in a car, ready to be attached to any tire that gets mired in snow or mud (Fig. 12.6*b*).

Even when the car is parked and locked, safety is an important concern—the worry here is not about accidents, of course, but about theft. In many areas of the country, car stealing has become almost as common as illegal parking. All kinds of new products have been developed to combat this problem, but the perfect protection has yet to be found. The automakers themselves have brought in new systems, including the locking steering wheel and transmission. Jonas Gorman of New York took a new approach to the problem and invented a housing for the steering wheel (Fig. 12.7). This is sufficiently complicated to remove so that the typical thief, not being particularly bright or resourceful, will simply move on to some easier target. A persistent punk would of course tow the car away, but that is a good deal riskier than getting in and driving off.

There are all kinds of auto thieves—joy-riding kids, specialists

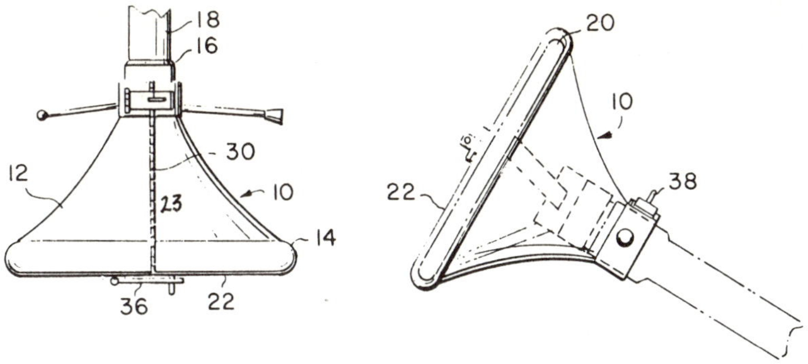

figure 12.7 Plan view and side view of the automobile antitheft device described in Patent No. 3,982,602 (1976).

The housing (10) is formed in two halves (12) and (14), having a separation along a line that is within a plane extending through the axis of the housing. At the axial extremity of the truncated cone that is smallest is a collar (16), configured to fit snugly with the steering column (18). The truncated conical housing is placed about the steering wheel (20) of an automobile. Carried across the larger axial extremity of the truncated cone is a generally planar top (22). Along the separation line (23) between the halves of the truncated conical housing is a hinge (30). Carried along the mating surface of the truncated conical housing are hasps (36) and (38) for selectively securing the housing about the steering wheel.

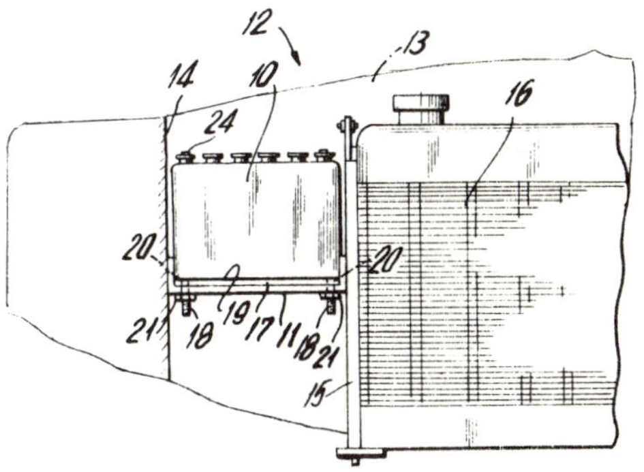

figure 12.8 Partially cutaway side view of a plate holding a battery in place, as a deterrent to theft. The illustration is from Patent No. 3,821,998 (1974). The battery housing (10) is secured to a mount (11) so that the battery cannot be removed from a motor compartment (13) within a vehicle (12). This usually occupies the space between the mudsplash guard (14) at the rear of a fender and the support bar (15) of a radiator (16). A rigid holding plate (17) can have threaded studs (18) molded into, embedded within, or welded or swedged to the bottom of the holding plate from which the studs extend down. The battery housing can be bonded to the holding plate by applying a satisfactory adhesive (19) to either the underside of the battery housing or the top surface of the holding plate. (The adhesive may be any suitable epoxy adhesive or other satisfactory bonding means which will prevent the battery housing from being easily separated from the holding plate.) Battery cradle (11) has holes which will align with the studs extending from the bottom of the holding plate. Nuts (21) can then be screwed onto the end of each of the studs until the nuts are tight against the underside of the battery cradle. Since the underside of the battery cradle is accessible only from beneath the vehicle and the vehicle frame is low slung, battery housing can be secured to and removed from the vehicle frame only by, in effect, lifting the vehicle off the ground so that the nuts can be placed onto or removed from studs from the underside of the vehicle.

The top surface of holding plate can have the same substantial area dimension as the underside of battery housing. Similarly, the outer edges of the holding plate can extend vertically upward from the top surface of the holding plate so as to surround the lower side portions of the battery housing and thus prevent the wedging of a sharp instrument between the battery housing and the holding plate.

in stealing expensive models to order, cannibalizers, and others. Many thieves don't want the whole car, just a part of it, like the battery. In cars not equipped with good hood locks, the battery is quite vulnerable. Figure 12.8 shows a plate that locks a battery in place. It's simple enough that it can be installed (and removed) easily, but only if the car is lifted off the ground. This maneuver would be too risky and time-consuming for the ordinary thief.

The war between thieves and their victims offers inventors the chance to prove that they can outsmart the bad guys and help the regular guys by developing new and more effective antitheft devices. See Chapter 6, Security Equipment, for additional information.

Gadgets are almost as popular in the car as they are in the home. They generally come and go according to the public's fancy, but the market is always there for the right thing at the right time. One example of a specialty item that proved more durable than most is the toll coin gun, patented in 1965. The inventor, John Thew of Connecticut, saw that many motorists have a difficult time at the exact-change toll booths. They fumble around, come up with a nickel instead of a quarter, finally get the right change, and then flip it so that it lands on the roadway, just short of the bucket. With

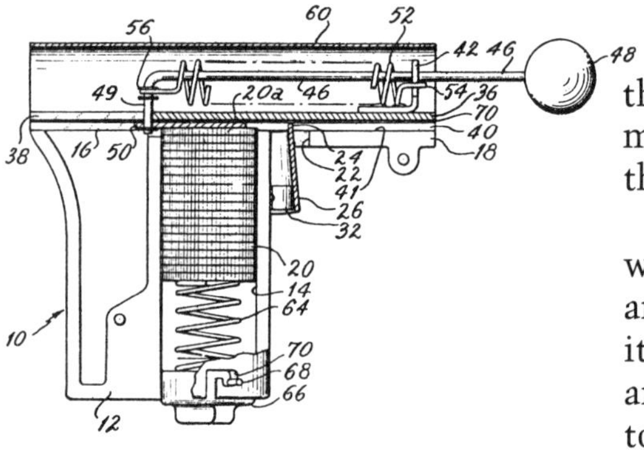

figure 12.9 Cross section of the coin gun illustrated in Patent No. 3,191,588 (1965).

The grip (10) is formed with a hollow, vertically extending coin-holding cylinder (14) at its forward edge, half of which is supplied by each of the grip halves (12). The grip is formed at its top edge, to the rear of the cylinder, with flat, outwardly extending flanges (16), whose surface is normal to the axis of the cylinder. The grip is formed forward of the coin cylinder, with a flat platform (18) extending laterally to each side of the cylinder and at a level below the flanges (16), a distance equal approximately to the thickness of a coin (20) or slightly greater.

The platform is provided, immediately in front of the cylinder, with an opening (22) through which projects a reduced end portion (24) of a wider, preferably arcuate trigger or coin release member (26), which may be formed with lateral extensions or ears (28) by which it is pivoted on the outer edges of the platform. The pivots are so arranged that when the trigger (26) is pressed toward the cylinder, the end thereof, projecting through the opening, is retracted and lowered to a position where it is below the upper surface of the platform. When the lowered end is moved away from the cylinder, the upper end of the trigger extension projects above the upper surface of the platform. Spring means, preferably in the form of a suitably curved leaf-spring (32) secured in suitable position on the coin cylinder, are provided for normally maintaining the trigger (26) away from the coin cylinder and the end of its upper projection above the surface of the platform; the plate (36) and platform form, respectively, the top and bottom walls of the discharge nozzle (41).

this gun, however, one shot and it's all over (Fig. 12.9). The gun mechanism is a bit complicated, and the product is not cheap, so there may be room for something new here.

No one likes a dirty job, even a little one such as checking the oil with a dipstick. William Denver of Illinois found it cumbersome and messy to remove the stick, wipe it clean, dip it, then withdraw it again to check the oil level. His invention, shown in Fig. 12.10, is an attachment for the conventional dipstick that makes it possible to check the oil in one clean swipe.

There are many other gadgets around, some good and some bad. You can get a pretty good idea of what's currently popular by snooping around your local auto-accessory store or department. Then think about what's needed and lacking in your own car. For instance, consider cold engines in winter—is there a way to heat them to a minimum temperature at a minimum cost? How about a radio antenna that can't be snapped off? Is there a simple way of changing a flat tire? Or an easy method of diagnosing an engine's problems? Is there any role for a small computer in automobile operation? (Computers are becoming ever more widely available for use by individuals. Their size and price are shrinking almost monthly. There are some predictions that mini- and microcomputers will be as widely used in the 1980s as calculators are today.) The questions noted above are only a few of those that are being asked (and answered) by inventors all over the country.

If you added up all the man- and boy-hours that have been put into customizing cars, you'd come pretty close to infinity. It's the great American indoor/outdoor all-the-year-round sport. And while most customizers take pride in doing things their way, as a group they are quite receptive to new designs and inventions that can help them in their pursuit of the perfect set of wheels. Four men from Michigan—Nadosi, Nadosi, Zelenkom, and Pero—developed a new construction and process for an ornamental car top (Fig. 12.11). Complete with opera windows, these vinyl units can be predesigned to fit the various basic body types made by U.S. auto manufacturers. Foreign cars can be customized, too, of course. Even the Volkswagen bug, one of the most popular cars of all time, seems to be—well, a little small. Many kits have been made that included a molded fiberglass body ready to be mounted on the chassis of a stripped-down VW. Conrad Weiffenbach of New York found these units generally complicated and expensive to install, so that only a comparatively few people can afford them. His invention (Fig. 12.12) extends the frame of a bug about two feet to accommodate a "long front hood type car body . . . without the necessity of relocating the driver's seat, etc." With this inexpensive addition, the customizer can install or have installed one of the VW body kits without having to go to the expense of having the car's body extended and adjusted by hand.

One of the most useful items for customizers and all those who work on cars in their home garages is the portable wheel ramp. Most models don't cost very much, and they lift the car just enough so that you can get down and under without flattening yourself out like some kind of worm. This ramp has generally been

figure 12.10 Cross section of the dipstick described in Patent No. 3,591,886 (1971). It comprises a unitary, tubular, pliable molding including a first tubular portion (22*a*) of an inner diameter approximately equal to the outer diameter of the pipe (14), a cup portion (24*a*) of a substantially larger inner diameter, and an internal shoulder (26*a*) intermediate those portions.

The first tubular portion (22*a*) is adapted to be telescoped over the upper end of the guide pipe (14), minor variations in the diameter of the pipe being accommodated by the resiliency or pliability of the material, and major variations being satisfied by models of different sizes, such as small, medium, and large. To secure the attachment in place, and to ensure sealed engagement between the portion (22*a*) and the pipe, a hose clamp (28*a*) is provided around the portion (22*a*).

The cup portion (24*a*) is adapted to extend above the upper end of the pipe, and the shoulder (26*a*) is intended to be located at substantially the level of the top of the pipe.

In use, the dipstick (16) is inserted into the pipe in the conventional manner, with the cap (18) of the stock closing the upper end of the pipe. By virtue of its large internal diameter, the cup portion

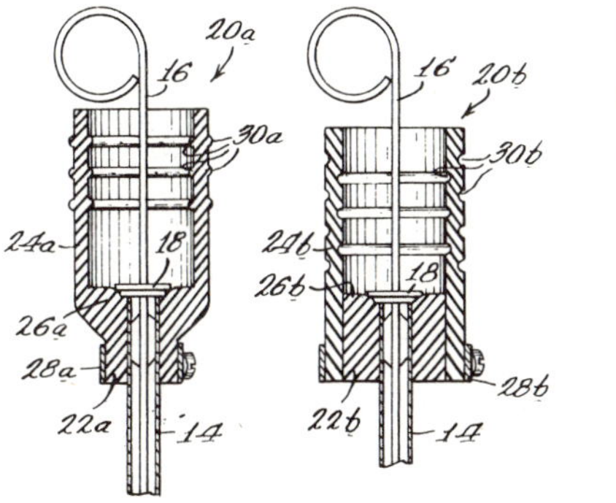

(24*a*-24*b*) of the attachment accommodates the conventional dipstick, including its cap, without modification. The operator grasps the finger grip of the dipstick in one hand, raises the stick slightly, and with the thumb and forefinger of the other hand squeezes inwardly on the resilient walls of the cup (24*a*-24*b*). The resilient material encompasses the dipstick, wiping it clean. The liquid wiped from the stick is captured and retained by the cup (24*a*-24*b*) and returned by the shoulder (26*a*-26*b*) directly to the crank case or reservoir through the interior of the pipe.

figure 12.11 Ornamental car top, as shown in Patent No. 3,926,471 (1976). The unit includes magnetic holding strips incorporated within the preformed vinyl shell for the ornamental car top. The various layers of insulation and ornamental cover are all fabricated into a unitary assembly prior to installation on a car. In the installation, because of the holding strength afforded on the interior part of the shell by the magnetic holding strips, only two anchoring bolts are required, and only two actual openings need be made in the car body itself. Additional holding force is provided by the structure, whereby one or more opera-type windows are formed with crimping edges at either side of the car top. These edges are designed to fit over the standard car windows and assist in holding the car top in place.

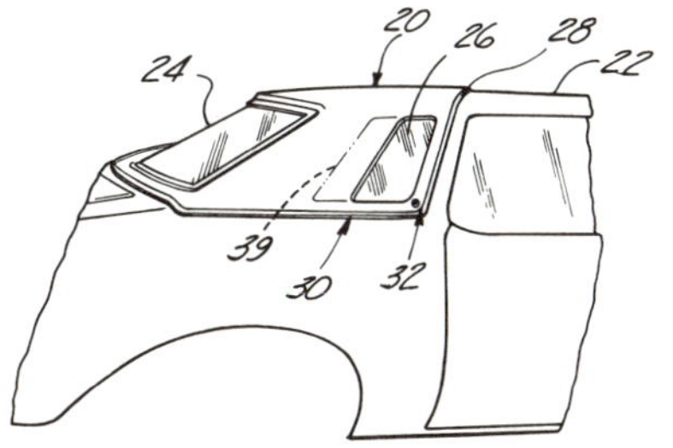

figure 12.12 Perspective view of an extension for the frame of a Volkswagen "bug" as illustrated in Patent No. 3,942,817 (1976). The extension includes a bracket or weldment (16) fixed to the front end of a backbone or tubular structure (18). The bracket is formed with upper and lower pairs of aligned, elongated concave mounting surfaces (22) and (24), which are sized and spaced to receive upper and lower axle tubes (26) and (28), respectively of front axle assembly (14). In a conventional VW construction, the assembly is clamped to the bracket by four bolts (30), which extend through bore openings (32) formed in a pair of vertical upstanding axle tube joining posts (34) and into threaded openings (36) formed in the bracket.

Extension unit (10) generally comprises (1) upper and lower rear attaching devices, which are preferably in the form of tubes (42) and (44), serving to define rearward-facing, convex attaching surfaces, sized and arranged to be received within bracket-mounting surfaces (22) and (24); (2) upper and lower front attachment devices, which are preferably in the form of pairs of aligned attachment channels or "split" tubes (46) and (48), serving to define forward-facing, concave front attachment surfaces, sized and arranged to receive upper and lower axle tubes (26) and (28), respectively; and (3) a pair of vertically disposed and horizontally spaced cantilever or extension plates (50), which serve to rigidly interconnect tubes (42) and (44) and channels (46) and (48) in a horizontally spaced, essentially parallel relationship.

Tubes (42) and (44) are best maintained in a vertically spaced, parallel relationship relative to each other by plates (50), which are welded to their opposite ends, and by a pair of parallel rear-attaching posts (54), which are vertically disposed and have their rear surfaces welded to the forward-facing surfaces of the attaching tubes. Attaching posts (54) are formed with through openings (56), which receive bolts (60) adapted to be fitted into bracket openings (36) for the purpose of clamping tubes (42) and (44) within and against casting mounting surfaces (22) and (24), respectively. Tubes (42) and (44) and attaching posts (54) may be considered as cooperating to define an attaching or first assembly.

Channels (46) and (48) are maintained in a vertically spaced, parallel relationship relative to one another by having their outer ends welded to a pair of vertically upstanding, generally U-shaped front attachment posts (62), which are in turn welded to facing surfaces of plates (50). Further, channels (46) and (48) are preferably welded adjacent to their inner ends to opposite ends of upper and lower angle-iron braces (66) and (68), respectively. Alternately, the braces may be omitted and each pair of channels may be replaced by a single channel or "split" tube having its opposite ends welded to front-attachment posts (62) or directly to facing surfaces of plates (50). Channels (46) and (48), attachment posts (62), and braces (66) and (68) may be considered as cooperating to define an attachment or second assembly.

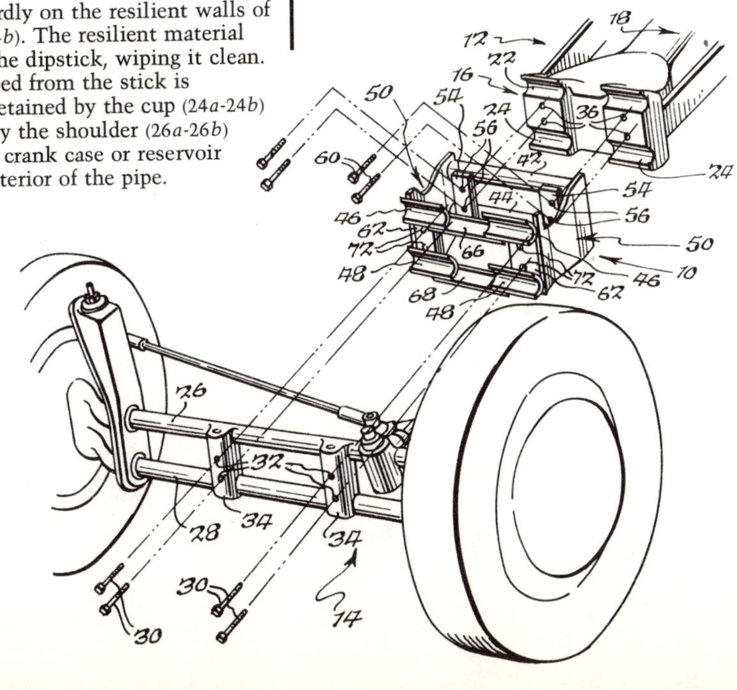

manufactured in two types: One comes with all the parts welded in place—rugged, but not easily shipped or handled. The other type is packed with the components detached, all ready to be screwed together—easy to ship, but often not as strong as you might like. An invention by Ronald Grewe and Eugene Hoersten of Ohio for the Huffman Company shows how even the common wheel ramp can be improved if you analyze its disadvantages and work to correct them as economically as possible (Fig. 12.13).

One of the most significant changes to affect American industry in recent years has been pollution control. This is not a fad. It has become a big business that will probably become even bigger. People are now worried about the purity of the air they breathe, the water they drink, and so on. Automobiles are acknowledged to be a major factor in the creation of air pollution. In recent years, the car makers and the federal government have engaged in a running battle as to how much can be done—and how soon—to reduce auto emissions. The rules keep changing, and the fight goes on.

The problem is basically one of efficiency. What the engine doesn't burn—its waste—is pumped out the exhaust pipe. The more waste product, the more pollution.

The air-injection system is one of the best ways discovered to re-

figure 12.13 A perspective view of a wheel ramp, Patent No. 3,863,895 (1975). The ramp (10) includes a pair of rigid, longitudinally extending side frames (12) extending in spaced parallel relationship to each other. Each of the side frames includes an upright front section (14), a substantially horizontal upper section (16), and an inclined section (18) extending rearward and downward from the rear end of the upper section (16).

Preferably, the sections (14), (16), and (18) are formed of a structural steel member of substantially rectangular tubular cross-sectional configuration. A lower section (20), which may be an angle member, extends from the lower end of the front section (14) to the rearward end of the inclined section (18) and is attached to each by welding. Additionally, a strut (22) of rectangular tubular construction extends from the lower section (20) upward to the inclined section (18) and is joined thereto by welding.

A pair of feet (24) extend transversely of the two side frames (12) and are attached to the bottom sections (20) by means of self-tapping screws (26). Each of the feet (24) extends transversely outwardly beyond the side frames and is provided with a pair of skid-resistant rubber pads (28) on its outer ends and a hole (29) for storage of the ramp on a hook or nail. It will also be noted that the lower rear end of the inclined section (18) is flattened slightly, as at (30), to provide a pair of feet at this point.

A plurality of separate treads (32) are attached to the upper surfaces of the upper and inclined sections (18) by means of self-tapping screws (34). The spacing between the side frames (12) is such that they are positioned inward of the outer ends of the treads (32). Thus, a wheel supported by the ramp imposes very little bending moments on its treads, and the major portion of the load is transmitted directly to the upper and inclined sections (16) and (18), depending upon where the wheel is positioned on the ramp.

In this regard, it will be noted that by utilizing a tubular construction for the load-supporting sections (16) and (18), a substantially lighter member can be used than would be the case if the sections (16) and (18) were angular in shape. The outer ends of the treads are turned upward to help center the wheel on the ramp and prevent it from running over the edge.

At its forward end the ramp is provided with a cross-bracing plate (40) which is bolted to the uprights (14) by a series of four bolts (42) extending through the plate and the forward and rear walls of the uprights. This provides a rigid cross brace that resists collapsing forces applied to the ramp.

Formed integrally with the plate is a stop member (43) which extends up over the front end of the ramp and has a wheel-engaging portion (44) positioned rearward of the front of the ramp. Self-tapping screws (46) extend through a flange (48) of the stop member and the upper surface of the upper member (16).

A safety stop (50) is provided with an opening (52) through a flange (54), and each of the treads (32) is provided with a centrally located opening (56) therethrough. Thus, the safety stop may be attached to the ramp after a wheel is positioned on its upper section by means of the bolt (57) passing through the openings (52) and (56) and secured in place by a wingnut (58).

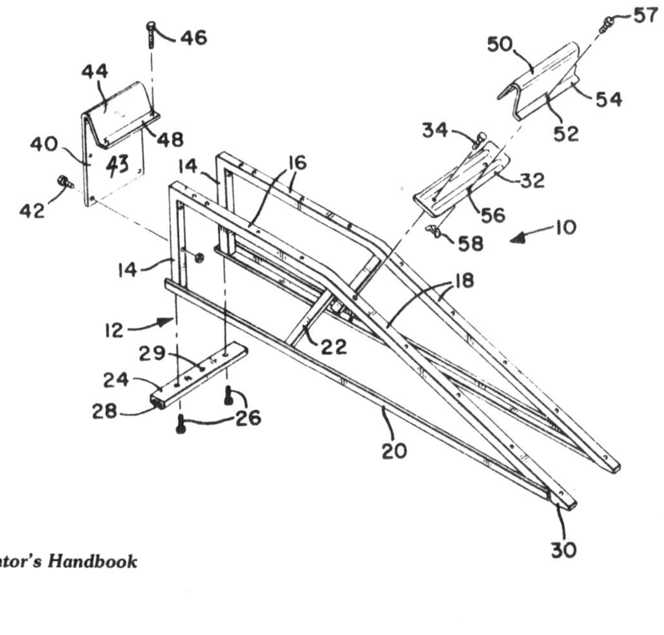

duce auto pollution. An engine-driven pump delivers air to the stream of hot exhaust gases as they are leaving the engine combustion chamber. Utilizing the heat of these exhaust gases, the injected air supports additional burning, thus reducing noxious emissions. Now, the air pump does not have to be used. A suitable tuned exhaust system can make for maximum air flow, induced by exhaust pressure pulsations. Patent No. 3,653,212, registered to Richard Gast and Harry Mitchell, describes an engine with a number of exhaust manifolds and pipes equal to half the number of combustion chambers. Each manifold has an air-induction unit with passages to direct the air from the atmosphere toward these chambers.

In a related development, Ernst Renft of New York and Kenneth Pfrengle of Georgia patented an air-induction valve, assigned to General Motors, which was adapted to prevent two events—engine backfire during deceleration and escape of exhaust gases from the air-induction unit. A number of check valves control the flow of air. These valves have thermostatic stops that close at the high temperatures that occur when the exhaust gases backflow (see Fig. 12.14).

An independent inventor from Michigan, William Bachmann, attacked the very heart of the pollution problem—the workings of the internal-combustion engine. According to his analysis, it is the design principles of present-day high-performance, relatively inexpensive engines that cause pollution. In his 1976 patent, Bachmann cites sixteen characteristics generally true of eight-cylinder, four-stroke engines that make it difficult to control emissions. They included the following:

1. High compression ratios
2. Dual or four-barrel carburetors
3. 300- to 400-cubic-inch breathing volume
4. Crank speed of 4,000-5,000 rpm
5. Intake and exhaust valves that can leak
6. Fuel-to-power conversion efficiency of less than 27 percent
7. Mechanical efficiency of not more than 85 percent

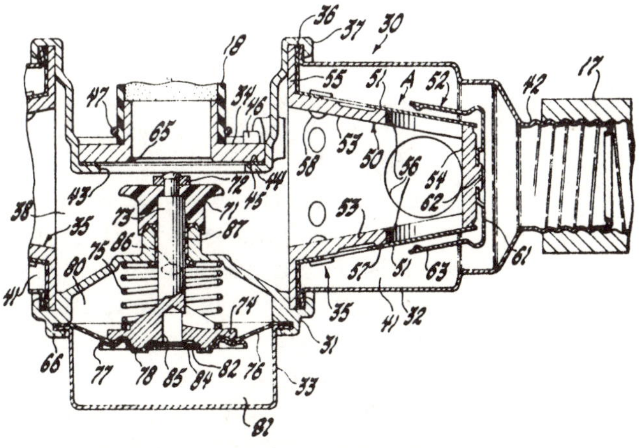

figure 12.14 Partially cutaway top view of an air-induction valve for an exhaust emission control system, Patent No. 3,751,915 (1973). In operation, air flows through air inlet hose (18), through flanged inlet duct (34) of the valve casing, past a normally open control valve (71), and into inlet chamber (38). From this chamber the air flows through reed-type check valves (51), which are adapted to open as the exhaust manifold pressure drops below atmospheric pressure.

In order to prevent backfire during engine deceleration, the high induction vacuum pulls the diaphragm up to close the control valve.

As illustrated, the air-induction valve (30) consists essentially of a multiple-piece housing having a central or main body (31), opposed discharge casings or reed valve covers (32), a bottom cover (33), and a flanged inlet duct (34). A pyramid reed valve assembly is mounted to each side of the main body. This valve assembly (35) is secured, together with a gasket (36), between flange (37) of the main body and the end flange of a discharge casing. Each discharge casing thus encloses a valve assembly, and, together with the main body, they form a central inlet chamber (38) and discharge chambers (41). These chambers are joined to the air manifold (15) by conduit couplings (17).

The inlet chamber has an aperture (43) in the upper flange (44) of the main body and by the flanged inlet duct (34), which is suitably secured to the upper flange (44) with a gasket (45) sandwiched between by means of positioning plugs (46). The flanged inlet duct is connected by hose (18) to the air cleaner (12), a wire hose clamp (47) being used to secure one end of the hose to the flanged inlet duct.

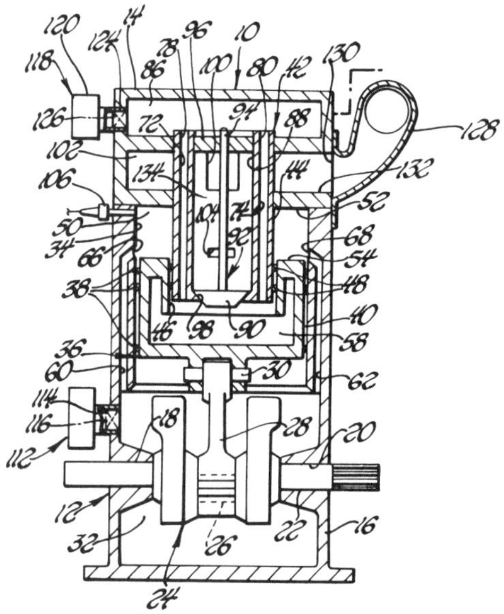

figure 12.15 Cross section of a new type of internal combustion engine, as illustrated in Patent No. 3,969,894 (1976). There is an engine housing with upper (14) and lower (16) sections joined together. The under part of section 16 has support bearings (18) and (20) that support a power output shaft (22). A crank has a rotatable throw (26) suitably journaled to one end of a connecting rod (28), which in turn has its upper end secured to a wrist pin (30).

The upper part of section 16 has a cylindrical wall (34) which receives a cuplike ring piston (36) for reciprocating movement. A cylinder (42) may be secured to section (14). It has a downward-depending portion with an outer cylindrical surface (44) that fits in piston (36). Inner piston rings (48) may be provided and carried by the piston to effect a seal between the outer surface and the inner surface (46).

Space (50), between surfaces (34) and (44) and between surface (52) and surface (54) of ring piston, is the combustion chamber.

Housing (42) also has passages (72), (74), and (76), which have their respective upper ends (78), (80), and (84) in communication with upper chambers (86) formed within upper engine housing section (14); their respective lower ends are in communication with variable chamber (58). This housing is also formed with an inner clearance passageway (88) to permit free movement of a head (90) of a valve comprising a stem (94) which may be within a closure (96). The lower part of member (42) is provided with a generally annular valve-seating surface (98), adapted to cooperate with valve head (90) to prevent communication between passageway (88) and chamber (58).

Radially directed ports (100) maintain communication between passageway (88) and a second chamber (102) formed within upper housing section (14). Other radially directed ports (104) are also formed through member (42) in order to provide for communication between conduit (88) and combustion chamber (50).

Angularly and equidistantly spaced plugs or nozzles (106), (108), and (110) are provided. The reason that these elements are referred to as "plugs" or "nozzles" is that they may be ignition spark plugs or igniters in an engine where such equipment is required.

Exhaust means (128) is provided and secured to the engine housing so as to have its inlet (130) in registry with the outlet orifice (132) of chamber (102). The purpose of conduit (128) is, of course, to convey the engine exhaust gases to any desired point away from the engine.

So Bachmann went ahead and developed his own internal-combustion engine (Fig. 12.15), best described by the abstract from his patent:

> Exhaust ports lead to an afterburner reactor situated medially of the annular cylinder; the exhaust gases are further burned within the after-burner section; to further reduce pollutants during the afterburning process, additional air is pumped into the afterburner section by the ring piston due to its special configuration enabling it to accomplish, for example, triple functions, such as: (a) to act as a power transmitter; (b) to act as a pump to supply the engine's combustion chamber with scavenging air and/or an air-fuel mixture; and (c) to act as an air pump to supply ambient air into the afterburner in timed intervals with a constant predetermined pressure and predetermined quantities in order to thereby maximize the degree of control of the temperature generated . . . and thereby more fully oxidize the unburned fuel residues.

Another shock to the American economy came from Energy Crisis I. Its effects are still rippling through all aspects of the country's life. Fuel-conservation devices are in great demand, though the public is still buying the big cars. The automakers are worried about the price and availability of fuel, however, and are trying to cut down the weight of their products so that they will use less gasoline. This trend will continue through the 1970s and into the '80s. There are other ways to save gas, too, and an independent inventor may have to lead the way. The big companies are reluctant to go along with any revolutionary change that would make their equipment obsolete.

Do you remember the fellow who was supposed to have found a cheap, readily available substitute for gasoline? His story has been told countless times in bars, gas stations, barbershops, and similar hangouts all over the country. The big oil and auto companies got wind of his discovery and paid him off to keep the secret to himself. Whatever happened to this durable folk hero in the 1970s, now that we really need him? The truth, of course, is that he was kidnapped by the Arabs, who gave him a small sheikdom on the Persian Gulf, where he and his harem have lived luxuriously ever since.

So far, despite the work and the hopes of inventors who have patented their fuel-conservation devices, no easy answers have been found. A good bet is a device that can be attached to a conventional engine. But that is easier said than done. The point is that if you can invent an accessory, something that is compatible with existing equipment, or if you can invent a substitute for an existing part—say, a new carburetor—you'll have an easier time selling it either directly to the public or to the automakers for attachment to their lines of cars.

Another accessory has come along to be just about the most popular new automotive item in recent years—the CB (citizens' band) radio. Within just a few years in the 1970s, aerials sprouted from millions of vehicles in both country and city. It's probably too late at this point for inventors to make any significant contributions to CB technology. However, there is still room for interesting accessories and byproducts such as the radio lock shown in Fig. 6.19 in the chapter on Security Devices.

The CB boom is interesting in retrospect because it shows how a product—whether a new one or something that has been around for awhile, perhaps in somewhat different form—can suddenly take off. For years, most people had been doing very well without CBs in their cars. Then, for some reason, millions suddenly decided that they needed one. CBs have their uses—to call help in an emergency, to communicate with the home base, to while away the hours on a long trip, and to catch warnings of Smokey Bear's radar setups—but none of these offer the real reason why most people bought CBs. When *A* buys a widget, and then *B* gets one, and then *C*, and they start talking about how nice their widgets are, *D* almost naturally follows along, as do *E* and *F* and many others, except *Q* and *Z*, who fancy themselves nonconformists. Can the average inventor capitalize in any way on this human characteristic? Perhaps. It takes a while for a boom to develop. If, for instance, you had been a CBer before the fad began, you might have sensed an increasing popularity. At that point, if you had anything to offer manufacturers, wholesalers, or retailers in the field, you would have had a better chance of getting a decent hearing because most of them were in an expansion phase, making more money, looking to a brighter future, and looking for new products. The only trouble with a boom like this is that you've got to act rather fast, or you'll be left behind.

Does the automobile have a good future, or are there so many problems that it may eventually be outmoded by another form of transportation? Rail enthusiasts have urged the federal government to invest billions of dollars to bring back good passenger service all over the country. To many travelers of a certain age who hope never again to have to take the Podunk local, these proposals to turn the clock back are not particularly welcome. What's needed, perhaps, are some fresh ideas. Inventors, as always, have

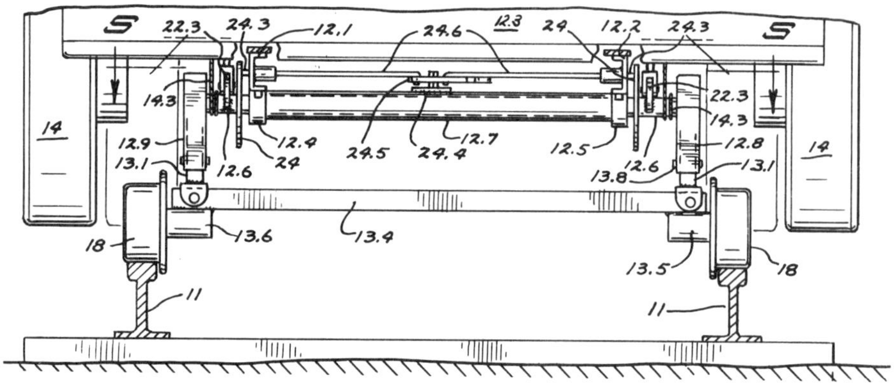

figure 12.16 Front cross section of the undersection of a convertible rail-highway vehicle, as its wheels rest on tracks, shown in Patent No. 3,980,025 (1976). There are front and rear sets of tires and of rail-guide wheels that are adjustable according to the desires of the operator.

Longitudinal beams (12.1) and (12.2) are extended beneath the front bumper (12.3) of the vehicle. To the front of the beams (12.1), (12.2) are attached bearings (12.4), (12.5). There is a rocker arm (12.6) within the bearings that extends to the front of the vehicle. If desired, a portion of the rocker arm between the bearings may be covered with a protective shield (12.7).

Support arms (12.8), (12.9) are attached to the respective ends of the rocker arm (12.6). Each support arm includes an outer tube having an open end within which is received a rod (13.1) in telescoping fashion, the other end of the tube being closed by a plate. The inner surface of the plate serves as a spring seat against which an internal helical spring bears at one end. The other end of the spring bears outwardly upon the confronting surface of the rod. At its outer end the rod of each of the support arms is welded or otherwise affixed to a transverse support bar (13.4), which is in turn being mounted to guide wheel axles (13.5) and (13.6), which mount the rail guide wheels (18). The transverse bar maintains the guide wheels the proper distance apart for contacting the rails (11). The outer wall of the tube of each of the support arms is slotted, and the inner rod (13.1) is provided with a pin which extends through the slot on either side to prevent the rod from separating from the tube. The springs (13.3) continuously urge the guide wheels (18) against the rails upon which they ride.

Means are provided to lock the front guide wheels in their rail-contacting, or retracted, positions. Rigidly mounted, as by welding, to each end of the rocker arm (12.6) are segment plates (24), which are parallel to the frame beams (12.1), (12.2). Holes are positioned in the segment plates to become aligned with holes formed in the frame beams as the rail guide wheels are moved into their rail-contacting position and as the guide wheels are fully retracted. Pins (24.3) pass through the aligned holes in the segment plates and vehicle frame to lock the rocker arm in position with the guide wheels retracted or in contact with the rails. Any forces tending to rotate the rocker arm when the rail guide wheels are in their rail-contacting positions are resisted by means of the locking pins (24.3), rather than by the hydraulic cylinders. To the surface of the rocker arm sheath (12.7), near the center of the latter, is welded a bracket (24.4). Pivotally attached to this bracket is a handle (24.5), and rods (24.6) are mounted to the handle fore and aft, respectively, of its pivotal mounting to the bracket, the rods extending transversely of the vehicle and ending in the pins (24.3), so that movement of the handle in a counterclockwise direction locks the pins (24.3) in the aligned holes of the frames and segment plates.

figure 12.17 Overall view of a suspension mass-transportation system, as illustrated in Patent No. 3,774,542 (1973). Shown are four tracks, consisting of hollow tubes that are supported by cables that extend along the right of way. These cables are supported by transverse bridges (12) [I-beams (28) and (29)] that in turn rest on towers positioned at suitable intervals. Rods (14) and collars (15) form the basis of the suspension system for the tracks, which are kept aligned by cross-bracing rods (16). Each vehicle (22) has a series of supports (23), which are attached to sleds, or runners, that glide along the track. These trains are propelled by jet engines (26) mounted on the locomotive (27).

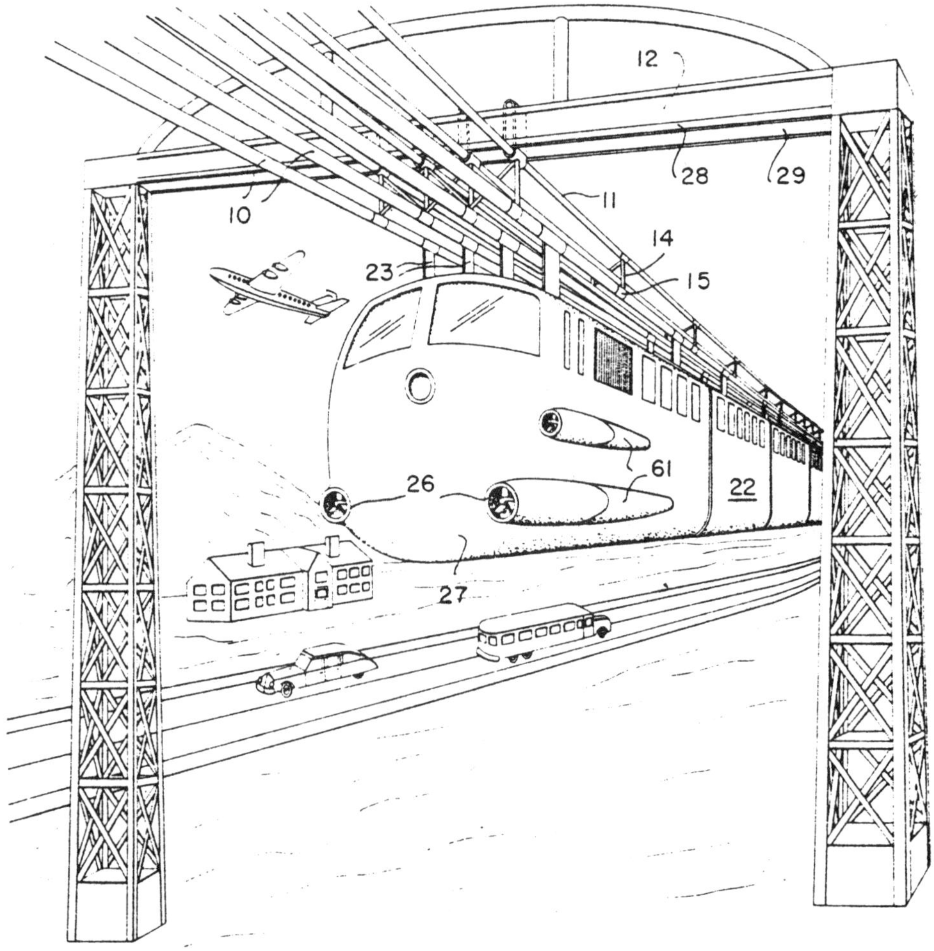

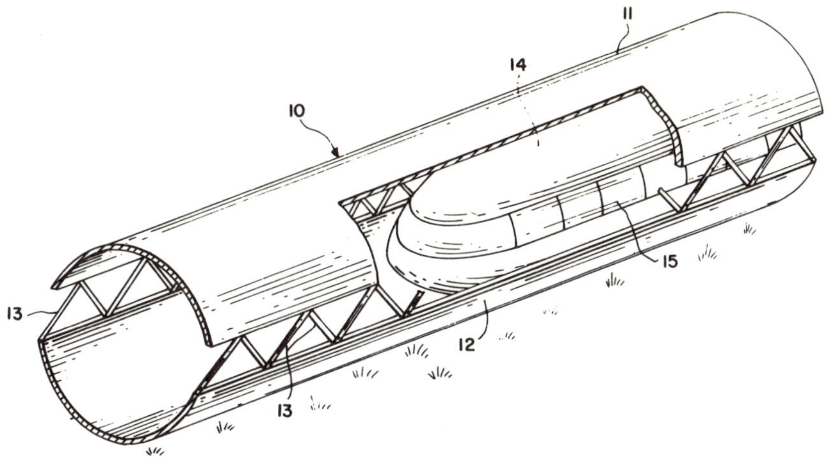

figure 12.18 Open-tube guideway for high-speed, air-cushioned vehicles, as shown in Patent No. 3,837,285 (1974).

The track (10) has top and bottom semicircular cross section portions (11) and (12), joined together by diagonal lacing (13) or other types of shear connectors, to effectively define a guideway having open sides. The air-cushioned vehicle (14) rides within guideway (10), and has windows (15) coincident in elevation with connectors (13) in order that passengers may view the outside world and to permit air to escape from the guideway to minimize the piston effect. The top portion is partially cut away to illustrate the position of the vehicle.

The guideway can be elevated upon pedestals with or without the addition of suspension cables or it can be placed on a grade. The top and bottom portions are relatively thin and may be made from a wide variety of materials, including a combination of corrugated and smooth sheet metal, precast concrete, pre-stressed concrete, etc.

Louvers may be added between top and bottom to offer protection from weather when needed.

these in plentiful supply. Three Minnesota men—Olsen, Nelson, and Davidson—have taken out two patents on flanged wheels and accompanying mechanism that allow an automobile or truck to use a railroad track (Fig. 12.16). Others have proposed more radical changes. Robert Walsh of Washington, D.C., developed a new rapid-transportation system similar to the monorail. His cars are suspended from runners that move in a tubular track and are supported on a right-of-way above the ground by cables. (See Fig. 12.17.) Jet engines provide the motive power and some of the braking action. This "train" is thus capable of much higher speeds than conventional equipment. The only trouble, of course, is the huge capital investment necessary to get a system of any size like this into operation.

Rodney Goering of Nebraska assigned to NASA his patent for an open-tube guideway that can accommodate air-cushioned vehicles (Fig. 12.18). This is one of the best ways now known for ground-level equipment to attain high speeds. Here again, however, the overall cost would be huge.

People are always going to want to get from here to there. Inventors are always going to try to help them do just that faster, more comfortably, and maybe even less expensively. Transportation is a good basic field for the application of inventive genius.

CHAPTER THIRTEEN

Tools for the Workshop

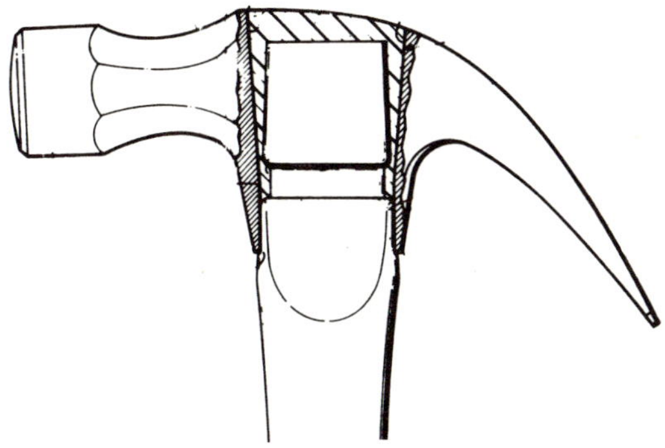

The basis of modern industrial society is tools. They are like levers that individuals can use to multiply their output by two, by ten, or, in extreme cases, by over a thousand. Without tools, all of mankind would still be living in the style of primitive tribes. There has been a strong link between the availability of more and better tools and the march of progress. As tools have become more numerous, faster, automatic, and specialized, society has undergone an increasing change of pace.

People have not always been happy about the results of new tools and new ways. In the early nineteenth century, for instance, English workers who came to be called Luddites, after an early leader, angrily smashed machine tools that they thought were responsible for the prevailing unemployment. In our own time, one of the causes of "future shock" has been the ever-increasing use of tools to do things that used to be done more slowly—and, some would say, more carefully and better—by hand.

There are two major categories of tools—machine and hand. The machine tools have, of course, been responsible for the most revolutionary changes in our lives. By enabling one machine to do the work of countless men, machine tools have been the basis of modern industrial mass production. They get more massive, more complicated, and more expensive with each year. To invent or improve one of these monsters lies beyond the reach of most independent inventors, who don't have the necessary capital or sophisticated equipment to deal with them. For that reason, machine tools will not be discussed except peripherally in this chapter. Our

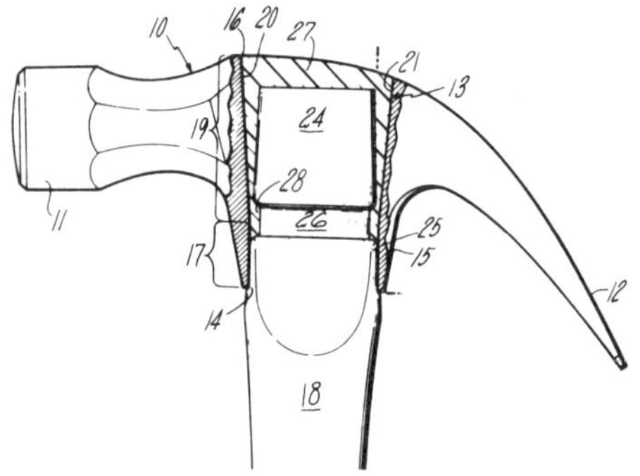

figure 13.1 Patent No. 3,877,826 (1975) is a handle connection for a tool such as this forged steel hammerhead. It has the usual poll and bell (11) at one end and claws (12) at the other. Centrally of the head (10) is a handle-receiving eye (13), which is a through opening having an entrance end (14) at the outer end of the boss (15) and an exit end (16) at the upper face of the head. In the specific embodiment here the cross section of the eye is shown as generally rectangular, although as will be apparent, the invention will apply equally to other cross-sectional configurations such as oval or round.

The entrance region (17) adjacent to end (14), which in the specific embodiment extends slightly less than one-third of the length of the eye, is of uniform cross section throughout or with a very minor flare or divergence toward the entrance end. The function of the entrance region (17) is to form a tight frictional engagement with the end of the handle (18) when it is driven in from the entrance end, the handle end being dimensioned so that it will be slightly compressed at the entrance region and so that the entrance end will dig slightly into the material of the handle when the handle end is fully inserted to form a tight seal. The remaining region (19) of the eye (13) gradually increases in cross section toward the exit end (16).

primary concern here is tools for the hands, tools that make the everyday tasks of construction and repair and so on easier and faster to accomplish.

Hammers, knives, axes, and other simple tools have been in use for thousands of years. Even before metals were discovered, the Stone Age people were able to do surprisingly fine work with their primitive implements. Until the eighteenth and nineteenth centuries, progress in toolmaking was relatively slow. Carpenters and other craftsmen worked in pretty much the same ways from generation to generation. The Industrial Revolution changed all that. What some historians have called the golden age of toolmaking occurred in England during the first half of the nineteenth century. A great many inventors, whose names would be unfamiliar to all but a few people today, participated in this explosive era. Machine tools were the principal vehicles of change, but hand tools were important, too.

In the early days most tools became more automatic but at the same time less versatile. Tools were made for just one or a few specialized jobs. It often didn't matter what level of skill was possessed by the workers; the tools could do the jobs. This was the beginning of automation, a process accelerated in the twentieth century by amazing advances in electronics.

With all the thought and effort (and time and money) that have been put into the development of tools, one might think that by now the field would be pretty well invented out. But that is not so. The new products keep coming into the market, and many of them meet with success. As in most other fields, the rules and boundaries keep changing. Before Energy Crisis I, there was emphasis on electrically and gasoline-powered tools. But now that fuel costs have increased, with the worst perhaps still to come, there has been a shift in attitude. Gimmicks in power tools are pretty much out. The use of productive and useful power tools continues to grow, through power units are now designed with a view toward the conservation of energy.

Safety is another concern that has become increasingly important to makers and users of tools. In recent years, tools have come to be used by more people with fewer skills than ever before. These do-it-yourselfers don't want anything that is difficult or dangerous to handle. Even for skilled people, the safer the tool, the more attention they can give to the job at hand.

Still another factor to be kept in mind is the expense of manufacture. Inventions that cost too much to produce relative to the price at which they can be sold in quantity have little interest for most manufacturers. This does not mean that cheaper is better in every instance. There are plenty of skilled craftsmen who want good tools and are willing to pay for them—if the value received is worth the price.

Inventors never seem able to leave well enough alone. Even when something seems not worth changing, they nevertheless persist in looking for a better way. This is especially true in the tool field, where very often new wrinkles are turned up in unlikely places. Take the hammer, for instance. It's been around almost as long as mankind, but people are still tinkering with it. Perfection has not been reached. One of the difficulties with hammers and other impact tools is the loosening and eventual separation of the metal head from the wooden or plastic handle. The traditional way of joining these two parts has been to provide the eye of the head with an outward flare or enlarged cross section toward its exit end to expand the end of the handle that has been inserted through the smaller entrance end of the eye into tight contact with the eye by wedging. Connections of this type are very tight at first, but tend to loosen as the tool is used. In recent years, improved glues have been used both with and without the wedges. This has improved performance, though fractures and splits still occur. Warren Shepherd, Jr., of New Jersey and Joseph Guarnaccia of Connecticut teamed up to patent a solution to this problem (Fig. 13.1). The patent was assigned to one of the most important manufacturers in tools, the Stanley Company.

There are other problems with the hammer, too. Lela Mae Hays of California developed the Bammer especially for work in cramped areas (Fig. 13.2). It has interchangeable handles and heads that enable the user to adapt it to different work conditions. The Bammer takes advantage of the principle of inertia to furnish a supplemental force to the usual hammer stroke. It's got a catchy name, too, which often helps get a manufacturer/marketer interested.

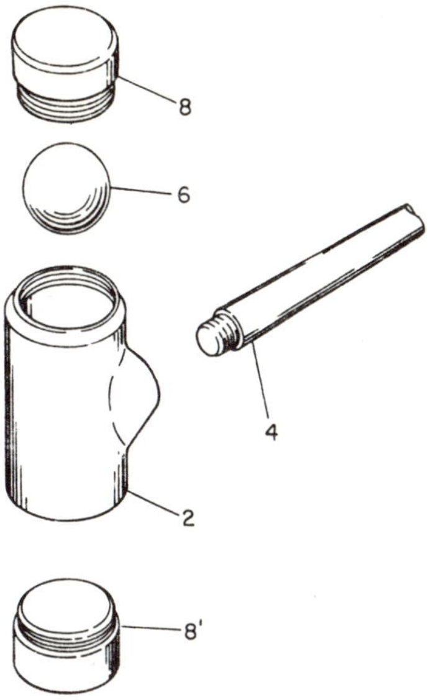

figure 13.2 The bammer, Patent No. 3,948,301 (1976), has only a few parts, as this drawing of a disassembly illustrates. The head (2) is joined to the handle shaft (4) by a threaded portion of the handle shaft, allowing the interchanging of handles. The construction of the head assembly, as shown, includes a hollow body with the cavity tapered at both ends to allow a positive seat for the movable ball (6), and threaded openings at either end, to accept the interchangable heads (8) and (8′).

Speed, not space limitations, was the main consideration in Bruce Malkin's work on his automatic hammer (Fig. 13.3). Such nailing machines, as they probably should be called, have been in use for many years, and significant improvements have been made along the way. These tools have often been powered by compressed air, which, according to Malkin, has several disadvantages. An electric tool such as the one illustrated here has the advantage of being more easily portable than a pneumatic device and of being easily plugged into a readily available source of power. Most builders bring electricity onto a construction site fairly early in the work cycle.

There are other alternatives in this area, many of which have not been fully explored. Swingline, the staple maker, has been assigned a patent for a "gun-type manually operable . . . driving means including a spring-loaded ram." There are two or more driving shots for each fastener, which can be used on jobs where additional force is required (Fig. 13.4). Using one of these machines, a workman can significantly increase productivity. In these days of high costs, that's of prime importance. It's evident that mass-production tools and techniques must be adapted for use in the construction industry—or else a large part of the industry will collapse under the weight of its own high costs, as it almost did during the 1974–1976 recession. Changes are in the wind, which means opportunities for inventors. Many young people cannot afford to buy a new house built and financed in the old way. Either new products and techniques are going to be used in home building, or a lot of people are going to be living in tents.

As illustrated in other chapters of this book, inventions need not be complicated, expensive machines or great departures from the prevailing state of the art. Figure 13.5 illustrates an accessory for a tack hammer. In "tackless" carpet installation there has been a need for a tool that a worker can use to apply the even force necessary to get the carpet into the proper position. Anyone who has handled a big piece of carpet knows that it is awkward and difficult to maneuver. Leonard Krell of Ohio figured that a simple gadget that could be attached to a tack hammer would be a useful thing. He developed and eventually patented the little device illustrated.

Carpenters carry more tools than just about any other craftsmen. On some complicated jobs they use more devices than the surgeon in an operating room. Not having a faithful nurse to stand by and slap tools into their hands as needed, carpenters are often tempted to try to combine two or more related tools into one. Sometimes this works quite well, as in the time-tested combination of hammerhead and claw nail remover. Ashby Cotter of Tennessee

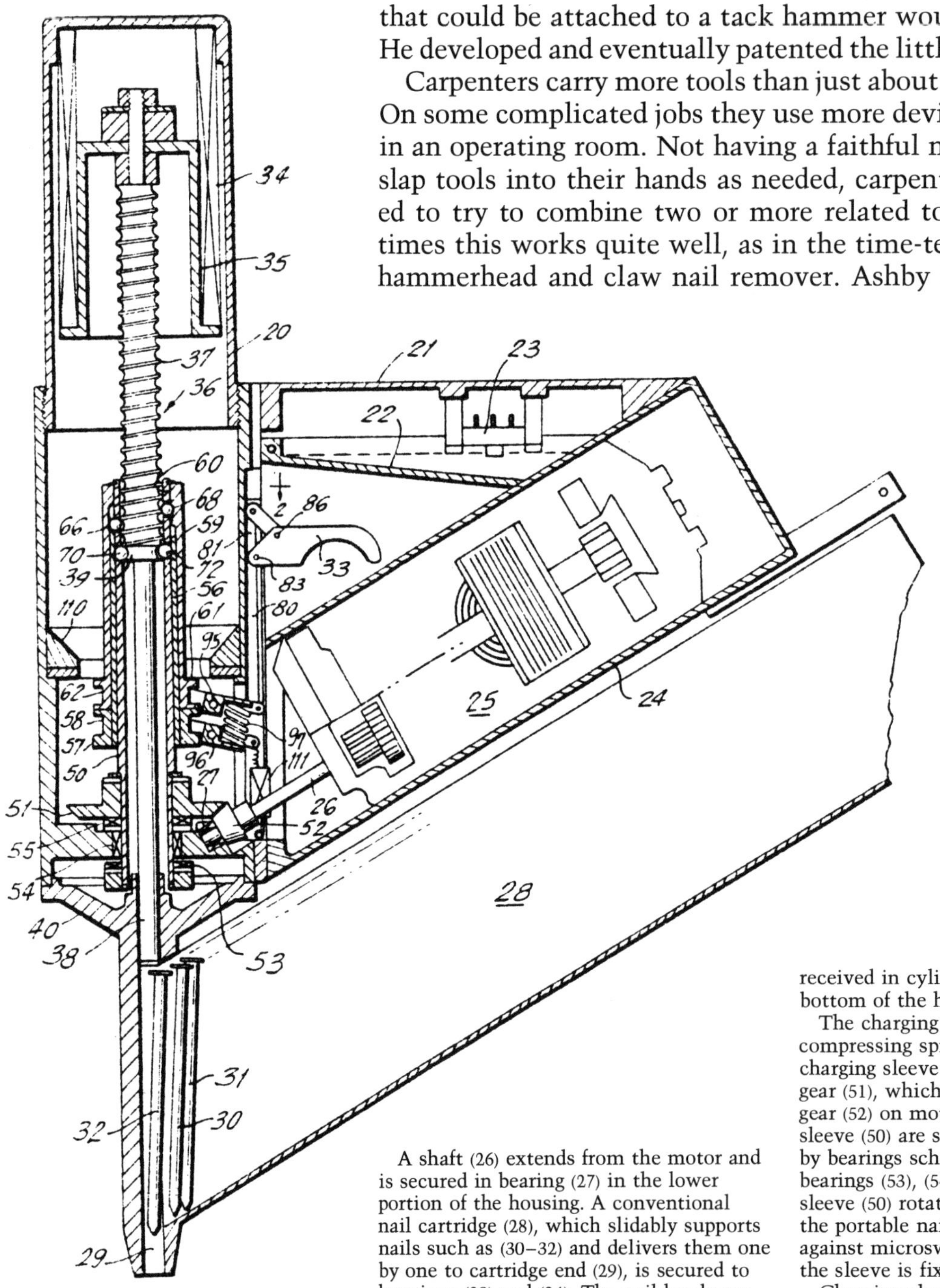

figure 13.3 A cross-sectional diagram of the electric nailer illustrated in Patent No. 3,810,572 (1974). A handgrip (21) in the main hammer housing (20) contains a movable plate (22) that actuates a microswitch (23) when a user picks up the tool. An electric motor housing (24) is connected to the housing and contains a suitable a-c motor (25) that is turned on when microswitch (23) is actuated by plate (22).

A shaft (26) extends from the motor and is secured in bearing (27) in the lower portion of the housing. A conventional nail cartridge (28), which slidably supports nails such as (30–32) and delivers them one by one to cartridge end (29), is secured to housings (20) and (24). The nail heads are kept within a suitable channel. [Nail (32) is shown in a delivered position, awaiting a hammer blow.]

A manual trigger (33) actuates the hammer blow. The upper portion of the housing contains a powerful compression spring (34) that is carried in a retaining cup (35). This cup is then connected to hammer (36), which consists of a threaded section (37) and elongated ram (38). The bottom end of the hammer is slidably received in cylindrical section (40) at the bottom of the housing.

The charging mechanism, for charging or compressing spring (34), includes rotatable charging sleeve (50), which is fixed to bevel gear (51), which is driven by cooperating gear (52) on motor shaft (26). Gear (51) and sleeve (50) are suitably rotatably mounted by bearings schematically shown as bearings (53), (54), and (55). Note that sleeve (50) rotates so long as the user holds the portable nailer and holds plate (22) against microswitch (23). Note further that the sleeve is fixed against axial movement.

Charging sleeve (50) is concentrically surrounded by an axially movable latching sleeve (56). The bottom of sleeve (56) is secured to a disk (57) which contains a channel (58) in its outer periphery.

Latching sleeve (56) is then surrounded by ball-drive latching sleeve (59), which has an internal restricted-diameter portion (60). Sleeve (59) is axially movable relative to sleeves (50) and (56) and is connected at its bottom to disk (61), which has a channel (62) in its outer periphery.

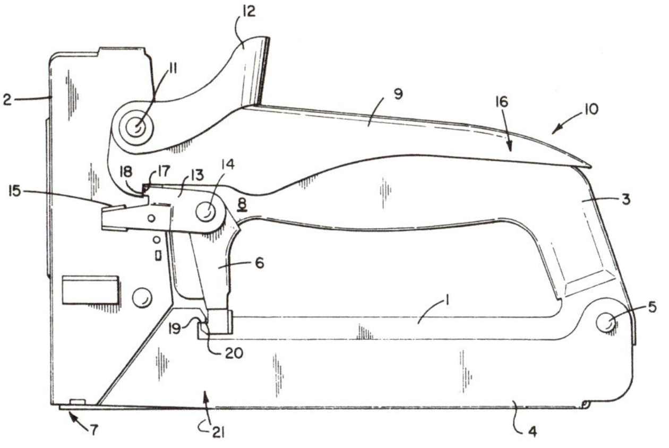

figure 13.4 The multiple-shot fastening gun of Patent No. 3,923,228 (1975) is best illustrated by this simple drawing. The frame (10), which includes a gripping handle (3) as shown, supports a housing for the operating components of the tool. The housing includes magazine (1) and sheath (2). The magazine cover (1) is pivotally secured to the rear end of the frame by hinge pin (5). The cover is swingable, upon release of cover latch (6), about hinge pin (5) to expose, for purposes of loading, the magazine-type fastening element feeding means contained within the magazine. This mechanism successively feeds individual fastening elements one at a time to the driver (17) in a sheath (2).

The forward portion (8) of the frame supports a specially constructed actuator that operates the drive mechanism. The actuator arm (9) is pivotally connected to frame (10) by hinge pin (11) and spring biased toward its open position. It includes a specially constructed linkage that acts to cock and fire the driving means. Arm (9) carries at its forward end a hand guard (12) that is also mounted on hinge pin (11) and fixed in place relative to the arm by rivets.

To lock the arm in the fully closed position, a trigger latch (13) is pivotally secured to the forward end (8) of gripping handle (3) by hinge pin (14). The latch has a thumb rest (15) that makes for convenient rotation and pivotal movement of the arm to its open position. To release the arm, it is helpful to depress it slightly from its locked position, as shown, in the direction of arrow (16). Such movement, it will be recognized, slightly separates contacting locking surfaces (17) and (18) of arm (9) and latch (13), to provide enough clearance between these locking surfaces, permitting easy release of the trigger. Similarly, it is helpful, for purposes of releasing cover (4), to depress it slightly from its closed position in the direction of arrow (21). This movement provides sufficient clearance between locking surfaces (19) and (20) on cover (4) and cover latch (6), to permit the latch to be easily pivoted from its locked position to its open position [as indicated by arrow (21)], thereby separating locking surfaces (19) and (20) to allow cover (4) to be swung open. Trigger latch (13), on the other hand, is held securely in its locked position with its locking surface (18) engaging locking surface (17) of trigger (9) by force of engagement between these locking surfaces created by the torsion spring biasing the trigger toward its open position. Thus the trigger is safely prevented from flying accidentally into open position when the tool is not being used.

---

figure 13.5 Conventional carpet installer's tack hammer with attachment from Patent No. 3,745,598 (1973). Inset is a close-up view of the invention (20) composed of a hollow cylindrical aft end (21), having a bored recess as at (22) and having secured to the opposite end a curvilinearly shaped, bladelike head (23). This bladelike extremity is about one-fourth to five-eighths inch and has a curved forward edge (24). The aft end (21) is provided with opposed threaded holes (25) and (25*a*) in which are threaded a pair of "Allen head" set screws (26). The holes (25) and (25*a*) are drilled and tapped transverse to the axis of the bored cylinder end (21).

A pair of half-sleeves, formed of metal and magnetized, are wrapped around the hammer's neck. Thereafter, the bore (21), is slid telescopically over the flared end of the tack hammer until the face of the tack hammer (18) abuts the inner end (40) of the cylindrical bore (22). The head (20) can be rotated until the elongate blade (23) is generally in alignment with the handle (12), whereupon the previously inserted set screws can be tightened, securing the aft end (21) onto the half-sleeve, squeezing it into secure surface abutment with the neck region of the tack hammerhead.

The user is thus supplied in a short time and with little effort a very familiarly shaped hand-operable tool which has, as its working end, a curved, tomahawk-shaped, relatively thin blade, which can be manipulated to accomplish the movement of the carpet into the crack or crevice to engage the tack strip as in the standard practice of the tackless installation technique.

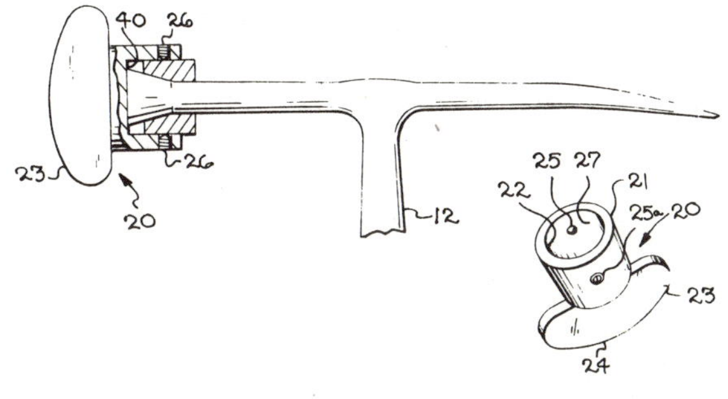

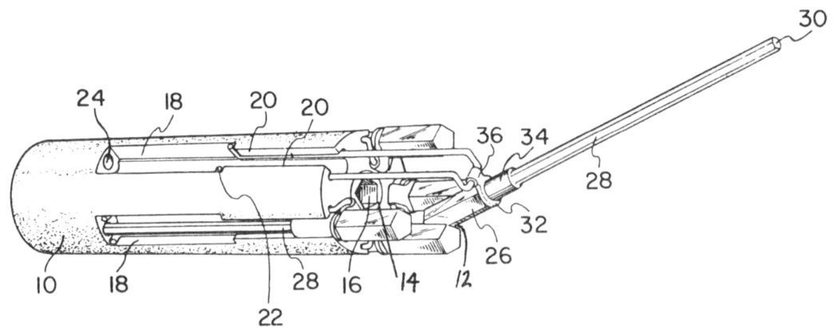

figure 13.6 Multiple hand tool with one attachment in partially open position, as shown in Patent No. 4,010,663 (1977). The handle (10) is provided with a central axial opening (12) extending into the handle from the front end. A steel socket (14) is tightly mounted in the opening (12) and is provided with a central axial opening (16). The selected tool must be mounted in this opening and locked against rotation. This can be done by providing the opening and tools with complementary locks such as five or six sides, tapers, keys, or other devices.

A number of grooves (18), as illustrated, extend rearward from the front end of the handle (10). Each groove is provided in its forward half with restricted portions (20) extending toward one another to provide overhanging portions. These portions form shoulders (22) at their rear ends. At the rear end of each groove, the handle is provided with a cut-out opening (24) for receiving the end of the tool when it is positioned in its groove. Each tool to be mounted on the handle comprises a base portion (26) having the same section as the axial opening (16). The shank of the tool (28) extends from the base and terminates at a free end (30), where it is shaped into the desired tool. A ring or loop (32) is slipped over the shank (28) until it abuts the inner end of the base. A lock collar (34) is slipped over the shank to lock the tool to the loop holder (32), allowing rotation of the tool for adjustment.

To mount the tool, the wire arms are pinched toward each other and inserted in one of the grooves (18). The arms will resiliently slip under the overhanging portions (20) to retain the arms in the groove. Furthermore, the ends of the arms will abut the shoulders (22) in the extreme outward position to prevent the wire arms from sliding out of the groove. The resilient arms will tend to engage the side walls of the groove to provide a friction fit.

invented a multiple hand tool that allows a variety of heads to be inserted and removed easily from a common handle (Fig. 13.6). Each tool head has a base that is adapted to lock into a grooved hole in the center of the handle. Another, earlier approach to this same idea is shown in Fig. 13.7. In this invention, Harry Condon of Michigan made a screwdriver that provides for easy movement of different heads into and out of the handle "without creating resistant suction forces and without providing an opening through the head end of the handle," which can be irritating to the user's hand whenever pressure must be applied.

The more jobs a tool can do, the fewer tools a craftsman will have to carry. Many different attachments have been developed for that most useful of tools, the electric drill. These are especially helpful for the special jobs that come up so seldom that it's not worthwhile to buy a tool just for them. And attachments are good for everyday jobs, too. William Atwater of North Carolina saw that most auto body shops used a separate, self-contained abrading tool when an attachment for an electric drill would be more efficient. There were a few accessory-type abrading tools available, but they had not come into widespread use because of certain deficiencies, most notably the failure to provide a practical means for expanding and contracting the drive cylinder over which the abrading sleeve is placed. Atwater's invention, illustrated in Fig. 13.8, consists of a drive cylinder that is connected through a universal joint arrangement to the chuck of an electric hand drill. A bracket extends between the tool and a bearing support so that the tool can be both driven and supported by the hand drill.

Atwater's tool is, of course, for a specialized market, and only someone in the auto-repair field could have invented it. But it points up the fact that the electric hand drill is made to receive all kinds of attachments for all kinds of uses. Anyone who has an idea for a practical time- or labor-saving device that can be powered by an electric drill is more likely to find a market for his device than the inventor of a new machine. Obviously, the former is a lot less expensive to manufacture and promote than the latter.

Many sawing and filing attachments for the drill have been invented, some with various disadvantages. Herbert Pfister of New York patented an improved jigsaw spindle sander in 1975 that attempted to correct two major faults in previous machines: first, converting rotary to reciprocating motion, and vice versa, formerly required external transmission devices; second, in cases where a composite movement (reciprocating-rotary) of the output shaft was provided, the speed of the output shaft proved to be excessive, especially for some sanding tasks. Pfister's invention (Fig. 13.9) provided variable speeds for different operations and a transmis-

figure 13.7 Screwdriver-type housing adapted to receive a series of different bits, illustrated in Patent No. 3,752,202 (1973). It is made up of a torque-magnifying handle (20) and an interchangeable and removable headed bit (30). This bit comprises a head (32) molded from a suitable plastic material and a metal bit (34). The bit has an elongate shank portion (36) and a fastener-engaging portion (38). At the end remote from bit (34), head (32) provides a smooth and rounded head end (40). Between the head end and the opposite end of the head a number of flats (42) are provided.

The torque-magnifying handle (20) defines an outer fluted longitudinal surface which provides a number of flutes (50). The handle defines a longitudinally oriented aperture or bore (52) opening at one end of the handle, and provides a smooth, nonperforated convex end (54) at the other end.

At one end, a portion of the bore is generally hexagonal in shape, each of the equilateral sides of the hexagon, for convenience, being designated as lands (56). Preferably, the lands are equal in transverse dimension. Pairs of opposed lands (56) are preferably spaced from each other a distance of not more than one ten-thousandth of an inch more than the spacing between confronting parallel flats (42) on the headed bit (30). Also, the lands incline very slightly inward from their outermost edges at one end of the bore (52) to their innermost edges intermediate the ends of the bore. It will be apparent, therefore, that the lands and the flats are of substantially the same size, but with the head defining a slightly smaller hexagon. Therefore, upon insertion of the head into the bore, there is provided a very slight space between the lands and the flats which defines an air passage extending from the outermost end of the bore to the innermost end, between the handle and the head.

Between the ends of the bore is a pair of projecting head-engaging means. These opposed gripper pads (58) project inwardly from a pair of opposed lands and are spaced apart a distance substantially equal to the space between an opposed pair of flats. When the headed bit is inserted into the aperture, the gripper pads snugly and frictionally grip the confronting pair of flats (42) thereby to retain the head, hence the headed bit, in removably assembled engagement with the torque-magnifying handle. Except in the vicinity of the gripper pads, the very slight spacing providing the air passage (57) between the head and the interior wall surface of the handle should be maintained to eliminate the suction effect caused when bits engage the walls of the bore in the handle.

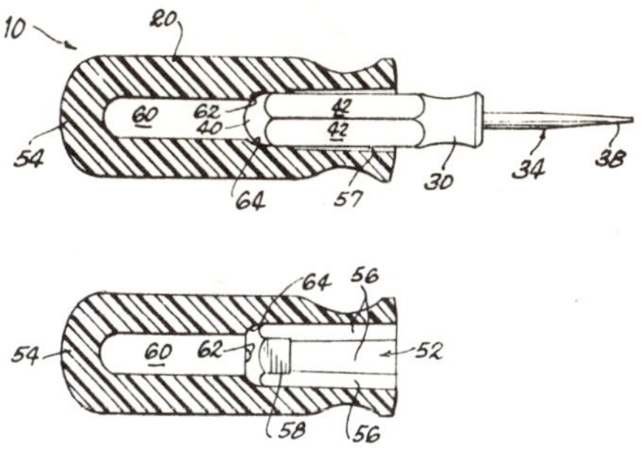

To maintain communication between chamber (60) and the other end of the bore (52), stops or spacers (62) are provided. These spacers, which are generally dimplelike, project from the rounded or concave intermediate portion (64) of the bore. Spacers (62) prevent the seating of head end (40) against the concave inner portion (64), thereby making it impossible for them to sealingly engage. This also provides constant communication, to the extent that it may be necessary, between suction-relief chamber (60) and the other end of bore (52). Spacers, by eliminating such sealing engagement, also eliminate the possibility of sweating or freezing of the plastic head end to the handle at the concave end portion. Spacers and gripper pads cooperate to locate the headed bit relative to the bore.

figure 13.8 Cross section of the abrading tool described in Patent No. 3,869,832 (1975), with the electric hand drill to which it is attached shown in dashed lines. The tool is designed to be operated by a conventional hand drill (20) having a conventional chuck (21) and a front housing (22) having a threaded aperture (25) adapted to receive a bolt (23) for securing a bent bracket (24). This aperture can be readily formed on the front housing, if not already part of the drill as originally manufactured. The bracket may be formed of a bent metal strip and is integrally secured to a bearing (30) having a hub portion (31) and a flange portion (32).

A universal joint (35) has a spindle portion (36), which is received by the chuck (21). A the opposite end, it has a socket portion (37) secured by a pin (38) that passes through a hole (39), which is provided in a reduced end portion (40) of drive shaft (41). Shaft (41) is rotatably received in the previously mentioned support bearing (30) and resides between socket (37) and a cam (50). The cam has a somewhat elliptical camming disc portion (51) and a hub portion (52), which receives shaft (41) and is pinned to it by a pin (55), which extends through a hole (56) in the shaft and holes (57) in hub (52) of the cam.

On the opposite end of shaft (41), there is mounted an auxiliary handle (60) having a pivotal pin connection (61) to a bearing (62), which receives the extreme end portion of shaft (41). A suitable locking ring (63) retains bearing (62) on shaft (41), which is further held in position by a sleeve secured to it by a set screw (71). The abrading sleeve (75) mounts on a split, thin-walled, metal cylindrical sleeve having a pair of longitudinal slots or slits extending from its driven end over a major portion of its length.

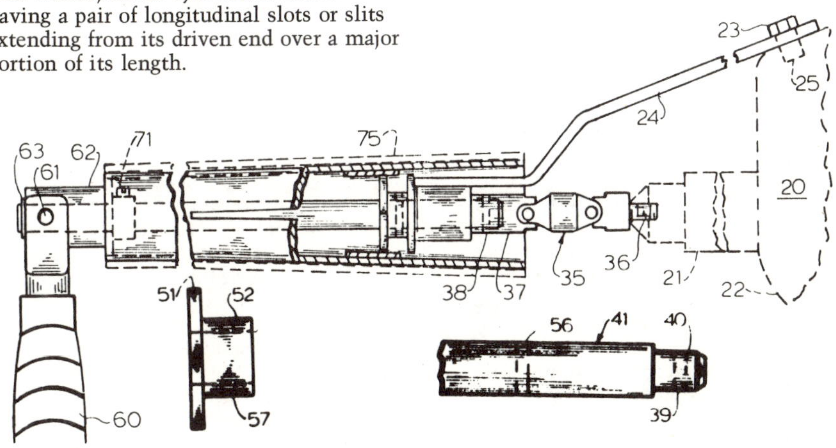

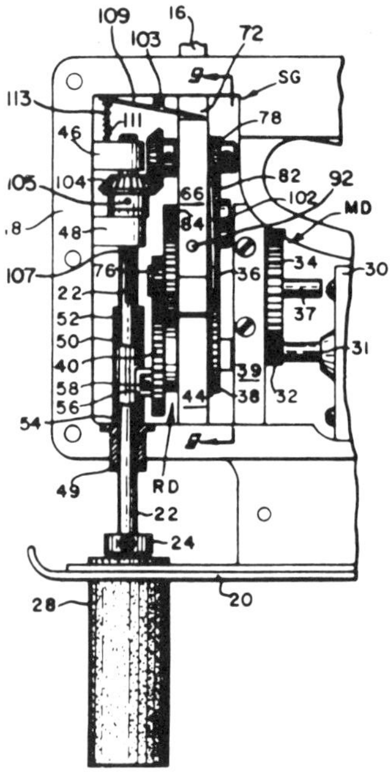

figure 13.9 Fragmentary side view of a portable jigsaw spindle sander with the cover removed, as illustrated in Patent No. 3,903,657 (1975). An electric motor (30) is suitably mounted within the motor housing (12) and is provided with an output shaft (31) to which a motor drive gear train (*MD*) is connected, such gear train including a pinion (32) arranged to drive gears (34) and (36), the latter being secured to a shaft (37) which is rotatably mounted in a housing part (39). Thus, it is seen, from this construction, that gear (36) of the motor drive (*MD*) will be rotated whenever the motor (30) is energized.

In order to effect reciprocating movement of the output shaft (22), the latter is connected to be driven by a reciprocating drive gear train (*RD*), which includes a pair of gears (38) and (40) secured to a shaft (42) rotatably mounted in a housing (44). The output shaft (22) is freely mounted for reciprocatory movement in the gear casing (18) by means of housing bracket supports (46) and (48) and bearing (49), and a drive sleeve (50) freely surrounds the shaft and is confined between collars (52) and (54) fixed thereto. The sleeve (50) is drivably connected with the gear (40) through a conventional "Scotch-yoke" device comprising a horizontally disposed channel (56) secured to the sleeve (50) and an eccentric drive pin (58) carried by the gear (40). Upon operation of the reciprocating drive gear train (*RD*), the Scotch-yoke device will effect reciprocation of the output shaft (22), and thus jigsaw operation of the portable tool.

An important feature of the invention is the means for connecting the motor drive with the reciprocating drive for effecting the reciprocation of the output shaft (22) as well as for connecting the motor drive with the output shaft for simultaneously rotating and reciprocating the latter so that the portable tool may be employed as a combined rotary and reciprocating drum sander. Such means comprises a shiftable gear train which includes three shafts that are rotably mounted in a shiftable carrier (66), the latter being slidably supported at one edge in a slot (68) formed in the gear case cover (70) and at the other edge in a slot (72) formed in the gear case (18). Gears (74) and (76) are affixed to shaft (60), while shaft (64) has a gear (78) secured at one end and a bevel gear that is secured at its opposite end. The intermediate shaft (62) has gears (82) and (84) secured to its opposite ends. These gears are positioned so that gears (78) and (82) are in constant mesh, as are gears (84) and (74). With the foregoing arrangement, it will be readily understood that the shiftable gear train (*SG*) comprises an input gear (82) and a pair of output gears (76) and (80), the latter functioning to impart a combined reciprocating and rotatable motion to the output shaft (22).

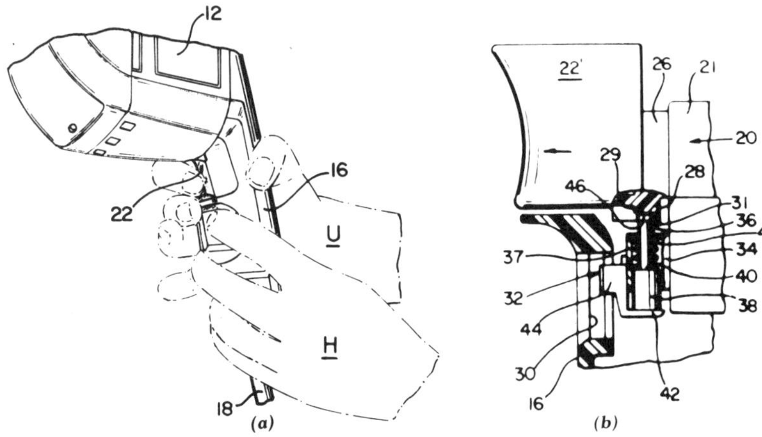

figure 13.10 Two views of a lock for an electric drill or similar tool as described in Patent No. 3,781,579 (1973). To turn the tool on, finger pressure is applied to the trigger (22); to turn it off, the pressure is released, the trigger returns to position, and switch (20) is opened.

The switch housing (21) has a guide sleeve (34) extending to its front. In the sleeve there is a plunger (36) with an enlarged portion (38) vertically and slidably placed in a bore (37). Its upper, flared end (46) is outside the sleeve. A compression spring (42) within the sleeve engages a shoulder (40) on the plunger (38) and normally biases it downward as seen in the figures, toward the position shown. Engagement between the flared end (46) and the sleeve limits this downward movement. An actuator (32) is provided for the plunger (36) and includes a foot (42) fixed to the enlarged portion (38), and a button (44) slidable on a face (34) of the sleeve. The button extends partly into an aperture (30) formed in the handle (16) below the trigger (22), and may be slid upwardly using one's index finger as shown in (*A*). When the button (44) is so moved, after the trigger has first been depressed using the index finger on one's other hand, also as shown in (*A*), the flared plunger end (46) moves into engagement with a boss (28) depending from the trigger.

Specifically, the illustrated switch (20) is of the "two speed" variety, such that partial depression of the trigger effects low speed tool operation, while full trigger depression effects high speed operation. The boss (28) has a pair of stepped shoulders (29) and (31) adapted to align with plunger (36) when the trigger is partially and fully depressed. Thus, when it is desired to lock the tool "on" for low-speed operation, the trigger is partially depressed and the button (44) raised [using the fingers in the manner shown in (*A*)]. Release of the trigger causes the shoulder (29) to bear against plunger end (46), holding the plunger in this raised position and the switch operator (26) in the low-speed position. To lock the tool in the high speed "on" position, the trigger is fully depressed and the button raised to locate plunger end (46) adjacent to shoulder (31). With the parts held in this position, shown in broken lines in (*B*), the trigger is first released and then the button (44), whereby biased engagement between shoulder (31) and plunger end (46) holds the trigger and plunger in this position. When it is desired to turn the tool "off," the user applies momentary pressure to the trigger releasing shoulders (29) (31) from plunger end (42), whereby plunger (36) returns to the full line position shown in (*B*) under the force of spring (42). Thereafter, release of pressure on the trigger allows it to return to the off position.

It will be appreciated that the "guarded" nature of button (44), shown here by the button's being recessed within the wall of handle aperture (30), prevents the actuator (32) from being accidentally or unintentionally actuated to a locking position, or, when in locked engagement with trigger boss (28), from being held in the locked position when it is desired to turn the tool off.

sion mechanism for switching between reciprocating and rotary/reciprocating motions. In addition, it was designed to be lightweight, relatively low in cost, and highly reliable.

The electric drill itself offers many possibilities for modification and improvement. Black & Decker has been one of the leaders in the competition to be the manufacturer with the best tools on the market. Figure 13.10 illustrates a feature that was planned to make their hand-held drills more convenient to use. Many of these tools have had a trigger or similar part capable of being locked in the on position, without continuous pressure from the operator. Prior to this invention by Francis Rosenthal, Jr., and Lewis Rohr, these devices could be locked in position accidentally.

Automatic sanders can do a job five or six times faster than it can be done by hand. They have turned a tedious task into something that's almost a pleasure. With faster sanding the dust comes faster, too. It's almost a necessity with one of these machines to have some kind of dust collector in order to keep the surrounding air clean, to prevent excessive mess, and to avoid clogging the sandpaper with dust. The typical small sander has a standard hose connection to a vacuum cleaner that sucks away the dust. Large machines, generally used for floors and similar surfaces, have a built-in vacuum cleaner. According to Melvin Glasser of New York arrangements such as the foregoing are expensive, cumbersome, and prone to clogging. His invention, which is for large machines, is a simple system that utilizes air generated by the sander to blow the small particles of wood away from the work area before they are ground into lighter-than-air dust. Thus, a simple device was designed to take the place of a more complicated one (Fig. 13.11). Could such a system, or something like it, work with a hand sander? Is there any other simple way to handle the dust? Glasser's experience thus far has been a frustrating one. He has not been able to get his invention on the market, despite the fact that he knows it is a good machine. This is in fact the common complaint among inventors—how do you reach the public with something you know is good if you have neither the manufacturing nor the marketing capabilities to do the job?

New tools and new versions of old tools are constantly being brought onto the market. It's not difficult for the craftsman to amass a large collection of tools in just a short time. But where do you put them all? All kinds of toolboxes and cases have been manufactured, some in large numbers. Back in 1951, Russell Phelan of Massachusetts patented two simple, useful devices that can still be found in many workshops—and elsewhere, too. His magnetic holders provide convenient, open storage for screwdrivers, knives, and other small, fairly lightweight tools (Fig. 13.12). Magnetism is one of a number of well-known scientific principles that can be applied to everyday uses—limited only by the ingenuity of the inventory.

A great many toolboxes have been designed and manufactured in the last hundred years, but not very many of them have been patented. They simply did not introduce anything new enough to merit a patent. Toolboxes for special purposes are something

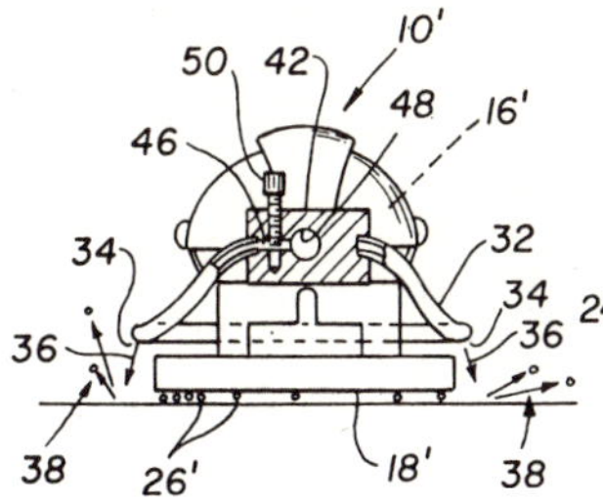

figure 13.11 Rear view of a sander with dust preventer, shown in Patent No. 3,906,682 (1975). A closed loop of conduit (32) is operatively arranged to encircle the body of the sander (10'). Conduit (32) is provided with a number of exit openings (34) in strategically spaced locations. The sander is also connected to a source of pressure air via conduit (14'), with the result that there is a continuous flow of pressure air in the form of streams (36) exiting from the openings (34) and directed against the work surface (24'). The important consequence of this is that the comparatively large particles (26'), being heavier-than-air, are removed from the work surface.

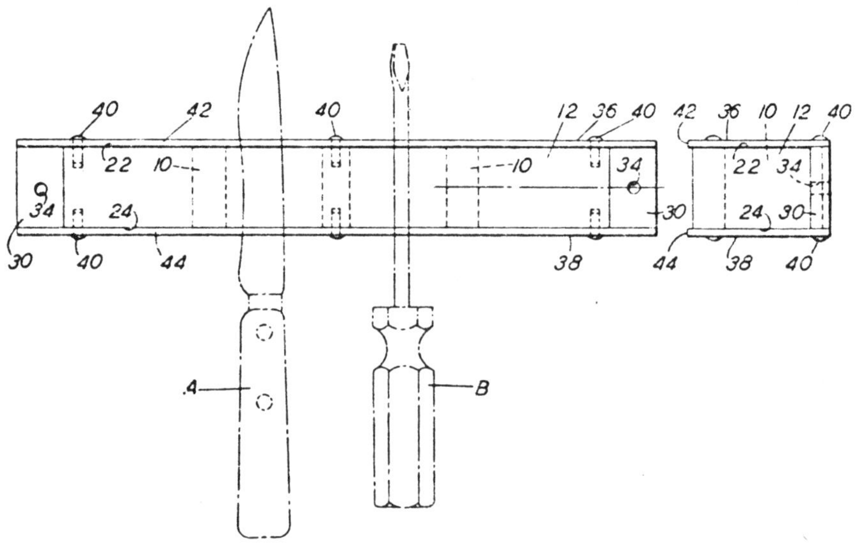

figure 13.12 Front view and (at right) end view of a magnetic holder, as illustrated in Patent No. 2,565,625 (1951). A body (12) of nonmagnetic material horizontally surrounds the magnets (10) and (10). The body has transversely spaced vertical front and rear walls (12) and (16) between which the magnets are located. The walls have longitudinally spaced pairs of separate interior notches for receiving the magnets. These notches have vertical faces. The front wall has two notches for receiving and fitting the front portions of the magnets, and the rear wall (16) has two notches for receiving and fitting the rear portions of the magnets.

The body (12) has initially exposed upper and lower horizontal faces (22) and (24), which are so spaced that they register with the opposite pole faces of the magnets. Inasmuch as the upper and lower body faces are intially exposed, the notches are open at the top and at the bottom and the magnets can be inserted vertically, either downward or upward.

Two relatively thin pole pieces (36) and (38) are provided, formed of magnetic material. They engage the upper and lower faces (22) and (24) of the body and the top and bottom pole faces of the magnets. The pole pieces are connected with the body by headed pins (40) and (40′) which extend through apertures in the pole pieces and into registering apertures in the body. The apertures in the body are initially slightly smaller than the pins so that the pins must be driven into place by them, being held frictionally so as to retain the pole pieces. By reason of their engagement with the magnets and their connection with the body, the pole pieces retain the magnets in place.

The pole pieces have vertically registering narrow edge faces (42) and (44), spaced outward or forward from the body. They span the magnets and are continuously curving convexly between points near the respective ends of the pole pieces. These edge faces (42) and (44) of the pole pieces have substantially greater curvature at the ends than in the center.

The narrow front faces (42) and (44) provide small areas of contact with the parts which are to be held. This makes it impossible to shunt so much magnetism through a few articles at one end of the holder that there is not enough left to support articles at the other end. The available amount of magnetism for each linear unit of length is limited by the saturation of the iron in the areas of contact with the articles.

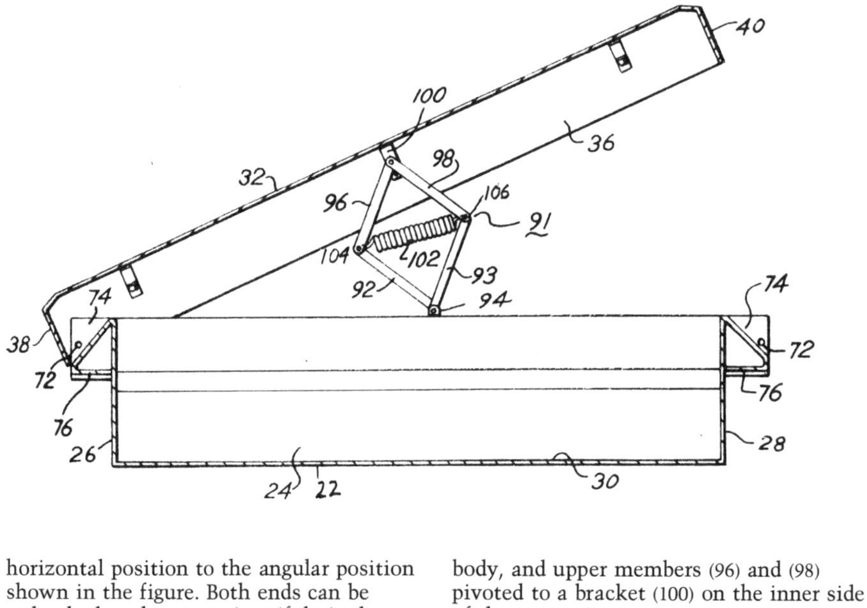

figure 13.13 A toolbox for a pickup truck is illustrated in Patent No. 3,640,423 (1972). The box has a lid (32) that extends downward to form an effective closure. The cover is pivoted at each end by a latch mechanism that permits either end to serve as a hinge when the other end is raised. The two latches are the same, consisting of two plunger assemblies mounted on the rear side of the two cover ends, each consisting of a bolt and a spring urging the bolt outward toward the side of the cover. The spring reacts between a collar and the end of a bracket that is rigidly secured to the inside surface of the cover end.

The outer end of the bolt locks into a fixture consisting of a hole (72) in an inner wall (74) joined to an extension (76) of the two sidewalls (22) and (24). The curvature of extension (76) forms a guiding surface for the end of the bolt and depresses it inwardly, permitting it to snap into hole (72) when the end of the lid is fully lowered. The bolt is retracted by an operating mechanism consisting of a cross member to which two cables are connected, the other end of the cable being connected to the respective bolt. A lever is rotatably mounted on a shaft which is operated by a handle. The rotation of the handle causes the lever to rotate and thereby move the cables inward, withdrawing the bolts from their respective holes (72), thus unlatching the respective end of the cover. Either end can be raised, and when the end is raised, the other end normally remains latched, thus permitting the cover to pivot from the horizontal position to the angular position shown in the figure. Both ends can be unlatched at the same time if desired.

In order to hold one end of the cover in an elevated position while material, tools, or other articles are being placed in or removed from the toolbox, yieldable supports (90) and (91) hold the cover in the position shown. The yieldable supports are mounted on opposite sides within the box, each support consisting of a frame formed by lower members (92) and (93) pivoted to a bracket (94) on the respective side of the body, and upper members (96) and (98) pivoted to a bracket (100) on the inner side of the cover. A spring (102) reacting between the pivot points (104) and (106) urges the four members inwardly in the direction to assist in lifting and supporting the cover. When the cover is to be lowered, the open end is pressed downward, thus causing the four members to move longitudinally in respect to the cover and body, thereby extending spring (102) and permitting the raised end to seat and latch.

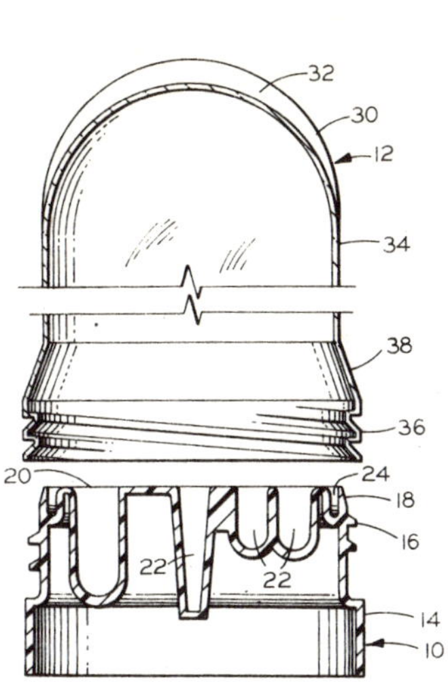

(a)

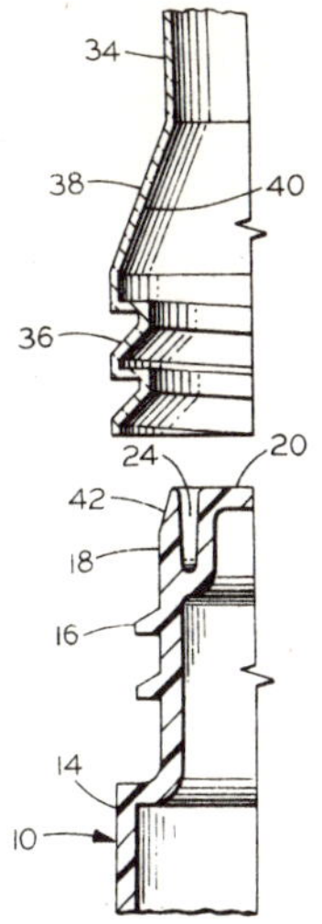

(b)

figure 13.14 At left is a perspective view of the container for boat tools described in Patent No. 3,937,324 (1976). At right is an enlarged cross section of part of the container. There is a base (10) and a cover (12). The base is circular and has a sidewall provided by the polygonal pedestal portion (14). An externally threaded portion (16) is above this pedestal and the upstanding lip (18) is at its upper end. The platform portion (20) of the base (10) extends transversely across the upper end and is spaced inwardly from the lip (18) so as to define a channel (24) between the parts. A multiplicity of upwardly opening recesses (22) is provided in the platform portion (20) to receive tools and the like.

The cover (12) is also of generally circular cross section and has a dome-shaped or spheroidal upper end portion (30) with a depressed rib (32) therein, and an elongated body portion (34) of generally cylindrical configuration extending downward. At the lower end the cover has an internally threaded portion (36) of larger diameter than the body portion (34), and a transitional portion (38) extends between the threaded portion and body portion, thus providing an interior shoulder (40) which slopes upward and inward.

As best seen in (*B*), the lip (18) of the base (10) is provided with an inward and upward sloping surface (42) about the upper circumference thereof to cooperate with the shoulder (40) of the cover. The dimensioning of the parts is such that when the cover is threaded onto the base, the lip will be cammed or deflected inwardly by the sloping shoulder of the cover. The channel (24) permits inward deflection of the lip, and the resilience of the lip ensures firm sealing engagement between its surface (42) and the shoulder of the cover. Thus, water about the container must penetrate past the threadably engaged portions (40) and (42) of the base and cover members (10) and (12), respectively.

else again. Many of them do have unique features. For instance, pickup trucks often have a toolbox mounted right behind the cab. Shirley Parker and John Rohrer of Indiana found certain disadvantages in the conventional lids for these boxes. These lids are generally hinged the length of the box, in the back, or in the middle. Parker and Rohrer found that this arrangement makes "removal and replacement of tools difficult . . . requires space beyond the edges of the box to open fully, or is not watertight." So they designed an improved toolbox to correct these and other deficiencies, with two latches, one at each end of the box cover (Fig. 13.13).

Waterproof tool storage is especially important aboard a boat. Carlton L. Whiteford of Connecticut developed a container for marine use that not only keeps out moisture but also will float if it is knocked overboard, even when weighted down with tools. A further advantage of this toolbox is that it can be made inexpensively from synthetic, corrosion-resistant material (Fig. 13.14).

One of the most onerous parts of a job can be the heavy lifting and pushing required to get something in place. An appliance dolly invented by Jan Macho and Richard Oxford of California is an example of how a simple device can be made to take a big load off your hands (Fig. 13.15). This dolly is unique in its steering mechanisms, which make it easy to maneuver heavy items through tight spaces.

Hardware and tools are so closely connected that it would be illogical not to include both categories in this chapter. As is true for most other areas, new products have mostly come out of the search for labor-saving components that can be manufactured at low cost. Four inventors employed by Bell Labs—John Andreini and Joseph Puccio of New Jersey and Edwin Borchard and Karl-Heinz Pohl of Colorado—were looking for new ways to mount and support electronic and switching equipment. The work they were doing lent it-

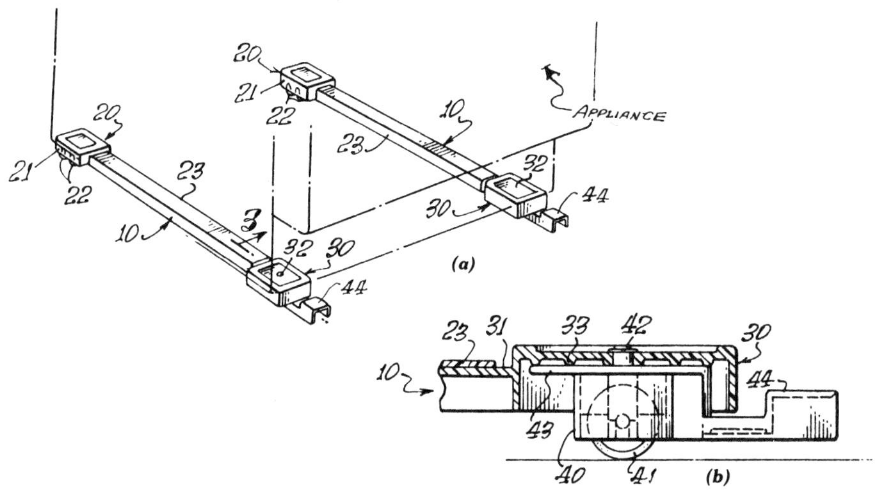

figure 13.15 Two views of the articulated appliance dolly described in Patent No. 3,918,733 (1976). At left are two of the dollies supporting a large appliance with four bearing points. At right is a detailed cross section of one load platform and wheel. A first load platform (20) is supported by a carriage (21) which rides on two or more wheels or tires (22) rigidly mounted to the carriage (21) in parallel plane or rotation. A second load platform (30) is preferably connected with the first platform by means of telescoping shafts (23) and (31), so that the distance between the two load platforms may be adjusted according to the size of the appliance to be moved. Carriage (21) is fitted with a female shaft (23) which telescopes with male shaft (31) fitted to the second load platform (30). The second load platform is supported by a truck (40) having two or more wheels or tires (41) rigidly mounted to the truck in parallel planes of rotation. Truck (40) is fitted with a vertical axle (42), which is rotatable within hole (32) of the second load platform (30). The underside of the second load platform forms a bearing surface supported by bearing surface (43) of truck (40). As shown in (*B*), the bearing surface of the second load platform consists of circular ribs (33) concentric about hold (32), which reduce the friction that would otherwise exist in the absence of such ribs. It is preferred that ribs (33) and bearing surface (43) be composed of materials having low surface friction properties, such as plastic or teflon. It will be understood that any means of securing truck (40) and load platform (30) in a horizontally rotatable relationship may be used. A tongue (44) extends from the truck to permit the operator to rotate the truck and thus steer the dolly. In normal operation the interplay of the ribs and bearing surface (43) establishes a load-leveling function, and the appliance weight is transmitted efficiently to the wheels.

figure 13.16 A perspective view of a hinge pintle described in Patent No. 3,689,115 (1972). This hardware is made up of a support (10), which includes a base (12) with a lateral extension (16) at the rear. In the front end of the base is a cavity (14) whose open end is surrounded by a planar surface (15).

The extension (16) has a flexible cantilevered arm (18) extending alongside the base (12) generally perpendicular to the plane of the face (15). The arm (18) may be a single, ribbonlike element as shown, or it may be bifurcated. In either case, the free end of the arm extends forward of the face and includes a catch (20) consisting of a cam surface (22) inclined to the plane of the arm and intersecting a detent surface (24) orthogonal to the plane of the arm.

These elements interact with a structural component (30) that includes a boss (32) adjacent to a lip (34). The surface of the lip (34) that is on the same side of the structural component as the boss is advantageously rounded, while the opposite surface of the lip is flat.

The elements of the hardware bear a particular relationship to the boss (32) and the lip. The detent surface (24) of the catch (20) is spaced from the face (15) of the base (12) a distance that is substantially the same as the distance between the flat surface of the lip and the surface surrounding the boss. In addition, the cavity (14) has a shape that conforms to the shape of the boss in sufficient detail that the boss is captured within the cavity. As illustrated, the boss has an "I" configuration, while the cavity has a cubelike configuration that has essentially the same height, width, and depth as the boss when measured across its outside corners. The corners of the boss cooperate with the corners of the cavity to prevent rotation of the base on the boss. However, the same result can be achieved where the boss and the cavity have a nonrectangular configuration by means such as a ridge on the boss that is received by a groove in the cavity.

With these relationships, the hardware is mounted on the structural component (30) by orienting the hardware so that the cavity (14) is in juxtaposition with the boss (32) and the arm (18) is in juxtaposition with the lip (34). The hardware is then moved toward the structural component whereby the cavity accommodates the boss and the cam surface (22) of the catch (20) engages the rounded surface of the lip. As the movement of the hardware continues, the arm (18) is deflected away from the base (12) and the catch (20) rides over the end surface of the lip. Then when the cam surface (22) of the catch (20) moves past the flat surface of the lip, the arm snaps back to its undeflected position wherein the detent surface (24) engages the flat surface. As a result, the walls of the cavity in the base embrace the boss to support the hardware, while the arm and catch grip the lip to secure the hardware in place.

The functional portion of the hardware extends from the support (10). In the embodiment shown, the hardware comprises a hinge pintle, and the functional portion consists of a tapered cylinder (40) extending from the base (12). This cylinder is slotted longitudinally to provide bifurcations (42) that have protruding lips (44) at the upper end, and the upper surface of each lip is curved while the under surface is flat.

The cylinder (40) is positioned within a hinge barrel (45) of a hinged member (46) by pressing the barrel against the curved upper surfaces of the lips (44). This deflects the lips and thereby the bifurcations (42) toward one another, permitting the lips and bifurcations to move into the barrel. When the barrel is moved past the lips, the bifurcations return to their undeflected position and the lips secure the barrel on the cylinder (40).

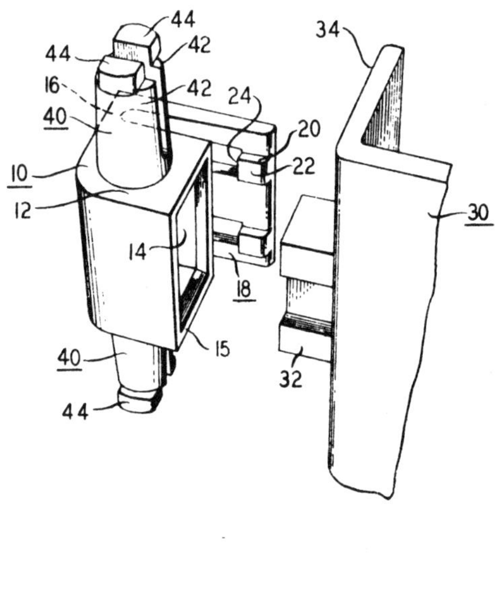

figure 13.17(a) The thumb tack of Patent No. 3,205,757 (1966) includes an oval base (12) which tapers into a parabolic configuration on the major axis of oval (12), this body comprising the head of thumb tack (10). A pair of opposite concave surfaces (14) along the minor axis of base (12) are provided with spaced-apart and elongated grooves (16), which provide a means for providing positive gripping with the fingers, particularly when removing the tack from a surface. An elongated pin (18) of circular cross-sectional configuration molded within the center of the tack is provided with a pair of equally spaced-apart annular recesses (20) which positively secure the pin within the tack. The pin is provided with a conical point (22) for easy insertion into a surface.

A person using the tack pinches the surfaces (14) between his fingers and urges the point (22) of the pin into the desired surface.

When it is necessary to remove tack (10), the operator pinches the surfaces (14) between his fingers as described and simultaneously pulls the head, which will in turn release the point (22), and while doing so, the fingers will grip into the triangularly configurated grooves (16), thus providing nonslip gripping of the surfaces when removing the thumb tack from the surface.

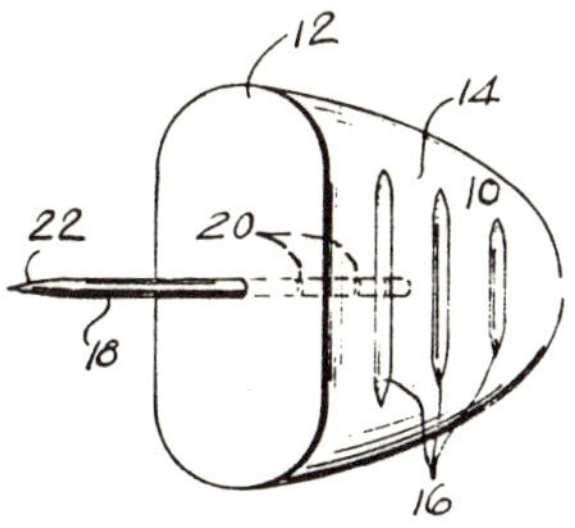

It will be noted that the pinch grip provided for by the concave surfaces (14) of thumb tack (10), in combination with the plurality of triangular configurated grooves (16), thus provides a maximum nonslip grip necessary for removing the thumb tack from a surface.

figure 13.17(b) The thumb screw tack of Patent No. 3,277,769 (1966) includes an enlarged generally planar head (10) and an elongated shank (12) secured thereto as by brazing at its connected end (14). Shank (12) is of a cylindrical configuration and at its other or free end (16) tapers to a sharp point (17). Shank (12) is approximately one-half inch long, and the planar head is approximately three-eighths inch in diameter. Between the ends of the shank, a thread (18) is provided. The thread terminates adjacent ends (14) and (16) and comprises one and one-half turns. The pitch of thread (18) decreases as it approaches its connected end (14).

The thread and the shank are integral. Head (10) is substantially planar and is circular in plane view. The upper surface of the head is roughened as by the provision of milling (24) having a first parallel component (26) and a second parallel component (28) intersecting the first. The depth of milling is considerably less than the dimension of the head normal to the plane of the upper surface.

The milling of the head provides for the frictional engagement of the upper surface of the head with an individual's thumb or finger, making it possible to rotate the tack into and out of threaded engagement with a tacking surface.

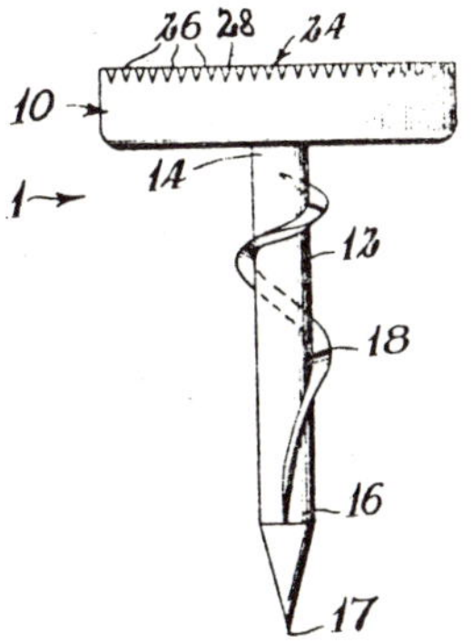

self nicely to the application of a relatively new engineering plastic, polycarbonate, which had a springiness that the inventors were able to use to advantage. The method of attaching hardware to a cabinet seemed ripe for change. For years this had been done by hand, using screws or nuts and bolts. With labor costs on a perpetual Up escalator, there was obviously opportunity for change. The snap-on hardware illustrated in Fig. 13.16 can save a significant amount of time and money in the assembling of a cabinet or similar structure. It is mounted on a structural component having a boss and a lip adjacent to the boss, and including a support portion with a cavity that is essentially the same size as the boss.

Hundreds of patents have been taken out on new types of nails, screws, and similar fastening devices. Unfortunately, most of them have not been successful. This is a difficult field that has brought disappointment to many inventors. Thumbtacks offer an interesting challenge that many have taken up: Is it possible to make a tack that will go in easily and come out reluctantly, except when you want to take it out? In 1965, Edward Kuennen of Iowa patented a tack with a head that was easy to grip and therefore easy to push into most surfaces (Fig. 13-17*a*). The next year, Charles Lippmeier of Illinois patented a tack for use in soft, porous areas. The principal features of his invention are an enlarged head and a lightly threaded shank that grips the material into which it is pressed (Fig. 13.17*b*). Both of these tacks are simple products designed to make an ordinary task easier to perform. Neither of them has managed to

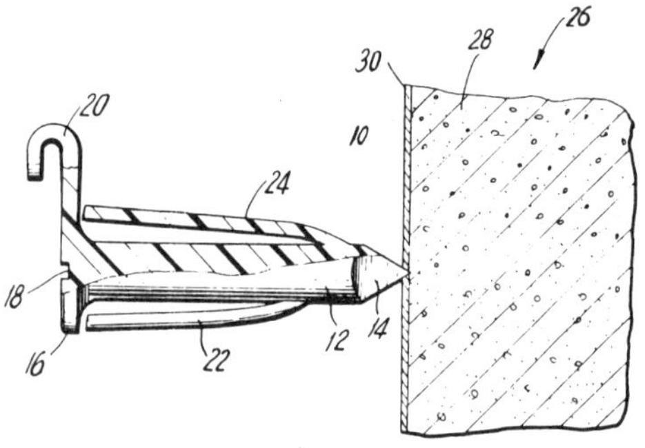

figure 13.18 Side view of the wall fastener in Patent No. 3,895,773 (1975), partially broken away. It is in position to be driven into a wallboard. The fastener (10) includes shank (12) having a point (14) at one end and a head at the other end. The head is engageable by the provision of a transverse slot (18) suitable for a screwdriver.

Support means in the form of a hook (20) is integrally formed with the head (16). The hook is suitable for supporting an object after the shank (12) is driven into the wall. The hook is slightly remote from the head itself in a radial direction with relation to an axis of the shank to prevent the hook from being damaged while the head is hit and the fastener is hammered into the wall.

At least one projection (22) extends from shank (12) proximate the pointed end (14) of the fastener and inclined relative to the shank (12) in the direction of the head (16). Two inclined projections (22) and (24) are provided, extending from substantially diametrically opposite portions of the shank. The projection (22) extends from a portion of the shank axially closer to the head (16) and further from the pointed end (14), while the projection (24) extends from a portion of the shank axially further from the head and closer to the pointed end. Accordingly, each of the projections (22) and (24) extends from the shank portions, which are axially displaced along the shank. These projections are integrally formed with the shank. The free ends of the projections are similarly disposed proximate to the head (16). Accordingly, the projection (22) is shorter than the projection (24). Such construction has resulted in improved performance and has facilitated penetration of the fastener into a wall surface.

The entire fastener (10) is formed integrally, including the integral connection between the hook (20) and the head (16), as well as the integral connection between the projections and the shank. The projections are made flexible by selection of the material from which the fastener is made as well as by selection of the dimensions thereof.

A plaster board or gypsum wallboard (26) is shown to include a plaster, gypsum, or other brittle material section (28), and a front paper-covering layer (30). The fastener (10) is positioned with its point (14) proximate to the paper covering layer (30), with the hook (20) in an upside down condition disposed above the head (16). It should be noted that the projections are disposed of in a plane which is substantially normal to the direction of the slot (18). The hook (20), on the other hand, is substantially aligned with and parallel to the projections.

displace the old reliable flathead, shortshanked tack that has been in use as long as most people can remember. This tack certainly isn't perfect, but it is so simple and so inexpensive that it is difficult to replace.

There is one area where the thumbtack has pretty much lost its usefulness—handing up pictures and other wall decorations. Various kinds of nails and adhesive-backed strips with hooks have been made for this purpose, and some of them have had great success. One of the latest entries in this competition is a molded wall fastener made of nylon that has been patented by Alan Solo of New York. This inventor found previous fasteners either too expensive to make or ineffective. His fastener is constructed so that it can be twisted into a plasterboard wall (Fig. 13.18).

In the twentieth century, many changes have been made in paints and related materials and tools. Hard epoxy paints, easy-to-apply latex types, and bright, fade-resistant colors are only a few of the major innovations in this field. The way in which most paint is applied has not changed much, with the notable exception of the roller that works so well with water-based paints. Inventors keep trying to replace the paintbrush, however, and they are having some success. Donald Seip of New Jersey has a patent on an adjustable-pad painter, which he has assigned to the Red Devil Company. Previous attempts at paint pads were mostly complicated gadgets that did not do the job very well. Seip's pad is designed to do clean work on window sashes, corners, and other areas where it's difficult to apply the paint without making a mess. His pad is removable for cleaning and adjustable to meet a variety of conditions (Fig. 13.19).

Painters who still use a brush for application have at least one major problem that inventors keep trying to eliminate—the drippiness of paint. The grooves on the top of the can often get clogged with paint, making it difficult to remove and replace the lid. And the paint often runs down the side of the can, making for a messy job. In these circumstances, leftover paint becomes more trouble than it's worth. An obvious solution is that practiced by most painters—to pour as much paint as is needed for a particular job into a separate container. But even when this is done carefully, some paint always seems to get into the grooves, there to harden and to make the replaced lid stick to the can. The grooves themselves could be eliminated only if some other way could be found whereby the lid could be sealed to the can's top.

George Beadles of California had an idea for something that he patented as a combination paint can spout and scraper that can be attached to any regular-size paint can (Fig. 13.20*a*). This eliminates the need for pouring, and with reasonably careful use it keeps all the paint out of the grooves. That patent was granted in 1956. Just twenty-one years later, John Ritter of Michigan patented a new paint can pour spout. It is designed for easy attachment to a paint can's lip, snapping down over the outer edge (Fig. 13.20*b*). This keeps the paint away from the grooves. There is also a square opening to facilitate brush wiping and a spout for pouring off some of the paint. These two inventions, decades apart, are de-

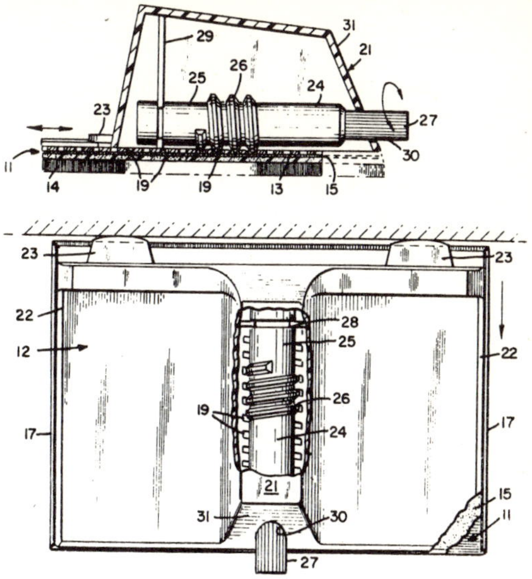

figure 13.19 Adjustable pad painter, from Patent No. 3,736,616 (1973), comprised of a pad (11) and a handle (12). The pad consists of a thin sheet of spongy material (13), with a layer of nylon filaments flocked thereon. (Other natural or synthetic bristles may be substituted for the nylon.) The filaments are adapted to receive paint for application to a surface. A layer of some suitable pressure-sensitive adhesive (15) is applied to the sheet of spongy material (13) on the surface opposite the nylon filaments. The sheet of spongy material (13) is secured to a rigid plate, preferably metal. The pressure-sensitive adhesive (15) makes it possible to replace the paint-receiving member.

The plate (16) is substantially rectangular in shape and provided with flanges (17) on each lateral edge, inwardly bent. A series of spaced slots (19) disposed midway between and parallel to the flanges (17) are provided in the plate (16), completing the structure of the paint-applying pad member.

The handle consists of a substantially rectangular horizontal support (20) and a hollow upstanding handle (21). The support is provided with lateral bevels (22) along each side. These bevels are adapted to receive the flanges of the plate in a freely sliding relationship. The front of the horizontal support (20) is formed with two outwardly extending spaced guides (23), which are adapted to bear against any surface not disposed in the plane of the surface being painted. The guides also establish a spaced relationship between the surface not being painted and the leading edge of the nylon filaments (14).

To change the space between the edge of the guides (23) and the paint-applying pad (11), a worm (24) is freely journaled within the upstanding handle part (21). This worm consists of a cylindrical body (25), threads (26), and an adjusting nob (27), all integrally formed. The body of the worm is provided with an annular groove (28), which is journaled within a wall (29) within the hollow handle part (21). The opposite end of the worm body (25), which is formed into the adjusting nob (27), is journaled within an opening (30) in the rear wall (31) of the handle part (21). The worm is so disposed within the handle part (21) that its threads (26) enter the spaced slots (19) in the plate (16). The slots (19) are offset to conform to the threads (26).

figure 13.20(a) Paint can with spout and scraper attached, as illustrated in Patent No. 2,767,891 (1956). The brush (35), indicated in dotted lines, is being scraped. The combination scraper/spout (*S*) has been attached to the paint can (*P*) by forcing it into the groove of the can's top rim. Walls (30) prevent any paint from spilling into the groove.

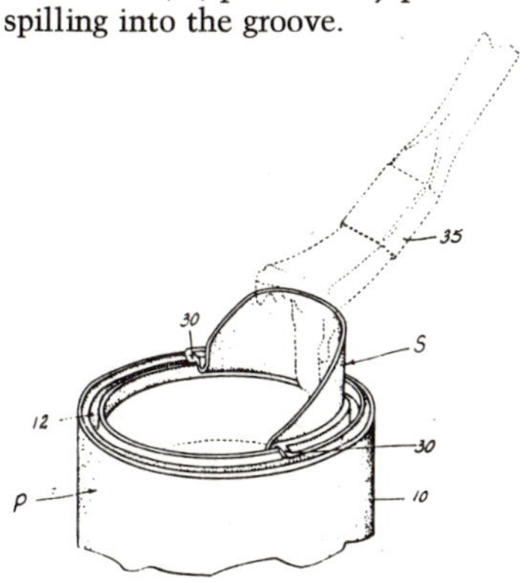

figure 13.20(b) Pouring spout, brush scraper, and cover attached to a paint can, as illustrated in Patent No. 4,014,465 (1977). The lid (10) can be made from plastic and is adapted to cooperate with can's rim so as to provide a seal around the rim to prevent paint from flowing into the circular groove. The flat area has an outer flange and an inner flange, engaging the outer and the inner edges of the can's rim. The square opening (26) permits entry of a paint brush. A pouring spout (28) is attached so that its bottom edges define two adjacent edges of the square. The remaining edges (29) and (30), define surfaces where the paint brush can be wiped.

The lid has a peripheral bead (38) that extends from the lower side edges of the spout and along the outer edge of the can, providing a raised barrier to prevent the spilling of paint. To facilitate separating lid and can, which are snap-locked together, the side wall of the lid is provided with parting apertures (40), which extend through the outer flange and into the channel defined between the outer and inner flanges. These apertures provide a simple means for inserting something like a screwdriver end through the lid and under the surface (20) to bear down against the paint can rim. A cover (12) permits the user to provide a temporary seal of the can and prevent evaporation of the paint when not in use. The cover comprises a top wall (44) having a handle (46) that permits the user to pick it up and place it in the opening (26). The cover has inclined side walls (48), formed to define a square opening that is so sized as to seal the square opening.

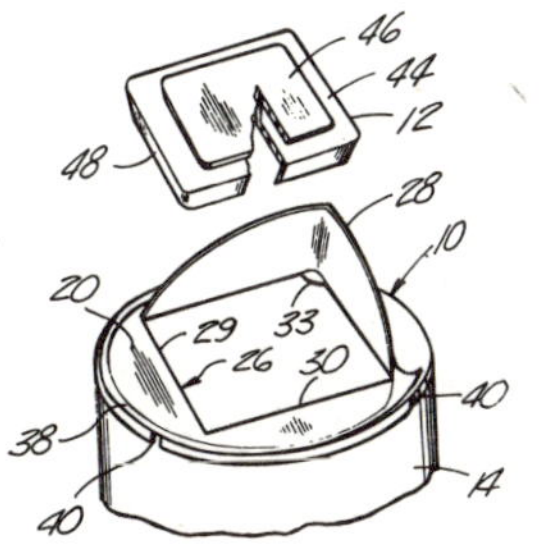

signed to accomplish the same thing in more or less the same way. Will the new one succeed where the old one did not? Time will give the answer.

Just a bit outside the main concerns of this chapter, but of great interest to craftsmen and like-minded readers, is the field of building materials. Richard Neilson of Iowa developed a mortarless system for building a wall out of masonry blocks (Fig. 13.21). Each block is coated on top, bottom, and ends with a material that has a relatively low melting point, but well above the maximum temperature that might be reached during the hottest day of summer. Each block also has matching thin shafts running from top to bottom. One hole in each set of blocks contains a connecting rod held

figure 13.21 Portion of a wall constructed with mortarless blocks, according to Patent No. 3,783,566 (1974). Each block can be partially hollow to reduce cost and make for easier handling. Each end wall has an opening (22) that extends vertically from top to bottom. In the center rib of each block there is a vertical opening (26), slightly larger than the opening (22), that extends halfway from top to bottom. The rod (28) is approximately the length of the distance from top to bottom of block.

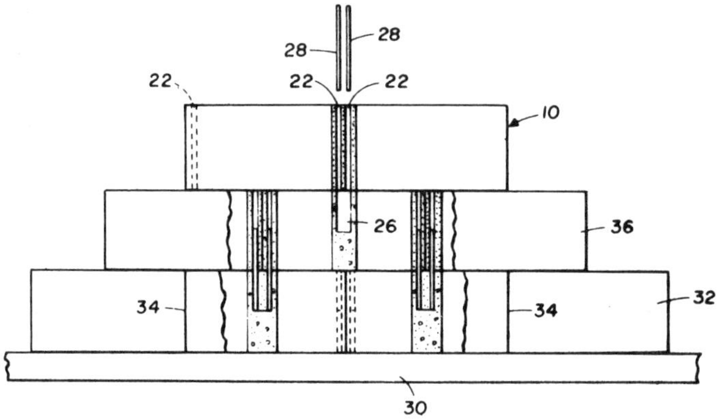

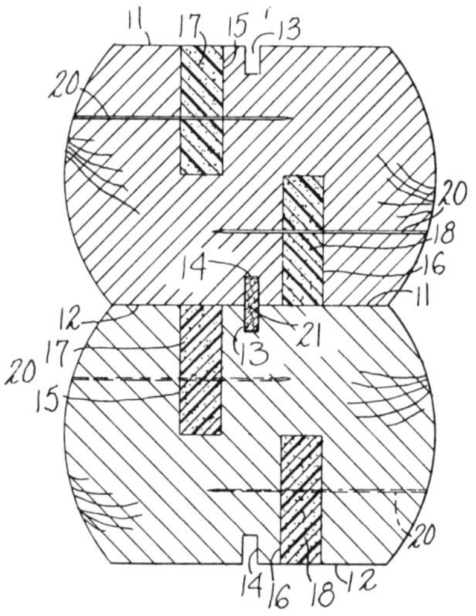

figure 13.22 Vertical transverse section of two insulated wall logs, one superimposed on the other, as illustrated in Patent No. 3,992,838 (1976). A natural wood log is de-barked, then sawed and planed to form flat parallel upper and lower faces (11) and (12), each provided with a medially-disposed longitudinal groove or kerf (13) and (14). A wider and deeper groove (15) and (16) is cut into each planed face, adjacent and parallel to the kerfs (13) and (14), the grooves (15) and (16) being on opposite sides of the vertical plane through the kerfs and each groove extending approximately halfway through the log. The wall logs usually average 9 inches in diameter, with variations between 7 inches minimum and 11 inches maximum, the kerfs (13) and (14) being one-half inch wide and the grooves (15) and (16) being one inch wide.

Each groove is filled with a suitable foamed plastic (17) and (18), preferably foamed in situ to ensure good adhesion and complete filling of the space. The stability and rigidity of the combination may be further aided by the provision of ties in the form of headless metal spikes or pins (20), driven horizontally into the sides of the log at suitable intervals and passing through the plastic at a level of slightly less than half the depth of the groove. Long screws could be substituted for the spikes with the same stabilizing effect but at a higher cost.

Proper vertical alignment of the logs, and basic sealing of the horizontal cracks between them, is effected by the provision of splines (21), designed to fit in the kerfs of adjacent logs.

in place by the same meltable adhesive material. After the blocks have been put in place to form the wall, heat is applied to (1) fuse the blocks together and (2) release the connecting rods so that they will all drop to interlock each layer with the next. This system is designed to save time and minimize the chance for error.

Interest in log cabins has increased in recent years, though interest in chopping down trees and shaping logs to fit one upon the other has remained pretty much where it has always been—not all that high. But the durability of logs, and their reasonable cost in comparison with some other materials, makes a log cabin an attractive investment. Two Canadians have proposed new methods of constructing a house from logs. Thomas Hisey developed prefabricated log sections that can be joined so as to be weathertight. The upper and lower sections are planed flat, and a narrow groove is cut lengthwise into each surface about halfway into the log. Then each groove is filled with foamed plastic, which is more effective than wood in providing insulation. The foam fills all the gaps between the logs quite naturally, and air penetration is virtually eliminated. (See Fig. 13.22.)

Ray Fell of British Columbia worked on simplifying log-cabin construction for many years before his death in 1976. He devised two machines, one that cut and shaped logs to be perfectly round, another that cut the logs to size and made the saddle corners. The structure made possible by these two machines—and by Fell's ingenious design—is shown in Fig. 13.23. Like Hisey's square-cut logs, Fell's round ones have grooves cut into one side to receive in-

figure 13.23 Vertical section through a part of a log cabin wall formed in accordance with the description in Patent No. 3,863,409 (1975).

A lower portion of one wall has a wall element that is cut essentially along a horizontal diameter so as to lie as shown on foundations of the building (not shown), as is common practice with log cabins of this construction. The log cabin wall element (13) has a central longitudinal axis (14) and is disposed upon a similar second wall element (16). The element (13) has a generally cylindrical surface (15) having a radius, the surface being concentric with the longitudinal axis. A lower portion of the cylindrical surface of the element (13) has a longitudinal groove (17), the groove extending along the element. A radial slot (18) extends from the groove to an end edge (20), radial length of the slot being approximately one-third of the radius of the element (13), the slot similarly extending along the element.

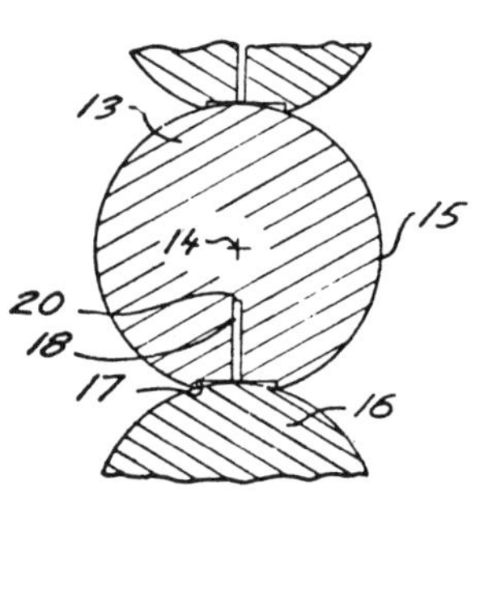

sulation and, if desired, electrical wiring. The grooved edges rest easily on the round surface of the log beneath. Since there are no grooves that must join up with each other at close tolerances, machining costs are kept to a minimum.

Tools, hardware, building materials, and the like will continue to offer excellent opportunities for the introduction of new ideas and new products. The challenge of the immediate future is to develop the capability to build soundly in the face of continually rising costs. The homebuilding field, in particular, will probably have to undergo radical changes. Do-it-yourself builders will continue to grow in both skill and numbers. Products that make it easier for them to do what they want to do will be in great demand. Patented inventions, many of them by people not affiliated with any manufacturer or distributor, will continue to play an important role in this vast, still-growing field.

Appendix

The statements, advisories, and suggested letters here are by way of general examples. No precise answers can be given for specific situations. If you wish to act or react to a specific set of circumstances, consult a patent counsel.

LETTER TO A PROSPECTIVE LICENSEE
(When a Patent Application Has Not Been Filed)

Dear __________:

I am in possession of confidential information with regard to certain processes and apparatus for *(the category, e.g., solar energy collectors).* I believe that I have a proprietary interest in this material, and I am willing to disclose it to *(name of company to which letter is addressed),* hereinafter referred to as A, and to license A to use the process and apparatus on the following terms:

1. A agrees to hold in confidence and not to divulge to anyone any and all information concerning processes and apparatuses for *(the category)* disclosed to A under this agreement, except:
   (a) information which at the time of disclosure is in the public domain or public knowledge;
   (b) information which, after disclosure, becomes part of the public knowledge by publication or otherwise, except by breach of this agreement by A;
   (c) information which A can establish by competent proof was in its possession at the time of disclosure by *(your name),* and was not acquired directly or indirectly from *(your name),* and
   (d) information which A receives in good faith from third parties, provided, however, that such information was not obtained directly or indirectly from *(your name).*
2. A agrees not to use any and all information as defined under 1(a) to 1(d) except under a license from *(your name).*
3. It is expressly understood between the parties hereto that A does not obtain any rights or privileges under any issued or to-be-issued patents of *(your name)* or others.
4. This agreement shall extend until such time as the confidential information referred to herein becomes part of the public domain or public knowledge by publication or otherwise, except by breach of this contract by A, or until written cancellation of the agreement is obtained from *(your name).*

If the foregoing terms and conditions meet with your approval, please show your acceptance and agreement by execution of the duplicate copies of this letter in the spaces provided below and return one copy to *(your name),* whereupon this letter shall have become the agreement between *(your name)* and A with respect to the proprietary interest of *(your name).*

Very truly yours,

(Your signature) __________

Dated:

Accepted and agreed to
(*Name of company*)

By: *(Signature of officer)* ________ *(Title)* ________

Dated:

[This suggested form can also be used by an officer or representative of a small business simply by inserting the name of the business wherever "your name" is indicated. Generally a short form of the name or an acronym is sufficient. The company's full name should be mentioned at the beginning of the letter and at the end, under "yours very truly."]

LETTER TO A PROSPECTIVE LICENSEE
(When a Patent Application Has Not Yet Been Filed)
Short Form

Dear __________:

I have certain information relating to *(very general description of information),* which I believe, if adopted and used, would result in substantial additional profits for your company.

(Optional) I have consulted my patent attorney, and it appears that the information may [or you may say "probably does not" depending on the attorney's advice] have patentable features.

I would be willing to disclose this information to you if you agree either to compensate me for use of the information or promise not to use the information or divulge it to others in the event that you do not elect to use this information.

Very truly yours,
(Your signature) ________

Dated:

LETTER TO A PROSPECTIVE LICENSEE
(After a Patent Application Has Been Filed)

Dear __________:

I have invented a device relating to *(description of invention in general terms),* which I believe you could add to your line of products.

A patent application has been filed on this invention.

If you would be interested in assessing the potential value of this invention to your organization, I would be willing to send you a copy of the patent application [see note below for what *not* to send] on which I will rely for my rights, provided, of course, that you inform me of your interest or lack of interest within a reasonable time and that you return the copy to me if you have no interest.

The invention is available for licensing on an exclusive or nonexclusive basis. If you should require a fairly long time to consider or to test it, terms can be arranged for an option of ninety days or longer.

Very truly yours,

(Your signature)

[*Note: When mailing a copy of a patent application to a prospective licensee, be sure to remove all data on the papers relating to serial number and filing date, including the attorney's file number.*]

LETTER TO A PROSPECTIVE LICENSEE
(After a Representative Has Agreed to See You)

Dear ___________:

I have agreed to show you certain of my disclosures and to give you the necessary detailed information and know-how with respect to them.

In return, you agree that you will keep this information in confidence and not divulge it to others and that you will not use it in any way, except with my written consent.

Please indicate your acceptance of this by signing a copy of this letter, in the space provided below, and returning such signed copy to me.

Very truly yours,

(Your signature)

Accepted and agreed to:
(Name of company)

By: *(Name of official)*

Dated:

LETTER TO AN INQUIRER
(From a Large Company to Someone Offering an Idea)

The following letter was quoted in a decision of the U.S. Court of Appeals, 9th Circuit, Nov. 5, 1965 (No. 19501), in the case of *Davies v. Carnation Co.* Miss Davies contended that she had given Carnation an idea for a possible use for its Instant Milk, that the company had subsequently adopted the idea and promoted it through its advertising, but had not compensated her for it.

Dear Miss Davies:

After studying the resume of your experience in the food industry, we can safely assume that you are aware of the dangers involved in giving consideration to unpatented and uncopyrighted ideas and suggestions which are submitted for our consideration and which are unsolicited by us. Because of this we have been forced to adopt a policy that we will not consider any suggestion or idea of this type unless the person wishing to submit the idea or suggestion first signs a release form of the type submitted to you by Mr. Taylor [who had met with

Miss Davies]. We also will not agree to pay any specified sum in advance for any suggestion submitted.

We realize, of course, that a program of this type undoubtedly causes many people to refuse to submit their idea or suggestion to us for our consideration. While we sincerely regret this, we have been forced, because of legal considerations, to adopt this rather harsh procedure. As I believe Mr. Taylor pointed out to you, we would be most happy to give consideration to your findings but only upon the conditions expressed in the release. We will not and cannot agree to pay any advance in the sum of $3500.00 [the amount Miss Davies requested] if we find your research findings to be of use to us. I might add that Carnation Company has always compensated people for any suggestion submitted which was new to the company and which was of value to the company. The amount of compensation, however, as expressed in our release, rests solely in our discretion and would naturally be based upon the value of the idea of suggestion which was submitted. . . .

Very truly yours,
R. D. Kummel

Several months later Miss Davies submitted her idea, saying, "To facilitate your evaluating the project and its timeliness with respect to your competition and to determine your position, I am enclosing a complete disclosure, as I have faith and trust in the fair dealings of your company." What the suggestion amounted to was "pouring or sprinkling dry milk into warm or hot liquids or food mixtures, during the usual cooking processes."

Mr. Kummel replied promptly, stating that Carnation had "no interest in acquiring whatever rights you may have to these ideas." He went on to point out, in effect, that the Company was aware that its Instant Milk could be used with heated liquids or foods, but had not yet chosen to advertise this property. Later, when they did, Miss Davies sued. She lost: As the court said, "Mere idea without novelty is not a property right to which one may claim exclusive ownership, especially where defendant . . . had already entertained the idea and shared it with the general public." In this particular case Carnation was able to prove that it did indeed know of the mixing properties that Miss Davies claimed as her own "idea."

A Corporation's Response to a Request That It Consider an Idea

YOUR IDEA

A Booklet Prepared by a Major American Company to Describe Their Policy in Considering Inventions and Ideas from the General Public

A. *About your idea*

We understand you have an idea that you think may be of interest to our Company. This booklet will tell you the basis on which we can consider your idea. Please read it carefully.

First of all, we appreciate your thinking about us when you think of new ideas. Our Company has been built on the principle that we will grow as we introduce new and useful products. To maintain this flow of new products, the company employs several thousand scientists and engineers. We think they do an outstanding job, but, even so, a few of the products which sparked our growth originated with outsiders.

However, this does not happen very often. Less than 1% of all ideas we receive from outsiders prove to be of interest.

At this point, it is only fair to say that we are not asking you to submit your idea. If you choose to do so, it will have to be according to the procedure set out in the following pages.

Also you must understand that it is the Company's policy not to accept ideas from the outside on any kind of confidential basis. If you decide to submit your idea to us, you must keep this policy in mind.

B. *We have some ground rules*

To avoid the possibility of misunderstanding, we have found it necessary to have some ground rules for considering ideas submitted by outside persons. These rules, which apply to the Company, its divisions and subsidiaries, cannot be modified or waived, except with the signed permission of one of the Company's officers. The general principle involved is that of fair play, that is the protection of both the outside inventor or author and the Company.

C. *Can you classify your idea?*

Read this book completely and then see if you can classify your idea. If your idea involves an invention of some kind (such as a device, an article or a design for an article, a process of manufacturing, a composition such as a formula or a chemical compound, etc.), Sections D, E, F, G, I, J, K, and L will probably be the most appropriate to your idea, and you should again read these pages carefully.

On the other hand, if your idea involves written, recorded, or pictorial materials such as instructions for games, designs for game boards, manuscripts, program scripts (e.g., for videocassette programs), recordings, film strips, or photographs for educational courses; or musical compositions for recorded background music, etc., your idea is probably based upon literary property, and you should be sure again to carefully read the information set out in Sections H through L.

D. *If your idea involves invention, it must pass some tests*

Speaking broadly, the Company will be interested in your idea only if it passes these three tests:

1. It is concerned with a working device, product, process, or formula and is not a sales promotion scheme or a bare idea for a new product. Additionally, please understand that we are not interested in ideas for possible uses for any of our own products. We rely on our own scientists, engineers, and marketing analysts for this.
2. It was not already old when you thought of it. By "old" we mean any of the following, but this list is only exemplary:
   —shown or described in a patent, catalog, magazine, encyclopedia, or some other printed publication.
   —commonly known to the public or the trade.
   —known to the Company
3. We feel it has attractive commercial possibilities.

E. *See your lawyer*

Before you present your idea to us, we urge you to consult an attorney, preferably a patent lawyer, of your choice. He can advise you whether your idea is patentable and, if appropriate, tell you what steps are necessary to provide you with patent protection.

F. *We like patents*

We prefer to consider only those ideas embodied in inventions or discoveries which are described and claimed in a patent or a patent application, since we are then in a position to know exactly how you have defined your invention. If you already have an issued patent covering your invention, just send a copy to us; nothing else is required.

G. *If you have no patent*

If you do not have an issued patent on your invention, the Company will consider your idea only under certain conditions.

1. Your idea must fall in one of the classes of subject matter which can be patented. (It must not, for instance, be concerned with advertising, slogans, selling, or how to run a business.)
2. You must complete and return to us the Idea Submission Agreement form which is part of this booklet. If you have already sent us a description

or other information about your idea, we will keep it sealed in a locked file unavailable to our employees until the signed form is returned.

3. You should accompany the completed Idea Submission Agreement form with a detailed written description of your invention, illustrating it with drawings or photographs, if you can. Please do not send models; we'll ask for them if we need them.

Be concrete and specific. In other words, don't merely make a suggestion such as "Why doesn't the Company develop a pill which you could add to a tank full of water and produce a motor fuel"—tell us what to make the pill out of!

(Incidentally, some ideas are hardy perennials—they come up every year. If your suggestion involves coating adhesive on the back of coarse grit sandpaper and cutting the resulting material into strips for use as temporary automobile tire chains, a great many other persons have already had the idea!)

4. To be certain that our files correspond, be sure to keep a copy of everything you send us. You must agree to let us keep whatever written material you submit, if we feel this is necessary.

H. *Ideas involving literary property*

Literary property such as game instructions; game board illustrations and designs; manuscripts, program scripts and illustrations for educational courses or programs; songs for background music systems; etc., may be copyrightable. As we stated earlier with respect to patents, we urge you to contact an attorney of your choice about copyrights, preferably a patent lawyer who will also be familiar with copyright problems. He can advise you as to whether or not your material may be copyrightable and how to secure a copyright. The Company cannot do this.

If your idea involves literary property, we must ask that you complete and return to us the Idea Submission Agreement form which is in the back of the booklet. Until you do so, any literary materials that you may have already sent to us will be kept sealed in a locked file which is unavailable to our employees.

If you have not already done so, send your literary materials to us at the same time that you return the signed Idea Submission Agreement form. If you have secured a copyright registration certificate from the Copyright Office in Washington for your material, send us a copy of the certificate along with the agreement form. Do not send the original certificate as we cannot return it. Also, be sure to indicate on the Idea Submission Agreement form whether or not your literary material has ever been published and, if it has, send us a copy of the publication.

Remember, as we said with respect to inventions, make a copy of everything you send to us as we must insist on the right to keep whatever we feel is necessary.

I. *What the Company will do*

After you have sent us either (1) your issued patent or (2) a completed Idea Submission Agreement form accompanied by the information you have checked and noted on the form, here's what we'll do:

We will submit your idea to those people in the Company best able to evaluate it. Any assistance in time or skill which you provide will, of course, be at your expense.

If we have any interest in your idea we will notify you as soon as we can—but this will probably take more time than you would suppose. We'll try to write you within six weeks; but some ideas (particularly ideas relating to games) require more time than this to evaluate. If you want to submit your idea to other companies while you're waiting, feel free to do so.

If we are interested in your idea after our evaluation, we may wish to enter into a formal written agreement which will, among other things, grant rights to the Company and provide for your compensation. We will not conduct any negotiations before we receive an issued patent, or a signed Idea Submission Agreement. A meeting may be arranged after our evaluation of the idea, if both you and the Company are interested in pursuing the matter further.

We will not assume any special obligation to you just because you make a disclosure to a Company officer or employee. If we consider your idea, we are not thereby admitting that it is new. Even if you obtain a patent, we can't promise that we will not contest its validity at a later date. This may seem harsh, but the Patent Office sometimes misses earlier publications which would have prevented the grant of a patent had these been apparent. There are many other reasons why a patent may be invalid. We are sure you will agree that anyone—including the Company—should have the right to challenge an invalid patent.

The same is true for copyright certificates. The Copyright Office simply registers your claim of copyright and does not take any steps to determine its validity.

As we said before, we can't accept your idea "in confidence," since that term has special legal meaning. Besides, a thorough study often requires the advice of several Company people, our attorneys, and perhaps even prospective customers or other outside people. To get their opinions, we may have to tell them about your idea.

We can't assume any obligation to act in a consulting capacity in regard to the patentability or copyrightability of your idea, the desirability of a patent or copyright, the commercial possibilities of your idea, or your rights against others.

If you are not an adult, we cannot deal with you unless your parent or guardian also signs the Idea Submission Agreement.

Nor can we consider any idea you submit unless you are the owner of the invention or literary material. If others have any rights in your materials, you must tell us who they are and the nature of their interest. You may, of course, appoint an attorney or agent to submit your idea; if you do, however, he should also submit proof of his appointment.

J. *In a nutshell*

Here is what we've tried to say:

1. We will consider your idea only at your request.

2. You must assure us that to the best of your knowledge you are the one who originated the idea, that you own the idea, and that you have the legal right to negotiate with us concerning it.

3. We assume no obligation to you whatsoever in the absence of a written agreement entered into after we evaluate your idea.

4. No confidential relationship between us is established when you submit your idea. The Company does not promise to keep it secret.

5. After we have studied your idea we will tell you

IDEA SUBMISSION AGREEMENT

My idea relates to: ______________________________

____

____

Check all appropriate boxes and fill in all pertinent blanks.

☐ I have sent to you ☐ I am enclosing ☐ I am sending under separate cover } information on this idea.

This information is in the form of:

- ☐ Patent Application (include serial number)
- ☐ Written description
- ☐ Photographs, Drawings, or sketches
- ☐ Unpublished manuscripts, scripts, etc.
- ☐ Published literary material
- ☐ Other (specify) ______________

Please indicate the number of pages of any of the above materials you send us on this line: ______________ pages.

If your idea involves literary material, please fill in the following:

1. Have your literary materials ever been published? Yes ____ No ____
2. If published, indicate the date of first publication: ______________
3. Was a copyright notice included in the publication? Yes ____ No ____
4. Did you secure a copyright registration certificate for the publication? If your answer is yes, please send a copy of the certificate. Yes ____ No ____
5. If your literary material was published, please send a copy of the publication.
6. If your idea involves a game not covered by an issued patent (some games may be patented, others copyrighted), send us a photograph of the game board and components and a copy of the instructions.

I am keeping a copy of this agreement and the other information sent to you as indicated above.

Name and address of present employer ______________________________

____

Name and address of employer when you first got the idea ______________________________

____

Does either of above employers have any right or equity in the idea? ☐ **Yes** ☐ **No**

List the names, addresses and interests of others having any rights in this idea and ask them to sign this agreement.

(If none, write "None")

____

____

____

____

I have read the "YOUR IDEA" booklet, and particularly the conditions set forth in Section J, immediately preceding this agreement. I agree to each of the conditions contained in the "YOUR IDEA" booklet with regard to the presently submitted idea and all related future suggestions.

______________ **Signature of submitter***	______________ **Date**	______________ **Name (please print)**
		______________ **Address (please print)**
______________ **Signature of parent or guardian if submitter is not an adult**	______________ **Date**	______________ **City, State and Zip (please print)**

***If this idea is submitted by an agent of the originator, the agent should sign and submit written proof of his authority.**

______________ **Signature of Agent** ______________ **Date**

only whether we are interested in it. We assume no obligation to tell you about anything we knew previously or have discovered since you submitted your idea.

6. By submitting your idea to us you are not giving us any rights under a patent or copyright you now have or may obtain in the future. On the other hand, the Company will not have any obligation to you, in respect to any such patent or copyright, unless and until a formal written contract has been entered into between you and the Company.

K. *It's your move*

If you would like to have the Company consider your idea, please complete the Idea Submission Agreement. A duplicate form is included on which you may make and retain a copy for your records. Tear out the original and mail it to the Company.

If your idea is described and claimed in your issued patent, you do not need to complete the Idea Submission Agreement—just send a copy of the patent to us.

L. *One idea to a form, please*

If you have more than one idea to submit, please make a copy of the Idea Submission Agreement or request another from us. Subsequent correspondence should cover only one idea or suggestion in any one letter.

Good luck in your creative career!

Addenda

STANDARD FORM OF RELEASE

A Mutual Release Regarding Suggestions or Inventions

Submitted to *(Name of Company)*

PURPOSE OF RELEASE

(Name of company) is constantly striving to improve its products and to expand its business, and welcomes the opportunity to negotiate with companies and individuals outside its own organization who may have worthwhile patent rights to sell or license. *(Name of company)* cannot receive or consider suggestions made in confidence and will consent to disclosure of suggestions to its representatives only with the understanding that reliance must be placed on such rights as are owned or acquired under the patent or copyright laws and not on any expectation of reward in any other circumstance.

(Name of company) may be working on a development similar to that submitted, or in a closely related field, and in the course of discussion (name of company) 's ideas may be disclosed. In that event (name of company) agrees that it will rely solely on such rights as it owns or acquires under the patent and copyright laws and not on any other expectation of reward.

No obligation of any kind is to be imputed to either party with reference hereto unless and until a formal written contract has been entered into, and then the obligation shall be only in accordance with the formal written contract.

(Name of company) has no intention of infringing anyone's patent rights and therefore must preserve records of material submitted to it. This privilege shall likewise be mutual. Accordingly, each party shall have the right to retain all material submitted to (it) by the other party or to make a copy. The inventor in *(his or her)* own interest should make and preserve duplicate copies of everything submitted and should consult *(his or her)* own patent attorney regarding the protection of the invention.

RELEASE

I have read and understand the above Purpose of Release. In consideration of *(name of company)*'s examining my suggestion or invention pertaining to and extending to me the same release as to any information *(name of company)* imparts to me with respect thereto, the undersigned hereby releases *(name of company)* from any liability in connection with any material now submitted or which may be submitted respecting the same, or any liability because of the use of any portion thereof, except such liability as may arise from unauthorized use of valid patents or copyrights now or hereafter issued. This release does not convey to either party any patent rights or copyrights in any circumstance. It is further agreed that as regards any improvement or variation, which is submitted or suggested in the future relating to the subject of this release, the same conditions shall apply unless they are expressly waived in writing signed by proper officials at the time such improvement or variation is submitted.

(Signature of owner of submitted material)

Date: __________

Address: ____________________

Witness: _________________________

DISCLOSURE DOCUMENT PROGRAM

(Date)

Hon. Commissioner of Patents
Washington, D.C. 20231
To Whom It May Concern:

The undersigned, being the inventor of the disclosed invention, requests that the enclosed papers be accepted under the Disclosure Document Program, and that they be preserved for a period of two years.

(Your signature)

Name
Address
City, State, Zip code

Send this letter along with the following:

1. *Fee of $10 (check payable to The Commissioner of Patents).*
2. *A copy of your disclosure, as complete as possible, (for illustrations of the invention, freehand sketches, rather than formal ink drawings, are permitted). Sign your disclosures as the inventor.*
3. *Self-addressed return envelope.*
4. *A copy of the letter (making a total of two).*

LIBRARIES WITH PATENT COLLECTIONS

State	*City*	*Library*
California	Los Angeles	Public
	Sunnyvale	Public
District of Columbia	Arlington, Va.	U.S. Patent Office
Georgia	Atlanta	Georgia Institute of Technology
Illinois	Chicago	Public
Massachusetts	Boston	Public
Michigan	Detroit	Public
Missouri	Kansas City	Linda Hall
New Jersey	Newark	Public
New York	Albany	State University of N.Y.
	Buffalo	Erie County
	New York	Public
Pennsylvania	Philadelphia	Franklin Institute
	Pittsburgh	Carnegie
Ohio	Cincinnati	Public
	Cleveland	Public
	Columbus	State University
	Toledo	Public
Oklahoma	Stillwater	State University of Agricultural and Applied Science

Rhode Island	Providence	Public
Virginia	Arlington	U.S. Patent Office
Wisconsin	Milwaukee	Public

The patents in these libraries are arranged numerically; there are various indexes to help readers find what they are looking for. Not all the collections are complete, especially for the early years. Other libraries around the country have information on patents, too, and some of them may offer just what you want. If none of the above-listed locations is close to you, check with your local library to find out if anything is available from nearby institutions.

Bibliography

Berle, Alf K.: *Inventions, Patents and Their Management,* Van Nostrand, Princeton, N.J., 1953.

Burlingame, Roger: *Machines That Built America,* Harcourt, Brace and World, New York, 1953.

Clarke, Arthur C.: *Profiles of the Future,* 2d ed, Harper & Row, New York, 1973.

Fenner, Terrence: *Inventor's Handbook,* Chemical Publishing Co., New York, 1969.

Hooper, Meredith: *Everyday Inventions,* Taplinger Publishing Co., New York, 1976.

Hope, Adrian: *Why Didn't I Think of It First?* Drake Publishers, New York, 1974.

Jones, Stacy V.: *The Inventor's Patent Handbook,* rev. ed., Dial Press, New York, 1969.

Kaempffert, Waldemar: *A Popular History of American Invention,* AMS Press, New York, 1974.

Kessler, Kenneth: *The Successful Inventor's Guide,* Prentice-Hall, Englewood Cliffs, N.J., 1965.

Lawrence, Lucas G.: *Inventor's Project Book,* Howard W. Sams & Co., Indianapolis: 1971.

McNair, Eric, and James Schwenck: *How to Become a Successful Inventor,* Hastings House, New York, 1973.

Moore, Arthur D.: *Invention, Discovery and Creativity,* Doubleday, Garden City, N.Y., 1969.

Neumeyer, Frederick: *The Employed Inventor in the United States,* MIT Press, Cambridge, Mass., 1971.

Patent and Trademark Office, U.S. Department of Commerce: *General Information Concerning Patents.* (Revised periodically.)

———: *General Information Concerning Trademarks.*

———: *Official Guide.* (Published weekly.)

———: *Patents and Inventions: An Information Guide for Inventors.*

Pressman, David: *Patent It Yourself!* McGraw-Hill, New York, 1979.

Stein, Ralph: *The Great Inventions,* Playboy Press, New York, 1976.

Taylor, Clarence: *How To Be a Successful Inventor,* Exposition Press, New York, 1972.

Index

NOTE: Page numbers in italic indicate patents